Astrometry for Astrophysics
Methods, Models, and Applications

The field of astrometry, the precise measurement of the positions, distances, and motions of astronomical objects, has been revolutionized in recent years. As we enter the high-precision era, it will play an increasingly important role in all areas of astronomy, astrophysics, and cosmology. This edited text starts by looking at the opportunities and challenges facing astrometry in the twenty-first century, from space and ground. The new formalisms of relativity required to take advantage of microarcsecond astrometry are then discussed, before the reader is guided through the basic methods required to transform our observations from detected photons to the celestial sphere. The final section of the text shows how a variety of astronomical problems can be solved using astrometric methods. Bringing together work from a broad range of experts in the field, this textbook is designed to introduce graduate students and researchers alike to the study of astrometry.

William F. van Altena is Professor Emeritus in the Astronomy Department, Yale University, and has taught undergraduate and graduate courses in introductory astronomy, astronomical observing, and astrometry for over 40 years. He served as Director of the Yerkes Observatory, Chair of the Yale Astronomy Department, President of the WIYN Consortium Board of Directors and the Yale Southern Observatory Board of Directors, and was Team Leader for the Hubble Space Telescope's Astrometry Instrument Definition Team and a member of the HST Astrometry Science Team. His current research involves studies of the kinematical structure of the Milky Way in the southern hemisphere, binary stars, and the use of high-technology cameras to study galaxy remnants that are merging with the Milky Way.

Cover illustration: the orbits of stars within the central arcsecond of our Galaxy. In the background, the central portion of a diffraction-limited adaptive optics image taken in 2008 is displayed. While every star in this image has been seen to move over the past 14 years, only a few of the stars with well-determined orbital parameters are highlighted. The annual average positions for these seven stars are plotted as colored dots, which have increasing color saturation with time. Also plotted are the best fitting simultaneous orbital solutions. These orbits provide the best evidence yet for a supermassive black hole with a mass of 4 million times the mass of the Sun. Note that the orbits have been offset from the Galactic center in the image to achieve balance in the cover illustration. See Fig. 10.2 for the correct placement of the orbits. Courtesy of Andrea Ghez.

Astrometry for Astrophysics

Methods, Models, and Applications

Edited by

WILLIAM F. VAN ALTENA

Yale University, Connecticut

CAMBRIDGE UNIVERSITY PRESS

CAMBRIDGE
UNIVERSITY PRESS

Shaftesbury Road, Cambridge CB2 8EA, United Kingdom

One Liberty Plaza, 20th Floor, New York, NY 10006, USA

477 Williamstown Road, Port Melbourne, VIC 3207, Australia

314–321, 3rd Floor, Plot 3, Splendor Forum, Jasola District Centre, New Delhi – 110025, India

103 Penang Road, #05–06/07, Visioncrest Commercial, Singapore 238467

Cambridge University Press is part of Cambridge University Press & Assessment,
a department of the University of Cambridge.

We share the University's mission to contribute to society through the pursuit of
education, learning and research at the highest international levels of excellence.

www.cambridge.org
Information on this title: www.cambridge.org/9780521519205

© Cambridge University Press & Assessment 2013

First published 2013

A catalogue record for this publication is available from the British Library

ISBN   978-0-521-51920-5   Hardback

Contents

Part V Applications of astrometry to topics in astrophysics

Contributors

ANDERSON Jay
Space Telescope Science Institute
3700 San Martin Drive
Baltimore MD 21218-2410
United States

BENEDICT G. Fritz
McDonald Observatory
University of Texas Austin
1 University Station C1402
Austin TX 78712-0259
United States

BROWN Anthony G. A.
Leiden Observatory
Leiden University
PO Box 9513
2300 RA Leiden
The Netherlands

CAPITAINE Nicole
SYRTE
Observatoire de Paris
61 Av de l'Observatoire
FR 75014 Paris
France

EASTHER Richard
Department of Physics
University of Auckland
Private Bag 92019
NZ Auckland 1142
New Zealand

FOMALONT Edward B.
NRAO
52 Edgemont Rd
Charlottesville
VA 22903-2475
United States

GHEZ Andrea
Department of Physics and Astronomy
University of California Los Angeles
430 Portola Plaza
Los Angeles CA 90095-1547
United States

GLINDEMANN Andreas
ESO
Karl-Schwarzschild-Str 2
DE 85748 Garching bei München
Germany

HORCH Elliott
Department of Physics
Southern Connecticut State University
501 Crescent St.
New Haven CT 06515
United States

HOWELL Steve B.
NASA Ames Research Center
M/S 244-30
Moffett Field CA 94035
United States

KLIONER Sergei
Lohrmann Observatory
Technical University Dresden
Mommsenstr 13
DE 01062 Dresden
Germany

LINDEGREN Lennart
Lund Observatory
Lund University
Box 43
SE 221 00 Lund
Sweden

LÓPEZ Carlos E.
Observatório Astronómico "Felix Aguilar"
Universidad Nacional de San Juan
San Juan
Argentina

MCARTHUR Barbara E.
McDonald Observatory
University of Texas Austin
1 University Station C1402
Austin TX 78712-0259
United States

MÉNDEZ René A.
Universidad de Chile
Dept de Astronomía & Observatorio
Astronómico Nacional
Casilla 36-D
Correo Central
Santiago
Chile

MIGNARD François
LAGRANGE
Observatoire de la Côte d'Azur
Bd de l'Observatoire
BP 4229
FR 06304 Nice Cedex 4
France

NÚÑEZ Jorge
Departament d'Astronomia i Meteorologia
Universitat de Barcelona and Observatorio
Fabra.
Av. Diagonal 647
ES 08028 Barcelona
Spain

PERRYMAN Michael
School of Physics
University of Bristol
Tyndall Avenue
Bristol BS8 1TL
United Kingdom

PLATAIS Imants
Dept. of Physics & Astronomy
Johns Hopkins University
3400 N. Charles St.
Baltimore MD 21218
United States

POURBAIX Dimitri
F. R. S. – FNRS
Inst d'Astronomie et d'Astrophysique
Université Libre de Bruxelles
CP 226
Bd du Triomphe
BE 1050 Bruxelles
Belgium

RABINOWITZ David L.
Center for Astronomy and Astrophysics
Yale University
PO Box 208121
New Haven CT 06520
United States

SCHROEDER Daniel J.
Beloit College
700 College St.
Beloit WI 53511
United States

SOZZETTI Alessandro
INAF – Osservatorio Astronomico
di Torino
Strada Osservatorio 20
10025 Pino Torinese
Italy

STAVINSCHI Magda
Astronomical Institute
Romanian Academy
Cutitul de Argint 5
RO 040557 Bucharest
Romania

TANG Zheng Hong
Shanghai Astronomical
Observatory CAS
80 Nandan Rd
Shanghai 200030
China PR

VAN ALTENA William F.
Astronomy Department
Yale University
PO Box 208121
New Haven CT 06520-8101
United States

WALLACE Patrick T.
Space Science & Technology Department
STFC
Rutherford Appleton Laboratory
Harwell Oxford
Didcot OX11 0QX
United Kingdom

ZACHARIAS Norbert
Astronomy Department
US Naval Observatory
3450 Mass. Ave. NW
Washington DC 20392
United States

Acronyms

2MASS	Two Micron All Sky Survey
A/D	analog-to-digital
AA3	third-order optical angular aberrations
AAS	angular astigmatism
AC	Astrographic Catalogue
ACS/WFC	Advanced Camera for Surveys/Wide-Field Channel
ADI	angular distortion
ADU	analog-to-digital unit
AGK	Astronomische Gesellschaft Katalog
AGK1	first AGK catalog
AGK2	second AGK catalog
AGK3	third AGK catalog
ALMA	Atacama Large Millimeter/submillimeter Array
ARC	Astrophysical Research Consortium
ASA	angular spherical aberration
ATC	angular tangential coma
ATCA	Australia Telescope Compact Array
BCRS	Barycentric Celestial Reference System
BIH	Bureau International de l'Heure
CCD	charge-coupled device
CCDM	Catalog of Components of Double and Multiple Stars
CDS	Centre de Données astronomiques de Strasbourg
CEO	Celestial Ephereris Origin
CEP	Celestial Ephemeris Pole
CGPM	Conférence Générale des Poids et Mésures
CHARA	Center for High Angular Resolution Astronomy
CIO	Celestial Intermediate Origin
CIP	Celestial Intermediate Pole
CMB	cosmic microwave background
CMOS	complementary metal oxide semiconductor
COAST	Cambridge Optical Aperture Synthesis Telescope
CTE	charge-transfer efficiency
CTI	charge-transfer inefficiency

CTIO	Cerro Tololo Interamerican Observatory
CTRS	Conventional Terrestrial Reference System
DENIS	Deep Near Infrared Survey of the Southern Sky
DI	distortion
DIVA	Deutsche Interferometer fur Vielkanalphotometrie und Astrometrie
DMSA	Double and Multiple Star Annex
DN	data number
DORIS	Doppler Orbitography and Radiopositioning Integrated by Satellite
DPAC	Gaia Data Processing and Analysis Consortium
DQE	detective quantum efficiency
dSph	dwarf spheroidal (galaxy)
EE	equation of the equinoxes
EIH	Einstein–Infeld–Hoffmann (equations)
EO	equation of the origins
EOPs	Earth orientation parameters
EPM2004	Ephemerides of Planets 2004
Eps	obliquity of the ecliptic
Eps0	obliquity of the ecliptic at J2000.0
ERA	Earth rotation angle
ERS	True Equinox and Equator of Date reference system
ESA	European Space Agency
ESO	European Southern Observatory
ET	Ephemeris Time
EVN	European VLBI Network
FAME	Fizeau Astrometric Mapping Explorer
FASTT	Flagstaff Astrometric Scanning Transit Telescope
FCN	free core nutation
FGS	Fine Guidance Sensor (units 1, 2 and 3)
FGS1R	replacement unit for FGS1
FITS	flexible image transport system
FK5	Fundamental Katalog 5
FOR	field of regard
FOV	field of view
FWHM	full width at half maximum
FWHP	full width at half-power
GAIA	Global Astrometric Interferometer for Astrophysics, now known as the Gaia mission
GAST	Greenwich Apparent Sidereal Time (GST)
GC	General Catalogue
GCRS	Geocentric Celestial Reference System

GLONASS	Global Navigation Satellite System (Russian GNSS)
GMST	Greenwich Mean Sidereal Time
GNSS	global navigation satellite system
GPS	Global Positioning System
GRF	Galactic rest frame
GSC	HST Guide Star Catalogue
GST	Greenwich Sidereal Time (GAST)
HIC	Hipparcos Input Catalogue
HIPPARCOS	HIgh Precision PARallax COllecting Satellite
HST	Hubble Space Telescope
IAU	Internationa Astronomical Union
IBPM	International Bureau of Weights and Measures
ICRF	International Celestial Reference Frame
ICRF2	ICRF second version
ICRS	International Celestial Reference System
IDS	International DORIS Service
IERS	International Earth Rotation and Reference Systems Service
IFOV	instantaneous FOV
IGS	International Service for GNSS
ILRS	International Laser Ranging Service
IMCCE	Institut de Méchanique Céleste et de Calcul des Éphémérides
ITRF	International Terrestrial Reference Frame
ITRS	International Terrestrial Reference System
IUGG	International Union of Geodesy and Geophysics
IVS	International VLBI Service
JASMINE	Japan Astrometry Satellite Mission for Infrared Exploration
JMAPS	USNO proposed astrometric satellite
JWST	James Webb Space Telescope
KAM	Kolmogorov–Arnold–Moser (theory)
KPNO	Kitt Peak National Observatory
LBA	(Australian) Long Baseline Array
LBT	Large Binocular Telescope
LCA	longitudinal chromatic aberration
LGSAO	laser guide star adaptive optics
LHC	Large Hadron Collider
L–K	Lutz-Kelker (corrections/bias)
LKH	Lutz–Kelker–Hanson corrections/bias
LLR	lunar laser ranging
LQAC	Large Quasar Astrometric Catalog
LSA	longitudinal spherical aberration

LSR	local standard of rest
LSST	Large Synoptic Survey Telescope
MAP	Multichannel Astrometric Photometer
MCF	mutual coherence function
MCMC	Markov Chain Monte Carlo
NCP	north celestial pole
NEO	near-Earth object
NLTT	New Luyten Two-Tenths catalogue
NOAO	National Optical Astronomy Observatory
NPM	Northern Proper Motion (survey)
NOVAS	(US) Naval Observatory Vector Astrometry Subroutines
NPOI	Navy Precision Optical Interferometer
OBSS	One Billion Star Survey
OCRS	Observer's Celestial Reference System
ODI	one-degree imager
OFAD	optical field-angle distortion
OPD	optical path difference
OT	Orthogonal transfer
QSO	quasi-stellar object
OTA	orthogonal-transfer array
PGC	Preliminary General Catalogue
PPM	Positions and Proper Motions
PPN	parameterized post-Newtonian
PRIMA	Phase Referenced Imaging and a Micro-arcsecond Astrometry
PSF	point-spread function
PTI	Palomar Testbed Interferometer
QUEST	Quasar Equatorial Survey Team
RAVE	RAdial Velocity Experiment
RQE	responsive quantum efficiency
SA	spherical aberration
SAOC	Smithsonian Astrophysical Observatory Catalogue
SDSS	Sloan Digital Sky Survey
SIM	Space Interferometer Mission
SKA	Square Kilometer Array
SLR	satellite laser ranging
SNR	signal-to-noise ratio
SOAR	Southern Astrophysical Research
SOFA	Standards of Fundamental Astronomy (service)
SOHO	Solar & Heliospheric Observatory
SPIE	International Society for Optics and Photonics

SPM	Southern Proper Motion (survey)
SPM2	second SPM catalog
SPM3	third SPM catalog
SPM4	fourth SPM catalog
STScI	Space Telescope Science Institute
SUSI	Sydney University Stellar Interferometer
TAI	International Atomic Time
TAS	transverse astigmatism
TC	tangential coma
TCA	transverse chromatic aberration
TCB	Barycentric Coordinate Time
TCG	Geocentric Coordinate Time
TDB	Barycentric Dynamical Time
TDI	transverse distortion
TDI	time-delay integration
TIO	Terrestrial Intermediate Origin
TIRS	Terrestrial Intermediate Reference System
TNO	trans-Neptunian object
TSA	transverse spherical aberration
TT	Terrestrial Time
TTC	transverse tangential coma
UCAC	USNO CCD Astrograph Catalog
UCAC2	second UCAC catalog
UCAC3	third UCAC catalog
UCAC4	fourth UCAC catalog
USNO	United States Naval Observatory
UT	Universal Time (UT, UT1)
UTC	Coordinated Universal Time
VERA	VLBI Exploration of Radio Astrometry
VIM	variability-induced movers
VISTA	Visible and Infrared Survey Telescope for Astrometry
VLA	Very Large Array
VLBA	Very Long Baseline Array
VLBI	very long baseline interferometry
VLT	Very Large Telescope
VST	VLT Survey Telescopes
VLTI	Very Large Telescope Interferometer
VO	Virtual Observatory
VSOP	secular variations of planetary orbits (french acronym)
WIYN	Wisconsin, Indiana, Yale, NOAO Observatory

WMAP	Wilkinson Microwave Anisotropy Probe
YORP	Yarkovsky, O'Keefe, Radzievskii and Paddack (effect)
YPC	Yale Parallax Catalogue
YPC95	1995 edition of the YPC
YSO	Yale Southern Observatory

Preface

Why do we need another text on astrometry?

Astrometry entered a new era with the advent of microarcsecond positions, parallaxes, and proper motions. Cutting-edge topics in science are now being addressed that were far beyond our grasp only a few years ago. It will soon be possible to determine definitive distances to Cepheid variables, the center of our Galaxy, the Magellanic Clouds and other Local Group members. We will measure the orbital parameters of dwarf galaxies and stellar streams that are merging with the Milky Way, define the kinematics, dynamics, and structure of our Galaxy and search for evidence of the dark matter that constitutes most of the Universe's mass. Stellar masses will be determined routinely to 1% accuracy and we will be able to make full orbit solutions and mass determinations for extrasolar planetary systems. If we are to take advantage of microarcsecond astrometry, we need to reformulate our study of reference frames, systems, and the equations of motion in the context of special and general relativity. Methods need to be developed to statistically analyze our data and calibrate our instruments to levels beyond current standards. As a consequence, our curricula must be drastically revised to meet the needs of students in the twenty-first century.

In October 2007, IAU Symposium 248 "A Giant Step: From Milli- to Micro-arcsecond Astrometry" was held in Shanghai, China. Approximately 200 astronomers attended and presented an array of outstanding talks. I was asked to present a talk on the educational needs of students who might wish to study astrometry in the era of microarcsecond astrometry and to organize a round-table discussion on the topic. In the process of preparing my talk and organizing the session, I realized that I had a unique opportunity to bring together the experts on virtually all topics that might be covered in an introductory text on astrometry. This book is the result of the advice from many individuals in the worldwide astrometric community, most of whom were present at that Shanghai meeting and in particular the 28 authors who wrote the 28 chapters.

Readership

This book is intended to fill a serious gap in texts available to introduce advanced undergraduates, beginning graduate students, and researchers in related fields to the science of astrometry. Several excellent books have been written in recent years that deal with the subject; however, they are written at a level suitable for a researcher in astrometry. What has been lacking to date is an introductory text designed to attract students to the field that might not otherwise be prepared to undertake the study of astrometry at an advanced level. This text gives advanced undergraduate and beginning graduate students an introduction to the field of astrometry, with examples of current applications to a variety of astronomical topics of current interest. It is hoped that the students' exposure at an introductory level will lead to more advanced study of this exciting field. For researchers in other fields, the goal is to provide sufficient background to understand the opportunities and limitations of astrometry.

Organization of the book

The text is intended for a one-semester introductory course that will hopefully lead to further study by students or serve as a primer on the field for researchers in related astronomical areas. To accomplish the above goals, the book is divided into five parts. Part I provides the impetus to study astrometry by reviewing the opportunities and challenges of microarcsecond positions, parallaxes, and proper motions that will be obtained by the new space astrometry missions as well as ground-based telescopes that are now yielding milliarcsecond data for enormous numbers of objects. Part II includes introductions to the use of vectors, the relativistic foundations of astrometry and the celestial mechanics of N-body systems, as well as celestial coordinate systems and positions. Part III introduces the deleterious effects of observing through the atmosphere and methods developed to compensate or take advantage of those effects by using techniques such as adaptive optics and interferometric methods in the optical and radio parts of the spectum. Part IV provides introductions to selected topics in optics and detectors and then develops methods for analyzing the images formed by our telescopes and the relations necessary to project complex focal-plane geometries onto the celestial sphere. Finally, Part V highlights applications of astrometry to a variety of astronomical topics of current interest that I hope will stimulate students and researchers to further explore our exciting field.

Acknowledgements

I am indebted to my 27 coauthors for devoting their time, energy, and expertise to preparing their contributions to this book and in particular I would like to acknowledge Sergei Klioner who helped with advice on numerous occasions. I would also like to express my gratitude to the many referees who carefully read each chapter and provided helpful suggestions. My colleagues at Yale, Terry Girard and Dana Casetti, provided help and advice as did the assistant editor, Claire Poole, the production editor, Chris Miller at Cambridge University Press and the copy-editor, Zoë Lewin. Finally, I would like to acknowledge four individuals who played very important roles in forming my professional career: George Wallerstein, who exposed me to stellar astronomy and astrophysics and guided me into the fascinating field of astrometry; Stanislaus Vasilevskis, who introduced me to the details of astrometry, set standards for the conduct of research, and guided me in so many ways during the formative stages of my career; William (Al) Hiltner, who introduced me to the fascinating field of instrumentation; and William Morgan, who helped me to develop a critical eye to judging quality and the importance of maintaining ethical standards in research. I dedicate this book to my wife Alicia Mora who encouraged me to undertake this project and provided the continuing support needed to bring it to completion.

William F. van Altena
Hamden, CT USA
November 2011

PART I

ASTROMETRY IN THE TWENTY-FIRST CENTURY

1 Opportunities and challenges for astrometry in the twenty-first century

MICHAEL PERRYMAN

Introduction

The fundamental task of measuring stellar positions, and the derived properties of trigonometric distances and space motions, has preoccupied astronomers for centuries. As one of the oldest branches of astronomy, astrometry is concerned with measurement of the positions and motions of planets and other bodies within the Solar System, of stars within our Galaxy and, at least in principle, of galaxies and clusters of galaxies within the Universe as a whole.

While substantial advances have been made in many areas of astrometry over the past decades, the advent of astrometric measurements from space, pioneered by the European Space Agency's Hipparcos mission (operated between 1989 and 1993), has particularly revolutionized and reinvigorated the field. Considerable further progress in space and ground measurements over the next decade or so collectively promise enormous scientific advances from this fundamental technique.

At the very basic level, accurate star positions provide a celestial reference frame for representing moving objects, and for relating phenomena at different wavelengths. Determining the changing displacement of star positions with time then gives access to their motions through space. Additionally, determining their apparent annual motion as the Earth moves in its orbit around the Sun gives access to their distances through measurement of parallax. All of these quantities, and others, are accessed from high-accuracy measurements of the relative angular separation of stars. Repeated measurements over a long period of time essentially provide a stereoscopic map of the stars and their kinematic motions.

What follows, either directly from the observations or indirectly from modeling, are absolute physical stellar characteristics: stellar luminosities, radii, masses, and ages. The physical parameters are then used to understand their internal composition and structure. Space motions, derived from a combination of proper motions, radial velocities, and distances, are used to infer the Galaxy's kinematic and dynamical properties and, eventually, to explain in a rigorous and consistent manner how the Galaxy was originally formed, and how it will evolve in the future.

Crucially, stellar space motions reflect dynamical perturbations due to all other matter, visible or invisible. In a highly simplified picture, most of the visible mass of our Galaxy

Astrometry for Astrophysics: Methods, Models, and Applications, ed. William F. van Altena.
Published by Cambridge University Press. © Cambridge University Press 2013.

resides in the form of disk stars. To first order, these stars display a bulk rotation, while more detailed observations reveal a whole host of substructure. This includes spiral arms and a central stellar bar, both of which appear to arise from intrinsic instabilities in a rotating population. In our immediate neighbourhood, the stellar motions reveal thin and thick disk populations, coherent groupings of open clusters and stellar associations, and various other dissolving remnants of earlier structures. How these various populations originated, and indeed how our Galaxy as a whole formed, can be inferred from these detailed space motions.

Although the large-scale acquisition of astrometric data provides the crucial foundation for studies of stellar structure and evolution on the one hand, and Galactic structure, kinematics, and dynamics on the other, astrometry is in principle capable of measuring a whole host of more detailed or "higher-order" phenomena: at the milliarcsecond level, binary star signatures, general relativistic light bending, and the dynamical consequences of dark matter are already evident; at the microarcsecond level, targeted by the next generation of space astrometry missions currently under development, direct distance measurements will be extended across the Galaxy and to the Magellanic Clouds.

Other effects will become routinely measurable at the same time: perspective acceleration and secular parallax evolution, more subtle metric effects, planetary perturbations of the photocentric motion, and astrometric micro-lensing; and at the nanoarcsecond level, currently no more than an experimental concept, effects of optical interstellar scintillation, geometric cosmology, and ripples in space-time due to gravitational waves will become apparent. The bulk of this seething motion is largely below current observational capabilities, but it is there, waiting to be investigated.

1.1 Some history

It is useful to place the current state-of-the-art milliarcsecond astrometric measurements in a brief, albeit highly selective, historical context. Chapman (1990) provides a fascinating historical account of the development of angular measurements in astronomy between 1500 and 1850, on which this summary is based.

After the remarkable achievements of the ancient Greeks, including their first estimates of the sizes and distances of the Sun and Moon, the narrative intensifies 300–400 years ago, when three main scientific themes motivated the improvement of angular measurements: the navigational problems associated with the determination of longitude on the Earth's surface, the comprehension and acceptance of Newtonianism, and understanding the Earth's motion through space. Even before 1600, astronomers were in agreement that the crucial evidence needed to detect the Earth's motion was the measurement of trigonometric parallax, the tiny oscillation in a star's apparent position arising from the Earth's annual motion around the Sun. The early British Astronomers Royal, for example, appreciated the importance of measuring stellar distances, and were very much preoccupied with the task. But it was to take a further 250 years until this particular piece of observational evidence could be secured.

In 1718, Edmund Halley, who had been comparing contemporary observations with those that the Greek Hipparchus and others had made, announced that three stars, Aldebaran, Sirius, and Arcturus, were displaced from their expected positions by large fractions of a degree. He deduced that each star had its own distinct velocity across the line of sight, or proper motion: stars were moving through space.

By 1725, angular measurements had improved to a few arcseconds, making it possible for James Bradley, England's third Astronomer Royal, to detect stellar aberration, as a byproduct of his unsuccessful attempts to measure the distance to the bright star γ Draconis. This was an unexpected result: small positional displacements were detected, and correctly attributed to the vectorial addition of the velocity of light to that of the Earth's motion around the Sun. His observations provided the first direct proof that the Earth was moving through space, and thus a confirmation both of Copernican theory, and Roemer's discovery of the finite velocity of light 50 years earlier. It also confirmed Newton's hypothesis of the enormity of stellar distances, and showed that the measurement of parallax would pose a technical challenge of extraordinary delicacy.

During the eighteenth century, the motions of many more stars were announced, and in 1783 William Herschel found that he could partly explain these effects by assuming that the Sun itself was moving through space. Attempts to measure parallax intensified. Nevil Maskelyne, England's fifth Astronomer Royal, spent seven months on the island of St. Helena in 1761, using a zenith sector and plumb-line, in an unsuccessful attempt to measure the parallax of the bright star Sirius.

Criteria for probable proximity were developed and, after many unsuccessful attempts, the first stellar parallaxes were measured in the 1830s.

Friedrich Bessel is generally credited as being the first to publish a parallax, for 61 Cygni (Piazzi's Flying Star), from observations made between 1837 and 1838. Thomas Henderson published a parallax for α Centauri in 1839, derived from observations made in 1832–1833 at the Cape of Good Hope. In 1840, Wilhelm Struve presented his parallax for Vega from observations in 1835–1837. Confirmation that stars lay at very great but nevertheless finite distances represented a turning point in the understanding of the Universe. John Herschel, President of the Royal Astronomical Society at the time, congratulated Fellows that they had *'lived to see the day when the sounding line in the universe of stars had at last touched bottom'* (quoted by Hoskin, 1997, p. 219).

The following 150 years saw enormous progress, with the development of accurate fundamental catalogs, and a huge increase in quantity and quality of astrometric data based largely on meridian-circle and photographic-plate measurements.

1.2 The Hipparcos mission

By the second half of the twentieth century, however, measurements from ground were running into essentially insurmountable barriers to improvements in accuracy, especially for large-angle measurements and systematic terms; a review of the instrumental status shortly in advance of the Hipparcos satellite launch in 1989 is given by Monet (1988).

Problems were dominated by the effects of the Earth's atmosphere, but were compounded by complex optical terms, thermal and gravitational instrument flexures, and the absence of all-sky visibility. A proposal to make these exacting observations from space was first put forward in 1967.

The Hipparcos satellite was a space mission primarily targeting the uniform acquisition of milliarcsecond level astrometry (positions, parallaxes, and annual proper motions) for some 120 000 stars. The satellite was launched in August 1989 and operated until 1993. The data processing was finalized in 1996, and the results published by the European Space Agency (ESA) in June 1997 as a compilation of 17 hard-bound volumes, a celestial atlas, and six CDs, comprising the Hipparcos and Tycho Catalogues. The original Tycho Catalogue comprised some 1 million stars of lower accuracy than the Hipparcos Catalogue, while the Tycho-2 Catalogue, published in 2000, provided positions and accurate proper motions for some 2.5 million objects, through the combination of the satellite data with plate material from the early 1900s. Details of the satellite operation, the successive steps in the data analysis, and in the validation and description of the detailed data products, are included in the published catalogue.

The Hipparcos astrometric results impact a very broad range of astronomical research, which can be classified into three major themes

1.2.1 Provision of an accurate reference frame

This has allowed the consistent and rigorous re-reduction of a wide variety of historical and present-day astrometric measurements. The former category include Schmidt plate surveys, meridian-circle observations, 150 years of Earth-orientation measurements, and re-analysis of the 100-year-old Astrographic Catalogue (and the associated Carte du Ciel). The Astrographic Catalogue data, in particular, have yielded a dense reference framework reduced to the Hipparcos reference system propagated back to the early 1900s. Combined with the dense framework of 2.5 million star-mapper measurements from the satellite, this has yielded the high-accuracy long-term proper motions of the Tycho 2 Catalogue.

The dense network of the Tycho 2 Catalogue has, in turn, provided the reference system for the reduction of current state-of-the-art ground-based survey data: thus the dense UCAC 3 and USNO B2 catalogs are now provided on the same reference system, and the same is true for surveys such as SDSS and 2MASS. Other observations specifically reduced to the Hipparcos system are the SuperCOSMOS Sky Survey, major historical photographic surveys such as the AGK2 and the CPC2, and the more recent proper-motion programmes in the northern and southern hemispheres, NPM and SPM. Proper-motion surveys have been rejuvenated by the availability of an accurate optical reference frame, and amongst them are the revised NLTT (Luyten Two-Tenths) survey, and the Lépine–Shara proper-motion surveys (north and south). Many other proper-motion compilations have been generated based on the Hipparcos reference system, in turn yielding large data sets valuable for open cluster surveys, common-proper-motion surveys, etc.

The detection and characterization of double and multiple stars has been revolutionized by Hipparcos: in addition to binaries detected by the satellite, many others have been revealed through the difference between the Hipparcos (short-term) proper motion, and the long-term photocentric motion of long-period binary stars (referred to as $\Delta\mu$ binaries). New binary systems have been followed up through speckle and long-baseline optical interferometry from ground, through a re-analysis of the Hipparcos Intermediate Astrometric Data, or through a combined analysis of astrometric and ground-based radial-velocity data. Other binaries have been discovered as common-proper-motion systems in catalogues reaching fainter limiting magnitudes. Various research papers have together revised the analysis of more than 15 000 Hipparcos binary systems, providing new orbital solutions, or characterizing systems which were classified by Hipparcos as suspected double, acceleration solutions, or stochastic ("failed") solutions.

Other studies since 1997 have together presented radial velocities for more than 17 000 Hipparcos stars since the catalogue publication, not counting two papers presenting some 20 000 radial velocities from the Coravel database, and more than 70 000 RAVE (Radial Velocity Experiment) measurements including some Tycho stars. These radial-velocity measurements are of considerable importance for determining the three-dimensional space motions, as well as further detecting and characterizing the properties of binary stars.

More astrophysically, studies have been made of wide binaries, and their use as tracers of the mass concentrations during their Galactic orbits, and according to population. Numerous papers deal with improved mass estimates from the spectroscopic eclipsing systems, important individual systems such as the Cepheid binary Polaris, the enigmatic Arcturus, and favourable systems for detailed astrophysical investigations such as V1061 Cygni, HIP 50796, the mercury–manganese star ϕ Her, and the spectroscopic binary HR 6046. A number of papers have determined the statistical distributions of periods and eclipse depths for eclipsing binaries, and others have addressed their important application in determining the radiative flux and temperature scales. Studies of the distributions of detached and contact binaries (including W Ursa Majoris and symbiotic systems) have also been undertaken.

The accurate reference frame has in turn provided results in topics as diverse as the measurement of general relativistic light bending; Solar System science, including mass determinations of minor planets; applications of occultations and appulses; studies of Earth rotation and Chandler wobble over the last 100 years based on a re-analysis of data acquired over that period within the framework of studies of the Earth orientation; and consideration of non-precessional motion of the equinox.

1.2.2   Constraints on stellar evolutionary models

The accurate distances and luminosities of 100 000 stars has provided the most comprehensive and accurate data set relevant to stellar evolutionary modeling to date, providing new constraints on internal rotation, element diffusion, convective motions, and asteroseismology. Combined with theoretical models the accurate distances and luminosities yield evolutionary masses, radii, and stellar ages of large numbers and wide varieties of stars.

A substantial number of papers have used the distance information to determine absolute magnitude as a function of spectral type, with new calibrations extending across the Hertzsprung–Russell diagram: for example for OB stars, AFGK dwarfs, and GKM giants, with due attention given to the now more quantifiable effects of Malmquist and Lutz–Kelker biases. Other luminosity calibrations have used spectral lines, including the Ca II-based Wilson–Bappu effect, the equivalent width of O I, and calibrations based on interstellar lines.

A considerable Hipparcos-based literature deals with all aspects of the basic 'standard candles' and their revised luminosity calibration. Studies have investigated the Population I distance indicators, notably the Mira variables, and the Cepheid variables (including the period–luminosity relation, and their luminosity calibration using trigonometric parallaxes, Baade–Wesselink pulsational method, main-sequence fitting, and the possible effects of binarity). The Cepheids are also targets for Galactic kinematic studies, tracing out the Galactic rotation, and also the motion perpendicular to the Galactic plane.

Hipparcos has revolutionized the use of red clump giants as distance indicators, by providing accurate luminosities of hundreds of nearby systems, in sufficient detail that metallicity and evolutionary effects can be disentangled, and the objects then used as single-step distance indicators to the Large and Small Magellanic Clouds, and the Galactic bulge. Availability of these data has catalyzed the parallel theoretical modeling of the clump giants.

For the Population II distance indicators, a rather consistent picture has emerged in recent years based on subdwarf main-sequence fitting, and on the various estimates of the horizontal branch and RR Lyrae luminosities.

A number of different methods now provide distance estimates to the Large Magellanic Cloud, using both Population I and Population II tracers, including some not directly dependent on the Hipparcos results (such as the geometry of the supernova SN 1987A light echo, orbital parallaxes of eclipsing binaries, globular cluster dynamics, and white-dwarf cooling sequences). Together, a convincing consensus emerges, with a straight mean of several methods yielding a distance modulus of $(m - M)_0 = 18.49$ mag. Through the Cepheids, the Hipparcos data also provide good support for the value of the Hubble constant, $H_0 = 72 \pm 8$ km/s/Mpc, as derived by the Hubble Space Telescope key project, and similar values derived by the Wilkinson Microwave Anisotropy Probe (WMAP), gravitational-lensing experiments, and Sunyaev–Zel'dovich effect.

A huge range of other studies has made use of the Hipparcos data to provide constraints on stellar structure and evolution. Improvements have followed in terms of effective temperatures, metallicities, and surface gravities. Bolometric corrections for the Hp photometric band have opened the way for new and improved studies of the observational versus theoretical Hertzsprung–Russell diagram.

Many stellar evolutionary models have, of course, been developed and refined over the last few years, and Hipparcos provides an extensive testing ground for their validation and their astrophysical interpretation: these include specific models for pre- and post-main-sequence phases, and models which have progressively introduced effects such as convective overshooting, gravitational settling, rotation effects, binary tidal evolution, radiative acceleration, and effects of α-element abundance variations. These models have been applied to the understanding of the Hertzsprung–Russell diagram for nearby stars, the reality of the Böhm–Vitense gaps, the zero-age main sequence, the subdwarf main sequence, and the properties of later stages of evolution: the subgiant, first-ascent and asymptotic giant branch, the horizontal branch, and the effects of dredge-up and mass-loss. Studies of elemental abundance variations include the age–metallicity relation in the solar neighbourhood, and various questions related to particular elemental abundances such as lithium and helium. Other studies have characterized and interpreted effects of stellar rotation, surface magnetic fields, and observational consequences of asteroseismology, notably for solar-like objects, the high-amplitude δ Scuti radial pulsators, the β Cephei variables, and the rapidly oscillating Ap stars.

Many studies have focused on the pre-main-sequence stars, both the (lower-mass) T Tauri and the (higher-mass) Herbig Ae/Be stars, correlating their observational dependencies on rotation, X-ray emission, etc. The understanding of Be stars, chemically peculiar stars, X-ray emitters, and Wolf–Rayet stars have all been substantially effected by the Hipparcos data. Kinematic studies of runaway stars, produced either by supernova explosions or dynamical cluster ejection, have revealed many interesting properties of runaway stars, also connected with the problem of (young) B stars found far from the Galactic plane.

Dynamical orbits within the Galaxy have been calculated for planetary nebulae and, perhaps surprisingly given their large distances, for globular clusters. In these cases the provision of a reference frame at the 1 milliarcsecond accuracy level has allowed determination of their space motions and, through the use of a suitable Galactic potential, their Galactic orbits, with some interesting implications for Galactic structure, cluster disruption, and Galaxy formation.

One of the most curious of the Hipparcos results in this area is the improved determination of the empirical mass–radius relation of white dwarfs. At least three such objects appear to be too dense to be explicable in terms of carbon or oxygen cores, while iron cores seem difficult to generate from evolutionary models. "Strange matter" cores have been postulated, and studied by a number of groups.

1.2.3 Galactic structure and dynamics

The distances and uniform space motions have provided a substantial advance in understanding of the detailed kinematic and dynamical structure of the solar neighbourhood, ranging from the presence and evolution of clusters, associations and moving groups, the presence of resonance motions due to the Galaxy's central bar and spiral arms, the parameters describing Galactic rotation, the height of the Sun above the Galactic mid-plane, the motions of the thin disk, thick disk and halo populations, and the evidence for halo accretion.

Many attempts have been made to further understand and characterize the solar motion based on Hipparcos data, and to redefine the large-scale properties of Galactic rotation in the solar neighbourhood. The latter has been traditionally described in terms of the Oort constants, but it is now evident that such a formulation is quite unsatisfactory in terms of describing the detailed local stellar kinematics. Attempts have been made to re-cast the problem into the nine-component tensor treatment of the Ogorodnikov–Milne formulation, analogous to the treatment of a viscous and compressible fluid by Stokes more than 150 years ago. The results of several such investigations have proved perplexing. The most recent and innovative approach has been a kinematic analysis based on vectorial harmonics, in which the velocity field is described in terms of (some unexpected) "electric" and "magnetic" harmonics. They reveal the warp at the same time, but in an opposite sense to the vector field expected from a stationary warp.

Kinematic analyses have tackled the issues of the mass density in the solar neighbourhood, and the associated force law perpendicular to the plane, the K_z relation. Estimates of the resulting vertical oscillation frequency in the Galaxy of around 80 Myr have been linked to cratering periodicities in the Earth's geological records. Related topics include studies of nearby stars, the stellar escape velocity, the associated initial mass function, and the star formation rate over the history of the Galaxy. Dynamical studies of the bar, of the spiral arms, and of the stellar warp, have all benefited. Studies of the baryon halo of the Galaxy have refined its mass and extent, its rotation, shape, and velocity dispersion, and have provided compelling evidence for its formation in terms of halo substructure, some of which is considered to be infalling, accreting material, still ongoing today.

New techniques have been developed and refined to search for phase-space structure (i.e. structure in positional and velocity space): these include convergent-point analysis, the "spaghetti" method, global convergence mapping, epicycle correction, and orbital back-tracking. An extensive literature has resulted on many aspects of the Hyades, the Pleiades, and other nearby open clusters, comprehensive searches for new clusters, and their application to problems as diverse as interstellar reddening determination, correlation with the nearby spiral arms, and the age dependence of their vertical distribution within the Galaxy: one surprising result is that this can be used to place constraints on the degree of convective overshooting by matching stellar evolutionary ages with cluster distances from the Galactic plane.

In addition to studying and characterizing open clusters, and young nearby associations of recent star formation, the Hipparcos data have revealed a wealth of structure in the nearby velocity distribution which is being variously interpreted in terms of open-cluster evaporation, resonant motions due to the central Galactic bar, scattering from nearby spiral arms, and the effects of young nearby kinematic groups, with several having been discovered from the Hipparcos data in the last 5 years.

The Hipparcos stars have been used as important (distance) tracers, determining the extent of the local "bubble", itself perhaps the result of one or more nearby supernova explosions in the last 5 Myr. The interstellar-medium morphology, extinction and reddening, grey extinction, polarization of star light, and the interstellar radiation field, have all been constrained by these new distance estimates of the Hipparcos stars.

1.2.4 Other applications

Superficially, it may seem surprising that the Hipparcos Catalogue has been used for a number of studies related to the Earth's climate. Studies of the passage of nearby stars and their possible interaction with the Oort Cloud have identified stars which came close to the Sun in the geologically recent past, and others which will do so in the relatively near future. Analysis of the Sun's Galactic orbit, and its resulting passage through the spiral arms, favour a particular spiral arm pattern speed in order to place the Sun within these arms during extended deep-glaciation epochs in the distant past. In this model, climatic variations are explained as resulting from an enhanced cosmic-ray flux in the Earth's atmosphere, leading to cloud condensation and a consequent lowering of temperature. A study of the Maunder Minimum, a period between 1645 and 1715 coinciding with the coldest excursion of the

"Little Ice Age," and a period of great hardship in Europe, was interpreted in the context of the number of solar-type stars out to 50–80 pc showing correspondingly decreased surface activity. Several studies have used the accurate distance data, accompanied by stellar evolutionary models, in an attempt to identify "solar twins" (stars which most closely resemble the Sun in all their characteristics, and which may be the optimum targets for searches for life in the future), and "solar analogs" (stars which resemble the Sun as it was at some past epoch or time in the future, and which therefore offer the best prospects for studying the Sun at different evolutionary stages).

Many studies have used the accurate photometric data, which can be derived from space observations in parallel with the positional information as part of the construction of absolute or bolometric magnitudes, or for their uniform colour indices. In addition, the extensive epoch photometry has been used for all sorts of variability analyses, including the rotation of minor planets, the study of eclipsing binaries, the complex pulsational properties of Cepheids, Mira variables, δ Scuti variables, slowly pulsating B stars, and many others.

In addition to all of these, the Hipparcos and Tycho Catalogues are now routinely used to point ground-based telescopes, navigate space missions, drive public planetaria, and provide search lists for programmes such as exo-planet surveys; one study has even shown how positions of nearby stars at the milliarcsecond level can be used to optimize search strategies for extraterrestrial intelligence.

More details of the Hipparcos scientific results are given by Perryman (2008), a book-length review covering the full range of the Hipparcos scientific findings. It offers an extensive summary and analysis of the scientific literature over the 10 years since the publication of the Hipparcos and Tycho Catalogues in 1997.

1.3 Other recent advances

While the Hipparcos satellite has represented a breakthrough in stellar astrometry in terms of numbers of stars and astrometric accuracy, many other projects have contributed to the advances in measurement and interpretation over the past 10–20 years. From space, the Fine Guidance Sensors of the Hubble Space Telescope have contributed accurate parallaxes of a small number of important objects. Various ground-based interferometers have also yielded small numbers of narrow-field astrometric measurements, which have been most notably targeted to the study of close binary systems showing orbital motion.

Radio astrometry has also provided a basic network of reference positions over the entire celestial sphere, as well as a small number of high-accuracy measurements of radio stars, masers, and the radio source in the centre of the Galaxy, Sgr A*. Digitized photographic-plate surveys and, more recently, large-scale CCD sky surveys, have provided a stellar reference system, with proper motions, for hundreds of millions of stars down to 20 mag. In turn, newly discovered high-proper-motion stars have been used as target lists for identifying low-luminosity stars in the solar neighbourhood.

1.4 The future

The stimulus given by Hipparcos has led to a wave of second-generation space astrometry missions, either studied but ultimately rejected (such as FAME, OBSS, and DIVA), or now approved and progressing towards launch (in the case of Gaia, JMAPS, and NanoJasmine).

A wider appreciation of the scientific importance of astrometric measurements has also led to astrometry being a driving force, or important byproduct, for many ground-based instruments, ranging from large-scale surveys such as the Sloan Digital Sky Survey (SDSS), VST/VISTA, Pan-STARRS, and LSST, to very-high-accuracy narrow-field astrometric measurements expected from future astrometric interferometers such as the Very Large Telescope Interferometer (VLTI).

These opportunities for research in astrometry and related sciences have revolutionized the field in a way that is probably even more dramatic than the application of photographic emulsions in the late 1800s. The prospects of future large-scale measurements, of significantly higher accuracy, will open up an even more exciting landscape for quantitative exploration in the future.

1.4.1 Gaia

The European Space Agency's Gaia mission builds on the principles of Hipparcos, employing two viewing directions and revolving sky scanning to build up a detailed map of positions, proper motions, and parallaxes over an observing period of 5 years. Through a combination of larger optics and a substantial charge-coupled device (CCD) focal plane, Gaia will dwarf the measurements made by Hipparcos. The astrometric measurements will extend to a completeness limit of $V = 20$ mag, yielding accuracies of some 10–20 micro-arcsec at 15 mag. This stereoscopic map will embrace some 1 per cent of the Galactic stellar population, providing direct distance estimates for vast numbers of stars extending throughout our Galaxy and beyond. The satellite includes dedicated instruments for providing multi-colour photometry for each object, and radial velocities for a significant number. This combination of large-scale astrometric and astrophysical characterization for a significant fraction of our Galaxy will provide a database to quantify the early formation, and subsequent dynamical, chemical, and star-formation evolution of our Galaxy. It will also have a colossal impact on studies of stellar structure and evolution, and binaries and multiple stars. Additional products include detection and orbital classification of thousands of extra-solar planetary systems, a comprehensive survey of some 10^5–10^6 minor bodies in our Solar System, through galaxies in the nearby Universe, to some 500 000 distant quasars. It will also provide a number of stringent new tests of general relativity and cosmology. The analysis of Gaia data on ground will also provide a correspondingly daunting challenge, ranging from the ingestion, storage, and processing of some 1 PetaByte of data, a numerical computation task of order 10^{21} floating-point operations, and the need for robust and highly efficient large-scale analysis and classification techniques to process the millions, or hundreds of millions, of binary and multiple systems, variable stars, microlensing events, and many other important classes of object that will appear in the data stream. More details of the science goals and measurements techniques are given by Perryman *et al.* (2001).

1.4.2 SIM

While the sheer scale of the Gaia space measurements promises a vast and unprecedented astrometric survey, the technique of continuous sky scanning results in one major limitation: the resulting astrometric accuracy for a given object is primarily a function of its apparent magnitude (and to a lesser extent its celestial coordinates), falling off from its highest accuracies of 10–20 microarcsec at 10–15 mag, down to some 0.2 milliarcsec at around 20 mag. The option to make more extensive observations of particularly important celestial objects simply does not exist. NASA's SIM (Space Interferometer Mission) PlanetQuest aimed to provide a very different approach to space astrometry, employing a 9 m baseline targeted Michelson interferometer. With a global astrometry accuracy of 3 microarcsec for stars brighter than $V = 20$ mag, it would measure parallaxes and proper motions of carefully selected stars, throughout the Galaxy, with unprecedented accuracy. Operating in a narrow-angle mode, it targeted a positional accuracy of 0.6 microarcsec for a single measurement, equivalent to a differential positional accuracy at the end of the nominal 5-year mission of better than 0.1 microarcsec. The SIM PlanetQuest mission would contribute to many astronomical fields, including stellar and galactic astrophysics, planetary systems around nearby stars, and the study of quasar and active galactic nuclei. Using differential astrometry, it would search for planets with masses as small as Earth orbiting in the habitable zone around the nearest stars, and would help to characterize the multiple-planet systems that are now known to exist. While not recommended in the latest US Decadal Survey, details of the intended science goals and measurements techniques are given by Unwin *et al.* (2008).

References

Chapman, A. (1990). *Dividing the Circle: The Development of Critical Angular Measurement in Astronomy 1500–1850*. London: Ellis Horwood.

Hoskin, M. (1997). *Cambridge Illustrated History of Astronomy*. Cambridge: Cambridge University Press.

Monet, D. G. (1988). Recent advances in optical astrometry. *ARAA*, **26**, 413.

Perryman, M. A. C. (2008). *Astronomical Applications of Astrometry: Ten Years of Exploitation of the Hipparcos Satellite Data*. Cambridge: Cambridge University Press.

Perryman M. A. C., de Boer, K. S., Gilmore, G., *et al.* (2001). Gaia: composition, formation and evolution of the Galaxy. *A&A*, **369**, 339.

Unwin, S. C., Shao, M., Tanner, A. M., *et al.* (2008). Taking the measure of the Universe: precision astrometry with SIM PlanetQuest. *PASP*, **120**, 38.

2 Astrometric satellites

LENNART LINDEGREN

Introduction

The launch of the Hipparcos satellite in 1989 and the Hubble Space Telescope in 1990 revolutionized astrometry. By no means does this imply that not much progress was made in the ground-based techniques used exclusively until then. On the contrary, the 1960s to 1980s saw an intense development of new or highly improved instruments, including photoelectric meridian circles, automated plate measuring machines, and the use of charge-coupled device (CCD) detectors for small-field differential astrometry (for a review of optical astrometry at the time, see Monet 1988). In the radio domain, very long baseline interferometry (VLBI) astrometry already provided an extragalactic reference frame accurate to about 1 milliarcsecond (mas) (Ma *et al.* 1990). Spectacular improvements were made in terms of accuracy, the faintness of the observed objects, and their numbers. However, there was a widening gulf between small-angle astrometry, where differential techniques could overcome atmospheric effects down to below 1 mas, and large-angle astrometry, where conventional instruments such as meridian circles seemed to have hit a barrier in the underlying systematic errors at about 100 mas. Though very precise, the small-angle measurements were of limited use for the determination of positions and proper motions, due to the lack of suitable reference objects in the small fields, and even for parallaxes the necessary correction for the mean parallax of background stars was highly non-trivial. Linking the optical observations to the accurate VLBI frame also proved extremely difficult. The advent of space astrometry removed many of these obstacles by providing a unified framework for optical small- and large-angle astrometry, whether on ground or in space, as well as an accurate link to the extragalactic and radio frames.

2.1 Advantages of making astrometric observations from space

The first proposals for making astrometric observations from space were put forward by Pierre Lacroute around 1966. Although the main advantages of space were recognized from the start (e.g. Bacchus and Lacroute 1974), it is only with the hindsight of the Hipparcos experience that they can be fully appreciated. Briefly, they can be stated as follows:

Astrometry for Astrophysics: Methods, Models, and Applications, ed. William F. van Altena.
Published by Cambridge University Press. © Cambridge University Press 2013.

- The absence of an atmosphere eliminates the systematic and random effects of refraction and turbulence (seeing) and gives sharper images.
- Weightlessness eliminates the differential mechanical deformation of the instrument as it is pointed in different directions.
- Space may provide a thermally and mechanically very stable environment.
- At any time, nearly the whole sky is simultaneously accessible from a single observatory, and over a few months the whole sky is accessible.
- Continuous observation over extended periods of time gives high throughput and facilitates calibrations and the uniform treatment of data.
- Immensely useful photometric data of uniform quality can be collected with a moderate additional effort.

In principle, all astrometric observations seek, in the first instance, to determine the direction of an incident wavefront of electromagnetic radiation in some instrument reference frame. The achievable precision is fundamentally limited by diffraction and the stochastic nature of the detection process (Lindegren 2010b). In space, the orientation of the instrument frame, and possibly other calibration data as well, must always be established as part of the observation procedure itself, and subsequently eliminated. For example, the instrument orientation can be eliminated by using simultaneous or quasi-simultaneous observations of several objects, and field distortion can be determined from multiple observations of the same configuration of objects in various parts of the field. After calibration, the measurements refer to the differential coordinates of the objects on the sky. Thus, space astrometry observations are by their nature always differential. A crucial difference compared to ground-based differential astrometry is that it is possible to measure long arcs with essentially the same accuracy as short ones. With an instrument like Hipparcos, arcs of the order of $90°$ are measured as accurately as the relative positions in a small field.

2.2 The Hubble Space Telescope (HST)

The main instruments on-board the HST used for astrometry are the Fine Guidance Sensors (FGS), the Wide Field and Planetary Camera 2 (WFPC2, 1993–2009, then replaced by WFC3), and the Advanced Camera for Surveys/Wide-Field Channel (ACS/WFC, since 2002). The primary purpose of the FGS is to provide an absolute pointing reference for other HST instruments. To this end, two of the three FGS are locked on suitable guide stars, while the third sensor can be used for differential astrometry in its $\sim 4 \times 17 \, \text{arcmin}^2$ quarter-annulus (pickle-shaped) field of view. Each sensor has an aperture of $5 \times 5 \, \text{arcsec}^2$, which can be accurately positioned anywhere in the pickle by means of a star selector assembly, and two white-light shearing interferometers, which track or scan the object in the aperture in orthogonal directions. Additional technical details can be found e.g. in Nelan *et al.* (1998). Benedict *et al.* (2008) give a brief history of astrometry with the FGS and outline some recent results on the Galactic Cepheid period–luminosity relationship and the determination of exoplanet masses. As a science instrument, the FGS can observe targets in the magnitude range from $V = 3$ to 17.5 mag. The FGS units used for astrometry (initially FGS3 and, after refurbishing in 1997, the improved FGS1R) have routinely produced small-angle astrometry with a precision of 1 mas or better per observation (Benedict *et al.* 2008 and Chapter 22).

Although not specifically designed for astrometry, the WFPC2 and ACS/WFC cameras have proved very useful for differential astrometry. Within their fields of view (150 and 200 arcsec, respectively) the achievable astrometric precision in a single frame is about 1 mas (0.01–0.02 pixel) including calibration errors (Anderson and King 2003, 2006). Moreover, this precision can be achieved for much fainter stars and in much denser fields than is typically possible from the ground. By combining exposures with an epoch difference of up to more than 10 years, very precise relative proper motions can thus be derived for faint objects. This is especially valuable for determination of cluster memberships and for studying the internal kinematics of clusters. Using faint background galaxies as reference, it is also possible to determine the absolute proper motions of the clusters (e.g. Kalirai *et al.* 2007). See Chapter 17 for a detailed discussion of the HST cameras.

2.3 The Hipparcos mission

The Hipparcos satellite, launched in August 1989, was the first space mission dedicated to astrometry. The Hipparcos and Tycho Catalogues (ESA 1997) consist of two catalogues, as the name suggests: the main (Hipparcos) catalogue gives astrometric and single-band photometric data for about 118 000 "entries" (i.e. single or multiple stars), 48 asteroids and 3 planetary satellites. The Tycho Catalogue, derived from the auxiliary star mapper data, gives less accurate astrometric data for over 1 million objects, including two-band photometry. A later re-reduction of the star mapper data and combination with a century of ground-based positional catalogues for proper motions resulted in the Tycho-2 Catalogue (Høg *et al.* 2000) of the 2.5 million brightest stars. A later re-reduction of the main mission data by van Leeuwen (2007) resulted in improved astrometric accuracies especially for stars brighter than ninth magnitude. Perryman (2009) comprehensively summarized the scientific achievements up to 2007 based on the Hipparcos data; his review based on some 5000 published papers convincingly demonstrates the importance of astrometry for a very broad range of astronomical problems. A summary of Hipparcos science results is given in Chapter 1.

The historical roots of Hipparcos are in the mid 1960s, when Pierre Lacroute, then Director of Strasbourg Observatory, proposed to use observations in space to construct an optical reference system of some 700 bright stars with a positional accuracy of about 10 mas (Lacroute 1967, Kovalevsky 2005). This concept already included key elements of the future Hipparcos instrument, such as the beam combiner and modulating slits. However, to transform these ideas into a feasible project required many years of innovative technical work. Crucial contributions were for example made by Erik Høg (Copenhagen) during the early studies conducted by the European Space Agency (ESA) from 1975, and subsequently by the industrial study partner, Matra Espace (Toulouse). When Hipparcos

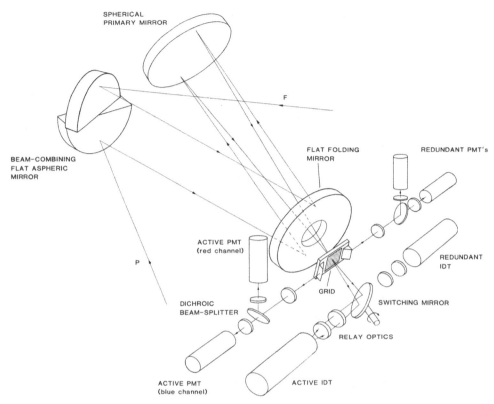

Fig. 2.1 Schematic illustration of the Hipparcos instrument, including the all-reflective Schmidt telescope (aspheric beam combiner and spherical primary), the two viewing directions (P = preceding field, and F = following field), modulating grid, main detector (IDT = image dissector tube), and star mapper detectors (PMT = photomultiplier). The solar panels are below the instrument as drawn.

was finally accepted within ESA's scientific programme in 1980, the aim was to observe 100 000 stars with a typical accuracy of 2 mas. The achieved accuracies for 118 000 stars were 0.8 mas for the positions (at the reference epoch J1991.25), 1.0 mas/yr for the proper motions, and 1.1 mas for the parallaxes (median values).

Hipparcos used an all-reflective Schmidt telescope of 1.4 m focal length, 0.29 m diameter circular pupil, and a square main field of view of $0.9° \times 0.9°$. In front of the telescope a beam combiner made of two plane mirrors, tilted in opposite directions and glued together, produced a fixed and very stable "basic angle" of $58°$ between the two light beams reflected into each half of the pupil (Fig. 2.1). The beam combiner also acted as a Schmidt corrector by the slight aspheric figuring of the mirror surfaces. Two fields on the sky, separated by the basic angle, were thus simultaneously imaged on the focal surface and, as the satellite slowly turned around an axis perpendicular to the two fields, they mapped out a $0.9°$ wide strip of the sky, approximately along a great circle. On the focal surface, a grid of parallel transparent slits produced a periodic modulation of the transmitted star light picked up by the detector. For its main observations, Hipparcos used a special kind of photomultiplier, an image dissector tube, with an electronically steerable "instantaneous field of view" (IFOV) of 30 arcsec diameter. Compared to an ordinary photomultiplier integrating the transmitted light over the whole field, the image dissector gave a vast improvement in signal-to-noise ratio and limiting magnitude. This was, however, gained at the expense of only being able to observe one star at a time and the need for an "input catalogue" and real-time attitude control to direct the IFOV to the correct positions. The required three-axis attitude was derived from gyroscopes and observations on the chevron-like star mapper grid (Fig. 2.1). The Hipparcos Input Catalogue (Turon *et al.* 1992), the list of objects to be observed with the main detector, was compiled from over 200 proposals solicited in the 1980s from the scientific community. It included a survey of about 55 000 stars complete to well-defined limits in magnitude, and a selection of fainter stars based on their relevance for the proposed studies.

The light modulation on the main grid essentially measured relative one-dimensional stellar positions in the "along-scan" direction normal to the slits. These positions were only determined up to a certain scale factor for the grid and, in the case of two images from different fields, relative to the actual value of the basic angle. Both the scale factor and the basic angle could, however, be determined from the closure condition when a sufficient number of relative measurements had been made along a complete great circle; had the wrong values been used for the scale factor or basic angle, the measured angles would not add up to 360°. Additional calibration parameters for optical distortion and grid imperfections were determined by a similar principle. To account for projection effects in the finite fields of view, the perpendicular ("across-scan") coordinates of the stellar images also had to be known, but only to a precision reduced in proportion to the size of the field, i.e. roughly by a factor 60. This information was indirectly provided by the star mappers via the across-scan-attitude determination.

The scanning law of Hipparcos ensured that the resulting one-dimensional measurements of any object could be combined to yield its two-dimensional position (α, δ), proper motion (μ_α, μ_δ), and absolute parallax (ϖ). The principle is illustrated in Fig. 2.2. Nominally, Hipparcos used a uniform revolving scanning law, in which the satellite spin axis $\mathbf{z}$ (perpendicular to the viewing directions) remains at a fixed angle ξ from the Sun while precessing around the solar direction, such that the inertial rate of change of $\mathbf{z}$ is roughly constant in size. In practice the quantity $S = |d\mathbf{z}/d\lambda_s|$ was kept constant, where λ_s is the ecliptic longitude of the Sun; the values used for Hipparcos were $\xi = 43°$ and $S = 4.481$.

Hipparcos was intended for a geostationary orbit, providing continuous contact with the ground station. This was necessary because of the small size of the on-board computer memory: the observations had to be transmitted to ground without delay, and the next portion of the observing programme uploaded at regular intervals. The Ariane-4 launcher successfully put the satellite in its geostationary transfer orbit, but when Hipparcos' own apogee boost motor failed to ignite, the satellite was left in its elliptical transfer orbit, with a perigee height of just a few hundred km, an apogee height of about 36 000 km, and a period of 10.7 hours. This unfortunate orbit had a very severe impact on the performance of the instrument and the possibility to operate the satellite. Three ground stations were required to maintain contact with the satellite, while its unintended journey through the Van Allen

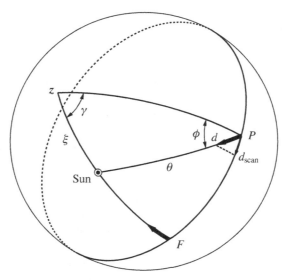

Fig. 2.2 Hipparcos was able to determine absolute parallaxes by measuring the small changes in the angles between stars in the two fields of view (P and F, separated by the basic angle γ) due to the parallax effect. See text for further explanations.

radiation belts several times per day temporarily saturated the detectors, upset the attitude control, progressively damaged sensitive electronics, and eventually set a premature end to the mission. Nevertheless, the original mission goals, in terms of astrometric accuracy, were in all respects fulfilled or even surpassed (Perryman 2009).

2.4 Global astrometry using scanning satellites

The advantages of space outlined in Section 2.1 are especially important for global astrometry, based on the direct measurements of large (~ 1 radian) angular separations on the sky. Global astrometry is essential for building a coherent reference frame over the whole sky as well as for the determination of absolute parallaxes. The recipe for global astrometry tried out by Hipparcos and adopted for the Gaia mission is as follows: (i) Make simultaneous observations in two fields separated by a large angle. (ii) Scan roughly along a great circle through both fields. (iii) Make (mainly) one-dimensional measurements, viz. along scan. (iv) Make sure the "basic angle" between the two fields is stable and known by appropriate calibration. (v) Repeat measurements as many times as required and with scans in varying orientations in order to cover, eventually, the whole sky.

Note that the measurements perpendicular to the scanning direction need not be as accurate as in the along-scan direction (where the large angles are obtained), because small orientation errors of the instrument only affect the along-scan measurements to second order. The astrometric instrument is therefore optimized for measurements in the along-scan direction. This explains the shape of the telescope pupil (elongated along scan) and of the pixels (elongated across scan), the method of CCD readout (drift-scanning or time-delayed integration [TDI] mode discussed in Chapter 15), as well as many other aspects of the design.

The basic one-dimensionality of the observations also motivates the scanning law. In order to obtain parallaxes as accurately as possible it is desirable to maximize the along-scan component d_{scan} of the displacement of the star image due to parallax (see Fig. 2.2). The total parallactic displacement of the star in the preceding field (P) is $d = \varpi R \sin \theta$, where ϖ is the parallax, R the Sun–spacecraft distance in AU, and θ the angle between the star and the Sun. But $d_{\text{scan}} = d \sin \varphi$, so using the sine theorem we have $d_{\text{scan}} = \varpi R \sin \xi \sin \gamma$. On the other hand, for the particular configuration shown in Fig. 2.2, the parallactic displacement of a star in the following field (F) is perpendicular to the scan. By measuring the relative along-scan displacements of the two stars it is therefore possible to get the absolute parallax of the star at P. To have the largest sensitivity to the parallax, it is clearly desirable to make both the spin axis to sun angle ξ and the fixed mirror angle γ as close to 90° as possible. The angles used for Hipparcos were $\xi = 43°$, $\gamma = 58°$. For Gaia, $\xi = 45°$ and $\gamma = 106.5°$ were chosen. For efficient stray-light suppression it is hardly possible to make ξ much greater than about 45° ($90° - \xi$ being the minimum angle between the Sun and the viewing direction during the scan), but it is important that it stays at this maximum value at all times during the mission, since any great-circle scan performed with a smaller Sun–spin axis angle would contribute less to the parallax determinations. This naturally leads to the revolving scanning law described above and adopted both for Hipparcos, Gaia, and some other proposed missions (cf. Table 2.1).

2.5 Step-stare observation mode

For a continuously scanning satellite, the data collection efficiency using CCDs in TDI readout mode can be close to 100%, since no overhead time is needed for re-pointing the instrument. However, the scanning imposes certain constraints on such a mission, e.g. the exposure time per object is fixed by the scan rate and detector size, and the readout mode could preclude the use of some promising alternative image sensor technologies. For example, CMOS (complementary metal oxide semiconductor) detectors and various hybrid sensors may provide significant advantages over CCDs in terms of readout noise, resolution, data throughput and radiation tolerance, as well as for applications in the infrared beyond 1 μm wavelength. See Chapter 14 for a more detailed discussion of CCDs. Abandoning the continuous scanning leads to the concept of a "stare-mode" (or "step-stare") astrometric space mission (Zacharias and Dorland 2006). The concept has some features in common with the ground-based plate overlap technique (Chapter 19), including the use of two-dimensional measurements, while benefiting from most of the advantages of a space platform outlined in Section 2.2. The step-stare mode has been adopted for some of the proposed future space astrometry missions (see Table 2.1). The calibration issues for these missions are fundamentally different from those of a continuously scanning satellite.

Table 2.1 A non-exhaustive list of (actual or proposed) space astrometry survey missions based on Hipparcos-type scanning or the step-stare observation mode. Several of the proposals evolved substantially over the years; the table generally gives the most recent known characteristics. An asterisk (*) indicates Fizeau-type interferometers (double-aperture telescopes). Note that the table does not include pointing space interferometers (Section 2.8). References: Hipparcos (Perryman *et al.* 1997); Tycho (Høg *et al.* 2000); AIST-Struve (Chubey *et al.* 1997); Roemer and Roemer+ (Høg 1993, 1995), GAIA (Lindegren and Perryman 1996), Gaia (Lindegren 2010a), DIVA (Bastian and Röser 2001), FAME (Johnston 2003), OBSS (Johnston *et al.* 2006), Nano-JASMINE (Kobayashi *et al.* 2008, Gouda 2010), JASMINE (Yano *et al.* 2008, Gouda 2010), JMAPS (Hennessy and Gaume 2010, Gaume 2010)

Project	Agency	Magnitude range	Number of stars	Typical accuracy (at mag.)	Observation mode (# FOV)	Status
Hipparcos	ESA	< 12	0.1 M	1000 µas (9)	Scanning (2)	1989–1993
Tycho		< 11.5	2.5 M	7000 µas (9)		
AIST-Struve	Russia	< 15	> 4 M	1000 µas (14)	Scanning (4)	Not developed
Roemer	ESA	< 18	400 M	140 µas (14)	Scanning (2)	Superseded
Roemer+		< 20	1000 M	35 µas (14)	Scanning (2)	Part of Gaia
GAIA (*)	ESA	< 15	50 M	6 µas (14)	Scanning (3)	Now Gaia
Gaia		6–20	1000 M	15 µas (14)	Scanning (2)	Launch 2013
DIVA (*)	Germany	< 15	35 M	900 µas (13)	Scanning (2)	Cancelled 2003
FAME	USA	5–15	40 M	50 µas (9)	Scanning (2)	Cancelled 2002
OBSS	USA	7–23	1000 M	10 µas (14)	Step-stare (1)	On hold
Nano-JASMINE	Japan	< 9	0.1 M	2000 µas (7.5)	Scanning (2)	Launch 2013
JASMINE	Japan	< 14	10 M	10 µas (14)	Step-stare (1)	Under study
JMAPS	USA	0–14	40 M	1000 µas (12)	Step-stare (1)	Launch 2014

Differences result from using only a single field of view, the two-dimensional measurements, and the need to tie observations locally to extragalactic sources in order to achieve global astrometry including absolute parallaxes.

2.6 The Gaia mission

Gaia is by far the most ambitious among the currently funded space astrometry missions. With a planned launch in 2013, it will observe a billion stars and other point-like objects, in particular quasars and minor planets, to a limiting magnitude around 20. The funding, construction, launch and operation of the satellite is under the responsibility of ESA, with EADS Astrium as the prime industrial contractor for building the satellite including the scientific instruments. Within a few months from launch Gaia will arrive at its orbit around the Sun–Earth L_2 Lagrange point, 1.5 million km from the Earth, after which the routine science operations start. This phase will last 5 years, with a possible 1 year extension. The scientific data processing will start immediately and continue for a few years after the end of the observations. The final catalog is thus expected around 2022, but with intermediate data releases produced during the operational phase. The exceedingly large and complex task of producing the Gaia catalog from the raw satellite data is entrusted to a consortium of scientists throughout Europe, the Gaia Data Processing and Analysis Consortium, DPAC (Mignard *et al.* 2008).

Gaia builds on the proven principles of Hipparcos to determine the astrometric parameters for a large number of objects by combining a much larger number of essentially one-dimensional (along-scan) angular measurements in the focal plane. As described in Section 2.4, the continuous scanning motion of the satellite is designed to ensure that every object is observed many times over the mission, in directions that are geometrically favourable for the determination of the astrometric parameters, including absolute parallaxes. The measurement principle also provides a globally consistent reference system for the positions and proper motions, which is tied to the extragalactic reference system (International Celestial Reference System, ICRS) through Gaia's observations of about half a million quasars. These are assumed to have, on the average, zero proper motion, which defines a non-rotating reference system; some of them also have accurate VLBI radio positions in the ICRS, which define the axes of the (α, δ) system.

The huge gain in accuracy, limiting magnitude and number of objects, compared with Hipparcos, is achieved by a combination of several factors: much larger optics (two entrance pupils, each of 1.45×0.5 m^2) gives much improved resolution and photon-collection area; the CCDs have a wider spectral range and much better quantum efficiency than the photomultipliers of Hipparcos; and there is a huge multiplexing advantage as the CCDs can observe tens of thousands of objects in parallel, while the main Hipparcos detector could only observe one object at a time. The CCDs of Gaia have a pixel size of 10×30 μm^2, matched to the optical resolution of the instrument and the effective focal length of 35 m. A schematic view of the optical instrument is shown in Fig. 2.3. The autonomous on-board processing system detects any point-like object brighter than $\sim$20 mag as it enters the skymappers at one end of the focal plane, then tracks the object across a sequence of CCDs dedicated to the astrometric, photometric, and radial-velocity measurements (Fig. 2.4). The tracking implies that only a small window of pixels centred on the projected path of the optical image is recorded and transmitted to ground. This is necessary in order to reduce the CCD readout noise (by spending more time per useful pixel) and data rate to acceptable values.

Photometric measurements of all detected objects are made using two slitless prism spectrometers, the blue and red photometers, with dedicated CCDs (BP and RP in Fig. 2.4) sampling the dispersed images at a very low spectral resolution ($\lambda/\Delta\lambda = 10$ to 30 depending on the wavelength). The spectral intervals covered are 330–680 nm (BP) and 640–1050 nm (RP). Similarly, radial-velocity measurements are made for the brighter objects using a slitless grating radial-velocity spectrometer (RVS). Its spectral resolution ($\lambda/\Delta\lambda = 10^4$) and wavelength range (847–874 nm) have been optimized to allow radial velocities to be measured, at the few km/s accuracy level, for a wide range of spectral classes.

The predicted accuracy of the astrometric parameters for single stars depends in a complex way on the magnitude, colour, and position on the sky, and to some extent on random factors such as unscheduled observation dead time. Sky-averaged values for unreddened solar-type (G2V) stars of different magnitudes are given in Table 2.2. The predicted parallax accuracies imply that extremely accurate stellar distances (relative error <1%) can be obtained for millions of stars within 1 kpc from the Sun, and less accurate distances

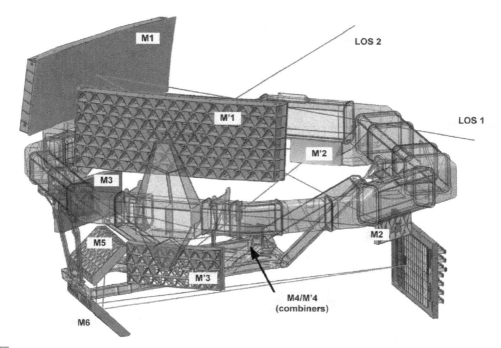

Fig. 2.3 The optical instrument of Gaia, comprising two telescopes (mirrors M1–M6 and M′1–M6), with a common focal plane, on a ring-shaped optical bench of ∼3.5 m diameter. M4/M′4 is the beam combiner situated at the exit pupil. LOS 1 and LOS 2 are the lines of sight for the two optical systems. (Image credit: EADS Astrium.)

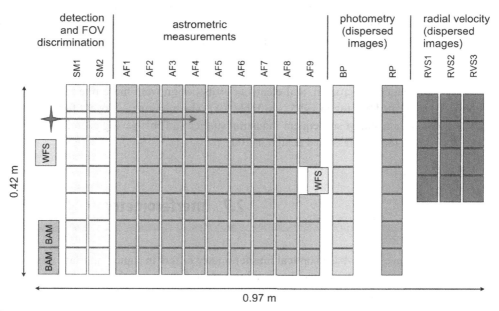

Fig. 2.4 Layout of CCDs in the focal plane of Gaia. Images travel from left to right, crossing in turn the skymappers (SM), astrometric field (AF), blue photometer (BP), red photometer (RP), and (some of them) the radial-velocity spectrometer (RVS). Also shown are the CCDs for the basic-angle monitor (BAM) and wavefront sensors (WFS, for in-orbit optical adjustment). Based on Lindegren *et al.* (2008).

Table 2.2 Predicted sky-averaged standard errors of the astrometric parameters for unreddened G2V stars observed with Gaia, assuming a science operations phase of 5 years (Lindegren 2010a)

V magnitude	6–13	14	15	16	17	18	19	20	Unit
Parallax	8	13	21	34	54	89	154	300	μas
Proper motion	5	7	11	18	29	47	80	158	μas/yr
Position at mid-epoch (≈2016)	6	10	16	25	40	66	113	223	μas

for many more stars out to tens of kiloparsecs. Transverse velocities to a few km/s are obtained for correspondingly large samples and distances. Combined with the photometric and spectroscopic information, these data will be vitally important for many areas of stellar and Galactic astrophysics. A comprehensive description of the science case for Gaia is found in the proceedings of the symposium *The Three-Dimensional Universe with Gaia* (Turon *et al.* 2005).

Technically, the Gaia mission as now implemented has few features in common with the original proposal, submitted to ESA in 1993, although the science goals are largely the same. At that time, it was believed that interferometry was the natural and perhaps necessary step to obtain the orders-of-magnitude improvement in accuracy required for astrophysical investigations on Galactic scales. Thus GAIA, an acronym for *Global Astrometric Interferometer for Astrophysics*, was conceived as a scanning satellite using three stacked Fizeau interferometers (Lindegren and Perryman 1996). At the same time, alternative mission concepts using CCDs in (non-interferometric) imaging mode were explored by Høg (1993, 1995) and others. Subsequent studies demonstrated unequivocally that interferometry was not the best solution for the next-generation astrometric-survey satellite. Indeed, the Gaia mission that was approved by ESA in 2000 was no longer an interferometer, its name no longer an acronym, and elements of its design were more like the Roemer+ concept described by Høg in 1995 (see Table 2.1). The mission underwent further significant changes as a result of subsequent industrial studies and cost-saving exercises before Gaia obtained its more or less final design in 2007.

2.7 Interferometry

Since space-based optical interferometry differs in some important ways from the ground-based interferometry discussed in Chapter 11, a repetition of the basic principles is necessary. Astrometric measurements made with space interferometers will benefit from the increased angular resolution obtained with baselines B that are much larger than any single telescope aperture. On the other hand, interferometers cannot easily be adapted to the systematic scanning principle described in Sections 2.3 and 2.4, and they are therefore better suited for making extremely precise differential measurements of a smaller number of pre-selected targets.

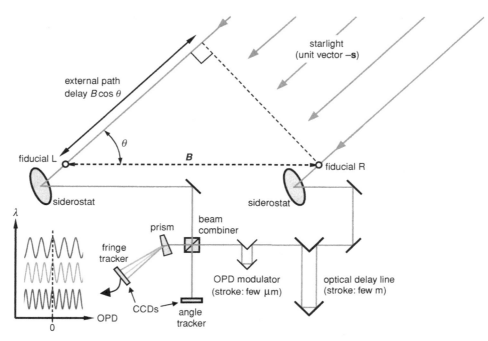

Principle of a space interferometer for astrometry. See text for explanation.

The principle of a space interferometer for astrometry is schematically illustrated in Fig. 2.5, modeled after the planned Space Interferometry Mission, SIM PlanetQuest (Unwin *et al.* 2008). This interferometer consists of three Michelson interferometers with roughly parallel baselines of 9 m length and 0.35 m apertures. Two of the interferometers are used for guiding on relatively bright stars, while the third moves among science targets within a field of regard (FOR) of 15° diameter. Observations can be made in two different modes. In the wide-angle (WA) mode, targets are observed relative to a full-sky grid of approximately 1300 grid stars, whose astrometric parameters are determined to 4 μas accuracy by repeated and systematic observations throughout the mission. Observations in the narrow-angle (NA) mode are made by chopping between targets within the same FOR and may reach a much higher precision of 1 μas for the *individual* measurements of bright (<10 mag) targets with respect to other targets in the same FOR. The extremely high precision of relative measurements in NA mode is primarily motivated by the search for terrestrial planets (Shao 2006).

Each interferometer consists of two steerable siderostats, directing the starlight along the two arms of the interferometer to a common beam combiner and detector. Constructive interference is obtained if the optical path length, from an incident plane wavefront to the detector, is very nearly the same in the two arms, i.e. if the optical path difference (OPD) is close to zero. This is achieved by adjusting the delay in one of the arms until fringes are observed. In order to determine the precise delay corresponding to OPD = 0, a saw-tooth modulation of the OPD by a few wavelengths is introduced, and the intensity of the combined beam measured by the fringe tracker as function of the delay. A dispersive prism allows tracking in quasi-monochromatic light for better fringe contrast (visibility) and a wider acquisition range.

With reference to Fig. 2.5 the two basic quantities that must be measured with the astrometric interferometer are the length of the baseline, $B = |\boldsymbol{B}|$, and the external path delay $d = \boldsymbol{B} \cdot \boldsymbol{s} = B \cos\theta$ for starlight propagating in the direction $-\boldsymbol{s}$. From these measurements, the polar angle θ of the star relative to the baseline vector follows. Physically, the baseline is defined by two fiducials (L and R, in reality situated centrally in the siderostat mirrors), consisting of corner cube retroreflectors. The length of the baseline is measured by the external metrology system sending laser beams directly between the fiducials. The external path delay equals the difference in optical length between the two interferometer arms, from the beam combiner to the two fiducials, when the OPD is zero for the starlight. It is measured by the internal metrology system sending laser beams through the interferometer arms to the fiducials and back.

Within NASA's Exoplanet Exploration Programme, the Jet Propulsion Laboratory is currently studying a mission called SIM Lite Astrometric Observatory. This is a more cost-effective version of SIM PlanetQuest, but with nearly the same science performance. The main science driver is the search for Earth-mass planets in the habitable zone of nearby solar-type stars, although the science case includes many other topics in Galactic astronomy, precision stellar astrophysics, and cosmology. The SIM Lite mission features two Michelson interferometers with parallel baselines: one for science observations, with 6-m baseline, 0.5-m apertures, and a 15° FOR; and a smaller guide interferometer with 4.2-m baseline, 0.3-m apertures, and very small FOR. Guiding around the axis pointing towards the science target is provided by a star tracker, consisting of a 0.3 m aperture telescope with a CCD-based pointing sensor. Although SIM Lite is technologically ready to enter its implementation phase, its future within the NASA program is uncertain at present.

2.8 Future prospects

In the two decades that separate Gaia from Hipparcos, space astrometry has matured from a highly original and risky experiment to established astrophysical technique. The expected gain in accuracy by a factor $\sim 10^2$, and in number of objects by $\sim 10^4$, derives partly from increased telescope size and detector quantum efficiency, but also reflects a unique technological transition from the single-channel sensor in Hipparcos to the large CCD mosaic of Gaia. Elementary considerations of resolution and photon noise show that, for a continuously scanning mission like Gaia, the astrometric precision scales as $\sigma \propto (DM)^{-1}$, where D is the linear size of the aperture, and M the number of resolution elements (pixels) across the field of view in the along-scan direction. Thus, a further increase in accuracy by a large factor requires either a much bigger telescope or a detector array with many more pixels, or both. It is likely that both could be improved by a moderate factor, especially with a new generation of image sensors, but a quantum leap similar to that from Hipparcos to Gaia seems unlikely, for survey-type missions, in the foreseeable future. The situation is different for pointed missions, making differential interferometric measurements on selected targets; here the resolution can in principle be increased without limit at the expense of sky coverage. There are clear scientific motivations for pursuing both directions.

References

Anderson, J. and King, I. R. (2003). An improved distortion solution for the Hubble Space Telescope's WFPC2. *PASP*, **115**, 113.

Anderson, J. and King, I. R. (2006). *PSFs, Photometry, and Astrometry for the ACS/WFC*. Space Telescope Science Institute, ISR-ACS 2006-01.

Bacchus, P. and Lacroute, P. (1974). Prospects of space astrometry. *Proc. IAU. Symp.*, **61**, 277.

Bastian, U. and Röser, S. (2001). DIVA, the next global astrometry and photometry mission. *ASP Conf. Ser.*, **228**, 321.

Benedict, G. F., McArthur, B. E., Bean, J. L. (2008). HST FGS astrometry – the value of fractional millisecond of arc precision. *Proc. IAU Symp.*, **248**, 23.

Chubey, M. S., Kopylov, I. M., Gorshanov, D. L., Kanayev, *et al.* (1997). The Aist-Struve space project sky survey. In *New Horizons from Multi-Wavelength Sky Surveys*, ed. B. J. McLean *et al. Proc. IAU Symp.*, **179**, 125.

ESA (1997). *The Hipparcos and Tycho Catalogues*. ESA SP-1200.

Gaume, R. (2010). Looking towards the future: testing new concepts. *EAS Publ Ser*, **45**, 143.

Gouda, N. (2010). Series of JASMINE missions. *EAS Publ Ser*, **45**, 393.

Hennessy, G. S. and Gaume, R. (2010). Space astrometry with the Joint Milliarcsecond Astrometry Pathfinder. *Proc. IAU Symp.*, **261**, 350.

Høg, E. (1993). Astrometry and photometry of 400 million stars brighter than 18 mag. In *Developments in Astrometry and Their Impact on Astrophysics and Geodynamics*, ed. I. I. Mueller and B. Kotaczek. *Proc. IAU Symp.*, **156**, 37.

Høg, E. (1995). A new era of global astrometry. II. A 10 microarcsecond mission. In *Astronomical and Astrophysical Objectives of Sub-Milliarcsecond Optical Astrometry*, ed. E. Høg and P. K. Seidelmann. *Proc. IAU Symp.*, **166**, 317.

Høg, E., Fabricius, C., Makarov, V. V., Urban, S., Corbin, T., Wycoff, G., Bastian, U., Schwekendiek, P., and Wicenec, A. (2000). The Tycho-2 catalogue of the 2.5 million brightest stars. *A&A*, **355**, L27.

Johnston, K. J. (2003). The FAME mission. *Proc. SPIE*, **4854**, 303.

Johnston, K. J., Dorland, B., Gaume, R., *et al.* (2006). The Origins Billions Star Survey: Galactic explorer. *PASP*, **118**, 1428.

Kalirai, J. S., Anderson, J., Richer, H. B., *et al.* (2007). The space motion of the globular cluster NGC 6397. *ApJ*, **657**, L93.

Kobayashi, Y., Gouda, N., Yano, T., *et al.* (2008). The current status of the Nano-JASMINE project. *Proc. IAU Symp.*, **248**, 270.

Kovalevsky, J. (2005). The Hipparcos project at Strasbourg Observatory. In *The Multinational History of Strasbourg Astronomical Observatory*, ed. A. Heck. New York; NY: Springer Astrophysics and Space Science Library, vol. 330, p. 215.

Lacroute, P. (1967). In *Transactions of the IAU*, **XIIIB**, 63.

Lindegren, L. (2010a). Gaia: Astrometric performance and current status of the project. In *Relativity in Fundamental Astronomy*, ed. S. Klioner, P. K. Seidelmann and M. Soffel. *Proc. IAU Symp.*, **261**, 296.

Lindegren, L. (2010b). High-accuracy positioning: astrometry. *ISSI Sci. Rep. Ser.*, **9**, 279.

Lindegren, L. and Perryman, M. A. C. (1996). GAIA: Global astrometric interferometer for astrophysics, *A&A Suppl. Ser.*, **116**, 579.

Lindegren, L., Babusiaux, C., Bailer-Jones, C., *et al.* (2008). The Gaia mission: science, organization and present status. *Proc. IAU Symp.*, **248**, 217.

Ma, C., Shaffer, D. B., de Vegt, C., Johnston, K. J., and Russell, J. L. (1990). A radio optical reference frame. I. Precise radio source positions determined by Mark III VLBI: observations from 1979 to 1988 and a tie to the FK5. *AJ*, **99**, 1284.

Mignard, F., Bailer-Jones, C., Bastian, U., *et al.* (2008). Gaia: organisation and challenges for the data processing, *Proc. IAU Symp.*, **248**, 224.

Monet, D. G. (1988). Recent advances in optical astrometry. *ARAA*, **26**, 413.

Nelan, E. P., Lupie, O. L., McArthur, B., *et al.* (1998). Fine guidance sensors aboard the Hubble Space Telescope: the scientific capabilities of these interferometers. *Proc. SPIE*, **3350**, 237.

Perryman, M. A. C. (2009). *Astronomical Applications of Astrometry. Ten Years of Exploitation of The Hipparcos Satellite Data*. Cambridge: Cambridge University Press.

Perryman, M. A. C., Lindegren, L., Kovalevsky, J., *et al.* (1997). The Hipparcos Catalogue. *A&A*, **323**, L49.

Shao, M. (2006). Search for terrestrial planets with SIM Planet Quest. In *Advances in Stellar Interferometry*, ed. J. D. Monnier, M. Schöller and W. C. Danchi. *Proc. SPIE*, **6268**, 1Z.

Turon, C., Gomez, A., Crifo, F., Creze, M., *et al.* (1992). The Hipparcos Input Catalogue. I – Star selection. *A&A*, **258**, 74.

Turon, C., O'Flaherty, K. S., and Perryman, M. A. C., eds. (2005). *The Three-Dimensional Universe with Gaia*. ESA SP-576.

Unwin, S. C., Shao, M., Tanner, A. M., *et al.* (2008). Taking the measure of the universe: precision astrometry with SIM PlanetQuest. *PASP*, **120**, 38.

van Leeuwen, F. (2007). *Hipparcos, the New Reduction of the Raw Data*. Astrophysics and Space Science Library, vol. 350, New York, NY: Springer.

Yano, T., Gouda, N., Kobayashi, Y., *et al.* (2008). Space astrometry JASMINE. *Proc. IAU Symp.*, **248**, 296.

Zacharias, N. and Dorland, B. (2006). The concept of a stare-mode astrometric space mission. *PASP*, **118**, 1419.

3

Ground-based opportunities for astrometry

NORBERT ZACHARIAS

Introduction

Chapter 1 surveys the opportunities and challenges for astrometry in the twenty-first century (van Altena 2008) while Chapter 2 discusses space satellites primarily designed for astrometry. We now review the situation for ground-based astrometry, since it is often mistakenly stated that there is no longer any need to pursue ground-based research once satellites are operating. It is certainly true that the levels of precision and accuracy projected for Gaia and others are far beyond what can be achieved from the ground. However, there are also consequences of the fairly small aperture size and short flight durations that impose constraints on the limiting magnitudes and our ability to study long-term perturbations. In this chapter we will explore those areas of research using astrometric techniques that will be able to make important contributions to our understanding of the Universe in the coming years, even with high-accuracy satellites such as Gaia operating.

3.1 Radio astrometry

It is likely that radio astrometry observations (see Chapter 12 for a detailed discussion of radio astrometry and interferometry) will continue to be made primarily from the ground due to the difficulty and cost of launching large objects into space. Since the diffraction limit of a telescope is inversely proportional to the wavelength and radio wavelengths are about 10^3 to 10^5 times longer than those of visible light, no high-resolution imaging or high-precision angular measures can be performed with a single radio telescope. The technique of interferometry and in particular very long baseline interferometry (VLBI) overcomes this problem (Thompson *et al.* 2001). VLBI connects radio telescopes across the globe and results in angular measures far more precise (order 0.1 mas) than obtained by traditional ground-based optical telescopes.

Following centuries of optically defined fundamental systems (Walter and Sovers 2000), a fundamental change was made for the International Celestial Reference Frame (ICRF) with its latest incarnation, the ICRF2 (see Chapters 7 and 8 for a detailed discussion of celestial coordinates and the ICRF). The modern reference system is now defined by VLBI

Astrometry for Astrophysics: Methods, Models, and Applications, ed. William F. van Altena.
Published by Cambridge University Press. © Cambridge University Press 2013.

observations of about 700 compact extragalactic sources (ICRF2 resolution, IAU General Assembly, 2009, Fey *et al.* 2009). VLBI observations are very accurate but reductions are complex since they have to solve for Earth orientation, plate tectonics, geophysical and other parameters simultaneously with the astrometric parameters (source positions). Since the reference sources are extragalactic, a mean zero motion and rotation of the ensemble of extragalactic sources is assumed. However, most of these radio sources are galaxies and the structure of their images has become an issue so VLBI observations now push to higher frequencies where this structure becomes less of a problem (Charlot *et al.* 2010).

A continuing problem with reference systems is that in order to obtain the highest accuracy, it is necessary to limit the reference objects included in the system to those objects with the highest-accuracy positions. As a consequence, there are few sources to which new observations may be referred. The next step is to incorporate more sources into the radio reference frame. Currently (www.vlba.nrao.edu/astro/calib, gemini.gsfc.nasa.gov/vcs/), about 4000 sources with median positional errors below 1 milliarcsecond (mas) are available. This densification of the VLBA calibrator catalog is a major activity today and will continue into the future. Quasars are strong emitters of radio energy and have the advantage of displaying nearly point-like images. They therefore are potential contributors to the reference system once their positions are accurately determined. The Large Quasar Astrometric Catalog (ftp://syrte.obspm.fr/pub/LQAC) contains over 100 000 quasars with the best available astrometric data, cross-correlating radio and optical information (Andrei *et al.* 2009).

The accuracy of VLBI radio observations can be improved even further to about 10 microarcseconds (μas) using phase-referencing techniques. Observations of radio stars (Boboltz *et al.* 2007), star-forming regions, and the Galactic center are leading to exciting results that are revolutionizing research on the luminosities of massive young stars and the Galactic distance scale (Reid 2008). Source structure analysis, proper motions, and trigonometric distance programs for specific targets are great opportunities for those interested in radio astrometry. Additional information and proposals to observe can be directed through the National Radio Astronomy Observatories NRAO (www.nrao.edu), the European VLBI Network (www.evlbi.org), or the Japanese VLBI Exploration of Radio Astrometry, VERA project (veraserver.mtk.nao.ac.jp/outline/index-e.html).

3.2 Optical astrometry

3.2.1 Differential astrometry

The atmosphere of the Earth acts like a corrugated wave front that distorts the image shape and displaces a star from its true position (see Chapter 9 for a detailed discussion of this topic). Fortunately, this image degradation acts nearly like a Gaussian broadening function, which makes it easier for us to model the shape of the stellar images that we observe. Due to the corrugated wave front, two stars that appear to be very close to each other have their positions displaced by nearly the same amount and therefore their relative positions are essentially unchanged. As a consequence, the accuracy of relative positions is much better than that of the individual, absolute positions. The error contribution to astrometric accuracy imposed by the atmosphere is to good approximation proportional to exposure time$^{-1/2}$ and separation$^{1/3}$ (Han 1989; see also Chapter 9). For example, 30 seconds of exposure on two stars separated by 10 arcminutes will yield an accuracy of about 7 mas in their separation, while if the two stars were separated by only 2 arcseconds the accuracy of their separation might be as small as 1 mas. Also, an increase of the integration time by a factor of four will cut the errors in half. This example offers a partial solution to our problem. If one of the pair of stars can be used as a guide star and we can guide sufficiently rapidly to track the atmospheric motions, then the second star and those surrounding it will have their relative positional shifts reduced. This approach is usually called "first-order adaptive optic," or "tip-tilt correction." Full adaptive optics (see Chapters 10 and 11) involves being able to map the shape of the corrugations in real time and distorting a flexible transfer mirror to compensate for the atmospheric corrugations; however, it is a very costly approach. At a good observing site the correlation between the atmospheric shifts is reduced to about 50% for two stars separated by 4 arcminutes, i.e. only half of the time will the two stars be shifted in the same direction. This limitation in the field size for the tip-tilt method means that in order to achieve larger fields of view we must use multiple guide stars, which is an approach developed by Tonry *et al.* (1997) and implemented on the orthogonal-transfer (OT) charge-coupled devices (CCDs) (see also Chapter 14). Large-scale implementation of OT CCDs into arrays have yielded orthogonal-transfer arrays, or OTAs (Tonry *et al.* 2004) that are being used with the Pan-STARRS telescopes (Kaiser *et al.* 2002) and the WIYN one-degree imager (ODI) as described by Jacoby *et al.* (2002). This new technology is revitalizing ground-based astrometry, especially since its use on medium-aperture telescopes yields much fainter limiting magnitudes than will be achievable from space, due to the limited aperture sizes that can now be built and launched. The earlier Mosaic cameras (8k by 8k pixels from eight CCD chips) have been the "workhorse" for many visiting astronomers at the KPNO and CTIO 4-meter telescopes. An astrometric evaluation is presented by Platais *et al.* 2002, 2003, as well as Chapter 19. These instruments were also used for the Deep Astrometric Standards project (Platais *et al.* 2006).

3.2.2 Wide-field astrometry

The concept of differential astrometry can also be applied to wide-field observing with the goal of obtaining extensive catalogs of positions, magnitudes, colors, and proper motions. Traditionally, this has been done with astrographs. Early epoch surveys are still important for astrometry due to the large "leverage" arm for determining accurate proper motions in combination with more recent observations. (A discussion and tables of old and new star catalogs may be found in Chapter 20.) The first all-sky astrometric survey at optical wavelengths with an electronic detector was published as the US Naval Observatory CCD Astrograph Catalog (Zacharias *et al.* 2004). Overlapping fields make it possible to establish a rigid celestial reference frame using block adjustment procedures (Zacharias 1992). Recent progress in detector technology has enabled the production of large single-chip CCDs rivaling the size of photographic plates (Zacharias 2008). Astrometric errors induced by the atmosphere are smaller for longer wavelengths (see Chapter 9). The 2-micron all-sky survey capitalized on this fact to obtain positions of over 400 million stars accurate to about 80 mas with short integration times, a small field of view, and telescope aperture (Zacharias *et al.* 2006).

3.2.3 Photometric surveys for astrometric research

Given the impressive increases in astrometric accuracy made possible by recent technology and its application to the large gigapixel cameras such as the WIYN ODI camera, we now look at some of the science that is possible. The starting points for many of these projects are the surveys, such as 2MASS, DENIS, and the SDSS recently made with relatively small telescopes. (See Chapter 20, Table 20.3, for details on those surveys.) The value of these primarily photometric surveys is that they enable one to search for groups of stars with possible common origin in color–magnitude arrays and then to determine their proper motions to ascertain if the stars are in fact moving with common proper motions, or determine the mean motion of the stars if we already have reason to believe that they are physically associated.

3.2.4 Studies in Galactic structure (see also Chapter 22)

Key for our understanding of our solar neighborhood, Galactic structure, luminosity function, and evolution of stars are the stellar distances. Unbiased, direct determination of distances can only be provided by trigonometric parallaxes. A century of observations with large refractors led to detailed color–absolute-luminosity (Hertzsprung–Russell) diagrams. The culmination of this effort is the 4th Yale General Catalog of trigonometric parallaxes (van Altena *et al.* 1995).

The Sagittarius dwarf galaxy is merging with the Galaxy as shown by radial velocity measurements. Photometric studies identified potential members of the Sagittarius dwarf and the radial velocities of those stars were determined leading to a number of possible orbit solutions. The determination of the proper motions of those stars (Dinescu *et al.* 2005) leads to a measurement of the tangential velocity of the Sagittarius dwarf and a definitive orbit. Several other possible merging dwarf galaxies are known and a determination of their proper motions would help us to understand the dynamics of their orbital evolution.

Currently accepted Lambda cold-dark-matter (CDM) cosmological models (see also Chapter 28) predict several hundred merging dwarf galaxies within 1 kpc of the Sun (Helmi and White 1999). Unfortunately, only a handful of possible candidates have been identified, which casts doubt on the validity of the model. An alternative possibility is that we have not been able to discover those faint objects. Johnston *et al.* (2002) predict that the velocity dispersion will be 5 km/s or less for the remnants of merging dwarfs. An intensive survey of about 100 square degrees should reveal the presence of about 20–30 streams within about 2 kpc, if they exist. If that number of streams is not identified, then something is clearly amiss with the Lambda CDM model.

The amount of "dark matter" in the local disk is still in dispute and we do not have a reasonably complete count of all stars in our solar neighborhood, a task the RECON project (Henry *et al.* 2006) is tackling. It is now possible to determine parallaxes to an accuracy of about 0.5 mas for stars as faint as magnitude 21 (see Chapter 21). This faint limiting magnitude will enable us to observe the faintest stars and obtain a complete inventory of the stellar density to an accuracy of 2%. We should then have a complete census of the local mass, which can then be compared to the dynamical mass to yield the amount of dark matter.

3.2.5 Using star clusters as laboratories for stellar evolution (see also Chapter 25)

Star clusters provide us with laboratories to study the formation and evolution of stars, since we have good reason to believe that the stars in a cluster formed from the same cloud of gas and all have approximately the same metal content and age. The only major variable left is the stellar mass, which gives us the perfect opportunity to study the luminosity and temperature changes as a function of mass for a given age and metalicity. Unfortunately, we observe the stars in a cluster against foreground and background contaminating stars. By measuring the relative proper motions of the stars in the field of the cluster we can identify stars not moving with the motion of the cluster. In the past it was necessary to wait 20 to 30 years after the first-epoch plates were taken to repeat them and determine the proper motions. With the new technology we can repeat the exposures after only 2 years. However, the short time needed to complete the study is not the only advantage. One of the main limitations in the past was that the old plates had a bright limiting magnitude and the interesting faint and low mass stars were beyond the old plate limit. Now we can carry out these studies in a 2-year time span to magnitude 22, or by extending the time span a bit to even fainter magnitudes.

3.2.6 Measuring the masses of black holes and stars

Black holes can be the final state of cataclysmic collapse of massive stars as well as a consequence of dynamical friction in the centers of galaxies where multitudes of stars spiral into a common gravitational center. What are the masses of these black holes? Estimates exist from spectroscopic measurements of the widths of spectral lines in the integrated spectra of the cores of those galaxies that are interpreted as the velocities of stars in dynamical equilibrium with the massive black hole. Astrometric measurements of the orbits of stars around the black hole in the center of the Galaxy have recently become available through the application of adaptive optics in the infrared. Over the course of the past decade Genzel *et al.* (1997) and Ghez *et al.* (2008) have measured the orbital parameters of several massive stars as they orbit the center of the Galaxy. Remarkably, by observing in the infrared they can see through 20 visual magnitudes of obscuration and directly determine the mass of the central object to be about one million solar masses, and with great certainty infer that the massive object is a black hole and not a cluster of neutron stars or some other kind of massive object (see also Chapter 10).

Similarly, the masses of stars can be measured with increasing accuracy in double-star systems. Using the technique of speckle interferometry (Labeyrie 1970, and Chapters 23 and 24), multiple very short exposures are combined in a computer to achieve diffraction-limited resolution and measure the separation and position angles of close binary stars (Hartkopf *et al.* 2008, Mason *et al.* 2009). Extension of that technique to simultaneous observations at two wavelengths has now made it possible to reach one-quarter of the diffraction limit of the telescope (Horch *et al.* 2006, 2009, 2011). Optical Michelson interferometer observations (see Chapter 11) from the Georgia State University's Center for High Angular Resolution Astronomy (CHARA) array have an even higher resolution (Ten Brummelaar *et al.* 2010). As a result, we can now determine accurate masses of stars for a comparison with stellar evolutionary models that is limited only by the accuracy of the parallaxes, i.e. we are waiting for the Gaia and similar missions to give us definitive parallaxes that will yield 1% masses (see Chapters 23 and 24).

3.2.7 Solar System astrometry

Astrometry could potentially become the life saver for the human race. Over geological timescales the Earth was hit by large asteroids leading to mass extinctions. The goal to detect and determine the orbits of all hazardous near-Earth objects (NEOs) led to significant funding for the Pan-STARRS and LSST projects (Ivezic *et al.* 2008). Astrometric observations of minor planets, natural satellites of major planets, and trans-Neptunian objects (TNOs) form the foundation for orbit determination and Solar System dynamics which eventually will allow us to understand the inventory, history, evolution, and formation of our Solar System (see Chapter 26).

3.2.8 Teaching of astronomy

In closing this chapter, it is important to note that ground-based telescopes offer the only practical solution for teaching observational astronomy and astrometry and training students in the methods of observation. It is of course possible to utilize observations taken from satellites as teaching materials, but the ability to experiment with different observational methods and to repeat those observations is greatly diminished. For these reasons there is a continuing need for ground-based astrometry even in the era of high-accuracy space astrometry. Last but not least, astrometry is technology driven. Ground-based telescopes provide the means of testing new hardware and software.

References

Andrei, A. H., Souchay, J., Zacharias, N., *et al.* (2009). The large quasar reference frame (LQRF). An optical representation of the ICRS. *A&A*, **505**, 385.

Boboltz, D. A., Fey, A. L., Puatua, W. K., *et al.* (2007). Very large array plus Pie Town astrometry of 46 radio stars. *AJ*, **133**, 906.

Charlot, P., Boboltz, D. A., Fey, A. L., *et al.* (2010). The Celestial Reference Frame at 24 and 43 GHz. II. Imaging. *AJ*, **139**, 1713.

Dinescu, D. I., Girard, T. M., van Altena, W. F., and Lopez, C. E. (2005). Absolute proper motion of the Sagittarius dwarf galaxy and of the outer regions of the Milky Way Bulge. *ApJ*, **618**, L25.

Fey, A., Gordon, D., and Jacobs, C. S., eds. (2009). *The Second Realization of the International Celestial Reference Frame by Very Long Baseline Interferometry*. IERS Technical Note No. 35.

Genzel, R., Eckart, A., Ott, T., and Eisenhauer, F. (1997). On the nature of the dark mass in the centre of the Milky Way. *MNRAS*, **291**, 219.

Ghez, A. M., Salim, S., Weinberg, N., *et al.* (2008). Probing the properties of the Milky Way's central supermassive black hole with stellar orbits. *Proc. IAU Symp.*, **248**, 52.

Han, I. (1989). The accuracy of differential astrometry limited by the atmospheric turbulence. *AJ*, **97**, 607.

Hartkopf, W. I., Mason, B. D., and Rafferty, T. J. (2008). Speckle interferometry at the USNO Flagstaff Station: observations obtained in 2003–2004 and 17 new orbits. *AJ*, **135**, 1334.

Helmi, A. and White, S. D. M. (1999). Building up the stellar halo of the Galaxy. *MNRAS*, **307**, 495

Henry, T. J., Jao, W.-C., Subasavage, J. P., *et al.* (2006). The solar neighborhood. XVII. Parallax results from the CTIOPI 0.9 m program: 20 new members of the RECONS 10 parsec sample. *AJ*, **132**, 2360.

Horch, E. P., Franz, O. G., and van Altena, W. F. (2006). Characterizing binary stars below the diffraction limit with CCD-based speckle imaging. *AJ*, **132**, 2478.

Horch, E. P., Veilliette, D. R., Baena Gallé, R., *et al.* (2009). Observations of binary stars with the differential speckle survey instrument. I. Instrument description and first results. *AJ*, **137**, 5057.

Horch, E. P., van Altena, W. F., Howell, S. B., Sherry, W. H., and Ciardi, D. R. (2011). Observations of binary stars with the Differential Speckle Survey Instrument. III. Measures below the diffraction limit of the WIYN telescope. *AJ*, **141**, 180.

Ivezic, Z., Axelrod, T., Brandt, W. N., *et al.* (2008). Large Synoptic Survey Telescope: from science drivers to reference design. *Serb. Astr. J.*, **176**, 1.

Jacoby, G., Tonry, J. L., Burke, B. E., *et al.* (2002). WIYN One Degree Imager (ODI). *Proc. SPIE*, **4836**, 217.

Johnston, K. V., Spergel, D. N., and Hayden, C. (2002). How lumpy is the Milky Way's dark matter halo? *ApJ*, **570**, 656.

Kaiser, N., Aussel, H., Burke, B. E., *et al.* (2002). Pan-STARRS: a Large Synoptic Survey Telescope Array. *Proc. SPIE*, **4836**, 154.

Labeyrie, A. (1970). Attainment of diffraction limited resolution in large telescopes by Fourier analyzing speckle patterns in star images. *A&A*, **6**, 85.

Mason, B. D., Hartkopf, W. I., Gies, D. R., Henry, T. J., and Helsel, J. W. (2009). High angular resolution multiplicity of massive stars. *AJ*, **137**, 3358.

Platais, I., Kozhurina-Platais, V., Girard, T. M., *et al.* (2002). WIYN open cluster study. VIII. The geometry and stability of the NOAO CCD Mosaic Imager. *AJ*, **124**, 601.

Platais, I., Kozhurina-Platais, V., Mathieu, R. D., Girard, T. M., and van Altena, W. F. (2003). WIYN open cluster study. XVII. Astrometry and membership to V = 21 in NGC 188. *AJ*, **126**, 2922.

Platais, I., Wyse, R. F. G., and Zacharias, N. (2006). Deep astrometric standards and galactic Structure. *PASP*, **118**, 107.

Reid, M. J. (2008). Micro-arcsecond astrometry with the VLBA. *Proc. IAU Symp.*, **248**, 141.

Ten Brummelaar, T. A., *et al.* (2010). An update of the CHARA Array. *Proc. SPIE*, **7734**, 773403.

Thompson, A. R., Moran, J. M., and Swenson, G. W. (2001). *Interferometry and Synthesis in Radio Astronomy*, 2nd edn. Chichester: Wiley.

Tonry, J., Burke, B. E., and Schechter, P. L. (1997). The orthogonal transfer CCD. *PASP*, **109**, 1154.

Tonry, J., Burke, B. E., and Schechter, P. L. (2004). The orthogonal parallel imaging transfer camera. In *Scientific Detectors for Astronomy: The Beginning of a New Era*, ed. P. Amico, J. W. Beletic and J. E. Beletic. Astrophysics and Space Science Library, vol. 300, New York, NY: Springer, p. 385.

van Altena, W. F. (2008). The opportunities and challenges for astrometry in the 21st century. *Rev. Mex. A&A* (Serie de Conf.), **34**, 1.

van Altena, W. F., Lee, J. T., and Hoffleit, E. D. (1995). *The General Catalogue of Trigonometric Parallaxes*, 4th edn. New Haven: Yale University Observatory.

Walter, H. G. and Sovers, O. J. (2000). *Astrometry of Fundamental Catalogues*: *The Evolution from Optical to Radio Reference Frames*. New York, NY: Springer.

Zacharias, N. (1992). Global block adjustment simulations using the CPC 2 data structure. *A&A*, **264**, 296.

Zacharias, N. (2008). Dense optical reference frames: UCAC and URAT. *Proc. IAU Symp.*, **248**, 310.

Zacharias, N., Urban, S. E., Zacharias, M. I., *et al.* (2004). The Second US Naval Observatory CCD Astrograph Catalog (UCAC2). *AJ*, **127**, 3043.

Zacharias, N., McCallon, H. I., Kopan, E., and Cutri, R. M. (2006). Extending the ICRF into the infrared: 2MASS-UCAC astrometry. In *JD16: The International Celestial Reference Ssytem: Maintenance and Future Realization*, ed. R. Gaume, D. McCharthy, J. Souchay. Washington, DC: USNO, p. 52.

PART II

FOUNDATIONS OF ASTROMETRY AND CELESTIAL MECHANICS

4

Vectors in astrometry: an introduction

LENNART LINDEGREN

Introduction

In astrometry, vectors are extensively used to describe the geometrical relationships among celestial bodies, for example between the observer and the observed object. Practical calculations using computer software are today mainly carried out with the help of vector and matrix algebra, rather than the trigonometry formulae typically found in older textbooks. It turns out that this often provides a better insight into the problem, and hence reduces the risk of errors in the derived algorithms, in addition to being advantageous in terms of computational speed and accuracy.

This chapter provides a brief introduction to the use of vectors and matrices in astrometry. It broadly uses the notational conventions from C. A. Murray's *Vectorial Astrometry* (1983), which seem to provide a particularly clear and consistent framework for theoretical work as well as practical calculations. By way of illustration, some useful transformations are explained in detail, while references to the general literature are provided for other applications. Only vectors in three-dimensional Euclidean space are considered.

4.1 What are vectors?

In this section we define classical vectors, unit vectors, matrices and present some important formulae for manipulating them.

4.1.1 Vectors and matrices

Classically, a vector is defined as a physical entity having both magnitude (length) and direction, as opposed to a scalar that only has magnitude. Vectors can be visualized as arrows that exist in space quite independently of any coordinate system. The usual vector operations – addition, subtraction, multiplication by a scalar, scalar (dot) product, and vector (cross) product – have simple geometrical interpretations that are independent of the coordinate system. For example, the sum of two vectors **a** and **b** can be constructed

Astrometry for Astrophysics: Methods, Models, and Applications, ed. William F. van Altena. Published by Cambridge University Press. © Cambridge University Press 2013.

by means of the usual parallelogram, and their scalar product can be obtained as $ab\cos\theta$, where a and b are the lengths of the vectors and θ the angle between them. Much of the theoretical development in astrometry results in vector expressions that can be interpreted in this general way without reference to any particular coordinate system.

When the vector expressions are being used for actual calculations, it is, however, necessary to agree on a numerical representation of the vectors and to map the vector operations into a corresponding set of numerical operations. Normally this is done by using Cartesian coordinates, arranged as column (3×1) matrices; thus:

$$\mathbf{a} = \begin{bmatrix} a_x \\ a_y \\ a_z \end{bmatrix}, \quad \mathbf{b} = \begin{bmatrix} b_x \\ b_y \\ b_z \end{bmatrix} \tag{4.1}$$

where a_x is the coordinate of $\mathbf{a}$ along the adopted x axis, etc. All the above-mentioned vector operations have simple and well-known equivalents in Cartesian coordinates, e.g.

$$\mathbf{a} + \mathbf{b} = \begin{bmatrix} a_x + b_x \\ a_y + b_y \\ a_z + b_z \end{bmatrix}, \quad \mathbf{a}'\mathbf{b} = a_x b_x + a_y b_y + a_z b_z, \quad \mathbf{a} \times \mathbf{b} = \begin{bmatrix} a_y b_z - a_z b_y \\ a_z b_x - a_x b_z \\ a_x b_y - a_y b_x \end{bmatrix} \tag{4.2}$$

Following Murray (1983), the prime (′) is here used to denote both the scalar product of vectors and the transpose of a matrix. Most of the time there is no need to worry about the distinction between the vector as a physical entity and its coordinate representation by means of a matrix. The $\mathbf{a}'\mathbf{b}$ in Eq. (4.2) can therefore be interpreted either as the scalar product of vectors $\mathbf{a}$ and $\mathbf{b}$, or as row matrix $\mathbf{a}'$ (the transpose of column matrix $\mathbf{a}$) multiplied by column matrix $\mathbf{b}$.

4.1.2 Unit vectors

The length of vector $\mathbf{a}$ is the non-negative scalar quantity $|\mathbf{a}| = \sqrt{\mathbf{a}'\mathbf{a}} = (a_x^2 + a_y^2 + a_z^2)^{1/2}$. Sometimes the corresponding variable in italic (a) is used to denote the length of a vector.

Unit vectors (of length 1) are ubiquitous in astrometry as representing directions, for example from the observer towards an object, the tangent direction of a light ray, or the direction of a coordinate axis. Constructing the unit vector from a given non-zero vector $\mathbf{a}$ is a common operation for which the special symbol $\langle\ \rangle$ is sometimes used,

$$\langle \mathbf{a} \rangle = \mathbf{a}|\mathbf{a}|^{-1} \tag{4.3}$$

4.1.3 Some important formulae

For arbitrary vectors or column matrices $\mathbf{a}$, $\mathbf{b}$, $\mathbf{c}$ we have

$$\mathbf{b}'\mathbf{a} = \mathbf{a}'\mathbf{b} \tag{4.4}$$

$$\mathbf{b} \times \mathbf{a} = -(\mathbf{a} \times \mathbf{b}) \tag{4.5}$$

$$(\mathbf{a} \times \mathbf{b})'\mathbf{c} = (\mathbf{b} \times \mathbf{c})'\mathbf{a} = (\mathbf{c} \times \mathbf{a})'\mathbf{b} = \mathbf{a}'(\mathbf{b} \times \mathbf{c}) = \mathbf{b}'(\mathbf{c} \times \mathbf{a}) = \mathbf{c}'(\mathbf{a} \times \mathbf{b}) \tag{4.6}$$

$$(\mathbf{a} \times \mathbf{b}) \times \mathbf{c} = \mathbf{b}\mathbf{c}'\mathbf{a} - \mathbf{a}\mathbf{b}'\mathbf{c} \tag{4.7}$$

4.2 Coordinate systems and triads

In Eqs. (4.1)–(4.3) the coordinate system is not explicitly defined but implied by the use of Cartesian coordinates a_x, etc. This is fine as long as we are consistently working in a single coordinate system. When more than one system is involved, for example in coordinate transformations, it helps to clarify the relation between the vectors and their different coordinate representations if an explicit notation is introduced for the systems. This can be done by means of coordinate triads.

A coordinate triad is a row matrix of three mutually orthogonal unit vectors, for example $\mathbf{Z} = [\mathbf{x\ y\ z}]$, representing the axes of a Cartesian coordinate system. From the orthogonality and unit length of the vectors, it follows that $\mathbf{Z'Z} = \mathbf{I}$, the 3×3 identity matrix.[1] Coordinate triads are normally right-handed, thus $\mathbf{x} \times \mathbf{y} = \mathbf{z}$ or $\det(\mathbf{Z}) = (\mathbf{x} \times \mathbf{y})'\mathbf{z} = +1$.

In the system defined by the coordinate triad $\mathbf{Z} = [\mathbf{x\ y\ z}]$ the Cartesian coordinates of the arbitrary vector $\mathbf{a}$ are given by the column matrix

$$\mathbf{Z'a} = \begin{bmatrix} \mathbf{x'} \\ \mathbf{y'} \\ \mathbf{z'} \end{bmatrix} \mathbf{a} = \begin{bmatrix} \mathbf{x'a} \\ \mathbf{y'a} \\ \mathbf{z'a} \end{bmatrix} = \begin{bmatrix} a_x \\ a_y \\ a_z \end{bmatrix} \tag{4.8}$$

The vector can be written in terms of its coordinates with respect to $\mathbf{Z}$ as

$$\mathbf{a} = \mathbf{x}a_x + \mathbf{y}a_y + \mathbf{z}a_z = \mathbf{Z} \begin{bmatrix} a_x \\ a_y \\ a_z \end{bmatrix} \tag{4.9}$$

Introducing now a second coordinate system, represented by the triad $\mathbf{K} = [\mathbf{i\ j\ k}]$, it is seen that the coordinates of $\mathbf{a}$ in the new system are given by the column matrix

$$\begin{bmatrix} a_i \\ a_j \\ a_k \end{bmatrix} = \mathbf{K'a} = \mathbf{K'Z} \begin{bmatrix} a_x \\ a_y \\ a_z \end{bmatrix} \tag{4.10}$$

Transforming the coordinates of $\mathbf{a}$ (or any other vector) from the $\mathbf{Z}$ to the $\mathbf{K}$ system is therefore accomplished through pre-multiplication with the 3×3 matrix $\mathbf{K'Z}$. Conversely, transforming the coordinates from $\mathbf{K}$ to $\mathbf{Z}$ is done through pre-multiplication with $\mathbf{Z'K}$. Note that the latter matrix is the transpose of $\mathbf{K'Z}$, and also its inverse, since they are orthogonal matrices.

4.3 Spherical coordinates

Celestial positions are often specified by means of spherical coordinates, for example (α, δ) in the equatorial system or (l, b) in the galactic system. Transformation of spherical

---

[1] It is a useful exercise to write out this equation in full. When transposing the matrix $\mathbf{Z}$ it is necessary to transpose its elements ($\mathbf{x}$, etc.) as well, as they are not scalar quantities; cf. Eq. (4.8).

coordinates from one system to another is conveniently done by means of a matrix multiplication as in Eq. (4.10). This involves some additional trigonometric operations, to be discussed below, for the conversions between spherical and Cartesian coordinates.

4.3.1 From spherical coordinates to vector

In the generic coordinate system $\mathbf{Z} = [\mathbf{x}\ \mathbf{y}\ \mathbf{z}]$, let (ϕ, θ) be the spherical coordinates representing the direction of the non-zero vector $\mathbf{a}$. The angles are defined in the usual astronomical sense, with ϕ the longitude-like angle (for example α or l) and θ the latitude-like angle (for example δ or b). The Cartesian coordinates of $\mathbf{a}$ in $\mathbf{Z}$ are

$$\mathbf{Z}'\mathbf{a} = \begin{bmatrix} a_x \\ a_y \\ a_z \end{bmatrix} = \begin{bmatrix} a\cos\theta\cos\phi \\ a\cos\theta\sin\phi \\ a\sin\theta \end{bmatrix} \qquad (4.11)$$

where $a = |\mathbf{a}|$ is the length of the vector. If $\mathbf{a}$ is the unit vector towards (ϕ, θ), then $a = 1$ in Eq. (4.11).

4.3.2 From vector to spherical coordinates

The inversion of Eq. (4.11) requires some care in order to avoid potential numerical difficulties and ambiguities. The following formulae are recommended because they work in all reasonable circumstances and give good numerical accuracy even for positions close to the poles:[2]

$$\phi = \operatorname{atan2}\left(a_y, a_x\right), \qquad \theta = \operatorname{atan2}\left(a_z, \sqrt{a_x^2 + a_y^2}\right) \qquad (4.12)$$

Here $\operatorname{atan2}(y, x)$ is the four-quadrant inverse tangent available in the mathematical libraries of many programming languages, including C++, Fortran, Java, MATLAB, Perl and Python. Equation (4.12) returns ϕ in the interval $[-\pi, \pi]$, which is usually fine for subsequent calculations, and there is normally no need to ensure that the angle falls in the "standard" range $(0, 2\pi)$ by adding 2π for negative values.

4.3.3 The normal triad

Associated with the coordinate system $\mathbf{Z}$ and spherical coordinates (ϕ, θ) are the three orthogonal unit vectors

$$\mathbf{p} = \mathbf{Z}\begin{bmatrix} -\sin\phi \\ \cos\phi \\ 0 \end{bmatrix}, \qquad \mathbf{q} = \mathbf{Z}\begin{bmatrix} -\sin\theta\cos\phi \\ -\sin\theta\sin\phi \\ \cos\theta \end{bmatrix}, \qquad \mathbf{r} = \mathbf{Z}\begin{bmatrix} \cos\theta\cos\phi \\ \cos\theta\sin\phi \\ \sin\theta \end{bmatrix} \qquad (4.13)$$

---

[2] The first formula in Eq. (4.12) fails if both a_x and a_y are exactly zero, in which case ϕ is undefined. If $a = 0$, both formulae fail and the spherical coordinates are completely undefined. In a computer implementation of Eq. (4.12) one has to decide whether these conditions should result in an error condition, or the angles being set to some conventional values. Both solutions have their pros and cons.

which form the so-called "normal triad" [**p q r**] at point (ϕ, θ) with respect to **Z**. The third vector **r** is just the unit vector towards the point, as seen from Eq. (4.11). Given **r**, the first two vectors can be computed using vector algebra,

$$\mathbf{p} = \langle \mathbf{z} \times \mathbf{r} \rangle, \qquad \mathbf{q} = \mathbf{r} \times \mathbf{p} \tag{4.14}$$

The significance of **p** and **q** becomes apparent when **r** is differentiated with respect to the spherical coordinates,

$$\frac{\partial \mathbf{r}}{\partial \phi} = \mathbf{Z} \begin{bmatrix} -\cos\theta \sin\phi \\ \cos\theta \cos\phi \\ 0 \end{bmatrix} = \mathbf{p}\cos\theta, \qquad \frac{\partial \mathbf{r}}{\partial \theta} = \mathbf{q} \tag{4.15}$$

resulting in the total differential

$$d\mathbf{r} = \mathbf{p}\, d\phi \cos\theta + \mathbf{q}\, d\theta \tag{4.16}$$

Therefore, **p** is the unit tangent vector in the direction of increasing longitudinal angle ϕ, and **q** the unit tangent vector in the direction of increasing latitudinal angle θ. (Note that $d\phi \cos\theta$ is the "true angle" representation of the longitude differential.)

4.3.4 The proper-motion vector

The normal triad is particularly useful in dealing with proper motions. The proper motion of a star is usually expressed as the time derivatives of the spherical coordinates (as viewed from the Solar System barycenter), $\mu_\phi = d\phi/dt$, $\mu_\theta = d\theta/dt$. For the longitudinal component, the $\cos\theta$ factor is usually included; in the Hipparcos and Tycho Catalogues (ESA 1997) this is indicated by an asterisk, $\mu_{\phi*} = (d\phi/dt) \cos\theta$. Then $\mu_{\phi*}$ and μ_θ can be regarded as the projections of the proper-motion vector $(d\mathbf{r}/dt)$ on **p** and **q**, respectively.

When ϕ and θ are functions of time, for example because of the object's proper motion, it means that the normal triad computed from Eq. (4.13) is also changing with time. It would, however, be highly inconvenient to express the proper-motion components in such a continually changing coordinate system. The practical solution is to use a fixed normal triad, usually the one computed for the values of ϕ and θ corresponding to the barycentric position at the reference epoch of the catalog. Let $\mathbf{u}(t)$ be the barycentric direction towards the star, which is then in general different from **r**. The proper-motion vector is given by

$$\boldsymbol{\mu} \equiv \frac{d\mathbf{u}}{dt} = \mathbf{p}\,\mu_{\phi*} + \mathbf{q}\,\mu_\theta \tag{4.17}$$

and conversely

$$\mu_{\phi*} = \mathbf{p}'\boldsymbol{\mu}, \qquad \mu_\theta = \mathbf{q}'\boldsymbol{\mu} \tag{4.18}$$

It is important to remember that **p** and **q** depend not only on the position of the object (i.e. on **r**) but also on the chosen coordinate system. Thus, when transforming from one coordinate system to another, the proper-motion vector in Eqs. (4.17) and (4.18) remains the same but its components along **p** and **q** will be different in the two systems.

4.4 Rotations

Rotating the arbitrary vector **a** by the angle ϵ about the unit vector **e** results in the new vector

$$\mathbf{b} = \mathbf{a}\cos\epsilon + \mathbf{e}\mathbf{e}'\mathbf{a}(1 - \cos\epsilon) + (\mathbf{e} \times \mathbf{a})\sin\epsilon \tag{4.19}$$

In particular, application of the above rotation $(\mathbf{e}, \epsilon)$ to the coordinate triad $\mathbf{Z} = [\mathbf{x}\ \mathbf{y}\ \mathbf{z}]$ results in the new triad

$$\tilde{\mathbf{Z}} \equiv [\tilde{\mathbf{x}}\ \tilde{\mathbf{y}}\ \tilde{\mathbf{z}}] = \mathbf{Z}\cos\epsilon + \mathbf{e}\mathbf{e}'\mathbf{Z}(1 - \cos\epsilon) + (\mathbf{e} \times \mathbf{Z})\sin\epsilon \tag{4.20}$$

Now let **r** be some vector fixed in space. According to Eq. (4.10), the coordinates $[r_{\tilde{x}},\ r_{\tilde{y}},\ r_{\tilde{z}}]'$ of the vector in the rotated system are obtained through pre-multiplication of the original coordinates $[r_x,\ r_y,\ r_z]'$ by the 3×3 matrix $\tilde{\mathbf{Z}}'\mathbf{Z}$. Using Eq. (4.20) we find

$$\tilde{\mathbf{Z}}'\mathbf{Z} = \mathbf{Z}'\mathbf{Z}\cos\epsilon + \mathbf{Z}'\mathbf{e}\mathbf{e}'\mathbf{Z}(1 - \cos\epsilon) + (\mathbf{e} \times \mathbf{Z})'\mathbf{Z}\sin\epsilon$$

$$= \mathbf{I}\cos\epsilon + \begin{bmatrix} e_x^2 & e_x e_y & e_x e_z \\ e_y e_x & e_y^2 & e_y e_z \\ e_z e_x & e_z e_y & e_z^2 \end{bmatrix}(1 - \cos\epsilon) + \begin{bmatrix} 0 & e_z & -e_y \\ -e_z & 0 & e_x \\ e_y & -e_x & 0 \end{bmatrix}\sin\epsilon \tag{4.21}$$

It may be noted that the components of **e** are the same in the two systems.

Equation (4.21) is much simplified when **e** coincides with one of the axes of **Z**. Putting **e** in turn equal to **x**, **y**, and **z** gives the elementary rotation matrices

$$\mathbf{R}_x(\phi) = \begin{bmatrix} 1 & 0 & 0 \\ 0 & c\phi & s\phi \\ 0 & -s\phi & c\phi \end{bmatrix}, \quad \mathbf{R}_y(\theta) = \begin{bmatrix} c\theta & 0 & -s\theta \\ 0 & 1 & 0 \\ s\theta & 0 & c\theta \end{bmatrix}, \quad \mathbf{R}_z(\psi) = \begin{bmatrix} c\psi & s\psi & 0 \\ -s\psi & c\psi & 0 \\ 0 & 0 & 1 \end{bmatrix}$$
$$\tag{4.22}$$

where ϕ, θ, and ψ are the rotation angles (using, for brevity, s and c for the sine and cosine). Since any rotation matrix can be decomposed as the product of three (or more) elementary rotation matrices, it is often convenient to express the relation between two coordinate systems by means of successive elementary rotations, rather than using $(\mathbf{e}, \epsilon)$. The minimum set of three successive rotation angles is sometimes referred to as the Euler angles, but there are many possible conventions in use depending on the order in which the axes are taken (Wertz 1978). An alternative way of representing general rotations is by means of unit quaternions (Wertz 1978), which are usually the preferred choice for example in spacecraft attitude control and computer games.

The rotation matrices (4.22) transform the coordinates of a fixed vector when the coordinate system is rotated. What happens if instead a vector is rotated in a fixed coordinate system? The answer is found by resolving Eq. (4.19) into the coordinates of a fixed triad, say $\mathbf{Z} = [\mathbf{x}\ \mathbf{y}\ \mathbf{z}]$:

$$\mathbf{Z}'\mathbf{b} = \mathbf{Z}'\mathbf{a}\cos\epsilon + \mathbf{Z}'\mathbf{e}\mathbf{e}'\mathbf{a}(1 - \cos\epsilon) + \mathbf{Z}'(\mathbf{e} \times \mathbf{a})\sin\epsilon$$

$$= \left(\mathbf{I}\cos\epsilon + \mathbf{Z}'\mathbf{e}\mathbf{e}'\mathbf{Z}(1 - \cos\epsilon) + \mathbf{Z}'(\mathbf{e} \times \mathbf{Z})\sin\epsilon\right)\mathbf{Z}'\mathbf{a} \tag{4.23}$$

since $\mathbf{e} \times \mathbf{a} = \mathbf{e} \times (\mathbf{ZZ'a}) = (\mathbf{e} \times \mathbf{Z})\mathbf{Z'a}$. This shows that the coordinates of $\mathbf{b}$ are obtained through pre-multiplication by a matrix which is the transpose of (4.21). Since the matrix is orthogonal, its transpose equals the inverse, which is also obtained by reversing the sign of the rotation angle. For example, if $\mathbf{a}$ is rotated an angle ϕ about the $\mathbf{x}$ axis, the relevant transformation matrix is $\mathbf{R}'_x(\phi) = \mathbf{R}_x^{-1}(\phi) = \mathbf{R}_x(-\phi)$.

In some literature, for example Seidelmann (1992, p. 552), elementary rotation matrices are defined in the opposite sense of (4.22), but since they are post-multiplied to the vector coordinates expressed as a row matrix, the end result is the same.[3] The reader should be aware of the several potential ambiguities involved in using rotation matrices, in particular whether they apply to column or row vectors, and to a rotation of the coordinate system or of the vector itself.

4.5 Example: conversion between equatorial and galactic coordinates

For transformation between the equatorial and galactic systems, the unit vector representing the barycentric direction to the star at the adopted reference epoch may be written

$$\mathbf{r} = \mathbf{E} \begin{bmatrix} \cos\delta\cos\alpha \\ \cos\delta\sin\alpha \\ \sin\delta \end{bmatrix} = \mathbf{G} \begin{bmatrix} \cos b\cos l \\ \cos b\sin l \\ \sin b \end{bmatrix} \qquad (4.24)$$

where $\mathbf{E}$ and $\mathbf{G}$ are the equatorial and galactic coordinate triads. Pre-multiplication by $\mathbf{G}'$ (remembering that $\mathbf{G'G} = \mathbf{I}$) gives the required formula for transformation from equatorial to galactic; pre-multiplication by $\mathbf{E}'$ gives the inverse transformation.

The rotation matrix relevant for converting from equatorial to galactic coordinates is therefore $\mathbf{G'E}$. Let (α_G, δ_G) denote the equatorial coordinates of the north galactic pole and l_Ω the galactic longitude of the first intersection of the galactic plane with the equator. It is then seen that a triad originally aligned with $\mathbf{E}$ can be brought into alignment with $\mathbf{G}$ by the following successive rotations: (i) by the angle $\alpha_G + 90°$ about the third axis; (ii) by the angle $90° - \delta_G$ about the resulting first axis; (iii) by the angle $-l_\Omega$ about the resulting third axis. Consequently

$$\mathbf{G'E} = \mathbf{R}_z(-l_\Omega)\mathbf{R}_x(90° - \delta_G)\mathbf{R}_z(\alpha_G + 90°) \qquad (4.25)$$

If $\mathbf{E}$ stands for the International Celestial Reference System (ICRS), as realized for example by the Hipparcos and Tycho Catalogues (ESA 1997), then we may adopt the following values as defining $\mathbf{G}$:

$$\alpha_G = 192.85948°, \qquad \delta_G = 27.12825°, \qquad l_\Omega = 32.93192° \qquad (4.26)$$

---

[3] The conventions used here seem to be the ones most commonly used in astronomical literature (for example Barbieri 2007, Eichhorn 1974, Kovalevsky and Seidelmann 2004, Taff 1981; also implicitly in Seidelmann 1992, pp. 103, 182, etc.).

(ESA 1997, vol. 1, p. 91), from which[4]

$$\mathbf{G}'\mathbf{E} = \begin{bmatrix} -0.054\ 875\ 560\ 416\ 215 & -0.873\ 437\ 090\ 234\ 885 & -0.483\ 835\ 015\ 548\ 713 \\ +0.494\ 109\ 427\ 875\ 584 & -0.444\ 829\ 629\ 960\ 011 & +0.746\ 982\ 244\ 497\ 219 \\ -0.867\ 666\ 149\ 019\ 005 & -0.198\ 076\ 373\ 431\ 202 & +0.455\ 983\ 776\ 175\ 067 \end{bmatrix}$$

(4.27)

In order to transform the proper-motion components, the proper-motion vector is written

$$\boldsymbol{\mu} = \mathbf{p}_E\, \mu_{\alpha*} + \mathbf{q}_E\, \mu_\delta = \mathbf{p}_G\, \mu_{l*} + \mathbf{q}_G\, \mu_b \tag{4.28}$$

where $[\mathbf{p}_E\ \mathbf{q}_E\ \mathbf{r}]$ and $[\mathbf{p}_G\ \mathbf{q}_G\ \mathbf{r}]$ are the normal triads at $\mathbf{r}$ with respect to $\mathbf{E}$ and $\mathbf{G}$. When transforming from equatorial to galactic components, use the first equality to compute the proper-motion vector in the $\mathbf{E}$ system, then obtain $\mu_{l*} = \mathbf{p}'_G \boldsymbol{\mu}$ etc as in Eq. (4.18). This last computation can be done in either system. If the $\mathbf{G}$ system is used, transform $\mathbf{r}$ and $\boldsymbol{\mu}$ into galactic coordinates and compute $\mathbf{G}'\mathbf{p}_G$ and $\mathbf{G}'\mathbf{q}_G$ by means of Eq. (4.14), noting that $\mathbf{G}'\mathbf{z}_G = [0,\ 0,\ 1]'$, where $\mathbf{z}_G$ is the third component of the coordinate triad $\mathbf{G}$. If $\mathbf{E}$ is used, then compute $\mathbf{E}'\mathbf{p}_G$ and $\mathbf{E}'\mathbf{q}_G$ from Eq. (4.14), noting that $\mathbf{E}'\mathbf{z}_G$ makes up the third column of $\mathbf{E}'\mathbf{G}$ or the third row of $\mathbf{G}'\mathbf{E}$.

References

Barbieri, C. (2007). *Fundamentals of Astronomy*. Boca Raton, FL: Taylor & Francis.

Eichhorn, H. (1974). *Astronomy of Star Positions – A Critical Investigation of Star Catalogues, the Methods of their Construction and their Purpose*. New York, NY: Ungar.

ESA (1997). *The Hipparcos and Tycho Catalogues*. ESA Special Publication SP-1200.

Kovalevsky, J. and Seidelmann, P. K. (2004). *Fundamentals of Astrometry*. Cambridge: Cambridge University Press.

Murray, C. A. (1983). *Vectorial Astrometry*. Bristol: Adam Hilger.

Seidelmann, P. K. E. (1992). *Explanatory Supplement to the Astronomical Almanac*. Mill Valley, CA: University Science Books.

Taff, L. G. (1981). *Computational Spherical Astronomy*. New York, NY: Wiley-Interscience.

Wertz, J. R. E. (1978). *Spacecraft Attitude Determination and Control*. Dordrecht: Reidel, Astrophysics and Space Science Library, vol. 73.

---

[4] The matrix in Eq. (4.27) is the transpose of the $\mathbf{A}_G$ given in the Hipparcos and Tycho Catalogues, vol. 1, Eq. [1.5.11]. This is consistent with the use of $\mathbf{A}'_G$ in Eq. [1.5.13]. $\mathbf{A}_G$ is only given to 10 decimals in that reference, which was sufficient for the data in the Hipparcos Catalogue tabulated to a precision of 0.01 mas. The result is here given to 15 decimals, which is adequate even for nanoarcsecond astrometry. Nevertheless, this matrix – or any other finite-precision matrix – cannot be taken as *defining* the ICRS/galactic relation; that should be done by means of a set of conventional angles as in Eq. (4.26).

5 Relativistic foundations of astrometry and celestial mechanics

SERGEI KLIONER

Introduction

Tremendous progress in technology during the last 30 years has led to enormous improvements of accuracy in astrometry and related disciplines. Considering the growth of accuracy of positional observations in the course of time, we see that during the 25 years between 1988 and 2013 we expect the same gain in accuracy (4.5 orders of magnitude) as that realized during the whole previous history of astrometry, from Hipparchus till 1988 (over 2000 years). It is clear that for current and anticipated accuracy requirements, astronomical phenomena have to be formulated within the framework of general relativity. Many high-precision astronomical techniques already require sophisticated relativistic modeling since the main relativistic effects are several orders of magnitude larger than the technical accuracy of observations. Consequently, many current and planned observational projects cannot achieve their goals without a properly relativistic analysis. In principle, some basic relativistic effects have been taken into account since at least the 1960s. However, for a long time relativity has been viewed as "one more small correction" to the standard Newtonian formulae, rather than the fundamental framework for the data analysis. For the observational accuracy expected from proposed space instruments like Gaia, we must include not only the main relativistic effects, but also a multitude of second-order corrections. In this case it is impossible to treat relativistic effects as small corrections to the standard Newtonian scheme of data reduction. Consequently, the whole modeling scheme should be formulated in a language compatible with general relativity. Many Newtonian concepts and ideas should be re-considered and replaced by mathematically rigorous relativistic alternatives, including time, celestial coordinates, parallax, proper motion, etc.

Space does not permit a detailed introduction to general relativity and its applications to the modeling of astronomical data. In addition to numerous textbooks for physicists (e.g. Schutz 1985), the books of Soffel (1989) and Brumberg (1991) can be recommended as introductory courses of relativity for astronomers. This chapter contains an explanation of the basic principles of relativistic modeling of observational data as well as a number of "recipes" for various relativistic effects.

Astrometry for Astrophysics: Methods, Models, and Applications, ed. William F. van Altena.
Published by Cambridge University Press. © Cambridge University Press 2013.

47

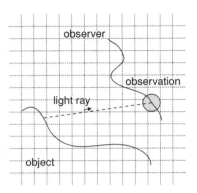

Fig. 5.1 Four parts of an astronomical event from the point of view of Newtonian physics: (1) motion of the observed object; (2) motion of the observer; (3) trajectory of an electromagnetic signal from the observed object to the observer, which is tacitly assumed to be a straight line in Newtonian astronomy; and (4) the process of observation responsible for Newtonian aberration. The coordinate grid in the background symbolizes a global inertial reference system.

5.1 Newtonian modeling of astrometric observations

Before proceeding to relativistic modeling of astronomical observations let us formulate the basic principles of Newtonian modeling from the physical point of view. The reduction of observations in Newtonian physics is rather simple. The existence of absolute Euclidean space and absolute time in Newtonian physics leads to the existence of a class of global reference systems that cover the whole space and time at once: inertial reference systems. Such a reference system is defined by up to 14 constants describing arbitrary linear transformations of time and three spatial axes (eight constants), arbitrary constant velocity of the origin of one inertial reference system relative to another one (three constants) as well as a time-independent rotation of spatial axes (three constants). Although one can introduce arbitrary curvilinear coordinates in Newtonian physics, inertial reference systems are physically preferred, since the laws of physics in these coordinates look drastically simpler than in non-inertial reference systems. Moreover, observable quantities – intervals of time and angles between directions – are immediately related to inertial coordinates. In fact, Newtonian space itself is often thought to be endowed with a global inertial reference system as illustrated in the famous lithograph "Cubic Space Division" by M. C. Escher.

Figure 5.1 sketches the four constituents of an astronomical observation from the point of view of Newtonian physics: (1) the observed object and its trajectory, (2) the observer and its trajectory, (3) propagation of an electromagnetic signal (photon) from the object to the observer, and (4) the process of observation, that is the process of interaction of a photon with the receiver. The models of the last two constituents are especially simple in Newtonian physics. It is usually assumed that in inertial coordinates the trajectories of photons are straight lines joining observed object and observer. Modeling of "the process of observation" reduces to a calculation of the velocity of a photon with respect to the observer. This is usually done by subtracting the velocity of the observer from the velocity of the photon in some chosen reference system. The difference in the direction of the photon's velocity in the chosen inertial coordinates and with respect to the observer is called aberration.

The goal of Newtonian reduction of astronomical observations is to model (that is, to predict on the basis of real observational data) the results of observations performed by a fictitious observer (normally situated at the origin of the chosen reference system, e.g. at the barycenter of the Solar System). We attempt here to correct for all the effects in observations that are induced by the motion of the real observer (aberration and parallax) and by the motion of the observed object (proper motion, orbital solution for binary stars or solar system objects and, possibly, light travel time effects). The structure of the Newtonian reduction scheme does not depend on the goal accuracy of reduction and can be described as subsequent calculation of (1) aberration, (2) parallax, and (3) proper motion and/or light travel time effects. For lower accuracies when only linear effects from aberration, parallax, and proper motion are of interest, one could apply the corresponding corrections in arbitrary order. On the contrary, for higher accuracies the order of these reductions is important. All parameters of the Newtonian model, i.e. the coordinates of the observer and the object as functions of time, are defined in the chosen inertial reference system. That is, the standard astrometric parameters of the object (right ascension α, declination δ, parallax π, proper motion in right ascension μ_α, and proper motion in declination μ_δ, and, possibly, radial velocity v_r) are also defined in the chosen reference system.

5.2 Why general relativity?

For many kinds of observations a Newtonian data reduction is not accurate enough: deviations of the Newtonian predictions from real observations can be several orders of magnitudes larger than the technical observational accuracy. Good examples here are (geodetic) very long baseline interferometry (VLBI), lunar laser ranging (LLR), radar location of planets, time dissemination, and satellite navigation systems (GPS, GLONASS, Galileo, etc.). In this way relativity has transformed from a purely academic discipline into an applied science. High-accuracy observational data confirm the validity of general relativity. The typical precision with which various predicted general-relativistic effects are observed is currently 0.001. The highest precision of 10^{-5} has been achieved using Doppler measurements of the Cassini spacecraft (Bertotti *et al.* 2003). As far as the fundamental idea of general relativity, the Einstein Equivalence Principle, is concerned, its three constituents – Weak Equivalence Principle (the equivalence of gravitational and inertial masses), Local Lorentz Invariance (the hypothesis that the local physical laws do not depend on the motion of observer), and Local Positional Invariance (the hypothesis that the local physical laws do not depend on the position of the observer in space and time) – are confirmed with precisions of 4×10^{-13}, $\simeq 10^{-21}$, and 2×10^{-4}, respectively. We should stress that general relativity is not the only theory of gravity that successfully describes all available observational data. As well as general relativity there are a number of alternative theories of gravity. However, general relativity is the simplest of all viable theories. The difference between general relativity and viable alternative theories is important for objects with strong gravitational fields and high velocities of motion (for example, binary pulsars or black holes). Direct detection of gravity waves will be a further important test of general relativity. Although this decisive test has not yet been done, general relativity has proved to be a reliable theory in the solar system with its slow motions and weak gravitational fields and can be used to formulate the standard scheme of relativistic modeling of observational data. Although the reference systems and reduction formulas given below are formulated in the framework of general relativity, it is possible to reformulate the whole scheme in the framework of the so-called parametrized post-Newtonian (PPN) formalism. This formalism formally combines a class of alternative theories of gravity in the first post-Newtonian approximation by defining a generic theory including a set of numerical parameters (e.g. the famous PPN parameters β and γ) so that each theory corresponds to certain values of those parameters. The main goal of the PPN formalism is to formulate the reduction models in such a way that these numerical parameters can be directly determined from observations. A detailed description of various tests of general relativity and of the PPN formalism can be found in Will (1993, 2006).

5.3 General schema of relativistic reduction of astronomical observations

Let us now outline the general principles of relativistic modeling of astronomical observations. It is interesting that in spite of a deep conceptual difference between Newtonian physics and general relativity, the structure of the reduction scheme changes, in principle, only in two points: (1) in curved spacetime of general relativity there are no preferred coordinates in which the laws of physics are drastically simpler than in other coordinates, and (2) light rays are no longer straight lines (in three-dimensional space) and should be carefully modeled. Any reference system covering the region of spacetime under consideration can be used to describe the physical phenomena in that region. Instead of Newtonian inertial coordinates we should choose some "curvilinear" reference system in curved spacetime (such a reference system is schematically shown by dashed curves on Fig. 5.2 that shows the four constituents of an astronomical observation in the relativistic framework).

The general scheme of relativistic modeling of astronomical observations of an arbitrary kind is shown in Fig. 5.3. Starting from general relativity, or any other theory of gravity or PPN formalism, it is possible to define at least one relativistic four-dimensional reference system, covering the region of spacetime where all physical processes constituting the astronomical observation are located. A typical astronomical observation from the point of view of general relativity is shown in Fig. 5.2 and consists of the same four constituents as in the Newtonian model: the processes of motion of the object and the observer, the process of signal propagation from the object to the observer, and the process of observation. Each of these constituents should be modeled in agreement with physical principles and mathematical language of general relativity or PPN formalism (Fig. 5.3). First, the equations

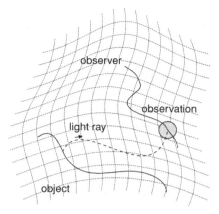

Fig. 5.2   Four parts of an astronomical event from the point of view of relativistic physics: (1) motion of the observed object; (2) motion of the observer; (3) trajectory of an electromagnetic signal from the observed object to the observer; (4) the process of observation. The grid of curved coordinates in the background symbolizes the chosen relativistic reference system.

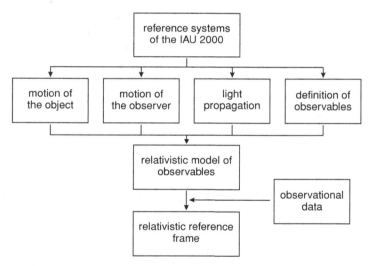

Fig. 5.3   General scheme of relativistic reduction of astronomical observations.

of motion for the observer and observed object relative to the chosen reference system and the way to solve these equations should be found. Normally, the equations of motion are ordinary second-order differential equations that can be solved numerically. The observer measures properties of electromagnetic signals coming from the observed object. Therefore, it is necessary to derive and solve the equations of light propagation in the chosen reference system. In order to solve all these equations some initial values have to be fixed for the position and velocity of each object at some initial moment of time. The motion of the object and the observer together with the laws of signal propagation allow the position and velocity of the observer, the observed object, and the photon (electromagnetic signal) to be computed in the chosen reference system at any moment of coordinate time. It is clear that these calculated positions and velocities depend on the reference system that we use for modeling. On the other hand, the results of observations cannot depend on any details of the modeling and, in particular, cannot depend on the reference system used. Therefore, one more step is required: relativistic description of the process of observation. This step allows the calculation of coordinate-independent values of observables from the coordinate-dependent position and velocity of the observer and, possibly, the coordinate velocity of the signal at the moment of observation.

These four components should be combined together into a single relativistic model of observables (Fig. 5.3). Such models are represented by some expressions for observable quantities as functions of a number of parameters. Numerical values of these parameters are estimated from observations using some statistical estimator (e.g. the least-squares estimator). Sets of these parameters, characterizing observed objects or observers, represent relativistic astronomical reference frames, that is, physical realizations of the chosen reference systems.

Let us recall that a reference system in astronomy is a mathematical concept, allowing us to ascribe four real numbers to any event in physical spacetime. On the contrary, a reference frame is some materialization (realization) of the corresponding reference system. Usually in astronomy such a materialization is given in the form of a catalog (or ephemeris) containing positions of some celestial objects relative to the selected reference system. Any astronomical reference frame (a catalog, an ephemeris, etc.) is defined only through the reference system(s) used to construct physical models of observations. It is also important to understand that the parameters that appear in a relativistic model are usually defined only in the reference system(s) used to construct the model and are, therefore, coordinate-dependent quantities. For example, the position and velocity of an observed object are obviously coordinate-dependent.

5.4 Relativistic reference systems

From the physical point of view all reference systems covering the region of spacetime being studied are equivalent and we can choose to work in any reference system. However, it is well known that a mathematical description of physical processes can be simpler in one reference system than in others and we naturally use the freedom to simplify the parametrization of the observables. Each reference system in general relativity is fully defined by its metric tensor. A metric tensor is a symmetric real 4×4 matrix, defined in each point of the region of spacetime covered by the reference system. The word "tensor" means that this matrix field can be transformed from one reference system to another using the so-called tensorial transformation rule. The matrix field is called "metric" because it can be used to calculate distances between any two infinitely close points. In two-dimensional Euclidean space $\mathbb{R}^2$ in rectangular Cartesian coordinates (x, y) the distance $\mathrm{d}s$ between two points with coordinates (x, y) and $(x + \mathrm{d}x, y + \mathrm{d}y)$ is given by the Pythagoras theorem $\mathrm{d}s^2 = \mathrm{d}x^2 + \mathrm{d}y^2$.

In polar coordinates (r, θ) the same distance can be written as $ds^2 = dr^2 + r^2 d\theta^2$. Both of these formulas can be written as a product of some matrix $g_{\alpha\beta}$, called the metric tensor or simply the metric, and a vector of coordinate differentials dx^α. For example, in $\mathbb{R}^2$ in rectangular Cartesian coordinates (x, y) we have $dx^1 = dx$ and $dx^2 = dy$, while in polar coordinates (r, θ) we get $dx^1 = dr$ and $dx^2 = d\theta$. Therefore, we have

$$ds^2 = g_{\alpha\beta} \, dx^\alpha dx^\beta. \tag{5.1}$$

Here the Einstein implicit summation rule is used: the repeated indices assume summation over all possible values of these indices. Traditionally, in general relativity the time coordinate has index 0 and three spatial coordinates have indices 1, 2, and 3, Greek indices run from 0 to 3, while Latin indices take values 1, 2, and 3 only.

Special relativity is based on the postulated constancy of the light velocity c. In inertial coordinates $(x^0, x^i) = (ct, x, y, z)$ this can be expressed as $dx^2 + dy^2 + dz^2 = c^2 dt^2$ or $ds^2 = 0$, where $ds^2 = -c^2 dt^2 + dx^2 + dy^2 + dz^2$. Hence, in the inertial coordinates of special relativity

$$g_{00} = -1$$
$$g_{0i} = 0 \tag{5.2}$$
$$g_{ij} = \delta_{ij}$$

where δ_{ij} is the unit matrix ($\delta_{ij} = 1$ for $i = j$; $\delta_{ij} = 0$ for $i \neq j$). As in Newtonian physics the inertial coordinates of special relativity are not defined uniquely. The transformations between inertial coordinates are given by the Lorentz transformations

$$t' = \left(t - \frac{1}{c^2} v^i x^i\right) \frac{1}{\sqrt{1 - v^2/c^2}}$$
$$x'^i = x^i - v^i t + \left(\frac{1}{\sqrt{1 - v^2/c^2}} - 1\right) \left(v^k x^k - v^2 t\right) \frac{v^i}{v^2} \tag{5.3}$$

This form of the Lorentz transformations involves three components of the relative velocity v^i as free parameters. The most general form of the Lorentz transformations (sometimes called Poincare transformations) also include arbitrary constant shifts (and, possibly, scalings) of all four coordinates (up to eight parameters), and arbitrary time-independent rotations of the spatial axis (three parameters).

In modern treatments of relativistic modeling two reference systems are used: the Barycentric Celestial Reference System (BCRS) and the Geocentric Celestial Reference System (GCRS). Both reference systems (referred to as IAU 2000) were adopted by the IAU (IAU 2001, Rickman 2001, Soffel *et al.* 2003). Several versions of the GCRS can be considered: reference system for the Earth (geocentric in the proper sense), analogous reference systems for other massive bodies of the Solar System, and a physically adequate reference system for a massless observer. The latter can be called the Observer's Celestial Reference System (OCRS). Let us describe the main properties of these reference systems.

5.4.1 Barycentric Celestial Reference System

A detailed analysis carried out by a number of working groups and commissions of the International Astronomical Union (IAU) and the International Bureau of Weights and Measures (IBPM) has shown that the reference system (t, x^i), which is most convenient for modeling the motion of Solar System bodies as well as for any observations performed from within the Solar System, is given by the metric tensor (Soffel *et al.* 2003):

$$g_{00} = -1 + \frac{2}{c^2} w - \frac{2}{c^4} w^2 + O(c^{-5})$$

$$g_{0i} = -\frac{4}{c^3} w^i + O(c^{-5}) \tag{5.4}$$

$$g_{ij} = \delta_{ij} \left(1 - \frac{2}{c^2} w \right) + O(c^{-4})$$

where the post-Newtonian gravitational potentials w and w^i read

$$w(t, \mathbf{x}) = G \int d^3x' \frac{\sigma(t, \mathbf{x}')}{|\mathbf{x} - \mathbf{x}'|} + \frac{1}{2c^2} \frac{\partial^2}{\partial t^2} \int d^3x' \, \sigma(t, \mathbf{x}') \, |\mathbf{x} - \mathbf{x}'| \tag{5.5}$$

$$w^i(t, \mathbf{x}) = G \int d^3x' \frac{\sigma^i(t, \mathbf{x}')}{|\mathbf{x} - \mathbf{x}'|} \tag{5.6}$$

Here σ and σ^i are the post-Newtonian density and vector density, related to the components of the energy-momentum tensor $T^{\alpha\beta}$ as follows:

$$\sigma = \frac{1}{c^2} \left(T^{00} + T^{kk} \right)$$

$$\sigma^i = \frac{1}{c} T^{0i} \tag{5.7}$$

The energy-momentum tensor $T^{\alpha\beta}$ fully characterizes the distribution of matter (and energy) in the Solar System as function of t and x^i. Usually, knowledge of $T^{\alpha\beta}$ itself is not required since the information on the distribution of matter is replaced by the multipole moments of gravitational fields of each required body (see below). The origin of spatial coordinates x^i coincides with the barycenter of the Solar System at any moment of time t. This reference system is called the Barycentric Celestial Reference System, BCRS. The BCRS is recommended by the IAU (IAU 2001, Rickman 2001, Soffel *et al.* 2003) for modeling of astronomical observations. The BCRS coordinates satisfy the so-called harmonic gauge conditions that make calculations especially simple. Straightforward calculations show that the metric tensor given above satisfies the Einstein field equations of general relativity in the first post-Newtonian approximation (up to the powers of c^{-1} shown in Eq. (5.4)). This approximation guarantees sufficient accuracy for modeling of modern astrometric observations, including microacrsecond (μas) positional observations, performed more than one degree from the Sun (this is true, for example, for Gaia). Higher-order (so-called post-post-Newtonian terms) can be added to the metric tensor (5.4) as it becomes necessary. The "celestial" in BCRS stresses that this reference system does not rotate with the Earth and that distant celestial objects (for example, quasars) are at rest with respect to the BCRS in some average sense. The coordinate time of the BCRS is called Barycentric Coordinate Time (TCB). The TCB and its linear function TDB are used to parametrize barycentric reference frames, that is, catalogs of celestial objects or ephemerides of the Solar System.

5.4.2 Geocentric Celestial Reference System

The other reference system defined by the IAU, the GCRS, can be constructed for any of the bodies of an N body system. Since at the present time the most important applications of such a reference system are related to the Earth, this reference system is called "geocentric." A body is simply some bounded region of space in which the energy-momentum tensor is not zero. It is assumed that there are N such bodies and that these bodies are separated from each other (so that between the bodies the energy-momentum tensor vanishes). The GCRS with coordinates (T, X^a) has two remarkable properties:

(a) The gravitational field of external bodies is represented only in the form of tidal terms, that have at least second order $O(X^2)$ relative to the spatial coordinates X^a and coincides with the Newtonian tidal potential in the Newtonian limit.
(b) The internal gravitational field of the central body coincides with the gravitational field of the corresponding isolated source provided that the tidal field of external masses is neglected.

Because of the so-called Strong Equivalence Principle that is satisfied in general relativity, one can construct a reference system that satisfies these two properties simultaneously (Soffel *et al.* 2003, Klioner and Soffel 2000). These two conditions impose certain constraints on the reference system (T, X^a) and on the coordinate transformations between the global BCRS and the local coordinates of the GCRS. These constraints together with the gauge conditions (again the harmonic gauge is applied to the GCRS) almost uniquely define both the metric tensor of the GCRS and the coordinate transformations between the global coordinates of the BCRS and the local coordinates of the GCRS. The coordinate transformations between the BCRS and GCRS involve both coordinate times and spatial coordinates and represent a generalization of the four-dimensional Lorentz transformation. Local reference systems like the GCRS are physically adequate for the modeling of physical processes located in the immediate vicinity of the central body: e.g. rotational motion of the central body, motion of its satellites, and its internal dynamics. The GCRS is constructed in such a way that fictitious relativistic effects (for example, fictitious changes of Earth's shape induced by Lorentz contraction) are avoided in the GCRS coordinates. The metric tensor of the GCRS is similar to the metric tensor of the BCRS:

$$G_{00} = -1 + \frac{2}{c^2}W - \frac{2}{c^4}W^2 + O(c^{-5})$$

$$G_{0i} = -\frac{4}{c^3}W + O(c^{-5})$$ (5.8)

$$G_{ij} = \delta_{ij}\left(1 - \frac{2}{c^2}W\right) + O(c^{-4})$$

with the gravitational potentials W and W^a consisting of three parts

$$W = W_E + Q_a X^a + W_T \tag{5.9}$$

$$W^a = W_E^a + \frac{1}{2}\varepsilon_{abc}C_b X^c + W_T^a \tag{5.10}$$

Here W_E and W_E^a are the post-Newtonian gravitational potentials of the central body (the Earth). These potentials are defined by Eqs. (5.5)–(5.7), but with the integration limited by the volume of the central body and all quantities referred to the GCRS coordinates. Potentials W_T and W_T^a represent the post-Newtonian tidal gravitational field of all external bodies. These potentials are of second order with respect to the local spatial coordinates X^a: $W_T = O(\mathbf{X}^2)$ and $W_T^a = O(\mathbf{X}^2)$. Two terms $Q_a X^a$ and $\frac{1}{2}\varepsilon_{abc}C_b X^c$ describe inertial forces. Here Q_a and C_a are arbitrary functions of time T. Function Q_a defines the acceleration of the GCRS origin relative to the geocentric, momentarily co-moving locally inertial reference system. In other words, an accelerometer placed at the GCRS origin measures acceleration Q_a (Klioner and Soffel 2000, Section VIII). For the Earth, Q_a is of the order of 4×10^{-11} m/s^2 and can usually be neglected. The function C_a defines the rotational motion of the spatial axes of the GCRS relative to the momentarily co-moving locally inertial reference system. Clearly, the equations of test particles relative to the GCRS with $C_a \neq 0$ contain Coriolis forces. The GCRS is kinematically non-rotating with respect to the BCRS, that is the coordinate transformations between BCRS and GCRS do not involve rotation of spatial coordinates. This condition requires a specific non-zero value for C_a which can be found (e.g. in Soffel *et al.* 2003). That specific value of C_a represents the angular velocity of the so-called relativistic precessions: geodetic precession, Lense–Thirring precession, and special-relativistic Thomas precession. The magnitude of that angular velocity is of the order of $2''$ per century. Coordinate time of the GCRS is called Geocentric Coordinate Time (TCG). This time and its linear function TT are used to parametrize geocentric reference frames (e.g. catalogs of positions of Earth-bound sites, theories of Earth rotation, and motion of artificial satellites).

One important role of the GCRS is the physically meaningful description of the structure of the gravitational field of the central body. With sufficient accuracy we have

$$W_E(T, \mathbf{X}) = \frac{GM_E}{R}\left\{1 + \sum_{l=2}^{\infty}\sum_{m=0}^{l}\left(\frac{R_E}{R}\right)^l P_{lm}(\cos\theta)\left[C_{lm}^E(T)\cos m\varphi + S_{lm}^E(T)\sin m\varphi\right]\right\}$$

$$W_E^a(T, \mathbf{X}) = -\frac{G}{2}\frac{(\mathbf{X}\times\mathbf{S}_E)^a}{R^3} \tag{5.11}$$

where C_{lm}^E and S_{lm}^E are, to sufficient accuracy, equivalent to the post-Newtonian multipole moments introduced by Blanchet and Damour (Damour *et al.* 1991, Soffel *et al.* 2003 and references therein), φ and θ are the polar angles corresponding to the spatial coordinates X^a of the GCRS, $R = |\mathbf{X}|$, and $\mathbf{S}_E$ is the total angular momentum vector of the central body. Note that although the first of Eqs. (5.11) looks exactly like its Newtonian counterpart, this is the post-Newtonian definition of the coefficients C_{lm}^E and S_{lm}^E.

5.4.3 Other versions of the GCRS

The technique developed to construct the GCRS can be directly applied for three other purposes. These three tasks differ only in the choice of the central body.

In the first case the whole Solar System is viewed as the central body. Galactic and extragalactic matter outside the Solar System are the source of the tidal gravitational potentials W_T and W_T^a in Eqs. (5.9) and (5.10). In this way one can evaluate and take into account, if necessary, the tidal influences of Galactic and extragalactic masses on the dynamics of Solar System. Sometimes the BCRS is claimed to be constructed using the assumption that the Solar System is isolated. However, a better argument is: construct a GCRS-like reference system for the whole Solar System as the central body, compute the external tidal potentials as well as the inertial forces and check that these potentials can be neglected for current purposes. This procedure results in the standard metric tensor of the BCRS given by Eqs. (5.4)–(5.7).

One further application of this technique is to construct GCRS-like reference systems for all other bodies of Solar System. These reference systems are physically adequate to describe processes in the vicinity of each of the bodies. For instance, the Selenocentric Celestial Reference System is used to model the rotational motion of the Moon and to represent the selenocentric coordinates of the retroreflectors that are used for LLR. Similar reference systems are used to model the dynamics of Mars and Mercury. The structure of the gravitational fields of each body is defined by multipole expansion (5.11) in its own celestial reference system.

Finally, considering an observer as the central body one can construct the Observer's Celestial Reference System, OCRS. The OCRS is a physically adequate GCRS-like reference system for an arbitrary "mass-less" observer. Mass-less means that the gravitational potential of the observer itself is so small that it can be neglected. A space vehicle, an observing site on the Earth etc. can be considered as such observers. The details of the construction of the OCRS can be found in Klioner (2004). The OCRS is very similar to the GCRS, but the internal gravitational potentials W_E and W_E^a vanish and the external tidal potentials W_T and W_T^a are generated by all external bodies (including the Earth). The inertial terms in Eqs. (5.9) and (5.10) can also play an important role. For example, for the OCRS constructed for an observer on the surface of the Earth $|Q_a| \approx 9.8 \text{ m/s}^2$. The value of C_a is again chosen in such a way that the OCRS does not rotate with respect to the BCRS (the coordinate transformations between these systems do not involve any rotation of spatial coordinates). The OCRS is physically adequate to describe processes located in the immediate vicinity of the observer: dynamics of the attitude of a space vehicle, the process of observation by its instruments etc. At the origin of the OCRS its coordinate time coincides with the proper time of the observer, being the time measured by an ideal clock moving together with the observer. Moreover, the coordinate basis of the OCRS at its origin represents a kinematically non-rotating tetrad. The latter is the standard tool to convert coordinate-dependent quantities into coordinate-independent observables in general relativity (see Section 5.7.2 below). Hence, the OCRS presents a very convenient tool to model any kind of observations performed by the observer (see Klioner 2004 for a detailed discussion).

Finally, the BCRS, GCRS and OCRS are given by IAU resolutions in the framework of general relativity. Variants of these reference systems in the framework of the PPN formalism are discussed in Will (1993), Klioner and Soffel (2000), Klioner (2004), and Kopeikin and Vlasov (2004).

5.4.4 Rigidly rotating relativistic reference systems

An important derivative of the GCRS-like reference systems is the local reference system rigidly rotating with respect to the corresponding celestial (i.e. non-rotating) reference system. Such a rotating reference system $(\bar{T}, \bar{X}^a)$ is defined from the GCRS-like non-rotating reference system (T, X^a) by the following Newtonian-like coordinate transformations:

$$\begin{aligned}
\bar{T} &= T \\
\bar{X}^a &= P^{ab}(T) X^b
\end{aligned} \tag{5.12}$$

where $P^{ab}(T)$ is an arbitrary time-dependent orthogonal matrix defining the rigid rotation of the spatial axes $\bar{X}^a$ relative to X^a. Such rotating reference systems can be used to formally separate the dynamics of the central body into a "rotation" and a "deformation" of any nature. Thus, $(\bar{T}, \bar{X}^a)$ for the Earth represents the relativistic model for International Terrestrial Reference System and Frame, ITRS/ITRF, where variations of coordinates attached to the solid surface of the Earth are only due to tectonic and tidal deformations of the Earth. The matrix $P^{ab}(T)$ in this case represents the rotation between the GCRS and the ITRS and contains all the components of Earth rotation (precession, nutation, and polar motion). The same sort of coordinates can be used also for other planets. In case of a space vehicle $P^{ab}(T)$ describes the attitude of the space vehicle with respect to the kinematically non-rotating OCRS.

5.5 Motion of observers and observed objects

Typically, for objects situated in the Solar System the equations of motion are second-order ordinary differential equations. The only qualitative change to the Newtonian N-body problem is that the forces depend now on the velocities of massive bodies and not just on their positions. Numerical integration with suitable initial or boundary conditions can be used to solve these equations of motion. For objects outside of the Solar System simple models like uniform and rectilinear motion in space are often used. More complicated models are needed for systems like binary stars. In any case, in the relativistic framework all these ad hoc models give positions and velocities of observed objects in the chosen relativistic reference system (usually BCRS). For binary pulsars we should use the relativistic equations of motion. The principal relativistic effects on the translational motion of Solar System bodies (including space vehicles) in the BCRS are contained in the so-called Einstein–Infeld–Hoffmann (EIH) equations of motion of N gravitating bodies, whose gravitational fields can be described by their masses M_A only:

$$\ddot{x}_A^i = -\sum_{B \neq A} GM_B \frac{r_{AB}^i}{r_{AB}^3}$$

$$+ \frac{1}{c^2} \sum_{B \neq A} GM_B \frac{r_{AB}^i}{r_{AB}^3} \left\{ (2\gamma + 2\beta + 1)\frac{GM_A}{r_{AB}} + (2\beta - 1)\sum_{C \neq B, A} \frac{GM_C}{r_{BC}} \right.$$

$$- (1 + \gamma)\dot{x}_B^k \dot{x}_B^k - \gamma \dot{x}_A^k \dot{x}_A^k + 2(1 + \gamma)\dot{x}_A^k \dot{x}_B^k$$

$$\left. + 2(\gamma + \beta)\sum_{C \neq A} \frac{GM_C}{r_{AC}} + \frac{3}{2}\frac{(r_{AB}^k \dot{x}_B^k)^2}{r_{AB}^2} - \frac{1}{2}\sum_{C \neq A, B} GM_C \frac{r_{AB}^k r_{BC}^k}{r_{BC}^3} \right\}$$

$$+ \frac{1}{c^2} \sum_{B \neq A} GM_B \frac{r_{AB}^k}{r_{AB}^3} \left\{ 2(\gamma + 1)\dot{x}_A^k - (2\gamma + 1)\dot{x}_B^k \right\} \left(\dot{x}_A^i - \dot{x}_B^i\right)$$

$$- \frac{4\gamma + 3}{2c^2} \sum_{B \neq A} \frac{GM_B}{r_{AB}} \sum_{C \neq A, B} GM_C \frac{r_{BC}^i}{r_{BC}^3} + O(c^{-4}) \tag{5.13}$$

where $r_{AB}^i = x_A^i - x_B^i$, $r_{AB} = |r_{AB}^i|$, indices A, B, and C enumerate the bodies of the Solar System, and x_A^i is the position of body A. Here we also used the PPN parameters β and γ. In general relativity $\beta = \gamma = 1$. The Newtonian part of these equations follows from the term of order c^{-2} in g_{00}. The relativistic contribution requires all other terms in the BCRS metric tensor specified above. If only one massive body is taken into account, the corresponding metric tensor and the equations of motion are the Schwarzschild metric and the equations of motion of a test (massless) body in that metric, respectively. Since the mass of the Sun is much greater than the masses of all other bodies in the Solar System, the Schwarzschild equations of motion describe the dominant relativistic effects in most cases. In general, various parts of Eq. (5.13) represent: (1) Schwarzschild perihelion advance due to the Sun ($\simeq 43''$ per century for Mercury, $\simeq 10''$ per century for Icarus, etc.); (2) geodetic precession ($\simeq 2''$ per century for the lunar orbit); (3) various periodic relativistic effects (important mostly for LLR and binary pulsar timing observations). Other effects not contained in the EIH equations are the relativistic effects due to rotational motion of the bodies (Lense–Thirring or gravitomagnetic effects) and those due to non-sphericity of gravitating bodies. These additional effects are marginal for the current accuracy of LLR and satellite laser ranging (SLR), but negligible for space astrometry. For space vehicles, the gravitational potential of which can be neglected, the EIH equation could be simplified (M_A can be put to zero in Eq. (5.13)). Although the main relativistic effects are the same as for massive bodies, the more complicated orbits of space vehicles lead to the necessity to investigate the magnitudes of various relativistic effects in each particular case.

The GCRS metric tensor also allows one to derive the equations of motion of artificial satellites of the Earth. The equations of motion of an artificial satellite in the GCRS are recommended by the International Earth Rotation and Reference Systems Service (IERS, 2010, Section 10.3) for high-accuracy modeling of satellite motion. The main effects here are the Schwarzschild effects due to the Earth, the Lense–Thirring terms coming from W_E^a and induced by the rotational motion of the Earth as well as the Coriolis forces coming from C_a and caused by geodetic precession. The GCRS equations of motion should be used for satellites on a geostationary orbit and below it. For space vehicles flying further from the Earth their motion can be adequately modeled directly in the BCRS.

The GCRS allows us also to model the rotational motion of the central body (i.e. of the Earth). The first attempt to construct a relativistic theory of Earth rotation has been undertaken in Klioner *et al.* (2009). In the same way the OCRS allows one to model the rotational motion of space vehicles (Klioner 2004).

5.6 Modeling of light propagation

In general relativity and in the PPN formalism the equations of light propagation coincide with the equations of geodetic lines in the chosen reference system. These are second-order ordinary differential equations, which could also be solved by numerical integration, but normally we prefer to use some approximate analytical solutions. Exact analytical solutions are only known in some special (normally, highly symmetrical) cases such as the Schwarzschild metric. In more realistic cases some sort of approximations are used. The solution of the BCRS equations of light propagation can be written as follows

$$x_p^i(t) = x_0^i + c\,\sigma^i(t - t_0) + \Delta x_p^i(t; \mathbf{x}_0, \boldsymbol{\sigma}) \tag{5.14}$$

where x_0^i and σ^i are the parameters of a straight line representing a Newtonian light ray, σ^i being a unit vector ($\sigma^k \sigma^k = 1$), and Δx_p^i are the relativistic corrections to the Newtonian straight line. For a post-Newtonian metric of BCRS or GCRS, Δx_p^i contains terms of order c^{-2} and c^{-3}. Terms of order c^{-2} in both g_{00} and g_{ij} are required to derive the terms of order c^{-2} in Δx_p^i, while the terms of order c^{-3} in Δx_p^i also require terms c^{-3} in g_{0i}. The higher-order effects, the so-called post-post-Newtonian effects, would require terms of order c^{-4} in both g_{00} and g_{ij} (the c^{-4} terms in g_{ij} are not in the current definition of the BCRS metric tensor, but can be added as soon as necessary).

Usually it is assumed that $\Delta x_p^i(t_0; \mathbf{x}_0, \boldsymbol{\sigma}) = 0$ and $\lim_{t \to -\infty} \Delta \dot{x}_p^i(t; \mathbf{x}_0, \boldsymbol{\sigma}) = 0$. This means that x_0^i is the position of the photon for $t = t_0$ that can be identified with the position of observer x_{obs}^i at the moment of observation $x_p^i(t_0) = x_p^i(t_{\text{obs}}) = x_0^i = x_{\text{obs}}^i(t_{\text{obs}})$, and σ^i is the unit direction of light propagation for $t \to -\infty$. The latter direction is unaffected by the gravitational light deflection due to the gravitational field of the Solar System. The unit coordinate direction of light propagation at the point of observation is the coordinate light velocity $\dot{x}_p^i(t_0) = c\,\sigma^i + \Delta \dot{x}_p^i(t_0; \mathbf{x}_0, \boldsymbol{\sigma})$ normalized to unity: $n^i = \dot{x}_p^i(t_0)/|\dot{\mathbf{x}}_p(t_0)|$. For stars and quasars the finite distance to them can be neglected while calculating the light deflection at the level of accuracy of up to 1 μas. For objects located within the Solar System their finite distance should be explicitly taken into account. For a Solar System object the unit direction from the source at the moment of signal emission t_{em} to the observer at the moment of reception t_{obs} plays the role of the unperturbed direction: $k^i = R^i/|\mathbf{R}|$, where $R^i = x_p^i(t_{\text{em}}) - x_p^i(t_{\text{obs}})$. Obviously, the position of the photon at the moment of emission coincides with the position of the observed source x_s^i at that moment: $x_p^i(t_{\text{em}}) = x_s^i(t_{\text{em}})$. The relations between n^i and σ^i, and between n^i and k^i allow the gravitational light deflection to be taken into account:

$$n^i = \sigma^i + \delta\sigma^i$$
$$n^i = k^i + \delta k^i \tag{5.15}$$

For many observations the most important effects in $\delta\sigma^i$ and δk^i come from the spherically symmetric part of the gravitational fields of the bodies. In this case we have

$$\delta\sigma^i = -\frac{1+\gamma}{c^2} \sum_A GM_A \frac{\hat{d}_A^i}{|\mathbf{d}_A|} (1 + \boldsymbol{\sigma} \cdot \hat{\mathbf{r}}_{oA}) + O(c^{-4}) \tag{5.16}$$

where for any vector $\hat{x}^i = x^i/|\mathbf{x}|$, $\mathbf{d}_A = \boldsymbol{\sigma} \times (\mathbf{r}_{oA} \times \boldsymbol{\sigma})$ is the impact parameter of the unperturbed light ray with respect to body A, and

$$r_{oA}^i = x_p^i(t_{\text{obs}}) - x_A^i(t^*) \tag{5.17}$$

where t^* is the retarded moment of time (Kopeikin and Schäfer 1999, Klioner 2003b) defined as

$$t^* = t_{\text{obs}} - \frac{1}{c} \left| \mathbf{x}_p(t_{\text{obs}}) - \mathbf{x}_A(t^*) \right| \tag{5.18}$$

For δk^i we have

$$\delta k^i = -\frac{1+\gamma}{c^2} \sum_A GM_A \frac{\hat{d}_A'^i}{|\mathbf{d}_A'|} \frac{r_{eA}}{R} (1 - \hat{\mathbf{r}}_{eA} \cdot \hat{\mathbf{r}}_{oA}) + O(c^{-4}) \tag{5.19}$$

where $\mathbf{d}_A' = \mathbf{k} \times (\mathbf{r}_{oA} \times \mathbf{k})$, and

$$r_{eA}^i = x_p^i(t_{\text{em}}) - x_A^i(t^*) \tag{5.20}$$

These formulas give the main relativistic effect of light deflection. The gravitational light deflection due to the Sun amounts to $1.75''$ for a grazing ray and attains 4 milliarcseconds (mas) if the angular distance between the source and the Sun is $90°$. The deflection due to Jupiter can be as large as 16 mas. Other effects include the influence of the quadrupole gravitational fields of giant planets (up to 240 µas) and higher-order Schwarzschild effects (up to about 20 µas). The corresponding formulas and detailed estimates for various bodies of the Solar System can be found in e.g. Klioner (2003a).

Let us consider a light ray propagating from point $\mathbf{x}_0$ to point $\mathbf{x}_1$. In the presence of a gravitational field the travel time $t_1 - t_0$ between these two points is slightly larger than the coordinate distance divided by the light velocity c. In the spherically symmetric gravitational fields of N bodies we have

$$c(t_1 - t_0) = |\mathbf{x}_1 - \mathbf{x}_0| + (1+\gamma) \sum_A \frac{GM_A}{c^2} \ln \frac{|\mathbf{x}_1| + |\mathbf{x}_0| + |\mathbf{x}_1 - \mathbf{x}_0|}{|\mathbf{x}_1| + |\mathbf{x}_0| - |\mathbf{x}_1 - \mathbf{x}_0|} + O(c^{-4}) \tag{5.21}$$

This additional retardation in light propagation is called Shapiro delay and can amount to 240 µs for observations of Venus in upper conjunction with the Sun.

5.7 Computation of observables

As mentioned above, the conversion of the coordinate-dependent quantities into coordinate-independent observables is an important part of relativistic modeling. Mathematically, the coordinate-independent quantities are scalars and standard mathematical techniques within general relativity let us perform these conversions.

5.7.1 Proper time

The simplest case is the transformation of intervals of coordinate time t of a reference system into the corresponding intervals of the proper time τ of an observer. The proper time τ of an observer having the trajectory $x^i_{\text{obs}}(t)$ relative to some reference system (t, x^i) is related to coordinate time t of that reference system by

$$
\frac{d\tau}{dt} = \left(-g_{00}(t, \mathbf{x}_{\text{obs}}(t)) - \frac{2}{c} g_{0i}(t, \mathbf{x}_{\text{obs}}(t)) \dot{x}^i_{\text{obs}} - \frac{1}{c^2} g_{ij}(t, \mathbf{x}_{\text{obs}}(t)) \dot{x}^i_{\text{obs}} \dot{x}^j_{\text{obs}} \right)^{1/2} \qquad (5.22)
$$

where $g_{\alpha\beta}(t, \mathbf{x}_{\text{obs}}(t))$ are the components of the metric tensor of the considered reference system evaluated along the trajectory of the observer. This formula is valid for any relativistic reference system including BCRS and GCRS. Substituting the form of the metric tensor in the BCRS and GCRS into Eq. (5.22) it can be seen that in order to compute terms of order c^{-2} on the right-hand side of this equation we need only terms c^{-2} in g_{00}, while terms c^{-4} in g_{00}, c^{-3} in g_{0i}, and c^{-2} in g_{ij} are required to get the terms of order c^{-4}. To lowest order in the BCRS we have

$$
\frac{d\tau}{dt} = 1 - \frac{1}{c^2} \left(\frac{1}{2} \dot{\mathbf{x}}^2_{\text{obs}} + w(t, \mathbf{x}_{\text{obs}}(t)) \right) + O(c^{-4}) \qquad (5.23)
$$

It can be seen that the relation between the proper time τ of a clock and coordinate time t involves both the velocity of the clock in the chosen reference system and the gravitational potential along the trajectory of the clock. Typically in the Solar System the terms of order c^{-2} in Eq. (5.23) are $\simeq 10^{-8}$, while the higher-order terms neglected in Eq. (5.23) are less than 10^{-16}. Detailed formulas for the transformations between proper time τ and coordinate times TCB and TCG are given e.g. by Soffel *et al.* (2003).

5.7.2 Proper direction

An important problem is the conversion of the coordinate direction of light propagation n^i defined above into the corresponding observable direction s^i, which is often called "proper direction" or direction relative to the proper reference system of the observer. A proper reference system is a mathematical model of an ideal clock and three orthogonal rigid rods that the observer uses to measure time intervals, distances, and directions in his vicinity. The coordinate basis of such a proper reference system at its origin is often referred to as a tetrad. The OCRS as described above represents such a proper reference frame. The coordinate basis of the OCRS at its origin gives a tetrad kinematically non-rotating with respect to the BCRS (Klioner 2004). In the special theory of relativity the proper reference frame of an observer is related to some background inertial reference system by a Lorentz transformation. It is therefore sufficient to use Lorentz transformations to convert n^i into s^i:

$$s^i = \left(-n^i + \frac{\Gamma}{c}\left\{1 - \frac{\mathbf{v}\cdot\mathbf{n}}{c}\frac{\Gamma}{\Gamma+1}\right\}\frac{v^i}{c}\right)\frac{1}{\Gamma\left(1 - \frac{\mathbf{v}\cdot\mathbf{n}}{c}\right)} \tag{5.24}$$

$$\Gamma = \frac{1}{\sqrt{1 - \frac{v^2}{c^2}}} \tag{5.25}$$

This is the well-known formula for special-relativistic aberration. The parameter of the Lorentz transformation v^i in this case coincides with the velocity of the observer relative to the chosen reference system. In general relativity it is also sufficient to use Eqs. (5.24) and (5.25), but the parameter v^i should be related to the BCRS velocity $\dot{x}^i_{\text{obs}}$ of the observer as

$$v^i = \dot{x}^i_{\text{obs}}\left(1 + \frac{1}{c^2}(1 + \gamma)\,w(t_{\text{obs}}, \mathbf{x}_{\text{obs}}(t_{\text{obs}}))\right) \tag{5.26}$$

where $w(t_{\text{obs}}, \mathbf{x}_{\text{obs}}(t_{\text{obs}}))$ is the BCRS gravitational potential evaluated at the location of the observer at the moment of observation (Klioner 2003a).

5.8 Relativistic model for positional observations

Having all these theoretical tools we can formulate the relativistic model for positional observations with microarcsecond accuracy. The relativistic model for Gaia is well documented (Klioner 2003a), and we outline its overall structure here. The model essentially consists of subsequent transformations between the following five vectors that were listed above (Fig. 5.4):

(a) $\mathbf{s}$ is the unit observed direction (the word "unit" means here and below that the formally Euclidean scalar product $\mathbf{s}\cdot\mathbf{s} = 1$ is equal to unity),
(b) $\mathbf{n}$ is the unit vector tangential to the light ray at the moment of observation,
(c) $\boldsymbol{\sigma}$ is the unit vector tangential to the light ray at $t \to -\infty$,
(d) $\mathbf{k}$ is the unit coordinate vector from the source to the observer,
(e) $\mathbf{l}$ is the unit vector from the barycenter of the Solar System to the source.

The vector $\mathbf{s}$ represents components of the observed direction relative to the OCRS of the observer and is equivalent to the projections of $\mathbf{n}$ on vectors of the kinematically non-rotating tetrad attached to the observer. The other four vectors are interpreted as sets of three numbers characterizing the position of the source with respect to the BCRS. All these vectors would change their numerical values if some other relativistic reference system is used instead of the BCRS. The model consists then in a sequence of transformations between these vectors as shown on Fig. 5.5. The physical meaning of each transformation can be summarized as follows (the numbering here coincides with the numbering on Fig. 5.5):

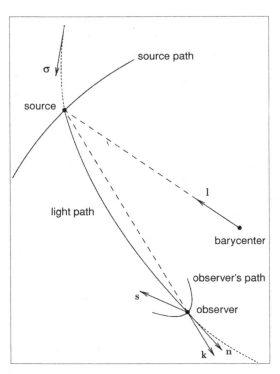

Fig. 5.4 Vectors involved in the model for positional observations (Klioner 2003a).

remote sources:

$$\mathbf{s} \xleftrightarrow{\ (1)\ } \mathbf{n} \xleftrightarrow{\ (2)\ } \boldsymbol{\sigma} \xleftrightarrow{\ (3)\ } \mathbf{k} \xleftrightarrow{\ (4)\ } \mathbf{l}, \pi \xleftrightarrow{\ (5)\ } \mathbf{l}(t_0), \pi(t_0), \boldsymbol{\mu}(t_0), \ldots$$

Solar System objects:

$$\mathbf{s} \xleftrightarrow{\ (1)\ } \mathbf{n} \xleftrightarrow{\ (2,3)\ } \mathbf{k} \xleftrightarrow{\ (6)\ } \text{orbit}$$

Fig. 5.5 Sequence of transformation in the model of positional observations.

(1) aberration (effects vanishing together with the barycentric velocity of the observer): this step is given by Eqs. (5.24)–(5.26) and converts the observed direction to the source **s** into the unit BCRS coordinate velocity of the light ray **n** at the point of observation;

(2) gravitational light deflection for the source at infinity: this step is given by Eqs. (5.15)–(5.18) and converts **n** into the unit direction of propagation $\boldsymbol{\sigma}$ of the light ray infinitely far from the Solar System at $t \to -\infty$;

(3) coupling of finite distance to the source and the gravitational light deflection in the gravitational field of the Solar System: this step converts $\boldsymbol{\sigma}$ into a unit BCRS coordinate direction **k** going from the source to the observer;

(4) parallax: this step converts **k** into a unit BCRS direction **l** going from the barycenter of the Solar System to the source;

(5) proper motion, etc: this step provides a reasonable parametrization of the time dependence of **l** (and, possibly, of the parallax π) caused by the motion of the source relative
to the barycenter of the Solar System;

(6) orbit determination process for Solar System objects.

These transformations have already been discussed in full detail (Klioner 2003a,b, Klioner
2004). The most complicated part of the model is the light-deflection model where the
effects of (a) monopole fields of all major Solar System bodies, (b) quadrupole fields of the
giant planets, and (c) gravitomagnetic fields due to translational motion of all major bodies
should be taken into account in order to attain the accuracy of 1 μas (Klioner and Peip 2003).
Moreover, each body with a mean density ρ and radius $R \geq \left(\frac{\rho}{1\,\mathrm{g/cm^3}}\right)^{-1/2} \times 624$ km produces
a light deflection of at least 1 μas. Therefore, a few tens of minor bodies (mainly, satellites of
the giant planets) should also be taken into account in certain rare cases (Klioner 2003a). For
objects situated at a distance greater than 1 pc from the Solar System, the transformation
from σ to **k** is trivial: **k** = σ plus terms smaller than 1 μas. The transformation from
k to **l** as well as the parametrization of **l**(t) and $\pi(t)$ have formally a Newtonian form
(Klioner 2003a), but it should be kept in mind that these parameters are only defined in
the BCRS.

5.9 Celestial reference frame

It is important to remember that all astrometric parameters (e.g. positions, proper motions,
parallaxes, radial velocities of stars, and orbits of binaries and Solar System objects) are
defined in the particular relativistic reference system used to model the observations,
namely, in the BCRS. These parameters represent the celestial reference frame, which is
a materialization of the BCRS. The celestial reference frame is, so to say, a model of the
Universe in the BCRS. Thus, the goal of astrometry in the relativistic framework is not to
find "the" barycentric inertial reference system, which is unique in Newtonian formulation,
but to find a materialization of a particular relativistic reference system.

5.10 Beyond the standard relativistic model

In the model described above the influence of gravitational fields generated outside of
the Solar System is ignored. For the majority of sources the external field can indeed
be fully neglected, but there are a number of cases when the external gravitational fields
produce observable effects. Several authors have discussed these additional effects in detail
(see e.g. Klioner 2003a, Kopeikin and Gwinn 2000). The main effects of this kind are:
(1) gravitational light deflection caused by masses situated outside of the Solar System,
which includes (a) weak microlensing on the stars of the Galaxy (Belokurov and Evans,
2002), (b) lensing on gravitational waves (both primordial ones and those from compact

sources), and (c) lensing of the companions of edge-on binary systems; (2) cosmological effects; and (3) more complicated models for the motions of observed objects in the BCRS, which are necessary for the case of binary stars, etc.

Note that all these effects can be easily taken into account by a simple additive extension of the standard model since at the required accuracy the external gravitational fields can be linearly superimposed on the Solar System gravitational field. The only exception could be the effects of the cosmological background, but preliminary studies by Cooperstock *et al.* (1998) and Klioner and Soffel (2005) show that even here the coupling of the local Solar System fields and the external ones can be neglected.

5.11   Astrometry as a laboratory for gravitational physics

An important application of astrometry is its ability to test special and general relativity theories. Although general and especially special relativity have been tested with high precision (see above), some physical theories predict deviations from general relativity at the level of 10^{-5}–10^{-8} (e.g. Damour *et al.* 2002). It is therefore very important to increase the accuracy of relativistic tests. It should be stressed that detection of deviations from general relativity (even if those deviations are very small) may have very profound consequences for the whole of physics and astrophysics. The most stringent relativistic test that can be expected from microarcsecond astrometry within the next decade is a measurement of gravitational light deflection with a precision of about 10^{-6} by Gaia. Since Gaia's astrometric measurements will be done in a wide range of angular distances from the Sun, not only the magnitude of light deflection, but also its dependence on the angular distance, will be precisely verified. Astrometric data from the BepiColombo project will allow measurements of the relativistic perihelion precession of Mercury with a precision of not worse than $5 \cdot 10^{-6}$. Besides these main tests a number of other tests will be performed and an overview of them can be found in Klioner (2007) and Mignard and Klioner (2008).

References

Belokurov, V. A. and Evans, N. W. (2002). Astrometric microlensing with the GAIA satellite. *MNRAS*, **331**, 649.

Bertotti, B., Iess, L., and Tortora, P. (2003). A test of general relativity using radio links with the Cassini spacecraft. *Nature*, **425**, 374.

Brumberg, V. A. (1991). *Essential Relativistic Celestial Mechanics*. Bristol: Adam Hilger.

Cooperstock, F. I., Faraoni, V., and Vollick, D. N. (1998). The influence of the cosmological expansion on local systems. *ApJ*, **503**, 61.

Damour, T., Soffel, M., and Xu, Ch. (1991). General-relativistic celestial mechanics. I. Method and definition of reference systems. *Phys. Rev. D*, **43**, 3273.

Damour, T., Piazza, F., and Veneziano, G. (2002). Violations of the equivalence principle in a dilaton-runaway scenario. *Phys. Rev. D*, **66**, 046007.

IAU (2001). *IAU Information Bulletin, 88* (errata in *IAU Information Bulletin*, 89).

IERS (2010). *IERS Conventions 2010*, eds. G. Petit and B. Luzum, IERS Technical Note, 36. Frankfurt am Main: Verlag des Bundesamts für Kartographie und Geodäsie.

Klioner, S. A. (2003a). A practical relativistic model for microarcsecond astrometry in space. *AJ*, **125**, 1580.

Klioner, S. A. (2003b). Light propagation in the gravitational field of moving bodies by means of Lorentz transformation I. Mass monopoles moving with constant velocities. *A&A*, **404**, 783.

Klioner, S. A. (2004). Physically adequate proper reference system of a test observer and relativistic description of the GAIA attitude. *Phys. Rev. D*, **69**, 124001.

Klioner, S. A. (2007). Testing relativity with space astrometry missions. In *Lasers, Clocks and Drag-Free: Exploration of Relativistic Gravity in Space*, eds. H. Dittus, C. Lämmerzahl, and S. G. Turyshev. Berlin: Springer, p. 399.

Klioner, S. A., Gerlach, E., and Soffel, M. (2009). Relativistic aspects of rotational motion of celestial bodies. In *Relativity in Fundamental Astronomy*, eds. S. Klioner, K. Seidelmann, and M. Soffel. Cambridge: Cambridge University Press, p. 112.

Klioner, S. A. and Peip, M. (2003). Numerical simulations of the light propagation in gravitational field of moving bodies. *A&A*, **410**, 1063.

Klioner, S. A. and Soffel, M. H. (2000). Relativistic celestial mechanics with PPN parameters. *Phys. Rev. D*, **62**, ID 024019.

Klioner, S. A. and Soffel, M. H. (2005). Refining the relativistic model for Gaia: cosmological effects in the BCRS. ESA Special Publication ESA SP-576, 305.

Kopeikin, S. M. and Gwinn, C. (2000). Sub-microarcsecond astrometry and new horizons in relativistic gravitational physics. In *Towards Models and Constants for Sub-Microarcsecond Astrometry*, eds. K. J. Johnston, D. D. McCarthy, B. J. Luzum, and G. H. Kaplan. Washington, DC: US Naval Observatory, p. 303.

Kopeikin, S. M. and Schäfer, G. (1999). Lorentz covariant theory of light propagation in gravitational fields of arbitrary-moving bodies. *Phys. Rev. D*, **60**, 124002.

Kopeikin, S. and Vlasov, I. (2004). Parametrized post-Newtonian theory of reference frames, multipolar expansions and equations of motion in the *N*-body problem. *Phys. Rep.*, **400**, 209.

Mignard, F. and Klioner, S. A. (2008). Space astrometry and relativity. In *Space Astrometry and Relativity, Proceedings of the Eleventh Marcel Grossmann Meeting on General Relativity*, eds. H. Kleinert, R. T. Jantzen, and R. Ruffini. Singapore: World Scientific, p. 245.

Rickman, H. (2001). Reports on Astronomy, *Trans. IAU*, **XXIVB**.

Schutz, B. (1985). *A First Course in General Relativity*. Cambridge: Cambridge University Press.

Soffel, M. (1989). *Relativity in Astrometry, Celestial Mechanics and Geodesy*. Berlin: Spinger.

Soffel, M., Klioner, S. A., Petit, G., *et al.* (2003). The IAU 2000 resolutions for astrometry, celestial mechanics, and metrology in the relativistic framework: explanatory supplement. *AJ*, **126**, 2687.

Will, C. M. (1993). *Theory and Experiment in Gravitational Physics*. Cambridge: Cambridge University Press.

Will, C. M. (2006). The confrontation between general relativity and experiment. *Living Rev. Relativity*, **9**, 3, http://www.livingreviews.org/lrr-2006-3 (Update of lrr-2001-4).

6 Celestial mechanics of the N-body problem

SERGEI KLIONER

Introduction

The dynamics of celestial bodies is an important topic for astrometry. First, as we have seen in the previous chapter, the position and velocity of the observer with respect to the BCRS as well as the positions and velocities of Solar System objects are necessary to reduce observations for aberration, parallax, and gravitational light deflection. Second, astrometric observations represent an important source of information allowing us to model the dynamical behavior of various celestial systems: Earth satellites, interplanetary stations, major and minor planets of the Solar System, binary and multiple stars, exoplanetary systems, etc.

A modern introduction in the various techniques of celestial mechanics can be found in the books of Murray and Dermott (1999), Beutler (2005) and Roy (2005). In this chapter we give a short overview of the most important results and concentrate on the practical aspects of the N-body problem relevant to obtaining and processing high-accuracy astrometric data.

6.1 Equations of motion and integrals of the N-body problem in Newtonian physics

Let us consider N bodies having positions $\mathbf{x}_A$ in an inertial reference system and characterized by their masses M_A. Here and below capital Latin indices A, B, etc. enumerate gravitating bodies (i.e. $A = 1, \ldots, N$, etc.). The Newtonian equations of motion of such a system read

$$\ddot{\mathbf{x}}_A = -\sum_{B \neq A} \frac{GM_B}{r_{AB}^3} \mathbf{r}_{AB} \tag{6.1}$$

where $\mathbf{r}_{AB} = \mathbf{x}_A - \mathbf{x}_B$ is the position of body A relative to body B, and G is the Newtonian gravitational constant. These equations can be also written in the form

$$M_A \ddot{x}_A^i = \frac{\partial}{\partial x_A^i} U \tag{6.2}$$

Astrometry for Astrophysics: Methods, Models, and Applications, ed. William F. van Altena.
Published by Cambridge University Press. © Cambridge University Press 2013.

where U is the potential of N gravitating bodies:

$$U = \frac{1}{2} \sum_A \sum_{B \neq A} \frac{GM_A M_B}{r_{AB}} = \sum_A \sum_{B < A} \frac{GM_A M_B}{r_{AB}} \qquad (6.3)$$

Clearly, these equations of motion have ten classical integrals: six integrals of the center of mass (t is the Newtonian time)

$$\sum_A M_A \dot{\mathbf{x}}_A = \mathbf{P} \qquad (6.4)$$

$$\sum_A M_A \mathbf{x}_A = \mathbf{P}t + \mathbf{Q}, \quad \mathbf{P} = \text{const}, \ \mathbf{Q} = \text{const} \qquad (6.5)$$

three integrals of the angular momentum

$$\sum_A M_A \mathbf{x}_A \times \dot{\mathbf{x}}_A = \mathbf{c} = \text{const} \qquad (6.6)$$

and one integral of energy

$$\frac{1}{2} \sum_A M_A \dot{\mathbf{x}}_A^2 - U = h = \text{const}. \qquad (6.7)$$

These ten integrals can be used to decrease the order of the system (6.1) or to check the accuracy of numerical integrations. Barycentric coordinates of the N-body system in which $\mathbf{P} = 0$ and $\mathbf{Q} = 0$ are often used. In this case Eqs. (6.4) and (6.5) can be used to compute the position and velocity of one arbitrary body if the positions and velocities of other $N - 1$ bodies are known. This procedure can be used to compute initial conditions satisfying (6.4) and (6.5) with $\mathbf{P} = 0$ and $\mathbf{Q} = 0$. Alternatively, one body can be completely eliminated from the integration, so that at each moment of time the position and velocity for that body are calculated using (6.4) and (6.5) with $\mathbf{P} = 0$ and $\mathbf{Q} = 0$ and the corresponding equation is excluded from Eqs. (6.1) or (6.2). Four remaining integrals are usually used to check the accuracy of the integration, the integral of energy being especially sensitive to numerical errors of usual (non-symplectic) integrators. Vector $\mathbf{c}$ calculated in Eq. (6.6) defines an invariant (time-independent) plane of the N-body system. This plane is also called the Laplace plane.

If the mass of one body is much larger than other masses in the system it is sometimes advantageous to write the equations of motion in the non-inertial reference system centered on that dominating body. In the Solar System the Sun is obviously dominating, having a mass about 1000 times larger than the planets. The heliocentric equations of motion read

$$\ddot{\boldsymbol{\rho}}_A + \frac{G(M_\odot + M_A)}{\rho_A^3} \boldsymbol{\rho}_A = \sum_{B \neq \odot, A} GM_B \left(\frac{\boldsymbol{\rho}_{AB}}{\rho_{AB}^3} - \frac{\boldsymbol{\rho}_B}{\rho_B^3} \right) \qquad (6.8)$$

where $\boldsymbol{\rho}_A = \mathbf{x}_A - \mathbf{x}_\odot$ and $\boldsymbol{\rho}_{AB} = \boldsymbol{\rho}_A - \boldsymbol{\rho}_B = \mathbf{x}_A - \mathbf{x}_B$. Here and below index "$\odot$" denotes the Sun. These equations can be also written as

$$\ddot{\rho}_A^i + \frac{G(M_\odot + M_A)}{\rho_A^3} \rho_A^i = \frac{\partial}{\partial \rho_A^i} R \qquad (6.9)$$

where

$$R = \sum_{B \neq \odot, A} GM_B \left(\frac{1}{\rho_{AB}} - \frac{\boldsymbol{\rho}_A \cdot \boldsymbol{\rho}_B}{\rho_B^3} \right) \tag{6.10}$$

The heliocentric equations can be directly integrated numerically or analyzed analytically to obtain the motion of planets and minor bodies with respect to the Sun. It is also clear that if we have only two bodies R vanishes and the remaining equations of motion describe the two-body problem (see below). Therefore, the forces coming from R can be considered as perturbations of the two-body problem (especially in the case when $M_\odot + M_A \gg M_B$ and the heliocentric motion of body A is close to the solution of the two-body problem). For this reason, R is called the *disturbing function*. The idea to treat any motion of a dynamical system as a perturbation of some known motion of a simplified dynamical system is natural and widely used in many areas of physics and astronomy. For the case of the dynamics of celestial bodies a suitable simplification is two-body motion, which is simple and given by analytical formulas. As perturbations, one can consider not only N-body forces as given above, but also non-gravitational forces, relativistic forces, etc.

The motion of N bodies is a very complicated problem. Since its formulation, the N-body problem has led to many new branches in mathematics. Here let us only mention the Kolmogorov–Arnold–Moser (KAM) theory that proves the existence of stable quasi-periodic motions in the N-body problem. A review of mathematical results known in the area of the N-body problem is given in the encyclopedic book of Arnold *et al.* (1997).

The main practical tool to solve the equations of the N-body problem is numerical integration. Three different modes of these numerical integrations can be distinguished. The first mode is integrations for a relatively short time span and with the highest possible accuracy. This type of solution is used for the Solar System ephemerides and space navigation. Some aspects of these high-accuracy integrations are discussed in Section 6.7 below. The second sort of integration is of a few bodies over very long periods of time with the goal to investigate the long-term dynamics of the motion of the major and minor bodies of the Solar System or exoplanetary systems. Usually we consider a subset of the major planets and the Sun as gravitating bodies and investigate the long-term motion of this system or the long-term dynamics of massless asteroids. For this sort of solution, it is important to have a correct phase portrait of the motion and not necessarily high accuracy of the individual orbits. Besides that, usually the initial conditions of the problem are such that no close encounters between massive bodies should be treated. Symplectic integrators are often used for these integrations because of their nice geometrical properties (e.g. the symplectic integrators do not change the integral of energy). Resonances of various natures play a crucial role in such studies and are responsible for the existence of chaotic motions. A good account of recent efforts in this area can be found in Murray and Dermott (1999) and Morbidelli (2002). The third kind of numerical integration is integration with arbitrary initial conditions that do not exclude close encounters between gravitating bodies. Even small-N numerical integrations of Eq. (6.1) in this general case are not easy, e.g. because of possible close encounters of the bodies which make the result of integrations extremely sensitive to small numerical errors. During the last half-century significant efforts have been made to improve the stability and reliability of such numerical simulations. This includes both an analytical change of variables known as "regularizations" and clever tricks in the numerical codes. An exhaustive review of these efforts can be found in Aarseth (2003). To increase the performance and make it possible to integrate the N-body problem for large N special-purpose hardware GRAPE has been created on which a special parallel N-body code can be run. Nowadays, direct integrations of the N-body problem are possible with N up to several million. This makes it possible to use these N-body simulations to investigate the dynamics of stellar clusters and galaxies (Aarseth *et al.* 2008).

6.2 The three-body problem

Special cases of the N-body problem are the two- and three-body problems. The two-body problem is the basis of all practical computations of the motion of celestial bodies and will be reviewed in the next section and, as will be seen below, it can be solved completely. The three-body problem also has important practical applications. The real motion of the Moon is much better described by the three-body system, Sun–Earth–Moon, than by the two-body problem, Earth–Moon. The motion of asteroids and comets can often be approximated by the Sun–Jupiter–asteroid system. However, the three-body problem is so complicated that the general case cannot be solved in an analytically closed form. The motion of three attracting bodies already contains most of the difficulties of the general N-body problem. Nevertheless, many theoretical results describing solutions of the three-body problem have been established. For example, all possible final motions (motions at $t \to \pm\infty$) are known and many classes of periodic orbits have been found. The three-body problem has five important special solutions called Lagrange solutions. These are points of dynamical equilibrium: all three masses remain in one plane and have in that plane a Keplerian orbit (being a conic section) with the same focus and with the same eccentricity. Therefore, in this case the motion of each body is effectively described by the equations of the two-body problem. The geometrical form of the three-body configuration (i.e. the ratio of mutual distances between the bodies) remains constant, but the scale can change and the figure can rotate. In a reference system where the positions of two arbitrary bodies are fixed, there are five points where the third body can be placed (see Fig. 6.1). The three bodies are either always situated on a straight line (three rectilinear Lagrange solutions L_1, L_2 and L_3) or remain at the vertices of an equilateral triangle (two triangle Lagrange solutions L_4 and L_5). The three rectilinear solutions were first discovered by Leonard Euler and are sometimes called Euler's solutions.

A simplified version of the three-body problem – the so-called restricted three-body problem – is often considered. In that problem the mass of one of the bodies is considered to be negligibly small, so that the other two bodies can be described by the two-body problem and the body with negligible mass moves in the given gravitational field of two bodies. Clearly, in many practical situations the mass of the third body can indeed be neglected (e.g. for the motion of minor bodies or spacecrafts in the field of the Sun and

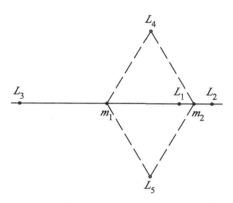

Fig. 6.1 Lagrange points L_1, L_2, $\ldots$, L_5 in the three-body problem. Two masses labeled with m_1 and m_2 are at the positions indicated. The third body has mass m_3. The positions of the Lagrange points depend only on the mass ratios between m_1, m_2, and m_3. For this plot masses with $m_2/m_1 = m_3/m_1 = 0.1$ were used.

one of the planets). Sometimes, it is further assumed that the motion of all three bodies is co-planar ("planar restricted three-body problem") and/or that the orbit of the two massive bodies is circular ("circular restricted three-body problem"). The five Lagrange solutions do also exist in these restricted versions of the three-body problem. In the circular restricted three-body problem the five configurations remain constant in the reference system co-rotating with the two massive bodies. These points are called libration or equilibrium points. Oscillatory (librational) motion around these points has been investigated in detail. In the linear approximation, such librational orbits around L_4 and L_5 in the circular restricted three-body problem are stable provided that the ratio of the masses of the two massive bodies is less than $1/2 - \sqrt{23/108} \approx 0.03852$. The orbits around L_1, L_2, and L_3 are unstable. Interestingly, librational motions around L_4 and L_5 are found in the Solar System; e.g. the asteroid family called the Trojans has orbits around L_4 and L_5 of the Sun–Jupiter–asteroid system. These librational orbits are stable since the ratio of the masses of Jupiter and the Sun is about 10^{-3}, which is smaller than the limit given above. The rectilinear Lagrange points also have practical applications, since librational orbits around these points – the so-called Lissajous orbits – are very attractive for scientific space missions. Lissajous orbits around L_1 and L_2 of the Sun–Earth–spacecraft system are used for such space missions as WMAP, Planck, Herschel, SOHO, Gaia, and the James Webb Space Telescope. The points L_1 and L_2 of the Sun–Earth–spacecraft system are situated on the Sun–Earth line at the distance of about 1.5 million kilometers from the Earth (see Fig. 6.1). Although the Lissajous orbits are unstable, the maneuvers needed to maintain these orbits are simple and require very limited amounts of fuel. On the other hand, placing a spacecraft in an orbit around L_1 or L_2 guarantees almost uninterrupted observations of celestial objects, good thermal stability of the instruments, and optimal distance from the Earth (too far for the disturbing influence of the Earth's figure and atmosphere, and close enough for high-speed communications).

One more important result in the circular restricted three-body problem is the existence of an additional integral of motion called the Jacobi integral. This integral can be used to recognize a minor body (e.g. a comet) even after close encounters with planets after which the heliocentric orbit of the minor body changes drastically. This is the so-called Tisserand criterion: the Jacobi integral should remain the same before and after the encounter even if the heliocentric orbital elements of the body (comet) have substantially changed. The value of the Jacobi integral also defines (via the so-called Hill's surfaces of zero velocity) the spatial region in which the massless body can be located. Details of the Jacobi integral can be found e.g. in Roy (2005).

Finally, let us note that although the N-body problem in general and the three-body problem in particular are among the oldest problems in astronomy, new results in this area continue to appear. A good example here is a remarkable figure-of-eight periodic solution of the three-body problem discovered by Chenciner and Montgomery (2000).

6.3 The two-body problem and the Keplerian elements

The two-body problem is a special case of the N-body problem for $N = 2$, which has been extensively applied to binary stars for the determination of their masses (see Chapter 23). Using the integrals of the center of mass or considering the relative motions, the equations of motion of the two-body problem can be written as

$$\ddot{\mathbf{r}} + \kappa^2 \frac{\mathbf{r}}{r^3} = 0 \tag{6.11}$$

where κ^2 is a constant depending on the masses M_1 and M_2 of the bodies and the kind of motion under consideration. If relative motion is considered, then according to Eq. (6.9) and the discussion afterwards, $\kappa^2 = G(M_1 + M_2)$. For the motion of M_i with respect to the common center of mass, $\kappa^2 = GM_{3-i}^3/(M_1 + M_2)^2$. Eq. (6.11) has four classical integrals: three integrals are analogous to Eq. (6.6)

$$\mathbf{r} \times \dot{\mathbf{r}} = \mathbf{c} \tag{6.12}$$

and one integral is analogous to Eq. (6.7)

$$\frac{1}{2}\dot{\mathbf{r}}^2 - \frac{\kappa^2}{r} = h = \text{const} \tag{6.13}$$

Eq. (6.12) immediately proves that the motion is planar and that the corresponding invariant plane contains the trajectory of motion. Sometimes, we consider one more integral of the two-body problem that has no counterpart in the general N-body problem: the so-called Laplace vector (sometimes also called Laplace–Runge–Lenz vector) defined as

$$\frac{\dot{\mathbf{r}} \times \mathbf{c}}{\kappa^2} - \frac{\mathbf{r}}{r} = \mathbf{e} = \text{const} \tag{6.14}$$

It is straightforward to see that the vector $\mathbf{e}$ remains constant during the motion. Because of relations $\mathbf{e} \cdot \mathbf{c} = 0$ and $2hc^2 = \kappa^4\left(e^2 - 1\right)$, where $e = |\mathbf{e}|$ and $c = |\mathbf{c}|$, among the three components of $\mathbf{e}$ only one is independent of the other integrals. For real motion, $2hc^2 \geq -\kappa^4$, so that e is always real. Note that e can be computed from $\mathbf{c}$ and h. Since $\mathbf{e}$ lies in the orbital plane ($\mathbf{e} \cdot \mathbf{c} = 0$), its only independent component defines a constant direction in the orbital plane (see below). Using the integrals it is easy to demonstrate that the trajectory of motion is always a conic section, which, in suitably defined polar coordinates (r, v) in the orbital plane, can be parametrized as

$$r = \frac{a(1 - e^2)}{1 + e \cos v} \tag{6.15}$$

where $a = -\kappa^2/h$ and e are the constants representing the semi-major axis and the eccentricity of the orbit, respectively. Angle v is called true anomaly. Equation (6.15) describes an arbitrary conic section. This conic section can be an ellipse ($0 \leq e < 1$), a parabola ($e = 1$), a hyperbola ($e > 1$), or, in the degenerate case $\mathbf{c} = 0$, a part of a straight line. In the following, only the case of the ellipse will be considered. Introducing the so-called eccentric anomaly E related to the true anomaly as

$$\tan \frac{E}{2} = \sqrt{\frac{1 - e}{1 + e}} \tan \frac{v}{2} \tag{6.16}$$

Eq. (6.15) can be rewritten as

$$r = a(1 - e \cos E) \tag{6.17}$$

In arbitrary inertial coordinates the components of vector $\mathbf{r}$ can be computed as

$$\mathbf{r} = \mathbf{R} \begin{pmatrix} X \\ Y \\ 0 \end{pmatrix} \tag{6.18}$$

$$\mathbf{R} = \begin{pmatrix} \cos\Omega\cos\omega - \sin\Omega\cos i\sin\omega & -\cos\Omega\sin\omega - \sin\Omega\cos i\cos\omega & \sin\Omega\sin i \\ \sin\Omega\cos\omega + \cos\Omega\cos i\sin\omega & -\sin\Omega\sin\omega + \cos\Omega\cos i\cos\omega & -\cos\Omega\sin i \\ \sin i\sin\omega & \sin i\cos\omega & \cos i \end{pmatrix}$$
$$\tag{6.19}$$

where

$$X = r\cos v = a(\cos E - e)$$
$$Y = r\sin v = a\sqrt{1 - e^2}\,\sin E \tag{6.20}$$

The angle Ω is the longitude of the ascending node ($0 \leq \Omega < 2\pi$), and i is the inclination of the orbit ($0 \leq i \leq \pi$). These angles Ω and i define the orientation of the orbital plane in space. The angle ω defines the orientation of the conic section within the orbital plane ($0 \leq \omega < 2\pi$). Namely, ω defines the direction towards the pericenter – the point where the distance $r = |\mathbf{r}|$ is minimal. It can be demonstrated that this direction is given by the integral $\mathbf{e}$ discussed above. The angle ω is called the argument of pericenter. The true anomaly v is the polar angle reckoned from the pericenter. Figure 6.2 illustrates the definitions of the angles. The eccentric anomaly is related to time t by the Kepler equation

$$E - e\sin E = M \tag{6.21}$$

where M is the mean anomaly

$$M = n(t - t_0) + M_0 \tag{6.22}$$

and $n = \kappa a^{-3/2}$ is the mean motion. The period of motion is $P = 2\pi/n$ and depends only on the semi-major axis a or on the energy constant h. Constant $M_0 = M|_{t=t_0}$ is the value of the mean anomaly at some moment of time t_0. The components of the velocity of the body can, therefore, be computed using

$$\dot{\mathbf{r}} = \mathbf{R} \begin{pmatrix} \dot{X} \\ \dot{Y} \\ 0 \end{pmatrix} \tag{6.23}$$

where

$$\begin{aligned} \dot{X} &= -an\, \frac{\sin E}{1 - e\cos E} \\ \dot{Y} &= an\, \sqrt{1 - e^2}\, \frac{\cos E}{1 - e\cos E} \end{aligned} \tag{6.24}$$

Six constants $(a, e, i, \omega, \Omega, M_0)$, called Kepler elements, together with two additional parameters κ and t_0 fully define the motion of the two-body system. The Kepler elements can be used to compute the position and velocity of the body. Vice versa, the position and velocity given at any moment of time can be used to compute the Kepler elements.

Thus, we see that the motion in the two-body problem can be computed analytically except for the solution of the Kepler equation (6.21). This equation is a transcendental equation having one and only one solution for E for any eccentricity $0 \leq e < 1$ and any real value of M. Many algorithms to solve this equation have been formulated.

The solution of the Kepler equation and any function of the coordinates of two-body motion can be represented analytically in the form of some infinite series. The principal types of these series are (a) expansions in powers of time, e.g.

$$\mathbf{r}(t + \tau) = F\,\mathbf{r}(t) + G\,\dot{\mathbf{r}}(t), \quad F = \sum_{k=0}^{\infty} \frac{1}{k!} F_k \tau^k, \quad G = \sum_{k=0}^{\infty} \frac{1}{k!} G_k \tau^k \tag{6.25}$$

where F_k and G_k are functions of $\mathbf{r}(t)$ and $\dot{\mathbf{r}}(t)$ and can be computed by elegant recurrent formulas; (b) expansions in powers of eccentricity, e.g.

$$E = M + \sum_{k=1}^{\infty} a_k(M)\, e^k \tag{6.26}$$

where $a_1 = \sin M$, $a_2 = \frac{1}{2}\sin 2M$, etc.; and (c) Fourier expansions in multiples of the mean anomaly, e.g.

$$E = M + \sum_{k=1}^{\infty} \frac{2}{k} J_k(k\,e)\sin kM \tag{6.27}$$

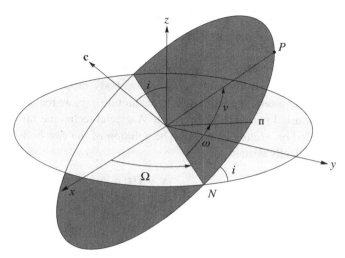

Fig. 6.2   Definition of the Keplerian elements. The dark gray plane is the orbital plane, while the light gray one is the xy-plane of the chosen reference system. Vector **c** is the vector normal to the orbital plane and defined by Eq. (6.12). The point P is the position of the body at some moment of time, Ω is the longitude of the ascending node N, i is the inclination of the orbit, ω is the argument of the pericenter Π and v is the true anomaly.

where $J_k(x)$ is the Bessel function of the first kind. The last type of expansion is also used in the form

$$\left(\frac{r}{a}\right)^n \exp(\iota\, mv) = \sum_{k=-\infty}^{\infty} X_k^{n,m}(e)\, \exp(\iota\, kM) \tag{6.28}$$

where ι is the imaginary unit ($\iota^2 = -1$) and $X_k^{n,m}(e)$ are the Hansen coefficients depending only on the eccentricity of the orbit. The Hansen coefficients can be computed using several methods for any n, m, k, and e. Further details on expansions in the two-body problem can be found in any classical textbook on celestial mechanics, e.g. Brouwer and Clemence (1985) and Roy (2005). Nowadays these techniques are only used for theoretical investigations or for some toy models.

The simplest numerical method to solve the Kepler equation is given by simple iterations $E_i = M + e \sin E_{i-1}$, $i = 1, 2, \ldots$ with the initial value $E_0 = M$. These iterations converge for any M and $0 \le e < 1$. Newton iterations $E_i = (M + e \sin E_{i-1} - e E_{i-1} \cos E_{i-1}) / (1 - e \cos E_{i-1})$ with the same initial value are more efficient. The most efficient numerical algorithm to solve the Kepler equation was published by Fukushima (1996). This algorithm does not require the use of trigonometric functions during the main calculations and is very easy to implement. It gives more than an order of magnitude performance improvement compared to the simple iterations and a speed-up of a factor of 3 compared to the Newtonian iterations, making this algorithm very attractive when the Kepler equation needs to be solved many times (e.g. for massive processing of binary stars or asteroid orbits).

6.4 Perturbation theory and osculating elements

The simplicity of the two-body motion and the fact that many practical problems of celestial mechanics are sufficiently close to two-body motion make it practical to use two-body motion as a zero approximation to the motion in more realistic cases and treat the difference by the usual perturbative approach. A special technique for the motion of celestial bodies is called *osculating elements*. The solution of the two-body problem given above can be symbolically written as

$$\mathbf{x} = \mathbf{f}(t; a, e, i, \omega, \Omega, M_0) \tag{6.29}$$

where the Keplerian elements $(a, e, i, \omega, \Omega, M_0)$ are constants. This representation gives a solution of the differential equation (6.11) with suitable initial conditions. Using the standard variation of constants approach, one can represent a solution of

$$\ddot{\mathbf{r}} + \kappa^2 \frac{\mathbf{r}}{r^3} = \mathbf{F} \tag{6.30}$$

where $\mathbf{F}$ is an arbitrary disturbing force, in the same form, but with constants depending on time:

$$\mathbf{x} = \mathbf{f}(t; a(t), e(t), i(t), \omega(t), \Omega(t), M_0(t)) \tag{6.31}$$

This is always possible since $\mathbf{F}$ has three degrees of freedom (three arbitrary components) and the representation (6.31) involves six arbitrary functions of time. The three "redundant" degrees of freedom can be used to compute not only the position $\mathbf{x}(t)$, but also the velocity $\dot{\mathbf{x}}(t)$ of the solution of Eq. (6.30) using standard formulas of the two-body problem given in the previous section. This can be done if the elements $(a(t), e(t), i(t), \omega(t), \Omega(t), M_0(t))$ satisfy the following condition

$$\frac{\partial \mathbf{f}}{\partial a} \frac{da}{dt} + \frac{\partial \mathbf{f}}{\partial e} \frac{de}{dt} + \frac{\partial \mathbf{f}}{\partial i} \frac{di}{dt} + \frac{\partial \mathbf{f}}{\partial \omega} \frac{d\omega}{dt} + \frac{\partial \mathbf{f}}{\partial \Omega} \frac{d\Omega}{dt} + \frac{\partial \mathbf{f}}{\partial M_0} \frac{dM_0}{dt} = 0 \tag{6.32}$$

This condition guarantees that the time derivative of Eq. (6.31) is given by the partial derivative with respect to time,

$$\dot{\mathbf{x}} = \frac{\partial}{\partial t} \mathbf{f}(t; a(t), e(t), i(t), \omega(t), \Omega(t), M_0(t)) \tag{6.33}$$

that is, velocity $\dot{\mathbf{x}}$ can be calculated by the standard formulas of the two-body problem. The elements satisfying these properties are called *osculating elements*. The osculating elements are functions of time. To compute position $\mathbf{x}(t)$ and velocity $\dot{\mathbf{x}}(t)$ at any given moment one first has to calculate the values of the six osculating elements for this moment of time and then use the standard equations (6.18)–(6.24).

In order to compute the osculating elements for a general disturbing force $\mathbf{F}$ one can use the *Gaussian perturbation equations* (which were also derived by Leonard Euler some time before Gauss and are sometimes called the Euler equations):

$$
\begin{aligned}
\frac{\mathrm{d}}{\mathrm{d}t}a &= 2a^2 e \sin v \left(\frac{S}{\kappa\sqrt{p}}\right) + 2a^2 \frac{p}{r}\left(\frac{T}{\kappa\sqrt{p}}\right) \\
\frac{\mathrm{d}}{\mathrm{d}t}e &= p \sin v \left(\frac{S}{\kappa\sqrt{p}}\right) + p(\cos v + \cos E)\left(\frac{T}{\kappa\sqrt{p}}\right) \\
\frac{\mathrm{d}}{\mathrm{d}t}i &= r\cos(v+\omega)\left(\frac{W}{\kappa\sqrt{p}}\right) \\
\frac{\mathrm{d}}{\mathrm{d}t}\omega &= -\frac{p\cos v}{e}\left(\frac{S}{\kappa\sqrt{p}}\right) + \frac{r+p}{e}\sin v \left(\frac{T}{\kappa\sqrt{p}}\right) - r\sin(v+\omega)\cot i \left(\frac{W}{\kappa\sqrt{p}}\right) \\
\frac{\mathrm{d}}{\mathrm{d}t}\Omega &= \frac{r\sin(v+\omega)}{\sin i}\left(\frac{W}{\kappa\sqrt{p}}\right) \\
\frac{\mathrm{d}}{\mathrm{d}t}\bar{M}_0 &= \frac{\sqrt{1-e^2}}{e}\left((p\cos v - 2er)\left(\frac{S}{\kappa\sqrt{p}}\right) - (r+p)\sin v \left(\frac{T}{\kappa\sqrt{p}}\right)\right)
\end{aligned}
\tag{6.34}
$$

Here $p = a(1 - e^2)$ and components (S, T, W) of $\mathbf{F}$ can be calculated as

$$
\begin{aligned}
S &= \frac{\mathbf{r}}{r}\cdot\mathbf{F} \\
T &= \frac{(\mathbf{r}\times\dot{\mathbf{r}})\times\mathbf{r}}{|\mathbf{r}\times\dot{\mathbf{r}}|\,r}\cdot\mathbf{F} \\
W &= \frac{\mathbf{r}\times\dot{\mathbf{r}}}{|\mathbf{r}\times\dot{\mathbf{r}}|}\cdot\mathbf{F}
\end{aligned}
\tag{6.35}
$$

Therefore, S is the component of $\mathbf{F}$ in the radial direction, T is in the momentary orbital plane (defined by vectors $\mathbf{r}$ and $\dot{\mathbf{r}}$), normal to $\mathbf{r}$ and pointing approximately in the direction of motion, and W is normal to the momentary orbital plane. Numerous alternative ways of computing the components of the disturbing force $\mathbf{F}$ and writing the Gaussian perturbation equations are given in Beutler (2005). Note that for osculating elements, instead of M_0 from Eq. (6.22) we use $\bar{M}_0$ related to the mean anomaly M as

$$
M = \int_{t_0}^{t} n\,\mathrm{d}t + \bar{M}_0
\tag{6.36}
$$

where the mean motion $n = n(t)$ is time-dependent and related to the time-dependent osculating semi-major axis $a = a(t)$ by usual formula $n = \kappa a^{-3/2}$.

For the special case when the perturbing force has a potential R so that

$$
F^i = \frac{\partial R}{\partial r^i}
\tag{6.37}
$$

the Gaussian perturbation equations can be modified so that the disturbing potential R appears in the equations instead of the components of the perturbing force $\mathbf{F}$. In this way we gets the *Lagrange perturbation equations*

$$\frac{d}{dt}a = \frac{2}{na}\frac{\partial R}{\partial \bar{M}_0}$$

$$\frac{d}{dt}e = \frac{1-e^2}{ena^2}\frac{\partial R}{\partial \bar{M}_0} - \frac{\sqrt{1-e^2}}{ena^2}\frac{\partial R}{\partial \omega}$$

$$\frac{d}{dt}i = \frac{\cot i}{na^2\sqrt{1-e^2}}\frac{\partial R}{\partial \omega} - \frac{\csc i}{na^2\sqrt{1-e^2}}\frac{\partial R}{\partial \Omega}$$

$$\frac{d}{dt}\omega = \frac{\sqrt{1-e^2}}{ena^2}\frac{\partial R}{\partial e} - \frac{\cot i}{na^2\sqrt{1-e^2}}\frac{\partial R}{\partial i} \qquad (6.38)$$

$$\frac{d}{dt}\Omega = \frac{\csc i}{na^2\sqrt{1-e^2}}\frac{\partial R}{\partial i}$$

$$\frac{d}{dt}\bar{M}_0 = -\frac{2}{na}\frac{\partial R}{\partial a} - \frac{1-e^2}{ena^2}\frac{\partial R}{\partial e}$$

where R is considered as a function of osculating elements: $R = R(a, e, i, \omega, \Omega, \bar{M}_0)$. Clearly, the Lagrange perturbation equations are fully equivalent to the Gaussian perturbation equations. The Lagrange perturbation equations have many remarkable properties and are widely used in theoretical investigations (Murray and Dermott 1999 and Beutler 2005).

Osculating elements are very convenient for analytical assessments of the effects of particular perturbations. They are also widely used for practical representation of orbits of asteroids and artificial satellites. Such a representation is especially efficient when the osculating elements can be represented using simple functions of time requiring a limited number of numerical parameters (i.e. when a relatively "lower-accuracy" representation over a relatively "short" period of time is required; the exact meaning of lower-accuracy and short depends on the problem). For example, the predicted ephemerides (orbits) of GPS satellites are represented in the form of osculating elements using a simple model for these elements (linear drift plus some selected periodic terms). The numerical parameters are broadcast in the GPS signals from each satellite. Osculating elements are also used in many cases by the Minor Planet Center of the International Astronomical Union (IAU) to represent orbits of asteroids and comets.

6.5 The *N*-body problem in the relativistic framework

In the post-Newtonian framework, the equations of motion of a system of N point-like bodies are called Einstein–Infeld–Hoffmann (EIH) equations and are given by Eq. (5.13). Different forms of these equations (e.g. Damour *et al.* 1991, Will 1993) can be found in the literature. All these forms are, however, equivalent. The EIH equations also have ten classical integrals (Brumberg 1972, 1991, Damour and Vockrouhlický 1995). Let us discuss these equations from the point of view of Lagrangian mechanics. It can be shown by explicit calculations that the EIH equations can be derived using the Lagrange–Euler equations

$$\frac{\mathrm{d}}{\mathrm{d}t}\frac{\partial L}{\partial \dot{x}_A^i} - \frac{\partial L}{\partial x_A^i} = 0 \tag{6.39}$$

with the following Lagrange function (Will 1993)

$$L = \frac{1}{2}\sum_A M_A \dot{\mathbf{x}}_A^2 \left(1 + \frac{1}{4c^2}\dot{\mathbf{x}}_A^2\right) + \frac{1}{2}\sum_A\sum_{B\neq A}\frac{GM_A M_B}{r_{AB}}\left(1 + \frac{2\gamma+1}{c^2}\dot{\mathbf{x}}_A^2\right.$$

$$\left. - \frac{4\gamma+3}{2c^2}\dot{\mathbf{x}}_A\cdot\dot{\mathbf{x}}_B - \frac{1}{2c^2}\frac{\dot{\mathbf{x}}_A\cdot\mathbf{r}_{AB}}{r_{AB}}\frac{\dot{\mathbf{x}}_B\cdot\mathbf{r}_{AB}}{r_{AB}} - \frac{2\beta-1}{c^2}\sum_{C\neq A}\frac{GM_C}{r_{AC}}\right) \tag{6.40}$$

Using the Lagrange function we can easily derive all the conservation laws that we also have in the Newtonian case. The technique to do this is described by e.g. Landau and Lifshitz (2007, Ch. 2). We have the conservation laws of momentum, energy and angular momentum, respectively:

$$\sum_A \frac{\partial L}{\partial \dot{x}_A^i} = P^i = \text{const} \tag{6.41}$$

$$\sum_A \frac{\partial L}{\partial \dot{x}_A^i}\dot{x}_A^i - L = h = \text{const} \tag{6.42}$$

$$\sum_A \varepsilon_{ijk}x_A^j\frac{\partial L}{\partial \dot{x}_A^k} = c^i = \text{const} \tag{6.43}$$

where $\varepsilon_{ijk}a^j b^k = (\mathbf{a}\times\mathbf{b})^i$ is the i-th component of the cross product of vectors $\mathbf{a}$ and $\mathbf{b}$. The conservation of the momentum reads

$$\sum_A M_A\dot{\mathbf{x}}_A\left(1 + \frac{1}{2c^2}\left(\dot{\mathbf{x}}_A^2 - \sum_{B\neq A}\frac{GM_B}{r_{AB}}\right)\right)$$

$$-\frac{1}{2c^2}\sum_A\sum_{B\neq A}\frac{GM_A M_B}{r_{AB}^3}(\mathbf{r}_{AB}\cdot\dot{\mathbf{x}}_A)\,\mathbf{r}_{AB} + O(c^{-4}) = \mathbf{P} = \text{const} \tag{6.44}$$

The integral defining the center of mass of the system can be derived by integrating (6.44) and reads

$$\sum_A M_A\mathbf{x}_A\left(1 + \frac{1}{2c^2}\left(\dot{\mathbf{x}}_A^2 - \sum_{B\neq A}\frac{GM_B}{r_{AB}}\right)\right) + O(c^{-4}) = \mathbf{P}t + \mathbf{Q}, \quad \mathbf{Q} = \text{const} \tag{6.45}$$

The integral of energy takes the form

$$\frac{1}{2}\sum_A M_A\dot{\mathbf{x}}_A^2\left(1 + \frac{3}{4c^2}\dot{\mathbf{x}}_A^2\right) - \frac{1}{2}\sum_A\sum_{B\neq A}\frac{GM_A M_B}{r_{AB}}\left(1 - \frac{2\gamma+1}{c^2}\dot{\mathbf{x}}_A^2 + \frac{4\gamma+3}{2c^2}\dot{\mathbf{x}}_A\cdot\dot{\mathbf{x}}_B\right.$$

$$\left. + \frac{1}{2c^2}\frac{\dot{\mathbf{x}}_A\cdot\mathbf{r}_{AB}}{r_{AB}}\frac{\dot{\mathbf{x}}_B\cdot\mathbf{r}_{AB}}{r_{AB}} - \frac{2\beta-1}{c^2}\sum_{C\neq A}\frac{GM_C}{r_{AC}}\right) + O(c^{-4}) = h = \text{const} \tag{6.46}$$

This expression also gives the Hamiltonian function associated with the EIH equations. Besides theoretical investigations, the canonical form of the EIH equations can be used, e.g. with symplectic integrators. Finally, the conservation of angular momentum reads

$$
\sum_A M_A \, \mathbf{x}_A \times \dot{\mathbf{x}}_A \left(1 + \frac{1}{2c^2} \dot{\mathbf{x}}_A^2 + \frac{2\gamma+1}{c^2} \sum_{B \neq A} \frac{GM_B}{r_{AB}} \right)
$$

$$
- \frac{1}{2c^2} \sum_A \sum_{B \neq A} \frac{GM_A M_B}{r_{AB}} \left((4\gamma+3) \, \mathbf{x}_A \times \dot{\mathbf{x}}_B - \frac{\mathbf{x}_A \times \mathbf{x}_B}{r_{AB}} \frac{\dot{\mathbf{x}}_B \cdot \mathbf{r}_{AB}}{r_{AB}} \right) \tag{6.47}
$$

$$
+ O(c^{-4}) = \mathbf{c} = \mathrm{const}
$$

The validity of these ten integrals can be demonstrated by explicit calculations using the EIH equations. The expressions for the integrals are not exact integrals of the EIH equations, but can be calculated only up to terms $O(c^{-4})$. For numerical calculations this means that the corresponding quantities are not exactly constant, but should be considered as "almost constant" during the integrations, the numerical level of their variations depending on the particular problem. For the same reason the solution of the Hamiltonian equations with Eq. (6.46) as the Hamiltonian function is not exactly equivalent to the corresponding solution of the EIH equations. However, both solutions are valid in the post-Newtonian approximation.

Let us note that the integrals (6.44) and (6.45) define the barycenter of the considered system. When the EIH equations are applied for the Solar System, this gives the position and velocity of the Solar System barycenter. Usually we use the BCRS (see Chapter 5) and, therefore, set the constants $\mathbf{P}$ and $\mathbf{Q}$ to zero.

The acceleration of the body A in the post-Newtonian approximation depends on its own mass M_A. All terms in the post-Newtonian force that are proportional to M_A are explicitly shown in Eq. (5.13) of the previous chapter (this is the first term in the braces). If the motion of a spacecraft or minor body is considered and its mass can be neglected, the EIH equations can be slightly simplified by dropping this term (or, equivalently, by putting $M_A = 0$).

The EIH equations have formed the basis for computations of the motion of the Solar System since the early 1970s. The main relativistic effect here comes from the two-body problem Sun–planet: it is the famous perihelion advance that amounts to $43''$ per century for Mercury. This secular effect can be very easily derived and taken into account using osculating arguments. For high-accuracy applications such as binary pulsar timing the post-Newtonian orbit of the binary system should be represented exactly. To this end one can use osculating elements (Brumberg 1972, 1991; Soffel 1989), but such an explicit representation is rather cumbersome. More elegant and compact representations have been suggested by several authors. Klioner and Kopeikin (1994) compare several such representations and give explicit relations between them.

While the relativistic equations of motion for the N-body problem are only known in the first post-Newtonian approximation where all terms of order c^{-2} are taken into account, the equations of motion for the two-body problem have been derived also in higher post-Newtonian approximations. The explicit form of the equations of motion of the two-body problem in the 3.5 post-Newtonian approximation (equations including terms of order

c^{-7}) can be found, e.g. in Blanchet (2006). Note that starting with the 2.5 post-Newtonian approximation (terms of order c^{-5}), the motion ceases to be conservative because of the gravitational radiation emitted by the system of moving bodies. The orbit of the relativistic two-body problem starting from this approximation is characterized by a secular decrease of the semi-major axis due to the loss of energy by gravitational waves. The strongest reason to develop the equations of motion in higher post-Newtonian approximations is the necessity to predict the gravitational wave signal of inspiralling binary systems of compact objects (neutron stars and black holes). In the extreme case of coalescing black holes and neutron stars even those equations of motion cannot adequately describe the reality since the neglected terms become too large. In this case direct numerical integration of the general-relativistic hydrodynamic equations is the only possibility. Tremendous advances in computer hardware and computational techniques have quite recently made such direct numerical solutions feasible.

6.6 Beyond the N-body problem: oblateness of the central body and non-gravitational effects

The problem of the motion of N "point-like" bodies is often not sufficient to describe the real motion of celestial bodies. In this section we will consider the effect of the oblateness of the central body on the motion of the two-body problem as well as review the most important non-gravitational forces.

The oblateness of certain bodies should be taken into account to construct modern Solar System ephemerides (see the next section) as well as to model the motion of spacecraft that come relatively close to massive bodies (e.g. Earth's satellites or planetary orbiters). The additional force due to the oblateness of the gravitating body can be considered as a small perturbation to the main force coming from the spherically symmetric gravitational field of the body. We will use this problem as an example of the use of the Gaussian perturbation equations. In general, the gravitational field of a non-spherical body at an external point given by its spherical coordinates (r, λ, φ) can be written as

$$U(r, \lambda, \varphi)$$
$$= \frac{\kappa^2}{r} \left(1 - \sum_{n=2}^{\infty} J_n \frac{L^n}{r^n} P_{n0}(\sin \varphi) + \sum_{n=2}^{\infty} \sum_{k=1}^{n} \frac{L^n}{r^n} (C_{nk} \cos k\lambda + S_{nk} \sin k\lambda) P_{nk}(\sin \varphi) \right)$$

$$(6.48)$$

where $\kappa^2 = GM$, L is the radius of a sphere encompassing the body, and $P_{nk}(x)$ are the associated Legendre polynomials $(P_{n0}(x) \equiv P_n(x))$. Here J_n are the zonal harmonics characterizing axially symmetric gravitational fields, and C_{nk} and S_{nk} are the tesseral and sectorial harmonics describing the deviation of the gravitational field from the axial symmetry. For an oblate, axially symmetric body only coefficients J_n are not zero, the second zonal harmonic J_2 typically giving the largest effect. Since J_2 itself is typically small (among the major bodies of the Solar System Saturn has the largest value, $J_2 = 0.0163$), we confine ourselves to the variations of the orbital elements linear in J_2. Since the perturbing force due to J_2 comes from a part of the gravitational potential U as given by Eq. (6.48) the Lagrange perturbation equations (6.38) can be used to model the corresponding orbital perturbations. The perturbing potential can be taken as

$$R(r, \varphi) = -J_2 \frac{\kappa^2 L^2}{r^3} P_2(\sin \varphi) \tag{6.49}$$

Taking the equatorial plane of the central body as the xy-plane of our coordinates (so that the inclination i of the orbit is the inclination with respect to the equatorial plane of the body), we can see that $\sin \varphi = \sin i \ \sin(v + \omega)$ and, therefore,

$$R(a, e, i, \omega, v) = \frac{J_2 \kappa^2 L^2}{4a^3} \left(\frac{1 + e \cos v}{1 - e^2}\right)^3 (2 - 3 \sin^2 i + 3 \sin^2 i \ \cos(2v + 2\omega)) \tag{6.50}$$

Since the perturbing potential does not depend on Ω, $\sqrt{a(1 - e^2)} \cos i = $ const. Using Eq. (6.38) it can be seen that the derivative of $\sqrt{a(1 - e^2)} \cos i$ is proportional only to $\partial R/\partial \Omega$ and indeed vanishes in the case under study. Now, using Eq. (6.50) we can derive the partial derivatives of R with respect to the Keplerian elements as needed in Eq. (6.38) and integrate the equations using

$$dt = \frac{a^{3/2}(1 - e^2)^{3/2}}{\kappa (1 + e \cos v)^2} dv \tag{6.51}$$

This procedure results in rather complicated equations that are given e.g. in Roy (2005, Sect. 11.4.1). Here we are only interested in secular variations of the Keplerian parameters. Let us represent the perturbing potential as a Fourier series in M:

$$R = R_0 + \sum_{k=1}^{\infty} \left(R_k^{(c)} \cos kM + R_k^{(s)} \sin kM\right) \tag{6.52}$$

where the coefficients R_0, $R_k^{(c)}$, and $R_k^{(s)}$ depend only on a, e, i, and ω. Clearly, the terms with $R_k^{(c)}$ and $R_k^{(s)}$ are responsible only for periodic terms in the osculating elements. We are, therefore, interested only in R_0. Using Eq. (6.28) and noting that $X_0^{-3,0} = (1 - e^2)^{-3/2}$ and $X_0^{-3,2} = 0$ we get

$$R_0(a, e, i, \omega) = \frac{J_2 \kappa^2 L^2}{4a^3} (1 - e^2)^{3/2} (2 - 3 \sin^2 i) \tag{6.53}$$

Now, substituting R_0 as the perturbing potential in Eq. (6.38) one concludes that the semi-major axis a, the eccentricity e and the inclination i have no secular drifts, and

$$\hat{\omega} = \frac{3J_2 L^2 \kappa}{a^{7/2}(1 - e^2)^2} \left(1 - \frac{5}{4} \sin^2 i\right) t + \text{const}$$

$$\hat{\Omega} = -\frac{3J_2 L^2 \kappa}{2a^{7/2}(1 - e^2)^2} \cos i \ t + \text{const} \tag{6.54}$$

$$\hat{M}_0 = \frac{3J_2 L^2 \kappa}{2a^{7/2}(1 - e^2)^{3/2}} \left(1 - \frac{3}{2} \sin^2 i\right) t + \text{const}$$

Hats over the osculating elements stress that we deal with their averaged values. The last equation describes a constant change in the period of the motion, while the first two demonstrate that the orbit precesses with constant angular velocity around the z-axis (changing Ω) and around the vector $\mathbf{c}$ (changing ω). From Eq. (6.54) one can see that for orbits perpendicular to the equatorial plane ($i = 90°$) the averaged orbital plane remains constant in space ($\hat{\Omega} = $ const and $\hat{i} = $ const), and for the specific value of inclination $i = \arcsin\frac{2}{\sqrt{5}} \approx 63°\,26'$, sometimes called the *critical inclination*, the orbit does not precess in the orbital plane ($\hat{\omega} = $ const). Precession rates for typical Earth satellites can reach some degrees per day. The same formulas can be applied for the planetary orbits in the field of the oblate Sun. Since the solar second zonal harmonics $J_2^\odot$ are $\simeq 2 \cdot 10^{-7}$, typical precession rates of planetary orbits are about 20 µas/yr.

Let us now turn to non-gravitational forces. They play an important role for high-accuracy modeling of the motion of celestial bodies. We will briefly discuss these forces for different classes of celestial objects. For Earth satellites the main sources of non-gravitational forces are the atmosphere of the Earth and radiation pressure. Atmospheric forces are the most important non-gravitational forces acting on low-altitude satellites. In order to calculate these forces we need a reasonable model of the density of the upper atmosphere and a detailed model of the interaction of various parts of the satellites with both neutral gas and charged particles. The main atmospheric effect is a force acting in the direction opposite to the velocity of the satellite and often called *atmospheric drag*. The atmospheric drag attempts to decelerate the satellite and that results in the decreasing of both the semi-major axis and the eccentricity of the orbit. The motion through the atmosphere also causes lift and binormal forces (in a way, similar to the forces acting on an airplane), but they are usually much smaller and can often be neglected. Atmospheric forces can also influence the spatial orientation of the satellites. *Radiation pressure* is the effect coming from reflection and absorption of photons by the surface of the satellite. The photons can come directly from the Sun or can be reflected by nearby extended bodies (e.g. by the Earth in the case of Earth's satellites). A detailed model of the reflection and absorption is required here to have a good accuracy of the prediction of these forces. One more source of non-gravitational forces is the thrust forces intentionally applied to space vehicles to correct their orbital and/or rotational motion.

A detailed introduction to modern practice of modeling these and other forces acting on Earth's satellites is given e.g. by Montenbruck and Gill (2000). The models of all these non-gravitational forces are in part empirical and include parameters that should be fitted to observations. This empirical nature of the models often limits the accuracy of orbital modeling. The so-called drag-free satellites allow us to compensate for all non-gravitational forces acting on the satellite by actively correcting the motion so that a freely flying test mass placed in a vacuum chamber inside the satellite is kept at the same position with respect to the body of the satellite. Such drag-free satellites promise to play an important role in all scientific applications, where extreme accuracy of the knowledge of orbital motion is required.

Non-gravitational forces are also important for the motion of other bodies. In addition to the radiation-pressure effects, that are only important for small bodies, the so-called Yarkovsky and YORP effects play an important role for the dynamics of asteroids. The

Yarkovsky and YORP effects are thermal radiation forces and torques that cause a drift of the semi-major axis and changes of the spin vector of the bodies (Bottke *et al.* 2006) and influence the long-term dynamics of smaller asteroids. In addition, thrust forces due to the natural non-isotropic outgassing of comets substantially influences their motion. These outgassing forces are often modeled by empirical forces with free parameters fitted to observational data (e.g. Marsden *et al.* 1973). Finally, a number of non-gravitational forces should be taken into account for the Earth–Moon system (see below).

6.7 Modern ephemerides of the Solar System

Modern ephemerides of the Solar System bodies are numerical solutions of the differential equations of motion. These solutions are obtained by numerical integration of those differential equations and by fitting the initial conditions of these integrations and other parameters of the force model to observational data.

The equations of motion used here are the EIH equations discussed above augmented by a number of smaller forces. These forces include Newtonian forces due to asteroids, the effects of the figures (non-sphericity) of the Earth, Moon, and the Sun as well as some non-gravitational forces. For the Sun it is sufficient to consider the effect of the second zonal harmonics $J_2^\odot$. The zonal harmonics J_n of the Earth and the Moon are usually used up to $n \leq 4$. In most cases one only needs forces coming from the interaction of these zonal harmonics with other bodies modeled as point masses. The dynamics of the Earth–Moon system requires even more detailed modeling since the translational motion of the Earth and the Moon are coupled with their rotational motions and deformations in a complicated way. For the Moon, even more subtle effects due to tesseral harmonics C_{nk} and S_{nk} again with $n \leq 4$ (see Eq. (6.48)) should be taken into account. Tidal deformations of the Earth's gravitational field influence the translational motion of the Moon and should be also taken into account. Rotational motion of the Earth is well known and obtained from dedicated observations by the International Earth Rotation and Reference Systems Service (IERS). Rotational motion of the Moon is often called physical libration and its modeling is an important part of the process of construction of Solar System ephemerides. Physical libration is modeled as the rotational motion of a solid body with tidal and rotational distortions, including both elastic and dissipational effects. A discussion of all these forces coming from the non-point-like structure of the gravitating bodies can be found in Standish and Williams (2012).

Asteroids play an important role for high-accuracy modeling of the motion of the inner Solar System, the motion of Mars being especially sensitive to the quality of the modeling. Since masses of asteroids are poorly known for most of them, the modeling is not trivial. Usually, asteroids are treated in three different ways. A number of "big" asteroids are integrated together with the major planets, the Moon, and the Sun. For these big asteroids the masses are estimated from the same observational data that are used to fit the ephemeris. Among these big asteroids are always "the big three" – Ceres, Pallas, and Vesta – and, sometimes, up to several tens of asteroids which influence the motion of Mars more than other asteroids. For some hundreds of asteroids their masses are estimated using their taxonomic (spectroscopic) classes and their estimated radii that are determined by photometry, radar data, or observations of stellar occulations by asteroids. For each of the three taxonomic classes – C (carbonaceous chondrite), S (stony) and M (iron) – the mean density is determined as a part of the ephemeris construction. The cumulative effect of other asteroids is sometimes empirically modeled by a homogeneous massive ring in the plane of ecliptic. The mass of the ring and its radius are again estimated from the same data that are used for the ephemeris (Krasinsky *et al.* 2002; Kuchynka *et al.* 2010).

Since the equations of motion are ordinary differential equations any method for numerical integration of such equations can be used to integrate them. A very good practical overview of numerical integration methods is given in Montenbruck and Gill (2000, Ch. 4). Even more details can be found in Beutler (2005, Part I, Ch. 2). In practice, for planetary motion, we use either multistep Adams (predictor-corrector) methods (Standish and Williams 2012, Fienga *et al.* 2008) or the Everhart integrator (Everhart 1985, Pitjeva 2005). The latter is a special sort of implicit Runge–Kutta integrators. Numerical round-off errors are an important issue for the integrations of planetary ephemerides. Usual double-precision (64 bit) arithmetic is not sufficient to achieve the goal accuracy and quadruple-precision (128 bit) arithmetic is often used at least for part of the computations. Since the beginning of the 1970s the JPL ephemeris team has used the variable-stepwise, variable-order multistep Adams integrators called DIVA/QIVA (Krogh 2004). Fienga *et al.* (2008) have shown that only a few arithmetical operations in the classical Adams integrator of order 12 must be performed with quadruple precision to achieve an acceptable accuracy over longer integration intervals. This substantially increases the performance of numerical integrations.

Observational data used for planetary ephemerides include radar observations of Earth-like planets, radar and Doppler observations of spacecraft (especially planetary orbiters), VLBI observations of spacecraft relative to some reference quasars, lunar laser ranging data, and, finally, optical positional observations of major planets and their satellites (especially important for outer planets with very few radiometric observations).

A total of 250 parameters are routinely fitted for the construction of planetary ephemerides. These parameters include: initial positions and velocities of the planets and some of their satellites, the orientation of the frame with respect to the ICRF, the mass parameter $GM_\odot$ of the Sun (or the value of the astronomical unit in meters for older ephemerides), the parameters of the model for asteroids (see above), various parameters describing rotational and translational motion of the Earth–Moon system, and various parameters used in the reduction of observational data (phase corrections for planetary-disk observations, corrections to precession and equinox drift, locations of various relevant sites on the Earth and other bodies, parameters of the solar corona, parameters describing the geometrical figures of Mercury, Venus, and Mars, etc.). A useful discussion of some models used for data modeling is given by Moyer (2003) and Standish and Williams (2012). The masses of the major planets can be also fitted from the same data, but, when available, they are taken from the special solutions for the data of planetary orbiters. However, the masses of the Earth and the Moon are often determined in the process of the construction of planetary ephemerides.

Modern ephemerides are represented in the form of Chebyshev polynomials. The details of the representation can vary from one ephemeris to another, but the principles are the same: each scalar quantity is represented by a set of polynomials p_i of the form

$$p_i(t) = \sum_{k=0}^{N_i} a_k^{(i)} T_k(x), \quad x = \frac{2t - t_i - t_{i+1}}{t_{i+1} - t_i} \tag{6.55}$$

where $T_k(x)$ are the Chebyshev polynomials of the first kind (given by the recurrent relations $T_0 = 1$, $T_1 = x$, and $T_{k+1} = 2x\,T_k - T_{k-1}$), and coefficients $a_k^{(i)}$ are real numbers. Each polynomial $p_i(t)$ is valid for some interval of time $t_i \leq t \leq t_{i+1}$. The representation (6.55) is close to the optimal uniform approximation of a function by polynomials of given order (Press *et al.* 2007, Section 5.8), and, thus, gives a nearly optimal representation of a function using a given number of free parameters. The orders of polynomials N_i do not usually depend on the time interval, but do depend on the quantity to be represented. Sometimes (e.g. for the JPL ephemerides) one polynomial represents both the position and the velocity of a body. The velocity can be then calculated as a derivative of Eq. (6.55):

$$\frac{d}{dt} p_i(t) = \frac{2}{t_{i+1} - t_i} \sum_{k=1}^{N_i} k a_k^{(i)} U_{k-1}(x) \tag{6.56}$$

where $U_k(x)$ are the Chebyshev polynomials of the second kind (given by the recurrent relations $U_0 = 1$, $U_1 = 2x$, and $U_{k+1} = 2xU_k - U_{k-1}$). At the boundaries t_i of the time intervals, the polynomials p_i must satisfy conditions like $p_{i-1}(t_i) = p_i(t_i)$, so that the whole approximant is continuous. Additional constraints for the derivatives $\dot{p}_{i-1}(t_i) = \dot{p}_i(t_i)$ can be applied to make the approximant continuously differentiable. An efficient technique to compute the coefficients $a_k^{(i)}$ starting from values of the quantity to be represented is described by Newhall (1989).

There are three sources of modern planetary ephemerides: Jet Propulsion Laboratory (JPL, Pasadena, USA; DE ephemerides), Institut de Méchanique Céleste et de Calcul des Éphémérides (IMCCE, Paris Observatory, France; INPOP planetary ephemerides) and the Institute of Applied Astronomy (IAA, St. Petersburg, Russia; EPM ephemerides). All of them are available from the Internet:

- http://ssd.jpl.nasa.gov/?ephemerides/ for the DE ephemerides,
- http://www.imcce.fr/inpop/ for the INPOP, and
- ftp://quasar.ipa.nw.ru/incoming/EPM/ for the EPM ephemerides.

Different versions of the ephemerides have different intervals of validity, but typically these are several hundred years around the year 2000. The longest readily available ephemerides are valid for a time span of 6000 years. Further details on these ephemerides can be found in Pitjeva (2005), Fienga *et al.* (2008), Folkner (2010), Standish and Williams (2012).

It should also be mentioned that for the lower-accuracy applications, semi-analytical theories of planetary motion called VSOP (French acronym for "secular variations of planetary orbits") are available (Bretagnon and Francou 1988, Moisson and Bretagnon 2001). The semi-analytical theories are given in the form of Poisson series $\sum_k c_k t^{n_k} \cos(a_k t + b_k)$, where a_k, b_k, and c_k are real numbers, and $n_k \geq 0$ are integer powers of time t. Formally any values of time can be substituted into such series, but the theory is meaningful only for several thousand years around the year 2000.

6.8 Can an observer determine its own velocity from positional observations?

In view of the planned and future high-accuracy astrometric missions we face the problem that the BCRS velocity accuracy requirements of those missions are rather difficult to satisfy. Indeed, an error δv of the velocity of an observer directly translates to an error $\delta v/c$ in the data reduction because of Newtonian aberration. For an accuracy of $\varepsilon = 1$ μas the maximum allowed error of the velocity is $\delta v_{\max} = c\varepsilon \approx 1.4$ mm/s, but it should be even lower to avoid systematic errors. For a hypothetical mission with accuracy $\varepsilon = 0.01$ μas, $\delta v_{\max}$ should be ≈ 0.01 mm/s. Even 1 mm/s is a relatively high accuracy for a big satellite flying sufficiently far from the Earth. Special care must be taken to reach this accuracy. Moreover, to avoid systematic errors we must be sure that the velocity of the satellite is given in the same (relativistic) reference system in which the model for astrometric observations is formulated and in which we would like to obtain the parameters of the observed objects (see the previous chapter). This naturally brings us to the question as to whether the data processing can be made self-calibrating, so that the velocity of the observer performing positional observations can be determined or improved from its own observational data.

Clearly, such a determination means that some correction $\delta \mathbf{v}(t)$ to the a priori "ephemeris" velocity $\mathbf{v}_{\mathrm{eph}}(t)$ is fitted to observations together with parameters of the sources, other parameters of interest and, possibly, some calibration parameters. The velocity correction is the difference between the unknown true velocity and some ephemeris velocity known a priori: $\delta \mathbf{v}(t) = \mathbf{v}_{\mathrm{true}}(t) - \mathbf{v}_{\mathrm{eph}}(t)$. For each source at least five parameters should be determined: two parameters giving the position at some epoch, two parameters of the proper motion, and the parallax. Here we consider the simple case when the velocity correction $\delta \mathbf{v}$ should be determined together with the source parameters only.

A detailed analysis of the problem has been done in the framework of the preparations for the European Space Agency mission Gaia (Butkevich and Klioner 2008 and references therein) and has revealed that the problem can be solved only partially. Indeed, there are three types of signals in the velocity error that are fully equivalent to certain changes in the astrometric parameters of the sources.

1. A small velocity correction $\delta \mathbf{v}$ leads to a change of the observed direction $\mathbf{s}$ that can be described as $\delta \mathbf{s} = \mathbf{s} \times (\delta \mathbf{v} \times \mathbf{s})/c$. Here higher-order terms coming from the relativistic aberration are neglected. The constant velocity error $\delta \mathbf{v}(t) = \mathbf{v}_0 = \text{const}$ obviously leads to shifts in the observed direction given by $\delta \mathbf{s} = \mathbf{s} \times (\mathbf{v}_0 \times \mathbf{s})/c$. This shift is constant for each source (for each point with a given direction $\mathbf{s}$). Therefore, a constant error in the velocity cannot be distinguished from a change of position of each source. Since we do

not know these positions a priori, this constant error in the velocity cannot be determined from positional observations. The effect is similar to that of secular aberration.

2. The same applies to any linear function of time in the velocity. If $\delta\mathbf{v}(t) = \mathbf{a}_0\,(t - t_0)$ with $\mathbf{a}_0 = \text{const}$, we have $\delta\mathbf{s} = \boldsymbol{\mu}\,(t - t_0)$, where $\boldsymbol{\mu} = \mathbf{s} \times (\mathbf{a}_0 \times \mathbf{s})/c$. This shift in $\mathbf{s}$ is indistinguishable from a change in individual proper motions of each source. Again, since the proper motions are unknown a priori, such an error in the velocity cannot be found from positional observations.

3. Finally, a velocity error of the form $\delta\mathbf{v}(t) = \alpha_0\,\mathbf{r}_{\text{obs}}(t)$, where $\alpha_0 = \text{const}$ and $\mathbf{r}_{\text{obs}}(t)$ is the barycentric position of the observer, gives the shift of the observed direction $\delta\mathbf{s} = \mathbf{s} \times (\alpha_0\,\mathbf{r}_{\text{obs}} \times \mathbf{s})/c$ which is indistinguishable from a constant change $\delta\pi = -\alpha_0\,\text{AU}/c$ in the parallaxes π of all sources. This is clear from the fact that a change of parallax $\delta\pi$ corresponds to the change of observable direction of the form $\delta\mathbf{s} = -\mathbf{s} \times \left(\dfrac{\delta\pi}{\text{AU}}\,\mathbf{r}_{\text{obs}} \times \mathbf{s}\right)$. Here AU is the astronomical unit of length.

Therefore, the velocity correction $\delta\mathbf{v}_{\text{fit}}$ that can be fitted from positional observations should not contain these three signals and the full (unknown) velocity correction $\delta\mathbf{v}$ can be written as

$$\delta\mathbf{v} = \delta\mathbf{v}_{\text{fit}} + \delta\mathbf{v}_{\text{bias}}$$
$$\delta\mathbf{v}_{\text{bias}} = \mathbf{v}_0 + \mathbf{a}_0(t - t_0) + \alpha_0\,\mathbf{r}_{obs}(t) \tag{6.57}$$

where $\delta\mathbf{v}_{\text{bias}}$ is the error in the velocity that cannot be obtained from positional observations and that leads to biases in the source parameters as described above. The velocity error $\delta\mathbf{v}_{\text{bias}}$ is determined by seven scalar constants $\mathbf{v}_0$, $\mathbf{a}_0$, and α_0. It is now clear that the velocity of the observer can be improved from positional observations performed by that observer if and only if one can demonstrate that $\delta\mathbf{v}_{\text{bias}}$ is negligibly small for the given observational accuracy and for the given ephemeris velocity $\mathbf{v}_{\text{eph}}$. Various calibration parameters that should be fitted along with the source parameters may bring additional degeneracies with the velocity correction and can make the problem more complicated. These additional degeneracies can, however, be avoided by a suitable reparametrization of the calibration parameters and the velocity correction. Whether $\delta\mathbf{v}_{\text{bias}}$ is negligibly small or not depends on particular circumstances of each astrometric project. It is therefore advisable to plan the future astrometric projects taking this aspect into account.

References

Aarseth, S. J. (2003). *Gravitational N-body Simulations: Tools and Algorithms*. Cambridge: Cambridge University Press.

Aarseth, S. J., Tout, Chr.A., and Mardling, R. A. (2008). *The Cambridge N-Body Lectures*. Berlin: Springer.

Arnold, V. I., Kozlov, V. V., and Neishtadt, A. I. (1997). *Mathematical Aspects of Classical and Celestial Mechanics*. Berlin: Springer.

Beutler, G. (2005). *Methods of Celestial Mechanics*. Berlin: Springer.

Blanchet, L. (2006). *Living Rev. Relativity*, **9**, 4, http://www.livingreviews.org/lrr-2006-4.

Bottke, Jr., W. F., Vokrouhlický, D., Rubincam, D. P., and Nesvorný, D. (2006). *Ann. Rev. Earth Planet. Sci.* **34**, 157.

Bretagnon, P. and Francou, G. (1988). *A&A*, **202**, 309.

Brouwer, D. and Clemence, G. M. (1985). *Methods of Celestial Mechanics*, 2nd edn. Orlando, FL: Academic Press.

Brumberg, V. A. (1972). *Relativistic Celestial Mechanics*. Moscow: Nauka (in Russian).

Brumberg, V. A. (1991). *Essential Relativistic Celestial Mechanics*. Bristol: Adam Hilger.

Butkevich, A. G. and Klioner, S. A. (2008). In *A Giant Step: from Milli- to Microarcsecond Astrometry*, eds. W. J. Jin, I. Platais and M. A. C. Perryman, Cambridge: Cambridge University Press, p. 252.

Chenciner, A. and Montgomery, R. (2000). *Ann. Math.* **152**, 881.

Damour, T. and Vockrouhlický, D. (1995). *Phys. Rev. D*, **52**, 4455.

Damour, T., Soffel, M., and Xu, Ch. (1991). *Phys. Rev. D*, **43**, 3273.

Everhart, E. (1985). In *Dynamics of Comets: Their Origin and Evolution*, eds. A. Carusi and G. B.Valsecci. Dordrecht: Reidel, Astrophysics and Space Science Library, vol. 115, p. 85.

Fienga, A., Manche, H., Laskar, J., and Gastineau, M. (2008). *A&A*, **477**, 315 (see also arXiv: 0906.2860).

Folkner, W. M. (2010). In *Relativity in Fundamental Astronomy*, eds. S. A. Klioner, P. K. Seidelmann, and M. H. Soffel. Cambridge: Cambridge University Press, p. 155.

Fukushima, T. (1996). *AJ*, **112**, 2858.

Klioner, S. A. and Kopeikin, S. M. (1994). *AJ*, **427**, 951.

Krasinsky, G. A., Pitjeva, E. V., Vasilyev, M. V., and Yagudina, E. I. (2002). *Icarus*, **158**, 98.

Krogh, F. T. (1994). *Annals of Numerical Mathematics*, **1**, 423 (the DIVA/QIVA software is available from http://mathalacarte.com).

Kuchynka, P., Laskar, J. , Fienga, A., and Manche, H. (2010). *A&A*, **514**, A96.

Landau, L. D. and Lifshitz, E. M. (2007). *Course of Theoretical Physics*. Amsterdam: Elsevier, Butterworth-Heinemann, vol. 1.

Marsden, B. G., Sekanina, Z., and Yeomans, D. K. (1973). *AJ*, **78**, 211.

Montenbruck, O. and Gill, E. (2000). *Satellite Orbits: Models, Methods and Applications*. Berlin: Springer.

Moisson, X. and Bretagnon, P. (2001). *Cel. Mech. Dyn. Astron.*, **80**, 205.

Morbidelli, A. (2002). *Modern Celestial Mechanics: Aspects of Solar System Dynamics*. London: Taylor & Francis.

Moyer, T. D. (2003). *Formulation for Observed and Computed Values of Deep Space Network Data Types for Navigation*. Hoboken: Wiley-Interscience.

Murray, C. D. and Dermott, S. F. (1999). *Solar System Dynamics*. Cambridge: Cambridge University Press.

Newhall, X. X. (1989). *Cel. Mech.*, **45**, 305.

Pitjeva, E. V. (2005). *Solar Syst. Res.*, **39**, 176.

Press, W. H., Teukolsky, S. A., Vetterling, W. T., and Flannery, B. P. (2007). *Numerical Recipes: The Art of Scientific Computing*, 3rd edn. Cambridge: Cambridge University Press.

Roy, A. E. (2005). *Orbital Motion*, 4th edn. Bristol: IoP Publishing.

Soffel, M. H. (1989). *Relativity in Astrometry, Celestial Mechanics and Geodesy*. Berlin: Springer.

Standish, E. M. and Williams, J. G. (2012). In *Explanatory Supplement to the Astronomical Almanac*, 3rd edn., eds. S. Urban and K. Seidelmann, Herndon, VA, University Science Books.

Will, C. M. (1993). *Theory and Experiment in Gravitational Physics*. Cambridge: Cambridge University Press.

7

Celestial coordinate systems and positions

NICOLE CAPITAINE AND MAGDA STAVINSCHI

Introduction

Astrometry of celestial objects is based on the determination and use of the coordinates of these objects with respect to appropriate space and time reference systems. Such coordinates are essential for many applications in astronomy and geodesy. This chapter provides the theoretical basis and definitions, as well as expressions to be used for applications related to this field. The astrometry textbooks by Green (1985) and Kovalevsky and Seidelmann (2004) are also useful references. Details on the corresponding algorithms can be found in Chapter 8.

7.1 Definitions

7.1.1 Coordinates of a celestial object

The coordinates of celestial objects and their time variations are required for expressing the positions and motions of celestial bodies, which are essential for the interpretation of astronomical observations in terms of physical and dynamical characteristics of the Universe. Those coordinates are used for Galactic and Solar System astrometry (optical and radio astrometry), for geodesy, celestial mechanics, astrophysics, and cosmology. The distance of a celestial object is accessible to direct measurement only for Solar System objects which can be observed by telemetry observations (e.g. with radar, laser ranging, etc.). An indirect way to access the distances of nearby celestial objects is the determination of their annual and/or diurnal parallaxes (see Section 7.2.4 and Chapter 22). When the distance is not considered, the "position" of a celestial object (also called "place"), as well as its "coordinates," will refer here to the apparent direction in which that object is seen from the observer, which can be represented by the unit vector in that direction. The positions of celestial objects are usually visualized on a conventional celestial sphere of unit radius, the center of which depends on the type of coordinates that are required (see Section 7.3). The corresponding space coordinates can then be derived in cases when the distances of

Astrometry for Astrophysics: Methods, Models, and Applications, ed. William F. van Altena. Published by Cambridge University Press. © Cambridge University Press 2013.

the objects are known. Distances and motions in the Solar System, in our Galaxy, and in the Universe, are fundamental parameters for the understanding of their origin and evolution.

7.1.2 Reference systems and their realization

Dynamical theories imply the existence of space and time "reference systems"; when applied to physical systems, these theories require physical representations of the reference systems. The definition of reference systems, their practical realizations, and their improvement and maintenance so that they match the quality of the observations, are a continuing preoccupation of fundamental astronomy.

A space reference system is a theoretical concept of a system of coordinates, including time and standards necessary to specify the bases used to define the position and motion of objects in time and space. A practical realization of a space reference system is called a "reference frame"; it is usually realized as a catalog of positions and motions of a certain number of fiducial points. A timescale is an accessible reference, which assigns a time coordinate, called the date, at any event. It is, in general, the realization of a theoretical time, as, for example, the Geocentric Coordinate Time (TCG), or its linear function, Terrestrial Time (TT), both as defined in Chapter 5).

For space reference systems and frames, the critical problem is that of the orientation of the non-rotating axes. Its empirical basis is the postulate that there is no global rotation of the Universe. Previous celestial reference systems were based on bright stars which are too close to allow a good application of this postulate. The current reference system (see Section 7.1.3) is based on astrometric application of very long baseline interferometry (VLBI), which enables the measurement of angular positions of the most distant bodies we know, the quasars, with uncertainties well below $0''.001$ (about one hundred times smaller than the uncertainties of the ground-based measurements in the visible domain). The angular proper motions of these objects are negligible, although some of them show a variable structure. For more details on radio interferometry, see Chapter 12.

Previous timescales based on astronomical observations, namely Universal Time and Ephemeris Time, were imperfect. The former was based on the variable rotation of the Earth and the latter on the revolution of the Earth around the Sun, but realized from inaccurate lunar observations. The construction of an adequate timescale became a reality when the first atomic frequency standard based on the cesium transition was begun in 1955. Finally, International Atomic Time (TAI) is the basis of the quasi-ideal representations of the various coordinate timescales needed in dynamics, now developed in a relativistic framework (see Section 7.1.4 and Chapter 5).

7.1.3 Essential reference systems and frames for astrometry

The International Celestial Reference System (ICRS), defined by the IAU 1997 Resolution B2,[1] is the idealized barycentric coordinate system to which celestial positions are referred.

---

[1] For IAU resolutions, see www.iau.org/administration/resolutions/general_assemblies/

This system is realized by the International Celestial Reference Frame (ICRF) through the adopted positions and uncertainties of a set of extragalactic objects.

Two systems of spacetime coordinates have been defined within the framework of general relativity (GR) by the IAU 1991 Resolution A4 and then specified by the IAU 2000 Resolution B1.3, namely the Barycentric Celestial Reference System (BCRS) for the Solar System and the Geocentric Celestial Reference System (GCRS) for the Earth. The latter resolution provides the full post-Newtonian coordinate transformation between the BCRS and the GCRS in the form of the corresponding metric tensors. This includes the transformation between the two time-coordinates in the BCRS and GCRS, respectively, i.e. the Barycentric Coordinate Time (TCB) and the TCG.

The IAU 1991 Resolution A4 specified that the relative orientation of barycentric and geocentric spatial axes in BCRS and GCRS are without any time-dependent rotation; then, the IAU 2006 Resolution B2 fixed the default orientation of the BCRS and GCRS as, unless otherwise stated, assumed to be oriented according to the ICRS axes.

As part of astrometry relies on ground-based observations, a terrestrial reference system and its corresponding terrestrial reference frame are also required. According to the IUGG 2007 Resolution 2, the Geocentric Terrestrial Reference System (GTRS) is the system of geocentric spacetime coordinates within the framework of general relativity, co-rotating with the Earth, and related to the GCRS by a spatial rotation which takes into account the Earth orientation parameters; the International Terrestrial Reference System (ITRS) is the specific GTRS for which the orientation is operationally maintained in continuity with past international agreements (Bureau International de l'Heure (BIH) orientation). The geocenter is understood as the center of mass of the whole Earth system, including oceans and atmosphere (International Union of Geodesy and Geophysics (IUGG) 1991 Resolution 2). The International Terrestrial Reference Frame (ITRF) is a realization of the ITRS by a set of instantaneous coordinates (and velocities) of reference points distributed on the topographic surface of the Earth.

7.1.4 Timescales for astrometry

In addition to the TCB and TCG defined in the previous section as the time-coordinates associated with the BCRS and GCRS, respectively, other coordinate timescales have been introduced with the purpose of facilitating some practical applications, e.g. for replacing Ephemeris Time (ET), which was the timescale used prior to 1984 as the independent variable in gravitational theories of the Solar System, with its unit and origin conventionally defined.

Terrestrial Time, which may be used as the independent time argument for geocentric ephemerides, is related to TCG by a conventional linear transformation provided by the IAU 2000 Resolution B1.9. In a very similar way, Barycentric Dynamical Time (TDB), which may be used to serve as an independent time argument of barycentric ephemerides and equations of motion, is related to TCB by a conventional linear transformation provided by the IAU 2006 Resolution B3. The slopes of both linear functions are such that the mean rate of TT or TDB is close to the mean rate of the proper time for an observer located on the rotating geoid. Note that in order to keep the equations of motion of celestial bodies and photons invariant, these scalings of coordinate times are accompanied by the corresponding scalings of spatial coordinates and mass parameters GM of celestial bodies. See Chapter 5 for detailed definitions and relationships between these timescales.

International Atomic Time is a widely used practical realization of TT with a fixed shift of 32.184 s from the latter due to historical reasons. TAI was adopted on an international basis by the CGPM (Conférence Générale des Poids et Mesures) in 1971. It is a continuous timescale, now calculated at the Bureau International des Poids et Mesures (BIPM), using data from some 300 atomic clocks in over 50 national laboratories in accordance with the definition of the SI second.

Unfortunately, TAI is not available prior to 1955, while old observations of planets and satellites are still valuable and need to be referred to a uniform timescale. For using these observations, and also to test relativistic theories, the realization of dynamical timescales, such as ET prior to 1984, and TT, TDB, etc. afterwards, remains essential.

7.2 The transformation from ICRS to observed positions

7.2.1 Requirements

The reduction of observations of celestial objects from ground stations requires coordinates of the celestial objects and ground stations with respect to a common reference system. As the coordinates of the celestial objects are referred to a celestial reference system and the ground stations are referred to a terrestrial reference system, this requires coordinate transformation between the ICRS and the ITRS. Since the observed positions are referred to a local reference system, this requires an additional transformation between the ITRS and the topocentric reference system.

7.2.2 The Earth orientation parameters

The transformation between the ICRS and the ITRS depends on the Earth's rotation, which can be represented by the time-dependent Earth orientation parameters (EOPs). Those parameters are for the motion of the celestial pole in the celestial reference system, the rotation of the Earth around the axis associated with the pole, and polar motion. The IAU 2000 and IAU 2006 resolutions have provided accurate definitions for the pole and origins on the equator to be used in the ICRS-to-ITRS transformation and have specified the form of the parameters as well as the process for taking them into account.

The Earth's instantaneous pole of rotation (IRP), which is moving both in spatial orientation (precession–nutation) and within the Earth (polar motion), is not directly observable by the current astro-geodetic techniques (i.e. VLBI, Global Positioning System (GPS), etc.; see Section 7.5). Those techniques are sensitive only to the GCRS orientation of the z-axis of the ITRS, whose variations with time covers the whole spectrum of frequencies, from

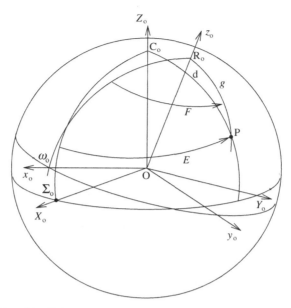

Fig. 7.1   The GCRS and ITRS position of the CIP (denoted P).

0 to infinity. The IAU 2000 Resolution B1.7 has defined a new pole for use in the GCRS-to-ITRS transformation. This pole, called the Celestial Intermediate Pole (CIP), is close to the IRP or the Earth's axis of figure, but it is defined through a convention separating nutation from polar motion in the frequency domain in order to best realize the pole in the high-frequency domain. According to this convention, the GCRS part of the CIP motion includes all the terms that are viewed in the ITRS in the retrograde diurnal band (i.e. all the terms with periods greater than 2 days in the GCRS), while its ITRS position results from the part of the motion which is outside the retrograde diurnal band.

The position of the CIP in the ITRS is provided by the ITRS x and y components of the CIP unit vector, or equivalently the ITRS polar coordinates g and F (with $x = \sin g \cos F$, $y = \sin g \sin F$; see Fig. 7.1). Note that the coordinates of the pole considered as representing the polar motion that is determined by observations, as in IERS publications, are $x_p = x, y_p = -y$.

The position of the CIP in the GCRS is provided by the coordinates X and Y, which are the GCRS x and y components of the CIP unit vector, or equivalently the GCRS polar coordinates d and E (with $X = \sin d \cos E$; $Y = \sin d \sin E$).

The coordinates X and Y can be written as (Capitaine 1990):

$$X = \xi_0 + \bar{X} - \mathrm{d}\alpha_0\, \bar{Y}$$
$$Y = \eta_0 + \bar{Y} + \mathrm{d}\alpha_0\, \bar{X} \tag{7.1}$$

where ξ_0 and η_0 are the celestial-pole offsets at the basic epoch J2000.0 and $\mathrm{d}\alpha_0$ the right ascension of the mean equinox of J2000.0 in the GCRS (i.e. frame bias in right ascension). The quantities $\bar{X}$ and $\bar{Y}$ are the celestial-pole coordinates in the mean equator and equinox

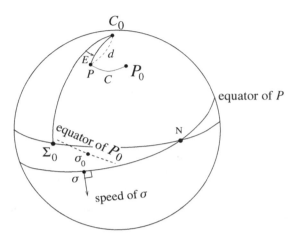

Fig. 7.2 Kinematical definition of the non-rotating origin (σ).

of J2000.0; they can be written as functions of the classical precession–nutation quantities as:

$$\bar{X} = \sin \omega \sin \psi,$$
$$\bar{Y} = -\sin \epsilon_0 \cos \omega + \cos \epsilon_0 \sin \omega \cos \psi \qquad (7.2)$$

where ω is the obliquity of the CIP equator on the J2000.0 ecliptic, ϵ_0 is its value at J2000.0, and ψ is the longitude, on the J2000.0 ecliptic, of the node of the CIP equator on the J2000.0 ecliptic. The parameters ω and ψ are such that:

$$\omega = \omega_A + \Delta\epsilon_1; \quad \psi = \psi_A + \Delta\psi_1 \qquad (7.3)$$

where ψ_A and $\Delta\psi_1$, and ω_A and $\Delta\epsilon_1$ are the precession and nutation quantities in longitude and obliquity referred to the ecliptic of epoch (see Fig. 7.4). The current expressions for X and Y as function of time have been derived from the IAU 2006 precession and IAU 2000A nutation (IAU 2006/2000A; see Section 7.2.4). The IAU 2006 value for ϵ_0 is $84381.406'' = 23° 26' 21.406''$.

For longitude origins on the CIP equator in the celestial and the terrestrial reference systems, the IAU 2000 Resolution B1.8 has recommended the "non-rotating origin" (Guinot 1979), which is such that for any displacement of the pole, the instantaneous motion of that origin is perpendicular to the equator (see Fig. 7.2), in contrast to the motion of the equinox (intersection of the equator and the equinox) due to precession–nutation (see Fig. 7.3). They were designated the Celestial Intermediate Origin (CIO) and the Terrestrial Intermediate Origin (TIO), respectively, by the IAU 2006 Resolution B2.

The positions of the CIO and TIO on the equator of the CIP are provided by two small quantities s and s', named "CIO locator" and "TIO locator," respectively. Their expressions are derived from the kinematical definition of "non-rotating" in the ICRS and ITRS, respectively when the CIP is moving due to precession–nutation and polar motion, respectively.

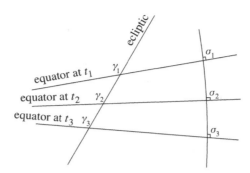

Fig. 7.3   Motion of the equinox (γ), and the CIO (σ) at successive dates, t_1, t_2, t_3.

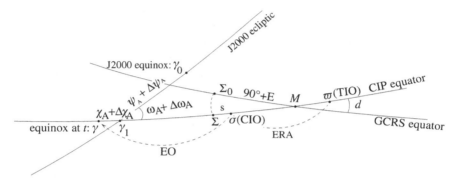

Fig. 7.4   The Earth rotation angle ERA ($= \widehat{\sigma\varpi}$) and its relation (GST $=$ ERA $-$ EO) to Greenwich Sidereal Time GST ($= \widehat{\gamma\varpi}$).

    The Earth rotation angle (ERA) is defined as the angle measured along the equator of the CIP between the CIO and the TIO (see Fig. 7.4) and the parameter UT1 is defined by a conventionally adopted linear proportionality to the ERA (see Section 7.4).

    Thanks to the kinematical definition of the CIO and TIO, the ERA accurately reflects the Earth's rotation around the CIP axis. It is determined by observations (currently from VLBI observations of the diurnal motions of distant radio sources).

    Greenwich Sidereal Time (GST), which is defined as the angle measured along the equator of the CIP between the equinox and the TIO, is related to the ERA, by the relation GST $=$ ERA(UT1) $-$ EO, where EO, called "equation of the origins," is the distance between the CIO and the equinox along the intermediate equator that is due to precession–nutation.

7.2.3 Transformation from ICRS to observed place

The procedure to be followed for accurate transformation from ICRS to the observed position (i.e. place) of a celestial object (not a planet) is illustrated by Figs. 7.5 and 7.6 from the IAU NFA Working Group documents (Capitaine *et al.* 2007). The chart summarizes the system, and the elements that are associated with that system, i.e. the name for the positions, the processes/corrections, the quantities required for that process, the origin to which the coordinates are referred, and the timescale to use. It provides the successive steps to be followed in order to be in agreement with the IAU 2000/2006 resolutions, noting at each step the space reference system and corresponding timescale. This shows in parallel the CIO- (see Section 7.2.2) and equinox based approaches (in the left and right of the chart, respectively) and specifies in what order to apply the usual corrections (annual aberration, precession-nutation etc.) when predicting apparent star directions for a ground-based observer.

The notation in the chart assumes that a rectangular coordinate system is being used, i.e. $v^T = [x, y, z]^T$ and $\mathbf{R}_1, \mathbf{R}_2, \mathbf{R}_3$, represent the standard rotation matrices about the x, y, and z axes, respectively.

In typical cases the BCRS-to-GCRS portion of the method, using existing annual aberration and light-deflection formulations (see Section 7.2.4), is accurate to a small fraction of a milliarcsecond (mas). However, the omission of light deflection by the planets could, in extreme cases, cause errors approaching 20 mas and there are various other missing terms at the sub-mas level. For very precise reductions it is necessary to use a fully GR-based approach (see Chapter 5). Note also that the transformation from the ITRS to observed place set out in the chart would require more complicated steps in the general relativity framework to achieve microarcsecond (μas) accuracy. Similarly, in demanding interferometer applications, such as VLBI, the geometry of the baselines requires a fully relativistic treatment.

The CIP (see Section 7.2.2) realized by the IAU 2006/2000A precession–nutation model defines its equator. The CIO based process makes use of an intermediate geocentric reference system in transforming to a terrestrial system. That system, defined by the CIP equator and the CIO on a specific date, is called the Celestial Intermediate Reference System (CIRS). The system obtained by applying to the CIRS a rotation of angle ERA around the z-axis is a geocentric reference system defined by the CIP equator and the TIO; it is called the Terrestrial Intermediate Reference System (TIRS).

The transformation to be used to relate the ITRS to the GCRS at the date t of the observation can be derived from the corresponding portions of the chart. It can be written, in the CIO based approach, as:

$$[\text{ITRS}] = \mathbf{W}(t)\mathbf{R}(t)\mathbf{C}(t)\,[\text{GCRS}] \tag{7.4}$$

where [ITRS] and [GCRS] denote coordinates in the reference systems ITRS and GCRS, respectively, and $\mathbf{W}(t)$, $\mathbf{R}(t)$, and $\mathbf{C}(t)$ are the transformation matrices arising from the ITRS motion of the CIP, from the rotation of the Earth around the axis of the CIP, and from the GCRS motion of the CIP, respectively:

$$W(t) = \mathbf{R}_1(-y_p) \cdot \mathbf{R}_2(-x_p) \cdot \mathbf{R}_3(s') \tag{7.5}$$

$$R(t) = \mathbf{R}_3(ERA) \tag{7.6}$$

$$C(t) = \mathbf{R}_3(-s) \cdot \mathbf{R}_3(-E) \cdot \mathbf{R}_2(d) \cdot \mathbf{R}_3(E) \tag{7.7}$$

The matrix transformation from the GCRS to the CIRS, $C(t)$ can be given in an equivalent form directly involving X and Y as:

$$C(t) = \mathbf{R}_3(-s) \cdot \begin{pmatrix} 1 - aX^2 & -aXY & -X \\ -aXY & 1 - aY^2 & -Y \\ X & Y & 1 - a(X^2 + Y^2) \end{pmatrix} \qquad (7.8)$$

with $a = 1/(1 + \cos d)$, which can also be written, with an accuracy of 1 μas, as $a = 1/2 + 1/8(X^2 + Y^2)$.

7.2.4 Description of the different effects and models to be used

The corrections to be considered for predicting apparent star directions for a ground-based observer (cf. Figs. 7.5 and 7.6) come from a number of effects that correspond to the motions of the observer and of the observed object, in combination with the finite velocity of light and the bending of a ray of light as it passes through the Earth's atmosphere and close to masses (relativistic light deflection due to the curvature of spacetime). The notations used in this section are those of Figs. 7.5 and 7.6.

Star's space motion

The space motion of a star at a TCB/TDB instant (see Fig. 7.5) from its catalog position is due to the relative space motion of the star and the Sun in the Galaxy. Proper motion is the component in the direction perpendicular to the radius vector from the Sun to the star. It results in an apparent angular motion of a star across the sky, which is usually given by the angular changes per year, μ_α and μ_δ in right ascension and declination, respectively. The most distant stars (i.e. with the smallest parallax, px) show the least proper motion. The second component of that space motion is the radial velocity, rv.

Parallax

The parallax (see also Chapter 22) is the difference in apparent direction of an object as seen from two different locations. The annual parallax refers to the difference in directions A and A' as seen from the Solar System barycenter and the geocenter (see Fig. 7.7), while diurnal parallax refers to the component of parallax due to the observer's separation from the geocenter. The computation of the effect of the annual parallax px involves the barycentric positions E_B and Q_B of the Earth and the object, respectively, evaluated at the required TCB/TDB instant (see Fig. 7.5); that of the diurnal parallax involves the equatorial horizontal parallax π, the geocentric latitude ϕ' and longitude λ, and geocentric distance ρ, of the observer at the TT instant (see Fig. 7.6).

The parallax is directly dependent on the distance between two different locations, so the parallax can be used to determine distances. This shows the importance of determining accurate annual parallaxes of stars (see also Chapter 22).

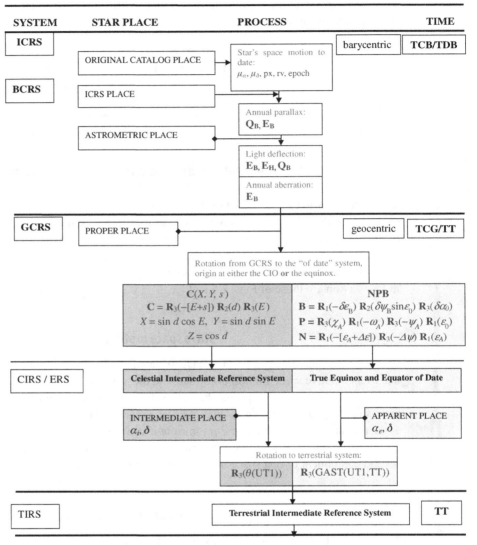

Fig. 7.5 Process to transform from the ICRS to the TIRS. Chart from the IAU Working Group on Nomenclature for Fundamental Astronomy (2006); see http://syrte.obspm.fr/iauWGnfa. This shows in parallel the CIO-, and equinox based-approaches, in the left and right part of the chart, respectively. The angle θ is the Earth rotation angle, ERA, that is related to UT1 by Eq. (7.9); GAST is the Greenwich apparent sidereal time that depends on both UT1 and TT.

Light deflection

Light deflection is the angle through which a ray of light is bent by the gravitational field of the Sun. The computation of this effect involves the barycentric and heliocentric positions E_B and E_H of the Earth, and the barycentric position Q_B of the object, all these quantities being evaluated at the required TCB/TDB instant.

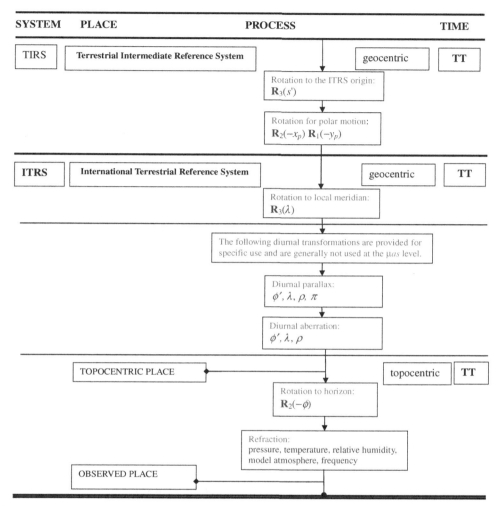

Fig. 7.6   Process to transform from the TIRS to the observed positions of a celestial object. Chart from the IAU Working Group on Nomenclature for Fundamental Astronomy (2006).

Aberration

Aberration is the apparent angular displacement of the observed position of a celestial object from its geometric position, caused by the finite velocity of light in combination with the motions of the observer and of the observed object. Annual aberration is due to the motion of the Earth around the Sun (see Fig. 7.8), while diurnal aberration is due to the Earth's rotation. Computation of the effect of annual aberration involves the barycentric velocity $\dot{E}_B$ of the Earth, evaluated at the required TCB/TDB instant, while that of the diurnal aberration involves the geocentric latitude ϕ' and longitude λ and geocentric distance ρ of the observer at the TT instant.

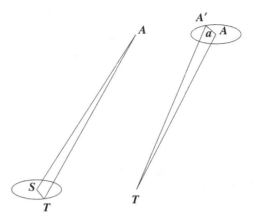

Fig. 7.7   Effect of the annual parallax: *A* and *A'* are the "true" and "apparent" directions of a celestial object; *S* and *T* are the positions of the Earth and the Sun, respectively and *a* is the distance *ST*.

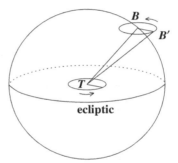

Fig. 7.8   Effect of the annual aberration: *B* and *B'* are the "true" and "apparent" directions of a celestial object.

Precession–nutation

Precession–nutation of the equator and of the CIP is the motion of the pole of rotation of a freely rotating body, undergoing torque from external gravitational forces. In the case of the Earth, precession–nutation of the equator is caused by Solar System objects acting on the Earth's equatorial bulge. This makes the pole of rotation describe a 26 000-year orbit around the ecliptic pole (precession) as well as a number of loops that constitutes the forced periodic part of the motion (nutation) (see Fig. 7.9).

The IAU 2000A nutation model, developed by Mathews *et al.* (2002) and denoted MHB2000, is based on the REN2000 rigid Earth nutation series of Souchay *et al.* (1999) for the axis of figure.

The rigid Earth nutation was transformed to the non-rigid Earth nutation by applying the MHB2000 "transfer function." This takes into account an Earth model composed of a deformable mantle, fluid core, and inner core; this also considers the existence of the anelasticity, ocean tides, electromagnetic couplings produced between the fluid outer core and the mantle as well as between the solid inner core and fluid outer core. It is based on

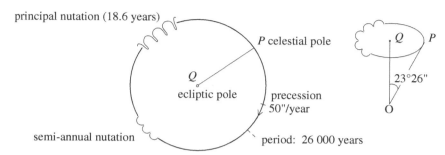

Fig. 7.9   Precession–nutation of the Earth's pole.

estimated values of seven of the parameters appearing in the theory, obtained from a least-squares fit of the theory to an up-to-date precession–nutation VLBI data set. Fundamental Earth parameters are the dynamical ellipticity e, dynamical ellipticity of the fluid core e_f, as well as parameters representing the deformabilities of the whole Earth and its fluid core, under tidal forcing.

The resulting nutation series includes 678 luni-solar terms and 687 planetary terms which are expressed as "in-phase" and "out-of-phase" components with their time variations. That model is expected to have an accuracy of about 10 μas for most of its terms. On the other hand, the free core nutation (FCN), being a free motion (of the fluid core) which cannot be predicted rigorously, is not considered a part of the IAU 2000A model, which limits the accuracy in the computed direction of the celestial pole in the GCRS to about 0.3 mas.

The sub-diurnal terms arising from the imperfect axial symmetry of the Earth are not part of this solution, so that the axis of reference of the nutation model is compliant with the definition of the CIP.

The IAU 2006 precession of the equator was derived by Capitaine *et al.* (2003) from the dynamical equations expressing the motion of the mean pole about the ecliptic pole. It includes the first- and second-order effect of the luni-solar torque and planetary torque acting on the Earth as well as the J_2 rate effect (i.e. $\dot{J}_2 = -3 \times 10^{-9}$ per century), mostly due to the post-glacial rebound.

The geodesic precession and nutation have been taken into account in the IAU 2006/2000A precession–nutation model. The GCRS being "dynamically non-rotating," Coriolis terms (that come mainly from geodesic precession) have to be considered when dealing with the equations of motion in that system.

Polar motion

Polar motion is the motion of the Earth's pole with respect to the Earth's surface. It is represented by the x_p and y_p coordinates of the ITRS direction of the CIP. It is quasi-periodic and essentially unpredictable. The main components are the Chandlerian free motion with a period of approximately 430 days, and an annual motion; there is, moreover, a drift of about 10 cm per year in the direction 80 degrees west. It also includes sub-daily

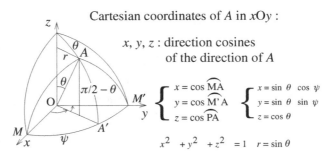

Fig. 7.10 Cartesian coordinates of a point A in the $Oxyz$ coordinates axes and spherical polar coordinates (polar distance, θ and azimuthal angle, ψ).

variations caused by ocean tides and periodic motions driven by gravitational torques with periods less than 2 days.

Earth's rotation

The variations in the Earth's diurnal angle of rotation are usually represented by the variations in UT1, or in length of day (LOD). The largest component is due to a deceleration of the Earth's angular velocity that is responsible for an increase of 2 milliseconds (ms) per century in LOD. There are also seasonal variations, decadal variations, and variations with periods from a few hours to a few tens of years.

7.3 Astronomical coordinates

Astronomical coordinates refer to the reference systems described above. They designate here the coordinates of the unit vector in the direction of a celestial object; they are represented (cf. Fig. 7.10 to Fig. 7.16) as coordinates on a conventional celestial sphere of unit radius. They are expressed either in cartesian coordinates in a right-handed coordinate frame, or in spherical coordinates. Depending on the case, the origin can be at the Solar System barycenter (barycentric), at the geocenter (geocentric), or at the observer (topocentric). In cases when the distance of the object is known, its space coordinates in the corresponding reference system can be deduced.

One coordinate is the angle from the pole of the reference plane, i.e. the pole of the horizon (the zenith), or that of the equator, or the ecliptic, or the Galactic plane. The second coordinate is an azimuthal coordinate, with an origin which can only be chosen by convention, the positive sense being also being a convention (see Fig. 7.10). The most common azimuth convention is clockwise from north, a left-handed system.

The reference plane is the equator for geographical, equatorial, hour angle, and declination coordinates; the ecliptic for ecliptic coordinates; and the horizon for horizontal coordinates.

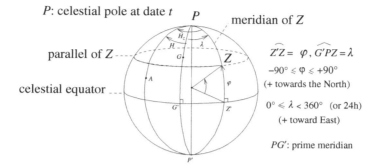

Fig. 7.11 Geographical coordinates, ϕ, λ, of Z and hour angle, H, of A.

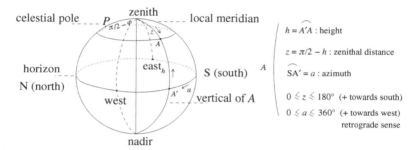

Fig. 7.12 Horizontal coordinates, z and a, of A.

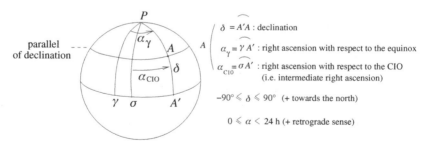

Fig. 7.13 Equatorial coordinates, α_γ and α_{CIO}, and δ, of A.

The "geographical coordinates," latitude φ and longitude λ (see Fig. 7.11), are referred to the equatorial plane with the center at the geocenter. The origin meridian is that of the ITRS.

The "horizontal coordinates," zenithal distance z and azimuth a (see Fig. 7.12), are referred to the local horizon plane, with the center at the observer (topocenter).

The "equatorial coordinates," right ascension α and declination δ (see Fig. 7.13), are referred to the equatorial plane with the center either at the barycenter of the solar system or at the geocenter.

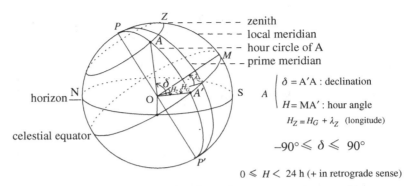

Fig. 7.14 Coordinates of A with respect to the equator and the meridian: declination δ, and hour angles H_Z and H_G with respect to the local meridian of Z (with longitude λ) and the prime meridian, respectively.

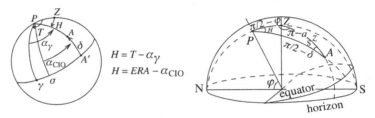

Fig. 7.15 Transformation between hour angle and right ascension on the left side and hour angle and declination and horizontal coordinates on the right side.

The equatorial coordinates can be referred to various equators (intermediate, GCRS) and various origins (true equinox of date or CIO on the CIP equator, mean equinox of epoch on the mean equator of epoch, GCRS origin on the GCRS equator, etc.). When referred to the CIO, it is called the "intermediate" right ascension (or CIO right ascension) and declination; this designation with respect to the CIO is analogous to the previous designation of "apparent right ascension and declination" when referred to the equinox.

Another set of celestial coordinates referred to the equator consists of using the declination as before, but the hour angle from the meridian instead of right ascension (see Fig. 7.14).

The observation of celestial objects requires a transformation between horizontal coordinates, which are referred to the horizon and meridian of the observer, and equatorial coordinates (see Fig. 7.15). Comparing coordinates and observations of the Solar System objects requires transforming equatorial to ecliptic coordinates and vice versa (see Fig. 7.16).

Intermediate right ascension and declination are the angular coordinates measured in the CIRS at a specified date. They specify a geocentric direction that differs from the ICRS direction by annual parallax, gravitational light deflection due to the Solar System bodies, except the Earth, annual aberration, and the time-dependent rotation describing the transformation from the GCRS to the CIRS. They are similar to apparent right ascension and

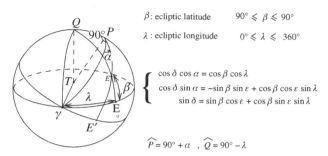

β: ecliptic latitude $90° \leqslant \beta \leqslant 90°$

λ: ecliptic longitude $0° \leqslant \lambda \leqslant 360°$

$$\begin{cases} \cos \delta \cos \alpha = \cos \beta \cos \lambda \\ \cos \delta \sin \alpha = -\sin \beta \sin \varepsilon + \cos \beta \cos \varepsilon \sin \lambda \\ \sin \delta = \sin \beta \cos \varepsilon + \cos \beta \sin \varepsilon \sin \lambda \end{cases}$$

$$\widehat{P} = 90° + \alpha \quad , \quad \widehat{Q} = 90° - \lambda$$

Fig. 7.16   Transformation between equatorial coordinates referred to the equinox and ecliptic coordinates.

declination when referring to the equinox based system. Note that intermediate declination is identical to apparent declination.

7.4 UT1, UTC and their relation to TAI

7.4.1 Definitions

UT1 is defined by its conventional relationship (IAU 2000 Resolution B1.7) to the ERA:

$$ERA(T_u) = 2\pi (0.779\,057\,273\,264\,0 + 1.002\,737\,811\,911\,354\,48\,T_u), \tag{7.9}$$

where $T_u =$ (Julian UT1 date $-$ 2451545.0), and UT1 $=$ UTC $+$ (UT1 $-$ UTC), or equivalently (modulo 2π), in order to reduce possible rounding errors,

$$\begin{aligned} ERA(T_u) = 2\pi\,(\text{UT1 Julian day fraction} \\ + 0.779\,057\,273\,264\,0 + 0.002\,737\,811\,911\,354\,48\,T_u). \end{aligned} \tag{7.10}$$

UT1, which is considered as an angle reflecting the Earth's rotation, can also be regarded as a time determined by the rotation of the Earth, which is a member of the family of Universal Time scales. It is an essential parameter for representing the Earth's orientation and is needed for celestial navigation and pointing telescopes.

TAI, which was shown in Section 7.1.4 to be the basis of quasi-ideal representations of time coordinates needed in dynamics, is also the basis of all civil timekeeping through the uniform timescale named Coordinated Universal Time (UTC). The UTC system in its present form was introduced on January 1, 1972, for the purpose of achieving worldwide coordination by means of adjustments in both time and frequency. UTC is the reference for civil (legal) time throughout the world; it was implemented firstly by the BIH and then by the BIPM and International Earth Rotation and Reference Systems Service (IERS).

UTC is defined as: UTC $=$ TAI $+ n$, where n is an integer number of seconds, such that $|$UT1 $-$ UTC$| < 0.9$ s, where UT1 is regarded as a time determined by the rotation of the Earth.

7.4.2 Relations

UTC provides to within 1 s the time corresponding to the Earth's variable rate of rotation (see Section 7.2.4) thanks to the introduction of leap seconds.[2] For more specialized applications, UTC provides the essential reference, with an uncertainty of less than 10 nanoseconds (ns), for many operational systems in science and technology.

UT1 can be obtained from the uniform timescale UTC through the relation: UT1 = UTC + (UT1 − UTC), using the quantity UT1−UTC, which is provided by the IERS.

7.5 Practical relationships and measurements

The accurate realization of the celestial and terrestrial reference systems as well as the celestial orientation of the Earth is essential for the reduction of high-precision astronomical observations and scientific exploitation.

Determining and providing that orientation is coordinated at an international level by the IERS. The IERS products, i.e. the ITRS, the ICRSs, and the EOP, are based on data provided by the IVS, ILRS, IGS and IDS international services. Those data are derived from observations by various modern techniques, namely VLBI of extragalactic radio sources for the IVS (International VLBI Service for Geodesy and Astrometry), laser ranging of artificial satellites and the Moon for the ILRS (International Laser Ranging Service), observations with the GNSS systems for the IGS (International Service for GNSS), and observations with the DORIS system for the IDS (International DORIS Service).

Each of these techniques have specific potentials for Earth orientation determination. The VLBI observations allow us to determine very accurately (i.e. at a 10 μas level) the ERA (and consequently UT1), as well as small adjustments to the celestial direction of the CIP as predicted by the a-priori precession–nutation model. GPS observations allow us to determine very accurately polar motion.

References

Capitaine, N. (1990). The celestial pole coordinates. *Celest. Mech. Dyn. Astr.*, **48**, no. 2, 127–143.

Capitaine, N., Wallace, P. T., and Chapront, J. (2003). Expressions for IAU 2000 precession quantities. *Astron. Astrophys.*, **412**, 567–586.

Capitaine, N., Andrei, A., Calabretta, M., *et al.* (2007). Proposed terminology in fundamental astronomy based on IAU 2000 resolutions. In *Transactions of the IAU*, XXVIB, ed. K. A. van der Hucht, p. 474.

---

[2] Proposals to discontinue leap seconds are currently under discussion. If and when this is done, UTC and UT1 will diverge.

Green, R. M. (1985). *Spherical Astronomy.* Cambridge: Cambridge University Press.

Guinot, B. (1979). Basic problems in the kinematics of the rotation of the Earth. In *Time and the Earth's Rotation*, eds. D. D. McCarthy and J. D. Pilkington. Dordrecht: D. Reidel Publishing Company, p. 7.

Kovalevsky, J., and Seidelmann, P. K. (2004). *Fundamentals of Astrometry.* Cambridge: Cambridge University Press.

Mathews, P. M., Herring, T. A., and Buffett B. A. (2002). Modeling of nutation and precession: New nutation series for nonrigid Earth, and insights into the Earth's interior. *J. Geophys. Res.*, **107**(B4), 10.1029/2001JB000390.

Souchay, J., Loysel, B., Kinoshita, H., and Folgueira, M. (1999). Corrections and new developments in rigid Earth nutation theory. III. Final tables "REN-2000" including crossed-nutation and spin-orbit coupling effects. *A&AS*, **135**, 111.

8

Fundamental algorithms for celestial coordinate systems and positions

PATRICK T. WALLACE

Introduction

This chapter offers practical advice to anyone developing, or supporting, astrometric computer applications related to celestial coordinate systems and positions. It identifies sources of ready-made software, useful in its own right or as benchmarks, and references to the underlying algorithms. This will be found useful in realizing many of the transformations detailed in Chapter 7, in which the theoretical bases and definitions of the transformations are provided. We shall limit ourselves to the sequence of transformations for ground-based astrometry of stars that links catalog coordinates with the direction of the incoming light as seen by the terrestrial observer. The catalog information with which the sequence begins typically comprises ICRS (International Celestial Reference System) right ascension and declination $[\alpha, \delta]$, and space motion in the form of proper motions in α and δ plus parallax and radial velocity. The final observed coordinates $[A, E]$ stand for azimuth and altitude, the latter informally called elevation. The sequence of transformations in this chapter are meant to aid the user who needs to take information in an astrometric catalog and predict the observed coordinates, in contrast to the situation in Chapter 19, where the observer wishes to transform in the opposite sense, i.e. from the observed coordinates to the final catalog positions.

Some special situations will not be dealt with due to limitations of space: the sequence for an observer in space, which is in most respects a subset of what we will examine; the Earth's attitude and rotation, which are dealt with in Chapter 7; Solar System targets (Chapter 26), which require certain additional steps (especially where surface features are involved), but can be treated in the same way as stars once the direction of the incoming light ray is known; binary stars which are dealt with in Chapters 23 and 24; variations in site coordinates, such as those caused by Earth tides and loading from oceans and atmosphere, which are beyond the scope of this text; and the relationship between the apparent directions of targets and the corresponding image positions in the focal plane of a camera, which are developed in Chapters 7 and 19.

Astrometry for Astrophysics: Methods, Models, and Applications, ed. William F. van Altena.
Published by Cambridge University Press. © Cambridge University Press 2013.

8.1 Useful references

The following publications provide astrometric methods, formulas and constants, as well as related material and links into the literature. The appendices in Wertz (1986) provide concise summaries of various coordinate transformation techniques, including spherical geometry, matrix and vector algebra, quaternions, rotation matrices, Euler-angle rotation, etc. The successive editions of IERS Conventions (McCarthy and Petit 2004) provide the latest recommendations on celestial and terrestrial coordinate systems and transformations, tables of constants, and associated topics. The IERS website also includes computer implementations of key algorithms.

Though it predates the ICRS and the IAU 2000 models, the 1992 *Explanatory Supplement to the Astronomical Almanac* (Seidelmann 1992, hereafter ES) is a useful source of algorithms. It includes astrometric transformation methods of milliarcsecond (mas) precision and presents a wide range of related topics. (A revised edition is in preparation.) The textbooks by Murray (1983), Green (1985), Kovalevsky and Seidelmann (2004), all include a fully relativistic treatment of astrometric topics. Section B of the annual *Astronomical Almanac* (hereafter AsA) provides up-to-date transformation procedures, including numerical examples. It also contains abridged transformations for dates close to the publication year. Both equinox-based and Celestial Intermediate Origin (CIO)-based methods are covered. The book's numerical tabulations are invaluable for checking software applications.

Standards Of Fundamental Astronomy (SOFA) is an IAU service that provides authoritative and up-to-date astronomical algorithms in the form of computer subroutines. Its "Tools for Earth Attitude" (SOFA 2007, 2010) is particularly relevant to the present discussion as it includes complete recipes for the conversion of GCRS to ITRS coordinates. The USNO NOVAS library[1] is a package of subroutines that provides a wide variety of common astrometric transformations at sub-mas accuracies.

Most of the references listed here are aimed mainly at ground-based applications with sub-mas-accuracy goals. For demanding space-astrometry applications of microarcseconds (μas) accuracy, see Klioner (2003) and Chapter 5. Kaplan (2005) describes ICRS-to-TIRS transformations, together with the relevant timescales and additional material. Volumes 1 and 3 of the Hipparcos and Tycho Catalogues (ESA 1997) provide detailed algorithms for aberration and light deflection. Capitaine and Wallace (2006) compare a number of different algorithms for generating the GCRS-to-CIRS matrix given the Celestial Intermediate Pole (CIP), and for locating the CIP and CIO. The paper includes accuracy tests and data on computational speed. Wallace and Capitaine (2006) set out algorithms for generating the full-accuracy bias–precession–nutation matrices (both equinox-based and CIO-based) and the GCRS-to-ITRS matrix. To assist implementers, high-precision numerical results are listed for a test case. Capitaine and Wallace (2008) show how the full-accuracy models

---

[1] NOVAS: *US Naval Observatory Vector Astrometry Subroutines*. See http://aa.usno.navy.mil/software/novas/novas_info.php.

CATALOG [α, δ]
Proper motion, catalog epoch to J2000
INTERNATIONAL CELESTIAL REFERENCE SYSTEM [α, δ] epoch J2000
Proper motion, J2000 to date
BARYCENTRIC CELESTIAL REFERENCE SYSTEM [α, δ] of date
Annual parallax
ASTROMETRIC [α, δ]
Light deflection
Annual aberration
GEOCENTRIC CELESTIAL REFERENCE SYSTEM [α, δ]
Frame bias
Precession
Nutation
CELESTIAL INTERMEDIATE REFERENCE SYSTEM [α, δ]
Earth rotation
TERRESTIAL INTERMEDIATE REFERENCE SYSTEM [λ, φ]
Polar motion
INTERNATIONAL TERRESTRIAL REFERENCE SYSTEM [λ, φ]
Site longitude
Diurnal aberration and parallax
TOPOCENTRIC [h, δ]
Equatorial to horizon
TOPOCENTRIC [A, E]
Refraction
OBSERVED [A, E]

Fig. 8.1   The astrometric transformations that take the "catalog position" of a star and predict its "observed place," where a perfect theodolite would see it. The items in upper case are coordinate systems, while the lower-case items are transformations and corrections. External inputs are time, observing wavelength, and local air conditions.

can be abbreviated, offering different compromises between speed and accuracy. Three examples are shown that in the modern era achieve 1, 20, and 400 mas results, with computing costs reduced by 1, 2, and 3 orders of magnitude respectively.

8.2   Notes on the transformation steps

We will suggest ways of implementing the transformations in Fig. 8.1, where the target is a star and we wish to predict the direction of the incoming light for a terrestrial observer. Vector methods are used throughout (see Chapter 4), as opposed to spherical trigonometry. Note that the formulation for the two directions (i.e. starting from, or ending with, celestial coordinates) will be subtly different in order to achieve precise reversibility. The rotations are straightforward, because the matrices concerned are inverted simply by taking the transpose. However, other effects, notably refraction but also aberration and light deflection, will require slightly different treatment, in some cases involving iteration.

The changes introduced in the IAU 2000 resolutions have little or no practical effect on the first few steps, namely proper motion, parallax, light deflection, and stellar aberration.

Consequently, the algorithms set out in the ES (q.v. for original references) and used in the AsA are sufficient for all applications except μas space astrometry. The latter application is addressed in Klioner (2003).

The AsA algorithm for **space motion** is non-relativistic but otherwise rigorous. This is sufficient given the accuracy of current catalogs, but if a treatment consistent with special relativity is preferred this is provided by the SOFA routine STARPM. Note that in any application dealing with proper motion great care is needed to interpret catalog data correctly. The tabulated values may for example be per year or per century, and either time derivatives of $[\alpha, \delta]$ or of the angles on the sky $[\alpha \cos \delta, \delta]$.

For **light deflection,** the AsA algorithm is adequate for mas applications, except close to planets. Applications need to detect cases behind the Sun's disk and take appropriate precautions. Note that solar-limb features are subject to a displacement of only a few mas, the 1.75 arcsec shift in an adjacent star notwithstanding, a counter-intuitive result. Klioner (2003) gives details of the light deflections that can be caused by planets and satellites. When calculating the position of the deflecting body the light-time should be taken into account in the usual way, i.e. giving the apparent direction.

For many practical applications, **stellar aberration** can be applied simply by subtracting from the light-deflected unit vector the Earth's barycentric velocity vector in units of c. However, the AsA algorithm includes relativistic terms that take the accuracy well beyond the 1 mas level and provides a good implementation of the GCRS. The Earth velocity vector can be obtained either from a numerical ephemeris such as the Jet Propulsion Library's, or by calling a closed-formula model such as that used in the SOFA routine EPV00, the latter enabling aberration to be predicted to a few microarcseconds.

The changes introduced by the IAU in 2000 mean that in the case of the GCRS-to-ITRS portion of the transformation sequence it is essential to use recent references. The AsA is a reliable guide, as are the IERS Conventions. Wallace and Capitaine (2006) describe two procedures for generating the **bias–precession–nutation** matrix, one based on Euler angles and the other on series that generate CIP $[X, Y]$ directly; the SOFA software offers both. Note that different implementations (angles, $[X, Y]$, IERS Conventions) of what is canonically a single model do in fact differ at the 1 μas level. This is of no practical significance, but it is unwise to develop applications in which a mixture of methods is used.

Great care must be taken in designing algorithms to calculate the **Earth rotation angle** if precision is not to be avoidably eroded. This is because the angle being calculated is the fractional part of a large number. The solution is to express the $\theta = f(t)$ equation in a manner which separates the integer and fractional components, and discards products of integers at the first opportunity. The SOFA routine ERA00 uses this technique.

Polar motion simply comprises three Euler rotations. The angles are very small, and so can be applied in any order without the results being noticeably affected. However, to ensure consistency, care must be taken to use the correct order, which is specified in the IERS Conventions (and in the SOFA routine POM00).

Corrections for **diurnal aberration** can be applied using the ES or AsA procedures. For stellar targets, diurnal parallax (maximum 55 μas) can be neglected for almost all ground-based applications.

There is no definitive **atmospheric refraction** algorithm, and the application developer must make a choice that depends on accuracy goals, altitude range, wavelength, and computational speed. An overview of refraction in the context of astrometry is given in Chapter 9. A general-purpose algorithm due to Hohenkerk and Sinclair (1985) for optical/IR applications can be found in Section 3.28 of the ES, after Auer and Standish (2000). This uses a ray trace through a model atmosphere and consequently is applicable even to large zenith distances ζ, but at some computational expense. The ES algorithm leaves room for further development, such as:

- extension to radio wavelengths;
- revised gas constants, e.g. from Murray (1983);
- a better model for saturated water-vapor pressure, e.g. from Buck (1981);
- a rigorous formula for the water-vapor pressure, given the saturation pressure and the relative humidity, such as (2.5.5) in Crane (1976);
- modern refractivity formulas, such as IAG (1999) for the optical/IR and Rueger (2002) for low-frequency radio; corrections for oxygen and water features, important in the mm/sub-mm, can be applied, such as the MPM93 model of Liebe *et al.* (1993); and
- variable tropopause height.

Applications requiring fast execution at moderate z can use the ray-trace method to sample two z values and solve for A and B in the familiar Oriani formula $\Delta\zeta_{\text{topocentric}} = A\tan\zeta_{\text{observed}} - B\tan^3\zeta_{\text{observed}}$, which typically gives good results down to about $\zeta = 75°$.

8.3   Computational efficiency

Many applications offer opportunities for great efficiency gains by limiting the per-target recomputation. Where many transformations are to be carried out over a short time interval it is often acceptable for the following quantities (inter alia) to be updated only infrequently:

- Earth barycentric position and velocity vectors;
- Earth heliocentric unit vector;
- bias–precession–nutation matrix;
- functions of geodetic latitude; and
- refraction constants A and B.

8.4   External inputs

The sequence of transformations in Fig. 8.1 depends on certain external inputs: site location, time, and the Earth orientation parameters (EOPs) and refraction information. Most of the transformations change relatively slowly, and in an undemanding application such as pointing a telescope the only one that requires accurate (millisecond) time is Earth rotation. Most of the other transformations strictly speaking require Barycentric Dynamical Time (TDB), but Terrestrial Time (TT) (not Coordinated Universal Time, UTC) is always an adequate substitute and avoids wasteful computation. The IAU SOFA document "Time Scale and Calendar Tools" presents methods for transforming times between scales.

The EOPs come from observations (using very long baseline interferometry VLBI) and cannot be predicted accurately in advance: they must be obtained from IERS publications. Where applications require only arcsecond accuracy it is permissible to neglect the polar motion (and also diurnal aberration), but [UT1 − UTC] (see Section 7.4.2) is a large effect and always needed. Very accurate (sub-milliarcseconds) applications require all the EOPs, including the celestial-pole offsets, which are due mainly to free core nutation but also in the long run imperfections in the precession–nutation model.

Ground-based applications that use observed directions need access to the ambient air conditions, namely pressure, temperature and relative humidity, as well as the observing wavelength. Humidity has a much bigger effect in the radio than in the optical/infrared, where for many applications it can be neglected. Conversely, wavelength has a big effect in the optical/infrared but not in the radio, at least between about 20 MHz and 100 GHz.

References

Auer, L. H. and Standish, E. M. (2000). Astronomical refraction: computational method for all zenith angles. *AJ*, **119**, 2472.

Buck, A. (1981). New equations for computing vapor pressure and enhancement factor. *J. Appl. Met.*, **20**, 1527.

Capitaine, N. and Wallace, P. T. (2006). High precision methods for locating the celestial intermediate pole and origin. *A&A*, **450**, 855.

Capitaine, N. and Wallace, P. T. (2008). Concise CIO based precession-nutation formulations. *A&A*, **478**, 277.

Crane, R. K. (1976). Refraction effects in the neutral atmosphere. In *Methods of Experimental Physics: Astrophysics*, ed. M. L. Meeks. New York, NY: Academic Press, vol. 12B, p. 186.

ESA. (1997). *The Hipparcos and Tycho Catalogues*. ESA Special Publication SP-1200 (17 volumes).

Green, R. M. (1985). *Spherical Astronomy*. Cambridge: Cambridge University Press.

Hohenkerk, C. Y. and Sinclair, A. T. (1985). *The Computation of Angular Atmospheric Refraction at Large Zenith Angles*. Greenwich: Royal Greenwich Observatory, NAO Technical Note No. 63.

International Association of Geodesy (IAG) (1999). Resolution 3, 22nd General Assembly, Birmingham, 19–30 July.

Kaplan, G. H. (2005). *The IAU Resolutions on Astronomical Reference Systems, Time Scales, and Earth Rotation Models*. Washington, DC: United States Naval Observatory, Circular No. 179.

Klioner, S. A. (2003). A practical relativistic model for microarcsecond astrometry in space. *AJ*, **125**, 1580.

Kovalevsky, J. and Seidelmann, P. K. (2004). *Fundamentals of Astrometry*. Cambridge: Cambridge University Press.

Liebe, H. J., Hufford, G. A., and Cotton, M. G. (1993). Propagation modeling of moist air and suspended water/ice particles at frequencies below 1000 GHz. In *Proceedings of the NATO/AGARD Wave Propagation Panel, 52nd Meeting*, Paper No. 3/1-10, Mallorca, 17–20 May.

McCarthy, D. D. and Petit G., eds. (2004). *IERS Conventions (2003)*. Frankfurt am Main: Verlag des Bundesamts für Kartographie und Geodasie, IERS Technical Note No. 32.

Murray, C. A. (1983). *Vectorial Astrometry*. Bristol: Adam Hilger.

Rueger, J. M. (2002). Refractive index formulae for electronic distance measurement with radio and millimetre waves. In *Unisurv Report S-68, School of Surveying and Spatial Information Systems*. Sydney: University of New South Wales.

Seidelmann, K. P., ed. (1992). *Explanatory Supplement to the Astronomical Almanac*. Sausalito, CA: University Science Books. Same edition published in paperback in 2006.

SOFA (2007, 2010). *Standards of Fundamental Astronomy Reviewing Board*. See www.iausofa.org.

Wallace, P. T. and Capitaine, N. (2006). *Astron. Astrophys.*, **459**, 981. See also *Astron. Astrophys.*, **464**, 793, 2007 (erratum).

Wertz, J. R. (1986). *Spacecraft Attitude Determination and Control*. Dordrecht: Reidel.

PART III

OBSERVING THROUGH THE ATMOSPHERE

9

The Earth's atmosphere: refraction, turbulence, delays, and limitations to astrometric precision

WILLIAM F. VAN ALTENA AND EDWARD B. FOMALONT

Introduction

The Earth's atmosphere imposes several limitations on our ability to perform astrometric measurements from the ground in both the optical and radio regions of the spectrum. First, we are limited to wavelengths where the absorption is not too great, i.e. the broad optical region from the ultraviolet to the near-infrared, scattered regions in the infrared and broad regions at radio wavelengths. The fundamental limitations imposed by the atmosphere are different in the optical and radio and in this chapter we will deal with those important for the optical; the radio part is largely dealt with in Chapter 12 on radio interferometry, except for a summary given here on the precision limitations imposed by the atmosphere. The second problem created by observing through the atmosphere is refraction of the light waves as they pass through different levels of the atmosphere. If it were only refraction through a stable medium, the problem would be very simple; however, the atmosphere is a turbulent medium that causes variations in the amount of refraction both spatially and temporally and it therefore limits the precision and accuracy of our observations. In this chapter we will deal with both effects using the developments in Schroeder (1987, 2000) as the basic reference.

9.1 Refraction through a plane-parallel atmosphere

When we are dealing with relative positions in fields of view less than several degrees, it is sufficient to adopt a plane-parallel atmosphere for the model. In cases where we need to consider the total displacement, such as with meridian circles, it is necessary to adopt atmospheric models that are substantially more complicated, such as those developed by Garfinkel (1967) and Auer and Standish (2000). For the purposes of this chapter we can safely use the plane-parallel atmosphere.

Assume that a light ray enters our plane-parallel atmosphere in a medium with index of refraction n, at a zenith angle z with respect to the vertical. The ray is refracted to an angle

Astrometry for Astrophysics: Methods, Models, and Applications, ed. William F. van Altena.
Published by Cambridge University Press. © Cambridge University Press 2013.

z' into the medium with an index of n', yielding Eq. (9.1):

$$n' \sin z' = n \sin z \tag{9.1}$$

Propagating Eq. (9.1) into deeper layers of the atmosphere we have

$$n'' \sin z'' = n' \sin z' = \text{const}$$

Differentiating Eq. (9.1) with respect to the distance, h, from the top of the atmosphere towards the Earth's surface, we have

$$(\mathrm{d}n/\mathrm{d}h) \sin z + n \cos z (\mathrm{d}z/\mathrm{d}h) = 0, \text{ or} \tag{9.2}$$

$$(\mathrm{d}z/\mathrm{d}h) = -n^{-1} \tan z (\mathrm{d}n/\mathrm{d}h) \tag{9.3}$$

We can now integrate Eq. (9.3) from the top of the atmosphere to the Earth's surface to calculate the total angle of refraction, R, as a function of the original angle of incidence and the index of refraction:

$$R = \int_z (\mathrm{d}z/\mathrm{d}h)\mathrm{d}h = -\int_z n^{-1} \tan z (\mathrm{d}n/\mathrm{d}h)\mathrm{d}h \tag{9.4}$$

Assuming $z \ll 1$, an angle of incidence at the top of the atmosphere equal to z_0 and the index of refraction $n \approx 1$, gives us

$$R \approx -n^{-1} \tan z_0 \int_z \mathrm{d}n \approx -(\delta n/n) \tan z_0 \tag{9.5}$$

At sea level, $\delta n/n \approx 0.00029 \, \mathrm{rad} \approx 60''$, so we have

$$R \approx -60'' \tan z_0 \tag{9.6}$$

If instead we let $\beta = -\delta n/n = f(\lambda, T, P, H)$, i.e. a function of the wavelength, and the atmospheric temperature, pressure and humidity, then we can write

$$R = \beta \tan z_0 \tag{9.7}$$

The function β can be written as

$$\beta \approx \beta_0(\lambda, T, P, H) + \beta' \tan^2 z_0 + \beta_c(B - V) + \dots, \tag{9.8}$$

where

$$\beta_0 = 16.272'' P \tan z_0/(273 + T) \tag{9.9}$$

with the atmospheric pressure, P, in millibars and the temperature, T, in degrees Celsius (see for example *Astronomical Almanac*, 1990, p. B62) and $\beta' \approx -0.082''$ (Konig 1962, p. 478). See also Stone (1996, Sec. 3.1). Substituting Eq. (9.8) into Eq. (9.7) yields

$$R \approx \beta_0 \tan z_0 + \beta' \tan^3 z_0 + \beta_c(B - V) \tan z_0 + \dots \tag{9.10}$$

Although atmospheric refraction constant, β_0, decreases in a continuous manner by about $3''$ over the range $320 \leq \lambda \leq 1000 \, \mathrm{nm}$ (see Table 3.1 in Schroeder 2000), the observed effective wavelength of a star can be an erratic and a discontinuous function of color. This is due to rapidly changing spectral features as a function of the star's surface temperature (see

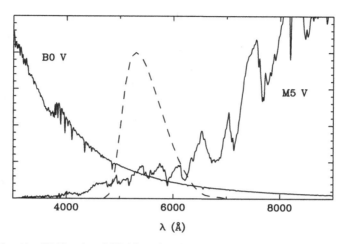

Fig. 9.1 The relative flux for a blue (B0 V) and a red (M5 V) star from Pickles (1998), as well as the transmission curve for the Bessell (1990) V-passband (dashed line), are illustrated, all as a function of wavelength. Note that when the flux of a particular star is convolved with the sensitivity of the detector, the resulting effective wavelength will change for different spectral types. Figure courtesy of D. Casetti and T. Girard.

Fig. 9.1). For that reason, β_c is a complicated function and a linear relation with $(B - V)$ is generally not adequate. A useful approximation between refraction in the blue and visual passbands is (Konig 1962):

$$\beta(\text{blue}) \approx 1.0155\beta(\text{visual}) \tag{9.11}$$

Stone (1996) gives a more precise calculation of the refraction constant β. He convolves the index of refraction with the spectral energy distribution of the star, the transmission of the filter, telescope, atmosphere, interstellar medium, and the quantum efficiency, all as a function of wavelength. Following Stone, we write the total refraction as

$$\beta = a/b \tag{9.12}$$

where

$$a = \int_0^\infty r(\lambda)[S(\lambda)F(\lambda)T(\lambda)A(\lambda)QE(\lambda)]\mathrm{d}\lambda \tag{9.13}$$

$$b = \int_0^\infty [S(\lambda)F(\lambda)T(\lambda)A(\lambda)QE(\lambda)]\mathrm{d}\lambda \tag{9.14}$$

with $r(\lambda) = $ refraction as a function of wavelength, $S(\lambda) = $ spectral energy distribution (sed) of the star, $F(\lambda) = $ filter transmission, $T(\lambda) = $ telescope transmission (lens, optics coatings, etc.), $A(\lambda) = $ atmospheric + intersteller transmission, and $QE(\lambda) = $ quantum efficiency of the detector.

This detailed approach brings us closer to being able to properly correct for atmospheric refraction, but it requires knowledge of the spectral-energy distribution of the object, which is not always available. An application of this approach to parallax observations at the Yerkes Observatory can be found in van Altena (1971). An alternative approach successfully used by Monet *et al.* (1992) for parallax observations at the US Naval Observatory in Flagstaff is to observe the field successively at increasing air masses. The deviation of the object's position with respect to the mean of the reference stars is plotted as a function of the air mass and that relationship used to correct the balance of the observations for that object. This method is especially useful when the object has a highly modulated spectral-energy distribution, such as found in the late-type low-mass stars.

9.2 Refraction in right ascension and declination

The zenith angle is easily calculated using Eq. (9.15), given the latitude of the observer, φ, the hour angle, HA, right ascension and declination of the object, (α, δ):

$$\cos z_0 = \sin \delta \sin \varphi + \cos \delta \cos \varphi \cos HA \tag{9.15}$$

To compute the components of refraction in right ascension and declination, we need to calculate the parallactic angle, χ, i.e. the angle between the celestial pole and the zenith as seen from the object being observed:

$$\sin \chi = \cos \varphi \sin HA / \sin z_0$$
$$\cos \chi = (\sin \varphi - \sin \delta \cos z_0) / \cos \delta \sin z_0 \tag{9.16}$$

In Eqs. (9.16) the subscript "zero" refers to the apparent zenith distance and the observed (α, δ) are corrected for refraction to yield (α', δ'):

$$\alpha' = \alpha - R \sec \delta \sin \chi$$
$$\delta' = \delta - R \cos \chi \tag{9.17}$$

where R is the refraction computed in Eqs. (9.07) or (9.10).

If we consider the developments up to here, then aside from the differences in spectral-energy distribution between objects, atmospheric refraction only results in a slight contraction of the field of view. At the zenith the contraction is equal in both coordinates, while on the meridian the contraction in azimuth is proportional to the field size and independent of the zenith distance and the contraction in the zenith coordinate is proportional to $(1 + z^2)$. To first order and for small fields-of-view these effects are similar to that produced by thermal variations in the telescope, so we may question why we need to correct for refraction since such changes in scale are easily dealt with in the process of the astrometric reductions described in Chapter 19. The answer depends on the end use of our astrometric analyses. Since the effective scale change due to differential refraction will be different in the orthogonal coordinates, allowing for different scale changes in each coordinate is one solution. However, for large field sizes and/or large zenith distances non-linear terms in the coordinates will be required to compensate for the effects of refraction. One important example for the need to correct for refraction is the pointing of fibers in multi-object spectrographs. If a fiber is not exactly pointed at the desired object, then some of the light will be lost and a penalty in observing time will be incurred. Equation 9.6 shows that the effects of refraction are very large, amounting to $60''$ at a zenith angle of $45°$. Clearly, the pointing of a telescope must be corrected for atmospheric refraction or we will completely miss our objects of interest. Even across the one-degree field of view of a multi-object spectrograph the differential effects of refraction amount to $1''$, which is the diameter of a typical fiber. Therefore, even if we can accurately center the field of view, by the time we move to the edge of the spectrograph field we will be missing most of the light in a misplaced fiber. Finally, since the effects of atmospheric refraction are calculable, they should be analytically removed from the observations so that the final relative positions can be analyzed and understood in terms of the instrumental characteristics without the interference of the atmosphere. Up to this point, the deviation in the light path has been calculable in an exact manner dependent only on the zenith angle of the object, its spectral-energy distribution, the atmospheric parameters, and their variation with height. In the next section we will explore the effects of turbulence in the atmosphere that introduce statistical fluctuations in the light-path direction.

9.3 Turbulence in the atmosphere

If we rewrite Eq. (9.3) and look only at the incremental change in the refraction when a light ray passes from one level to the next in the atmosphere and assume z is small, we have

$$\delta z \approx -z(\delta n/n) \tag{9.18}$$

We can now utilize Eq. (9.18) in a slightly different way. Turbulence in the atmosphere results from the largely chaotic temperature and pressure changes that occur from one element of the atmosphere to another. Temperature, as well as pressure and humidity, variations, produce corresponding changes in the index of refraction (δn), which, according to Eq. (9.18), result in a refraction change δz that we interpret as astronomical "seeing." As noted by Schroeder (2000), integrating Eq. (9.18) from the top of the atmosphere down to the surface yields a ray that random walks with an angular scale on the order of one arcsecond around the refracted position for a stable atmosphere. Depending on the duration of the exposure, the time average of the deviations will be the same as that for a stable atmosphere, while the time average of the squared deviation will yield a blur circle that we refer to as the seeing disk. Fortunately, for much of astronomical data analysis, the shape of that seeing disk is very close to Gaussian (see for example Fig. 16.6 in Schroeder 1987). In addition, the detailed integration (Roddier 1981) of the light-ray path though a turbulent atmosphere also yields a nearly Gaussian shape for the seeing disk, although since a Gaussian significantly underestimates the power in the wings, a Moffatt (1969) function gives a better fit to the observed seeing disk.

For a perfectly figured telescope without atmospheric degradation, the resolution would be limited by diffraction to the Rayleigh limit $\Delta\theta$, where $\Delta\theta$ is the *radius* of the first zero in the Airy disk. The usual interpretation is that two stars are just resolvable when they are separated by the Rayleigh limit, i.e. the second star is centered on the first zero of the primary's Airy disk:

$$\Delta\theta = 1.22\lambda/D \tag{9.19}$$

If we express $\Delta\theta$ in units of arcseconds, λ in μm and D in meters, then with $\lambda = 0.5$ μm and $D = 1$ m, we have $\Delta\theta \approx 0.25\lambda/D = 0.12''$. Conversely, if the atmosphere degrades the seeing disk to 1 arcsec, then this will be matched by the diffraction limit of a telescope with a mirror of diameter $D = 25$ cm. Larger telescopes will of course gather more light to reach fainter magnitudes, but different technologies are needed to achieve higher resolutions (see Chapters 10, 18 and 23). If we approximate the Airy disk by a Gaussian, then, very roughly, the full width at half maximum (FWHM) of the Airy disk is $\sim\lambda/D$.

Roddier calculates the point-spread function (PSF), or "degradation function" as he calls it, as

$$T(\nu) = \exp[-3.44(\lambda\nu/r_0)^{5/3}] \tag{9.20}$$

where λ is the wavelength, ν is the angular frequency in cycles per arcsec and r_0 is the scale length of the atmospheric turbulence, which is typically on the order of 25 cm. If the telescope diameter $D < r_0$, then the limiting resolution is set by the telescope, while if $D > r_0$, then the limiting resolution is set by the atmosphere. The nature of the seeing disk also differs for $D < r_0$, since the whole image is seen to move in the focal plane, while for $D > r_0$, many atmospheric cells with sizes characterized by r_0 fill the telescope aperture. The multitude of cells cause the light rays arriving at the focal surface to interfere and create independent images that are diffraction-limited according to the aperture diameter D and move more or less randomly in the focal plane with a typical time scale of tens of milliseconds. For exposures on such a timescale, and within a field of view of about 2 arcsec, it is possible to freeze the atmospheric motion and many individual "speckles" are observed within the seeing disk (see Chapters 10 and 23). Finally, Roddier calculates the turbulence scale length as

$$r_0 \approx 0.185\lambda^{6/5}\cos^{3/5}z \tag{9.21}$$

If we substitute (9.21) into (9.20), we find the seeing disk is proportional to $\lambda^{-1/5}$, i.e. the well-known observation that the seeing improves slowly with increasing wavelength and, as you observe at increasing zenith distances, r_0 decreases and you therefore get poorer seeing. Finally, for wavelengths longer than 10 microns, the resolution is set by the telescope diameter.

The distribution of cell sizes (r_0) is a function of the observing site and the instantaneous atmospheric conditions. In some cases the local conditions may be due to the wind passing over a nearby hill or obstruction that results in a very large-scale cell whose characteristics may change slowly with time. Such effects are commonly seen as systematic positional deviations in the course of meridian-circle observations. These deviations are correlated with the direction and speed of the wind and displace all of the observations in one direction instead of the effects being random as described above. We therefore have to contend with a variety of positional deviations that range from the random fluctuations about the mean position to non-random deviations from the mean.

If we consider the apparent angular separation between a pair of stars, then for stars separated by less than about 2″, the rays from the stars are so close together that they pass through essentially the same cells and they will be deviated by nearly equal amounts and there will be no image blurring. This is the "isoplanatic" region where we can achieve nearly diffraction-limited imaging using techniques such as speckle interferometry and adaptive optics as described in Chapters 23 and 10. When we consider stars separated by more than 2″, but less than about 250″, we are in the regime of rays that are confined more or less to what is called the "isokinetic" region, that is, the objects within that region are displaced approximately in the same direction by similar amounts. For objects separated by more than 250″ we lose coherence in the movements and except for the special circumstances of local disturbances in the wind-flow patterns described above, which cause large-scale deviations in the image position, the precision of an image position is governed largely by the integration time.

9.4 Atmospheric limitations on astrometric precision and accuracy

We divide the discussion of the limiting positional precision and accuracy at the various separation scales into two cases: first, the total motion of a single star, which is important for guiding telescopes and observations with meridian circles; and second, the relative motions of two or more stars. In both cases we must first consider the centering precision limit, which is set by the wavelength, telescope diameter, and signal-to-noise ratio (SNR) in the image. According to Lindegren (1978), the measuring precision for a diffraction-limited image is given by

$$\sigma_{\mathrm{meas}} = \frac{1}{\pi}\frac{\lambda}{D}\frac{1}{SNR}\,\mathrm{rad} = 0.361\left(\frac{\lambda}{0.55\,\mu\mathrm{m}}\right)\left(\frac{1\,\mathrm{m}}{D}\right)\left(\frac{100}{SNR}\right)\mathrm{mas} \qquad (9.22)$$

where we have scaled the relation to visible light at 0.55 μm for a 1 m diameter and $SNR = 100$. For the cases of a seeing-limited image with a nominal background of one-ninth of the peak and one with no background, Lindegren (1978) finds

$$\sigma_{\mathrm{meas}} = 1.335\frac{FWHM}{SNR}\,\text{or, with no background,}\,\sigma_{\mathrm{meas}} = \frac{FWHM}{SNR} \qquad (9.23)$$

9.4.1 Total motion of a star

In this case we are primarily interested in the total motion of a star in the context of guide stars for telescopes. Most telescopes employ dual-guiding systems, the first level being a general tracking system that drives the telescope according to the calculated diurnal rate, while a star tracker follows the residual motion of the guide star. The residual motion will include slow changes due to flexure in the telescope, refraction as the hour angle of the object changes and rapid excursions due to atmospheric turbulence and wind shaking the telescope. Hoeg (1968) and Lindegren (1978, 1980, 2010) studied the total motion of a single star based on visual observations with northern European meridian circles (apertures

~20 cm). These instruments observe the transit of an object through their field of view for about 40 seconds, during which time the instantaneous position of the image is monitored. Tatarski (1971) derived an analytical expression relating the precision to the wavelength, the characteristic atmospheric cell size, and the telescope diameter as

$$\sigma = 0.412'' \lambda r_0^{-5/6} D^{-1/6} \tag{9.24}$$

According to Kolmogorov turbulence, the characteristic cell size $r_0 \propto \lambda^{6/5}$, so that

$$\sigma \propto D^{-1/6} \tag{9.25}$$

If we incorporate the brightness of a seeing-limited object, the total variance of a guide star's excursions from its mean position will be

$$\sigma^2 = \left(\frac{FWHM}{SNR_T} \right)^2 + \left(\frac{0.412\lambda}{r_0^{5/6} D^{1/6} T^{1/2}} \right)^2 \tag{9.26}$$

where SNR_T is the signal-to-noise ratio accumulated in the target/guide star during the guider integration time T, assuming that there is negligible sky background. The guider integration time must be chosen in the context of the scientific field of view being tracked. For example, if the field is very small we may be able to successfully track all of the atmospheric motions using a time constant on the order of 10 milliseconds as with adaptive optics (see Chapter 10). If the field is somewhat larger, then time constants of tens of milliseconds will remove most of the coherent motion within a field of about 250 arcsec. Once we attempt to guide fields of view larger than about 4 arcmin, rapid guiding will track the field-center image motions but degrade the PSF farther from the center and complicate the data analysis. Wide fields can often be guided satisfactorily using time constants of 10–20 seconds allowing the use of fainter guide stars. This of course depends on the characteristics of the telescope's sidereal tracking system and the presence of wind shake, which may disturb the pointing of the telescope.

9.4.2 Observations in the isoplanatic region (adaptive optics and speckle interferometry)

For most astrometric applications, the precision of the separation between pairs of objects in the field of a telescope is of more interest. Lindegren (1980), Cameron *et al.* (2009), Helminiak *et al.* (2009), Fritz *et al.* (2010), and others have studied this topic in detail. Following Fritz *et al.* (2010) we write the variance of the angular separation differences for a pair of objects in the isoplanatic patch, i.e. for diffraction-limited images with a separation less than about ±2 arcsec, as a column matrix:

$$\left(\begin{array}{c} \sigma_{\parallel,TJ}^2 \\ \sigma_{\perp,TJ}^2 \end{array} \right) = 2 \left(\frac{1}{\pi} \frac{\lambda}{D} \frac{1}{SNR} \right)^2 + \alpha \left(\begin{array}{c} 3 \\ 1 \end{array} \right) \theta^2 D^{-7/3} \frac{\tau}{T} \tag{9.27}$$

where $\sigma_{\parallel,TJ}^2$ is the variance of the tilt-jitter (TJ) separation differences parallel to the line joining the pair of objects, while $\sigma_{\perp,TJ}^2$ is the variance perpendicular to the separation, α is related to the second moment of the atmospheric turbulence profile, θ is the separation of the stars in arcseconds, T is the integration time in seconds and τ is the transit time of the turbulence pattern across the telescope aperture. For the 8 m telescope they find $\tau = 0.4$ s, which corresponds to a wind velocity of about 20 m/s. When the separation θ is measured in arcseconds, σ in milliarcseconds (mas), and D in meters, the constant $\alpha \approx 3$. The first term in Eq. (9.27) is the image-centering error from Eq. (9.22) for a pair of diffraction-limited images assumed to have the same brightness.

9.4.3 Observations in the isokinetic region (CCD observations)

Finally, we consider what is often termed long-focus or narrow-field astrometry, where objects within the isokinetic region are studied. The isokinetic region is defined as the region where objects are displaced more or less in the same direction by atmospheric turbulence on timescales of some tens of milliseconds. This corresponds to the separation range from a few arcseconds to about 4 arcminutes in the visual and near-infrared parts of the spectrum. For a "poor" observing site, where the atmosphere is more turbulent, the isokinetic region will on average be substantially smaller. The relationship between measurement precision, field size, and exposure time has been studied by Lindegren (1980), Han (1989), Pravdo and Shaklan (1996), Vieira et al. (2005), and Monet (2010, private communication), among others. Han found that Eq. (9.28), where C is a constant to be determined, fit his data well and agreed with the prediction of Lindegren (1980) in that the precision was proportional to the cube root of the separation:

$$\sigma = C\theta^{1/3}T^{-1/2} \tag{9.28}$$

However, note that Eq. (9.27), which was derived for the regime of adaptive optics, predicts that the precision is directly proportional to the separation.

We now review a number of studies in an attempt to understand this discrepancy. In Table 9.1 the astrometric precision from a variety of ground-based investigations using CCD (charge-coupled device) detectors is summarized as a function of the type of study, i.e. whether the precision $\langle\sigma\rangle$ is estimated from an ensemble of reference stars (Ref) or from the separations (Sep) of pairs of stars. In the latter case the precision of the separation is reduced to that of a single star assuming that the pair have equal brightness. For estimates of the precision based on an ensemble of reference stars, the "separation" is taken as two-thirds the diameter of the circle enclosing the reference stars, i.e. an approximation to the mean diameter of the reference frame. The table also lists the exposure time (Exp) in seconds, the FWHM in arcseconds as well as the diameter of the telescope (Diam) in meters and a proxy for the precision Q, defined in Eq. (9.29b), which removes two of the parameters that determine the precision, i.e. exposure time and telescope diameter. If we rewrite Eq. (9.27), ignore the component of the precision due to the SNR in the images, i.e. the first term on the right-hand side, and set the time taken for a disturbance pattern to cross the telescope aperture, $\tau = D/v$, where v is the wind velocity, we get

$$\sigma^2 = \alpha \begin{pmatrix} 3 \\ 1 \end{pmatrix} \theta^2 D^{-4/3} \frac{1}{vT} \tag{9.29a}$$

Table 9.1 Astrometric precision measurements versus separation

Telescope	Type	Sep (")	⟨σ⟩ (mas)	Exp (s)	FWHM (")	Diam (m)	Q	Ref.
UCAC	Ref	2460	30.0	125	2.0	0.21	118	1
YSO	Ref	2160	25.0	120	1.8	0.51	174	2
MAP	Sep	1200	5.2	600	2.0	0.76	107	3
KPNO	Ref	928	10.1	100	1.8	0.9	94	4
UCAC	Ref	840	15.0	180	2.0	0.2	69	5
Subaru	Sep	800	15.6	10	0.5	8.2	201	6
Subaru	Sep	800	7.8	30	0.5	8.2	174	6
KPNO/4m	Ref	720	26.0	10	1.1	4.0	207	7
KPNO/4m	Ref	720	10.0	30	1.1	4.0	138	7
CTIO	Ref	544	8.2	100	1.4	0.9	76	4
Subaru	Sep	400	7.8	10	0.5	8.2	100	6
Subaru	Sep	400	5.7	30	0.5	8.2	127	6
MAP	Sep	330	3.4	600	2.0	0.76	69	3
USNO	Ref	256	2.7	600	1.5	1.5	87	8
Subaru	Sep	200	6.4	10	0.5	8.2	82	6
Subaru	Sep	200	5.0	30	0.5	8.2	111	6
WIYN/OPTIC	Ref	186	2.0	300	0.6	3.5	80	9
Subaru	Sep	100	4.8	10	0.5	8.2	62	6
Subaru	Sep	100	3.5	30	0.5	8.2	78	6
Hale P&S Dec	Ref	38	1.4	60	1.0	5.0	32	10
Hale C	Sep	24	7.2	1.4	0.2	5.0	25	11
Subaru	Sep	20	1.4	10	0.5	8.2	18	6
Subaru	Sep	20	1.4	30	0.5	8.2	31	6
Hale P&S RA	Ref	18	0.8	60	1.0	5.0	18	10
Hale C	Sep	12	4.9	1.4	0.2	5.0	17	11
VLT/NACO	Sep	8	0.42	34.4	0.078	8.2	10	12
Hale C	Sep	4	3.5	1.4	0.2	5.0	12	11
VLT/NACO	Sep	4	0.26	34.4	0.078	8.2	6	12
VLT/NACO	Sep	2	0.20	34.4	0.078	8.2	5	12
VLT/NACO	Sep	1	0.19	34.4	0.078	8.2	5	12
VLT/NACO	Sep	0.5	0.18	34.4	0.078	8.2	4	12
WIYN Speckle	Sep	0.5	1.1	1.6	0.84	3.5	3	13
CTIO Speckle	Sep	0.5	1.75	0.4	0.84	4.0	3	14

References: 1 = US Naval Observatory astrograph at CTIO, Chile, Zacharias *et al.*, 2004, Fig. 8 mean rms; 2 = Yale Southern Observatory double astrograph at El Leoncito, Argentina, Vieira *et al.* 2010 mean rms; 3 = Allegheny Observatory Multichannel Astrometric Photometer at Pittsburg, PA, Han 1989, Fig. 3; 4 = KPNO Mayall telescope, in Arizona, Zacharais 1996, average used; 5 = See 1, Zacharias 1997, average of test data used; 6 = Subaru telescope at Hawaii, Monet (2010, private communication), linear model; 7 = KPNO Blanco telescope in Arizona, Platais *et al.* 2002, data from his Fig. 6; 8 = USNO astrometric reflector at Flagstaff, AZ, Harris, H. (2010, private communication), mean rms; 9 = WIYN reflector at KPNO in Arizona, Vieira *et al.* 2005, OPTIC camera, FWHM = 0.6"; 10 = Hale reflector in CA, Pravdo and Shaklan 1996, all exp.; 11 = See 10, Cameron *et al.* 2009, data from his Fig. 3 average used; 12 = VLT reflectors at Paranol, Chile, Fritz *et al.* 2010, data from his Fig. 3 NACO average used; 13 = See 9, Horch *et al.* 2011, data from his Figs. 1 and 4 external error; 14 = Blanco telescope at CTIO, Chile, Tokovinin *et al.* 2010, Blanco and SOAR external error.

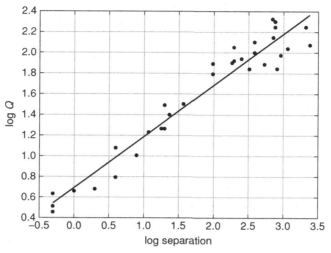

Fig. 9.2 The astrometric precision proxy Q as a function of the separation in arcseconds plotted on logarithmic scales for a variety of CCD imaging astrometric investigations listed in Table 9.1. See text for details. A linear regression of the data in Table 9.1 yields a slope $n = 0.49$ with a correlation coefficient $r = 0.969$. Therefore, the corresponding astrometric precision varies approximately as the square root of the separation from $0.5''$ to $2460''$, i.e. between the cube root and linear predictions of Eqns. (9.27) and (9.28), and independent of the type of investigation: classical reference stars, adaptive optics, or speckle interferometry.

The wind velocity will certainly vary from site to site, but if we assume to first order that v is a constant, then we can write our precision proxy Q as

$$Q \equiv \sigma D^{2/3} \sqrt{T} = c\theta^n \tag{9.29b}$$

where c is an unknown constant and our goal is to determine the exponent n, which is predicted by Eq. (9.27) to be unity in the isoplanatic region and by Eq. (9.28) to be one-third in the isokinetic region. Note that we are investigating estimates of the astrometric precision for a variety of CCD studies. This differs from the excellent and well-controlled study by Zacharias (1996) who separated the astrometric precision into the component due to the atmosphere and that due to the detector, optics, and electronics. In the present study we have no knowledge of the parameters that might make it possible to determine the atmospheric component, so we will study the total astrometric precision.

In Fig. 9.2 we plot our precision proxy Q versus the separation in arcseconds on a log-log scale. The separations range from $0.5''$ for speckle interferometry observations and VLT (Very Large Telescope) adaptive optics studies to over $2400''$ for the CCD astrometry used in the UCAC and SPM wide-field studies. Remarkably, a single square-root relation fits the data with a correlation coefficient of 97% as shown in Eq. (9.30). One important limitation of this analysis is that we have no information on the first term in Eq. (9.27), i.e. the SNR of the stars involved in the investigations. For example, the two data points with the largest separations refer to the YSO and UCAC, both of which are based on stars with well-exposed images, i.e. high SNR, and they lie below the mean regression line. Points that lie above the line may be due in part to the images having low SNR; however, the publications give us no information on the SNR. It is probable that, for investigations where the SNR is carefully controlled and a uniform analysis is used, a different relationship might exist and we look forward to new studies on this important topic. However, for the present, it appears that Eq. (9.30) is the best overall predictor of the precision as a function of the separation, telescope diameter, and exposure time:

$$\sigma = 5\,\theta^{0.49} D^{-2/3} T^{-1/2}, \tag{9.30}$$

where θ is measured in arcseconds, D in meters and T in seconds of time. Note that the constant derived in Eq. (9.30) was 4.83, but it has been rounded to 5.0 due to the many unknown factors in this analysis.

Beyond the limits imposed by the atmosphere we are constrained by several instrumental factors such as charge-transfer efficiency, camera electronics, irregularities in the CCD pixel layout, telescope, filter, and camera alignment. Surface irregularities in the telescope optics also impose a limit to the accuracy of relative positions. For example, if the section of the secondary seen by the target star differs by $\lambda/20$ from that seen by the mean of the reference stars, then the offset between the two will be 1 mas for an intermediate-aperture telescope (Shao 2010, private communication). One should therefore be cautious about believing quoted positional accuracies of less than 1 mas. In principle, the precision can be improved by accumulating more SNR in the images up to the point where the image begins to saturate. In practice, it has been found that reaching a centering precision better than 0.01 pixel is difficult due to pixel sampling, sensitivity variations across the individual pixels and variations of the PSF with location in the field of view. For cameras with a sampling ratio of more than 3 pixels/FWHM the location of the image center with respect to the pixel center (pixel-phase error) is relatively unimportant (Zacharias *et al.* 2004). However, in cases where the sampling is poor, such as with the Hubble Space Telescope (HST) imaging cameras, special techniques such as those outlined in Chapters 17 and 18 may have to be used. Repeated exposures, when combined with weights related to their SNR, are found to provide essentially the same precision as a single long exposure of the equivalent total exposure time.

Finally, it is important to remember that the position of an image is determined with respect to another star or a group of reference stars. We therefore need to consider the image-centering precision of each of the reference stars and in that case the observed luminosity function works against us. For example, if our target star is the brightest star in the field and has a measurement SNR of 100 or more, then the typical reference star will be a few magnitudes fainter and have its SNR about five to ten times lower. The effective relative precision of the target star is then dominated by the mean position of the reference stars, unless there are many of them. It is of course possible to increase the number of reference stars by using multi-chip cameras, but accurately relating the coordinate systems of the large cameras has proven to be very difficult, as discussed by Platais *et al.* (2002).

9.5 Optical interferometry

A detailed discussion of optical interferometry is given in Chapter 11, so we summarize here the basic limitation on the accuracy of those observations. In principle, the accuracy of interferometric observations increases with the baseline. However, in the optical region of the spectrum, if the baseline is much greater than 100 meters, then you begin to resolve the stellar disk and lose fringe visibility, which then decreases the measurement precision. A reasonable limit to the measurement precision for pairs of objects in optical interferometry for angles larger than about $10''$ seems to be about 10 microarcseconds (μas). For angles smaller than $1''$, since the precision scales with the angle between the objects, the limiting measurement precision could be about 10 times smaller, or on the order of 1 μas.

For the limits of space interferometry, the reader is referred to the discussions about SIM by Unwin *et al.* (2008).

9.6 Radio astrometry

Radio astrometric analysis can be divided into two regimes: (1) the narrow-field case where the relative position of sources within several degrees are determined; and (2) the wide-field case where the relative/absolute position of sources over the sky are determined. At both angular scales the atmospheric delay and turbulence (consisting of the wet troposphere, the dry troposphere, and the ionosphere) produce the major part of the positional error. The angular structure of the radio-source emission and unmodeled astrometric parameter errors also contribute to the error budget.

In the narrow-field case, it is sufficiently accurate to assume that the residual tropospheric, ionospheric, and astrometric parameters produce a quasi plane-parallel phase (delay) gradient over each antenna in the region of the two sources. In this case, the relative positional accuracy of sources scales roughly with their angular separation. At present, there are several catalogs of compact radio sources that contain the position, strength, and quality of over 3000 sources. Thus, at frequencies lower than 23 GHz, it is likely that a suitable calibrator can be found within $2°$ of any random target, except south of $\delta = -30°$ where the density of known calibrators is considerably less.

For narrow-field measurements, observations (scans) between a target source and a nearby compact calibrator source are alternated with a switching time from about 30 seconds to about 5 minutes. This type of radio observations is called *phase-referencing*. If the calibrator is not a point source, additional astrometric errors are introduced (see the end of this section). The technique assumes that the residual delay and phase measured for the calibrator applies to the target, and when transferred, the target structure and position can be determined from Fourier techniques or analytical modeling. However, the calibrator must be detectable, which means that the intensity of the source must be at least several times the root-mean-square (rms) noise for each baseline over the scan length that may be as short as 15 seconds. The switching timescale is determined by the stability of the short-term atmospheric-delay fluctuations, which must be less than 20% of the observing wavelength in order to maintain phase coherence between the source scans. At frequencies above 5 GHz, the scan times are about 20/frequency [GHz] minutes. Below 1 GHz, the ionospheric refraction increases markedly and switching times may become relatively short. During periods of poor weather, the necessary phase stability for baselines over 1000 km may not be achievable, especially above 10 GHz. For calibrators and targets that are closer than about 5 arcmin, or when observing with an array like the Japanese VERA array, which can observe two sources within $2.3°$ simultaneously, source switching is not needed.

Most observing sessions last about 1 hour, but if the target has weak emission, the session must be longer. Even though strong sources can be detected in a few minutes, longer observations are useful for two reasons. First, the imaging quality of the data (as given by the (u, v) coverage which depicts the synthesized aperture formed by the array data) improves with time, and non-point objects can be more accurately imaged. Second, systematic delay variations in the troposphere have timescales of many minutes and are often the main limitation to the astrometric accuracy. These changes are generally averaged out with observations that are considerably longer. These types of observations at appropriate intervals (about every 3 months) over several years can determine the target parallax, proper motion, or additional orbital motions. The astrometric effect of the change of the effective calibrator position caused by its variable emission is discussed below.

The observed phase difference between the target and calibrator contains several contributions from: (1) the short-term stochastic component; (2) the residual plane-parallel troposphere and ionosphere delay above each antenna toward the sources; (3) the modeling errors associated with the astrometric parameters of the array; and, of course, (4) the residual offset between the two sources. The stochastic component is caused by the turbulent component of the troposphere and ionosphere with a timescale between 2 seconds to 60 seconds in the radio regime. It is caused, mostly, by the wet component of the troposphere and is extremely location- and weather-related. The delay noise varies from 0.1 cm to 2.0 cm over about 30 seconds. The effect does not bias the determination of the angular separation, but can significantly degrade the resolution. If extremely severe, the detection of even a strong source may not be possible.

The longer-term residual tropospheric delay is the major limitation to obtaining accurate positions. Although a large part of the troposphere and ionosphere delay can be modeled using weather instrumentation and delay models from the GPS databases, a typical delay error of 4 cm at zenith remains in the raw experimental data, and this will produce a residual plane-parallel phase gradient above each antenna in the region of the two sources. This gradient varies from a timescale of tens of minutes, caused by troposphere/ionospheric cells approximately $10°$ in size, to a scale of many hours arising from an offset in the assumed spherical atmosphere delay model at zenith. The delay difference between two sources is also a strong function of zenith angle z and varies with $\tan z \sec z$, an increase of a factor of 3.5 and 6 at $z = 60°$ and $z = 70°$, respectively. The inclusion of special calibration observations during the main observing session can reduce the unmodeled delay to about 1 cm at zenith, and these calibrations are discussed in Chapter 12 on radio astrometry.

For frequencies less than 5 GHz, the ionospheric delay (frequency dependent) becomes dominant and residual delays at 1 GHz, for example, often exceed 10 cm. Simultaneous dual-frequency observations (see below) can remove the ionospheric refraction.

Any errors in the astrometric parameters associated with the array (Earth orientation and rotation, timing errors, nutation, polar motion, antenna location, and velocity) will also produce a phase gradient. However, with astrometric data available from the International Earth Rotation and Reference Systems Service (IERS) less than 1 week after the observations, residual delays associated with the astrometric models are usually less than 1 cm for baselines >3000 km.

The relevant resolution (FWHP = full-width at half-power in mas) that is obtained from the three major radio astrometric arrays, VLBA, EVN, and VERA, is the following:

$$FWHP \approx 8.0 \frac{\lambda}{D} \tag{9.31}$$

where λ is the wavelength in cm and D is the size of the array in units of 1000 km, more formally the two-dimensional size of the (u, v) coverage of the observations. The FWHP can vary by a factor of two depending on how the various baselines in the array are weighted.

The relative positional accuracy σ_{radio} in mas associated with a systematic delay error of $\Delta\tau$ in cm for a source pair separated by s degrees in an array size in units of 1000 km is

$$\sigma_{\text{radio}} \approx 0.1 \frac{s \Delta\tau}{D^{0.5}} \tag{9.32}$$

The square-root dependence on the array size reflects that the delay error tends to increase with baseline length. The loss of accuracy with the zenith angle of the observations has not been included.

Using the Very Long Baseline Array (VLBA) (maximum baseline of 8000 km) at 8 GHz as an example, the resolution for a source at declination $20°$ is 1.0×0.6 mas in position angle $0°$. The expected angular separation precision for two sources separated by $1°$, assuming a delay error of 1.4 cm (1.0 from the troposphere and 1.0 from other astrometric effects – added in quadrature), is about 0.03×0.02 mas. Positional accuracies of 0.01 mas have been reached with many VLBA experiments using careful calibrations with good weather conditions (Fomalont and Kopeikin 2005, Reid *et al.* 2009).

The astrometric accuracy is also a function of the strength of the calibrator and target, and is measured by the SNR, which is equal to the strength of the source divided by the rms noise expected integrated over all scans and baselines. The theoretical positional accuracy σ_{SNR} is given by

$$\sigma_{\text{SNR}} = \frac{FWHP}{2\,SNR} \tag{9.33}$$

where the appropriate SNR applies to the weaker of the calibrator or target. For weak target sources (SNR < 10), the astrometric accuracy is limited more by the observation sensitivity than the residual atmosphere delay. However, the short-term phase fluctuations (turbulence), rather than the brute-force sensitivity of the observations, often limit the SNR to less than about 50, so that a relative position accuracy less than 1% of the FWHP is rarely obtained even for strong sources.

Wide-field radio astrometric accuracy is more difficult to assess because of the complex nature of the observations and analysis. The main observational goal is the determination of a basic quasi-inertial frame on which the positions of a large number of bright, compact sources, located around the sky, are determined. This goal requires a completely different observational approach compared with that for narrow-field astrometry. For nearly all of the absolute astrometric observing sessions, two frequencies are observed simultaneously (8.4 GHz and 2.3 GHz) in order to remove the ionospheric refraction using an appropriate difference of two frequency data. Each session lasts 24 hours and contains several hundred scans of compact and strong sources, distributed over the sky in as short a time as possible (typically 30 to 60 min) in order to sample the tropospheric delay as a function of zenith angle as quickly as possible. The 24-hour schedule is critical in order to separate unambiguously the residual delay dependence associated with many astrometric and geodetic parameters (Earth-orientation, rotation, polar motion, nutation, antenna location and velocity, crustal tidal motions, source-position offsets), and the instrumental delay and phase variations.

The initial residual delay, even when the most accurate a-priori models of the session parameters are used, is about 10 cm for sources over the sky so that there is little phase coherence between scans. Hence, the fundamental observable is the group delay (rather than the phase delay used for narrow-field astrometry), which is the measured phase gradient with frequency. The group delay is determined by sampling simultaneously the data at many frequency channels after careful phase and delay calibrations remove the purely instrumental offsets and variations among the channels. The common observing frequencies are between 2.3 and 2.4 GHz, and between 8.1 and 8.6 GHz. The group delays measured at each frequency are then combined to remove that part caused by the ionosphere (Rogers 1970, Sovers *et al.* 1998).

The International Celestial Reference Frame ICRF2 catalog has been generated from hundreds of 24-hour observing sessions over several decades. The measured group delays from each session are analyzed using a complex least-squares method (Ma *et al.* 1998, Sovers *et al.* 2004, Fey *et al.* 2010) in order to obtain the relevant astrometric and geodetic parameters, the antenna troposphere model from hour to hour (zenith path delay plus gradients in the north–south and east–west directions), and a time correction associated with the independent clocks/oscillator at each VLBI station. Many sessions are then analyzed to obtain a global solution in which the source positions and a smooth time dependence of all relevant astrometric/geodetic parameters are obtained. The origin of the quasi-inertial frame determined by the observations is the barycenter of the Solar System, and the orientation of the three-coordinate axes are accurate to about 0.01 mas, with some minor corrections for solar motion and Galactic rotation.

For a typical 24-hour session, the residual delay error after obtaining the best solution is about 25 ps (0.75 cm), which is comparable to the 1.0 cm residual for narrow-field astrometry. In the narrow-field cases, the small source separation dilutes the effect of this error to about 0.02 cm. For the ICRF2 catalog, the dilution of the session error is obtained by the averaging of many good quality 24-hour sessions. The precision obtained for the most observed and strongest compact sources in the ICRF2 catalog is 0.04 mas, and corresponds to a delay error at 8 GHz of about 5 psec = 0.15 cm. This limitation is probably dominated by the subtle interaction of the troposphere modeling with the source positions, especially in declination (Fey *et al.* 2010; see Chapter 12).

Both narrow-field and wide-field astrometric precision are limited by the tropospheric variations. In an attempt to remove the short-term variations, radiometers, placed on or near radio telescopes, measure the water-vapor emission (at 22.4 or 183 GHz) in the direction of the radio sources. Recent implementation of radiometers at 183 GHz for the Atacama Large Millimeter Array (ALMA) in Chile show good reduction in the phase fluctuations caused by water vapor in the line of site (Jacobs *et al.* 2006, Nikolic *et al.* 2007).

Several new radio arrays will observe a much higher density of sources, so that the troposphere refraction can be modeled in periods much less than 30 minutes, perhaps as little as 5 minutes. These models may also include the $10°$–$50°$ refraction scales. The VLBI2010 project, a global VLBI array now under construction, will consist of small telescopes that can slew quickly between sources over the sky. The proposed Square Kilometer Array (SKA) will contain hundreds of telescopes, and it may be able to observe many sources simultaneously over the sky using many subarrays, again increasing the ability to model details of the tropospheric refractions. It is hoped that the accuracy of the absolute astrometric grid will increase by about a factor of 10 from the present day 0.04 mas source position accuracy and the 0.01 mas accuracy of the ICRF frame orientation with VLBI2010 and the SKA (see www.haystack.mit.edu/geo/vlbi_td/2010/index.html and www.skatelescope.org/).

The SKA will have unprecedented sensitivity so that the field of view centered on any target will contain several background calibrators that can be used as reference sources for a random target. Hence, the approximately plane-parallel-delay dependence above each antenna can be determined accurately with several calibrator probes. Relative positions between the target and the calibrators should be determined to 0.001 mas, and very weak targets will be detectable. However, the SKA must contain at least 20% of its collecting area associated with antennas up to 3000 km distant from the main collecting area that will be contained within a 100 km region, in order to realize this high astrometric precision.

Space-based observations avoid the troposphere problem, but only if there is more than one space-antenna (Hirabayashi *et al.* 2000). With three or more space-antennas, the angular precision would be limited by random motions of the antennas that could not be modeled directly with calibration observations. A large radio interferometer in space is not anticipated for several decades, but the resolution and angular precision would be an order of magnitude more than currently available.

Finally, radio calibrators are not point sources. They consist of a compact component (radio core) containing >20% of the source intensity, usually within an 0.05 mas region. The remaining emission is jet-like and usually emanates away from the radio core on one side to a distance that often exceeds 1 mas. The intensity and structure of most quasars are variable over timescales of days to years. At present, the positional error for narrow- and wide-field astrometry from tropospheric residuals is estimated to be about twice that from calibrator structure changes.

The calibrator structure and variation can lead to target position errors for three reasons that are discussed in detail in Chapter 12. (1) First, even if the calibrator does not vary with time, its precise location is ambiguous and resolution-dependent; (2) its variability changes the apparent position of the calibrator with time at the level of 0.02 to 2.0 mas;

(3) the calibrator structure has a complex dependence on frequency, and the location of the radio core can differ between 8.4 GHz and 23 GHz by about 0.05 mas, and is probably displaced from the nucleus of the galaxy or quasar (Fomalont *et al.* 2010).

The calibrator structure can be determined for any session using self-calibration techniques (Cornwall and Fomalont 1999) if the source has been observed with an array of many elements. However, at the 0.05 mas astrometric level, the effects of calibrator structure for narrow-field astrometry are not important, as long as the chosen calibrator is not particularly unstable. For more accurate precision, several calibrators for any target, besides aiding in the determination of the phase-gradient over the telescopes, will check if any of the calibrators are particularly unstable; however, these observations are time consuming.

For wide-field astrometry, the use of the most compact calibrators decreases the astrometric error associated with calibrator positions. The correction of the data for the calibrator structures can be made, but does not improve significantly the post-fit residuals that are still dominated by the troposphere. The position of the sources at frequencies other than 8.4 GHz (the ICRF frequency) can change by 0.05 mas, and it is likely that a catalog of sources at a higher radio frequency would decrease the effect of source structure and refer to a position that is more closely associated with the nucleus of the calibrator optical counterpart (Blandford and Konigl 1979, Charlot *et al.* 2010). Since the wide-field astrometric results are based on the average properties of hundreds of sources, the definition of the astronomical quasi-inertial grid is hardly affected by the structure changes of the calibrators.

9.7 Summary

As we have seen, the Earth's atmosphere limits our optical observations in three fundamental ways. First, by selectively absorbing various wavelengths of light; second, by distorting the images that we observe; and third, by displacing them from their true positions, thus limiting the astrometric precision of ground-based observations. Space astrometry offers the opportunity to determine significantly higher precision and accuracy, but at a very high cost and for limited periods of time. In spite of its limitations, ground-based astrometry will continue to play a very important role in the future in areas where long time baselines are required, such as binary-star orbits and Solar System observations. Extremely faint targets will continue be the domain of ground-based astrometry due to the prohibitive cost and difficulty of launching large imaging optics. Finally, ready access to ground-based telescopes is critical for the development of new instrumentation and the education of students.

In the radio spectrum, the troposphere limits narrow-field astrometric accuracy to 0.05 mas, but 0.01 mas can be reached if careful calibrations are made. The present accuracy of the ICRF (wide-field astrometry) is about 0.04 mas, based on several decades of observations. Until methods are found to decrease the tropospheric error by at least a factor of two, other astrometric limitations will remain smaller than those caused by the atmosphere.

The authors of this chapter wish to acknowledge the valuable input from a number of astronomers who have provided summaries of their experiences in pushing the limits of precision and accuracy in astrometric observations. In particular, they thank: Fritz Benedict, Andreas Glindemann, Elliott Horch, Lennart Lindegren, Chopo Ma, Hal McAlister, David Monet, Mark Reid, Steven Shaklan, Michael Shao, and Norbert Zacharias. William van Altena is indebted to his collaborators who worked with him to explore the limits of optical imaging astrometry: Dana Casetti, Terrence Girard, Imants Platais, and Katherine Vieira.

References

Astronomical Almanac for the year 1990, London: Her Majesty's Stationary Office, p. B62.

Auer, L. H. and Standish, E. M. (2000). Astronomical refraction: computational method for all zenith angles. *AJ*, **119**, 2472.

Bessell, M. (1990). UBVRI passbands. *PASP*, **102**, 1181.

Blandford, R. D. and Konigl, A. (1979). Relativistic jets as compact radio sources. *ApJ*, **232**, 34.

Cameron, P. B., Britton, M. C., and Kulkarni, S. R. (2009). Precision astrometry with adaptive optics. *AJ*, **137**, 83.

Charlot, P., Boboltz, D. A., Fey. A. L., *et al.* (2010). The Celestial Reference Frame at 24 and 43 GHz. II. Imaging. *AJ*, **139**, 1713.

Cornwell, T. and Fomalont, E. (1999). Self-calibration. *ASP Conf. Proc.*, **180**, 187.

Fey, L., Gordon, D., and Jacobs, C. S., eds. (2010). *The Second Realization of the International Celestial Reference Frame by Very Long Baseline Interferometry*. IERS Technical Note No. 35. See: http://www.iers.org/IERS/EN/Publications/TechnicalNotes/tn35.html.

Fomalont, E. B. and Kopeikin, S. M. (2005). The measurement of the light deflections from Jupiter: experimental results. *ApJ*, **598**, 704.

Fomalont, E. B., Johnston, K. J., Fey, A., Boboltz, D., Oyama, T., and Honma, M. (2010). The position/structure stability of four ICRF2 sources. *AJ*, **141**, 91.

Fritz, T., Gillessen, S., Trippe, S., *et al.* (2010). What is limiting near-infrared astrometry in the Galactic Center. *MNRAS*, **401**, 1177.

Garfinkel, B. (1967). Astronomical refraction in a polytropic atmosphere. *AJ*, **72**, 235.

Han, I. (1989). The accuracy of differential astrometry limited by the atmospheric turbulence. *AJ*, **97**, 607.

Helminiak, K. G., Konacki, M., Kulkarni, S. R., and Eisner, J. (2009). Precision astrometry of a sample of speckle binaries and multiples with the adaptive optics facilities at the Hale and Keck II telescopes. *MNRAS*, **400**, 406.

Hirabayashi, H., Fomalont, E. B., Horiuchi, S., *et al.* (2000). The VSOP 5 GHz AGN Survey I. Compilation and observations. *PASJ*, **52**, 997.

Hoeg, E. (1968). The mean power spectrum of star image motion. *Zeitschrift für Astrophysik*, **69**, 313.

Horch, E. P., Gomez, S. C., Sherry, W. H., *et al.* (2011). Observations of binary stars with the differential speckle survey instrument. II. Hipparcos stars observed in January and June 2010. *AJ*, **141**, 45.

Jacobs, C. S., Keihm, S. J., Lanyi, G. E., *et al.* (2006). *Improving astrometric VLBI by using water vapor radiometer calibrations.* IVS 2006 General Meeting Proceedings, Concepcion, Chile, January 9–11. See http://ivs.nict.go.jp/mirror/publications/gm2006/jacobs2.

Konig, A. (1962). Astrometry with astrographs. In *Astronomical Techniques*, ed. W. A. Hiltner. Chicago, IL: University of Chicago Press, ch. 20.

Lindegren, L. (1978). Photoelectric astrometry – a comparison of methods for precise image location. In *Modern Astrometry; Proceedings of the Colloquium*. Vienna: Universitaets-Sternwarte Wien, p. 197.

Lindegren, L. (1980). Atmospheric limitations of narrow-field optical astrometry. *A&A*, **89**, 41.

Lindegren, L. (2010). High-accuracy positioning: astrometry. *ISSI Sci. Rep. Ser.*, **9**, ch. 16.

Ma, C., Arias, E. F., Eubanks, T. M., *et al.* (1998). The International Celestial Reference Frame as realized by very long baseline interferometry. *ApJ*, **116**, 516.

Moffatt, A. F. J. (1969). A theoretical investigation of focal stellar images in the photographic emulsion and application to photographic photometry. *A&A*, **3**, 455.

Monet, D. G., Dahn, C. C., Vrba, F. J., *et al.* (1992). U.S. Naval Observatory CCD parallaxes of faint stars. I – Program description and first results. *AJ*, **103**, 638.

Nikolic, B., Hills, R. E., and Richer, J. S. (2007). *Limits on phase correction performance due to differences between astronomical and water-vapour radiometer beams. ALMA Memo Ser.*, **573**.

Pickles, A. J. (1998). A stellar spectral flux library: 1150–25 000 Å. *PASP*, **110**, 863.

Platais, I., Kozhurina-Platais, V., Girard, T. M., *et al.* (2002). WIYN open cluster study. VIII. The geometry and stability of the NOAO CCD mosaic imager. *AJ*, **124**, 601.

Pravdo, S. H. and Shaklan, S. B. (1996). Astrometric detection of extrasolar planets: results of a feasibility study with the Palomar 5 meter telescope. *ApJ*, **465**, 264.

Reid, M. J., Menten, K. M., Zheng, X. W., Brunthaler, A., and Xu, Y. (2009). A trigonometric parallax of SGR B2. *ApJ*, **705**, 1548.

Roddier, F. (1981). The effects of atmospheric turbulence in optical astronomy. *Progr. Opt.*, **19**, ch. 5.

Rogers, A. E. E. (1970). Very long baseline interferometry with large effective bandwidth for phase-delay measurements. *Radio Science*, **5**, 1289.

Schroeder, D. (1987). *Astronomical Optics*. San Diego, CA: Academic Press.

Schroeder, D. (2000). *Astronomical Optics*, 2nd edn. San Diego, CA: Academic Press.

Sovers, O. J., Fanslow, J. L., and Jacobs, C. S. (1998). Astrometry and geodesy with radio interferometry: experiments, models, results. *Rev. Mod. Phys.*, **70**, 1393.

Sovers, O. J., Jacobs, C. S., and Lanyi, G. E. (2004). *MODEST: a tool for geodesy and astrometry*. IVS 2004 General Meeting Proceedings, Ottawa, Canada, February 9–11. See: http://ivs.nict.go.jp/mirror/publications/gm2004/sovers.

Stone, R. C. (1996). An accurate method for computing atmospheric refraction. *PASP*, **108**, 1051.

Tatarski, V. I. (1971). *The Effects of the Turbulent Atmosphere on Wave Propagation.* Jerusalem: Israel Program for Scientific Translations.

Tokovinin, A., Mason, B. D., and Hartkopf, W. I. (2010). Speckle interferometry at the Blanco and SOAR telescopes in 2008 and 2009. *AJ*, **139**, 743.

Unwin, S. C., Shao, M., Tanner, A. M., *et al.* (2008). Taking the measure of the Universe: precision astrometry with SIM PlanetQuest. *PASP*, **120**, 38.

van Altena, W. F. (1971). Trigonometric parallaxes determined with the Yerkes Observatory 40-inch refractor. I. Methods of observation, measurement reduction, and first results. *AJ*, **76**, 932.

Vieira, K., van Altena, W. F., and Girard, T. M. (2005). Astrometry with OPTIC at WIYN. *ASP Conf. Series*, **338**, 130.

Vieira, K., Girard, T., van Altena, W., *et al.* (2010). Proper motion study of the Magellanic Clouds using SPM data. *AJ*, **140**, 1934.

Zacharias, N. (1996). Measuring the atmospheric influence on differential astrometry: a simple method applied to wide-field CCD frames. *PASP*, **108**, 1135.

Zacharias, N. (1997). Astrometric quality of the USNO CCD astrograph (UCA). *AJ*, **113**, 1925.

Zacharias, N., Urban, S. E., Zacharias, M. I., *et al.* (2004). The Second US Naval Observatory CCD Astrograph Catalog (UCAC2). *AJ*, **127**, 3043.

10

Astrometry with ground-based diffraction-limited imaging

ANDREA GHEZ

Introduction

The construction of large ground-based optical and infrared telescopes is driven by the desire to obtain astronomical measurements of both higher sensitivity and higher angular resolution. With each increase in telescope diameter the former goal, that of increased sensitivity, has been achieved. In contrast, the angular resolution of large telescopes ($D >$ 1 m), using traditional imaging, is limited not by the diffraction limit ($\theta \sim \lambda/D$), but rather by turbulence in the atmosphere. This is typically $1''$, a factor of 10–20 times worse than the theoretical limit of a 4-meter telescope at near-infrared wavelengths. This angular resolution handicap has led to both space-based and ground-based solutions. With the launching of the Hubble Space Telescope (HST), a 2.4-m telescope equipped with both optical and infrared detectors, the astronomical community has obtained diffraction-limited images. These optical images, which have an angular resolution of $\sim0.''1$, have led to exciting new discoveries, such as the detection of a black hole in M87 (Ford *et al.* 1994) and protostellar disks around young stars in Orion (O'Dell *et al.* 1993, O'Dell and Wen 1994). However, HST has a modest-sized mirror diameter compared to the 8–10 meter mirror diameters of the largest ground-based telescope facilities.

With the development of techniques to overcome the wavefront distortions introduced by the Earth's atmosphere, diffraction-limited observations from the ground have become possible. These techniques cover a wide range of complexity and hence expense. Speckle imaging, which provided the earliest and simplest approach, is described in Sections 10.1 and 23.3.1 and adaptive optics, which has more recently become scientifically productive and which is a much more powerful technique, is discussed in Section 10.2.

10.1 Speckle imaging

10.1.1 Method

Speckle imaging provides a "passive" approach to constructing diffraction-limited images from data that have been affected by atmospheric turbulence. By passive, we mean to

Astrometry for Astrophysics: Methods, Models, and Applications, ed. William F. van Altena.
Published by Cambridge University Press. © Cambridge University Press 2013.

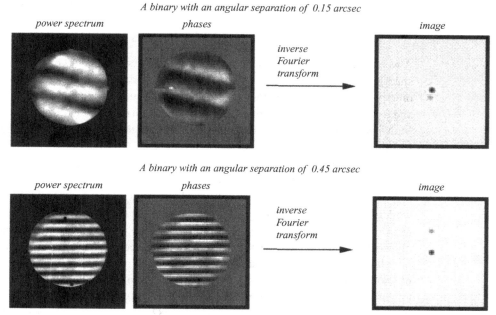

Fig. 10.1 Speckle-interferometric measurements of two binary-star systems. The recovered Fourier amplitudes (left) and Fourier phases (middle) are combined through an inverse Fourier transform to produce a final image, which has been convolved with a diffraction-limited point-spread function (right).

indicate that the Earth's atmosphere is compensated for in post-processing either via the shift-and-add method or speckle-interferometric methods. The basic input for both of these techniques is a large number of images which have exposure times that are short compared to the timescale over which the atmosphere changes ($\tau_0 \sim 100$ ms $\times (\frac{\lambda}{2.2\,\mu m})^{1.2}$; Fried 1966). These exposures are obtained both on the object of interest and a nearby point source, which is used to measure the average transfer function of the atmosphere and telescope. Via speckle interferometry, all the information in the images is used to recover a diffraction-limited image using Fourier-analysis techniques that are analogous to those used in radio interferometry (cf. Labeyrie 1970, Weigelt 1977, Lohmann *et al.* 1983). This works well for isolated objects, such as close binary stars (see Fig. 10.1), which naturally go to zero flux density at the edges of the field of view. In more complicated fields, such as stellar clusters that extend beyond the field of view, shift-and-add has often proven to be a more effective technique. As its name suggests, this technique is implemented by registering the short exposure images on the brightest speckle and then averaging over all the shifted images. While both of these techniques have the advantage of being computationally straightforward and having minimal imaging requirements for obtaining the necessary data, they have the disadvantage of being insensitive to faint objects since the source needs to be detectable in each individual short exposure. Currently, in the near-infrared this corresponds to $K_{\text{total}} < \sim 11$ mag and this limit is independent of telescope size, since the number of speckles scales as D^2.

10.1.2 Examples of astrometric speckle-imaging results

Stellar binaries

A wealth of astrometric measurements for binary stars within our Galaxy has resulted from speckle imaging. Several significant long-term efforts have used optical speckle interferometry to measure the relative positions of bright, nearby, double stars ($V < \sim 11$ mag) and to thereby derive their orbital elements; most notable among these efforts are those by McAlister *et al.*, using 4-m class telescopes around the world (e.g. McAlister *et al.* 1987, McAlistair 1977, Hartkopf *et al.* 1989), the US Naval Observatory's speckle interferometry program (e.g. Douglass *et al.* 1997, Mason *et al.* 1999, 2008), and Belegas *et al.*, using telescopes in Zlelnchuk (e.g. Belega and Belega 1988, Balega *et al.* 2002, 2006). In addition to nearby stars, many of these optical speckle-imaging programs also extended their survey work to include massive binary stars (e.g. Mason *et al.* 1998) as well as white dwarfs (e.g. McAlister *et al.* 1996). The advent of infrared arrays significantly increased the diversity of objects that could be studied with speckle-imaging techniques both for multiplicity, which is an important constraint on star-formation theories, and to determine the orbital elements of known binaries, which provides a direct measure of the systems' masses and distances. Of particular emphasis in the infrared speckle-imaging studies were samples of young stars in nearby T associations (e.g. Dyck *et al.* 1982; Ghez *et al.* 1993, 1995, 1997; Leinert *et al.* 1993; Brandner *et al.* 1996; Brandner and Köhler 1998; Köhler *et al.* 2000; Duchêne *et al.* 2006) and Orion – the nearest giant molecular cloud (e.g. Petr *et al.* 1998, Preibisch *et al* 1999, Weigelt *et al.* 1999), Herbig Ae/Be stars (e.g. Leinert *et al.* 1997a), members of young open clusters (e.g. Torres *et al.* 1997; Peterson *et al.* 1988; Patience *et al.* 1998, 2002), and M-dwarfs (e.g. Henry and McCarthy 1990, 1993; McCarthy *et al.* 1991; Leinert *et al.* 1997b).

Galactic center

Near-infrared speckle imaging has also revealed the presence of a supermassive black hole at the center of our Galaxy. Initially, velocity dispersion measurements based on proper motions demonstrated the existence of $\sim 3 \times 10^6 M_\odot$ of dark matter confined to within a radius 0.015 pc (e.g. Eckart and Genzel 1997, Ghez *et al.* 1998). This increased the implied minimum dark-matter density by three orders of magnitude, compared to earlier seeing-limited work (e.g. Lacy *et al.* 1980, McGinn *et al.* 1989, Sellgren *et al.* 1990, Haller *et al.* 1996, Genzel *et al.* 1997) and eliminated a cluster of dark objects as a possible explanation of the Galaxy's central dark-mass concentration (Maoz *et al.* 1998), but still left the fermion ball hypothesis (e.g. Tsiklauri and Viollier 1998, Munyaneza and Viollier 2002) as an alternative to a single supermassive black hole. Over time, significant curvature was detected (Ghez *et al.* 2000, Eckart *et al.* 2002) and now complete orbital solutions have been achieved providing the most definitive case for the existence of a supermassive black hole at the center of a normal-type galaxy (see Fig. 10.2; Schödel *et al.* 2002, 2003; Ghez *et al.* 2003, 2005a).

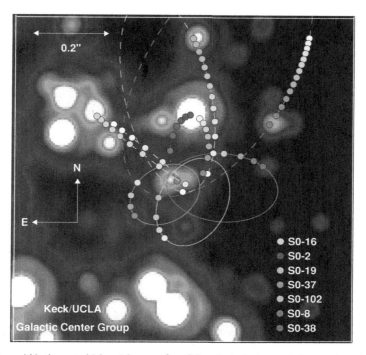

Fig. 10.2 The orbits of stars within the central 1.0 × 1.0 arcsec of our Galaxy. In the background, the central portion of a diffraction-limited adaptive optics image taken in 2008 is displayed. While every star in this image has been seen to move over 14 years (1995–2005 with speckle imaging; 2004–2008 with adaptive optics), only stars with well-determined orbital parameters are highlighted. The annual average positions for these seven stars are plotted as shaded dots, which have increasing color saturation with time (see color version of this figure on the cover of this book). Also plotted are the best-fitting simultaneous orbital solutions. These orbits provide the best evidence yet for a supermassive black hole, which has a mass of 4 million times the mass of the Sun. Measurements from data sets presented in Ghez *et al.* (2008).

10.2 Adaptive optics

10.2.1 Method

Adaptive optics provides an "active" approach to obtaining diffraction-limited images from light that has been affected by atmospheric turbulence. With this method, the wavefront errors are corrected in real time with optical components that are added to the telescope imaging system. Light from a guide star, either a natural one (i.e. a nearby star brighter than $R < \sim 14$ mag) or an artificial one created by a laser, is analyzed as a part of a feedback loop, which typically runs at a rate of few hundred hertz, in order to reconfigure a deformable mirror in a way that removes the effects of the atmosphere on the incoming wavefronts. This approach was originally proposed by Babcock (1953) and became scientifically productive

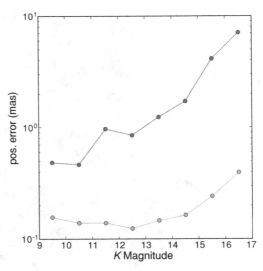

Fig. 10.3 Comparison of the positional uncertainties as a function of brightness for speckle imaging (upper curve) and laser
guide star adaptive optics (LGSAO) (lower curve) measurements of stars at the Galactic center from Ghez *et al.* (2008).
Because the very brightest stars ($K \sim 9$) are saturated in their cores in the adaptive optics images, there is a slight
rise in their centroid uncertainties compared to somewhat fainter sources. Overall, however, for bright sources
($K < 13$), the long exposure adaptive optics images achieve a centroiding uncertainty of just 0.17 mas, a factor of
$\sim$6 better than the earlier work done with speckle imaging.

in the 1990s, once the required computational power was easily available and work done in
this area by the military was de-classified (cf. e.g. Hardy 1991, Fugate *et al.* 1994, Roddier
et al. 1994, Max *et al.* 1994, 1997).

Adaptive optics is a much more powerful technique for doing relative astrometry compared to speckle imaging. It is orders of magnitude more sensitive, with a demonstrated
capability of astrometric measurements for stars as faint as $K \sim 20$ mag. For the brighter
stars, it also offers an order of magnitude improvement in astrometric precision (~ 0.2 vs.
1 mas) due to its lower observing overheads and higher strehl ratios[1] (see Fig. 10.3). In
general, the relative astrometric precision scales roughly as the square root of the observing
time, as we average over differential tilt error between sources (cf. e.g. Cameron *et al.*
2009).

10.2.2   Examples of astrometric results

With enhanced sensitivity, astrometric studies using adaptive optics have much greater
scientific reach than their speckle-imaging precursors. Below, just a few examples are
highlighted.

---

[1] For point-spread functions with diffraction limited cores, the strehl ratio is approximately the amount of energy
in the diffraction-limited core.

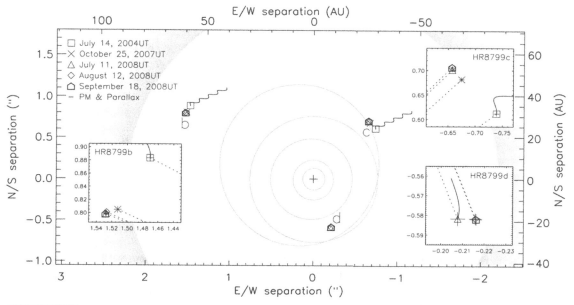

Fig. 10.4   HR 8799bcd astrometric analysis from Marois *et al.* (2008). The positions of HR 8799bcd at each epoch are shown in both the overall field of view and in the zoomed-in insets. The solid oscillating line originating from the first detected epoch of each planet is the expected motion of unbound background objects relative to the star over a duration equal to the maximum interval over which the companions were detected (4 years for b and c, 2 months for d). All three companions are confirmed as co-moving with HR 8799 to 98σ for b, 90σ for c and $\sim 6\sigma$ for d. Counter-clockwise orbital motion is observed for all three companions. The dashed lines in the small insets connect the position of the planet at each epoch with the star. A schematic dust disk – at 87 AU separation thought to be in 3:2 resonance with b while also entirely consistent with the far-infrared dust spectrum – is also shown. The inner gray ellipses are the outer Jovian-mass planets of our Solar System (Jupiter, Saturn, Uranus, and Neptune) and Pluto shown to scale. Credited to NRC-HIA, C. Marois & Keck Observatory.

Sub-stellar companions: planets and brown dwarfs

Adaptive optics has expanded the scope of companion-object studies, allowing much less massive components to be discovered and studied. One of the most exciting, recent examples of the benefits of astrometry with adaptive optics is what appears to be the first unequivocal cases of exoplanets being directly imaged and resolved around nearby stars (see Fig. 10.4 from Marois *et al.* 2008). Based on the low luminosity of the companions and the estimated age of the system, the authors estimate companion masses between 5 and 13 times that of Jupiter, which makes this system resemble a scaled-up version of the outer portion of our Solar System.

Astrometric adaptive optics work has also produced the first precision ($<10\%$) dynamical mass estimates of brown-dwarf binaries (Liu *et al.* 2008, Konopacky *et al.* 2010, Dupuy *et al.* 2009). These measurements allow the theoretical substellar evolutionary models to be tested and calibrated. Initial work suggests that the late-M and early-L spectral-type systems have higher dynamical masses than the models predict, while the mid-L system is

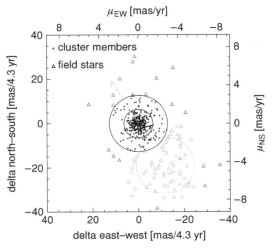

Fig. 10.5 Relative proper motions of stars in the central Arches field from Stolte *et al.* (2008) shows that the cluster's motion is easily measured with respect to the field. Extended proper-motion measurements over a few years will yield strong constraints on, or possible detection of, a central intermediate-mass black hole, and with increased radial coverage, a determination of the cluster's tidal radius.

consistent with model predictions and the one T-dwarf system has a lower dynamical mass than predicted (Konopacky 2010). If true, this result has important implications for masses inferred for directly imaged planets, which may currently be over-predicted.

Galactic center

The recent advent of adaptive optics has also significantly expanded the scientific reach of high-spatial-resolution infrared studies of the center of our Galaxy. This has led to a number of important results, including (1) the first infrared detection of SgrA*, the radio source associated with the central supermassive black hole, and its dramatic short-timescale infrared variations (Genzel *et al.* 2003; Ghez *et al.* 2004, 2005b, Eckart *et al.* 2006; Hornstein *et al.* 2007; Do *et al.* 2009a; Meyer *et al.* 2008); (2) the discovery that the stars within 0.01 pc (0."25) of the central supermassive black hole are young stars (B stars; Ghez *et al.* 2004, Eisenhauer *et al.* 2005, Gillessen *et al.* 2009, Do *et al.* 2009b); (3) the discovery that half the young stars beyond 0.04 pc (1″) reside in a stellar disk (Paumard *et al.* 2006, Lu *et al.* 2009); and (4) the first proper-motion measurement of the Arches, a massive young star cluster located at a projected galacto-centric distance of 30 pc (see Fig. 10.5, Stolte *et al.* 2008).

10.3 The future

Adaptive optics is a tremendously powerful technique for carrying out astrometric studies. We are only now reaping the first rewards of investing in this approach and much more exciting work is likely to come in the near future. Furthermore, as future planned adaptive optics systems come on-line, with higher strehl ratios and larger fields of view (e.g. Cady *et al.* 2009, Moretti *et al.* 2009, Frogel *et al.* 2009), the scientific potential will continue to blossom.

References

Babcock, H. W. (1953). The possibility of compensating astronomical seeing. *PASP*, **65**, 229.

Balega, I. I. and Balega, Y. Y. (1988). Binary star measurements using digital speckle interferometer of 6-m telescope. *JPRS Rep. Sci. Technol. USSR Space*, **6**, 508.

Balega, I. I., Balega, Y. Y., Hofmann, K.-H., *et al.* (2002). Speckle interferometry of nearby multiple stars. *A&A*, **385**, 87.

Balega, I. I., Balega, Y. Y., Hofmann, K.-H., *et al.* (2006). Orbits of new Hipparcos binaries. II. *A&A*, **448**, 703.

Brandner, W. and Koehler, R. (1998). Star formation environments and the distribution of binary separations. *ApJ Lett*, **499**, 79.

Brandner, W., Alcala, J. M., Kunkel, M., Moneti, A., and Zinnecker, H. (1996). Multiplicity among T Tauri stars in OB and T associations. Implications for binary star formation. *A&A*, **307**, 121.

Cady, E., Macintosh, B., Kasdin, N. J., and Soummer, R. (2009). Shared pupil design for the Gemini Planet Imager. *ApJ*, **698**, 938.

Cameron, P. B., Britton, M. C., and Kulkarni, S. R. (2009). Precision astrometry with adaptive optics. *AJ*, **137**, 83.

Do, T., Ghez, A. M., Morris, M. R., Yelda, S., *et al.* (2009a). A near-infrared variability study of the Galactic black hole: a red noise source with NO detected periodicity. *ApJ*, **691**, 1021.

Do, T., Ghez, A., Morris, M. R., Lu, J. R., *et al.* (2009b). High angular resolution integral-field spectroscopy of the Galaxy's nuclear cluster: a missing stellar cusp? *ApJ*, **703**, 1323.

Douglass, G. G., Hindsley, R. B., and Worley, C. E. (1997). Speckle interferometry at the US Naval Observatory. I. *ApJ S*, **111**, 289.

Duchêne, G., Beust, H., Adjali, F., Konopacky, Q. M., and Ghez, A. M. (2006). Accurate stellar masses in the multiple system T Tauri. *A&A*, **457**, L9.

Dupuy, T. J., Liu, M. C., and Ireland, M. J. (2009). Dynamical mass of the substellar benchmark binary HD 130948BC. *ApJ*, **692**, 729.

Dyck, H. M., Simon, T., and Zuckerman, B. (1982). Discovery of an infrared companion to T Tauri. *ApJ Lett*, **255**, 103.

Eckart, A. and Genzel, R. (1997). Stellar proper motions in the central 0.1 pc of the Galaxy. *MNRAS*, **284**, 576.

Eckart, A., Genzel, R., Ott, T., and Schödel, R. (2002). Stellar orbits near Sagittarius A*. *MNRAS*, **331**, 917.

Eckart, A., Baganoff, F. K., Schödel, R., *et al.* (2006). The flare activity of Sagittarius A*: new coordinated mm to X-ray observations. *A&A*, **450**, 535.

Eisenhauer, F., Genzel, R., Alexander, T., *et al.* (2005). SINFONI in the Galactic center: young stars and infrared flares in the central light-month. *ApJ*, **628**, 246.

Ford, H. C., *et al.* (1994). Narrowband HST images of M87: evidence for a disk of ionized gas around a massive black hole. *ApJ Lett*, **435**, L27.

Fried, D. L. (1966). Optical resolution through a randomly inhomogeneous medium for very long and very short exposures. *J. Optical Soc. America (1917–1983)*, **56**, 1372.

Frogel, J. A., Alcock, C., Bolte, M., *et al.* (2009). Frontier science and adaptive optics on existing and next-generation telescopes. Astro 2010: The Astronomy and Astrophysics Decadal Survey, Position Papers, no. 16.

Fugate, R. Q., *et al.* (1994). *J. Opt. Soc. Am. A*, **11**, 310.

Genzel, R., Eckart, A., Ott, T., and Eisenhauer, F. (1997). On the nature of the dark mass in the centre of the Milky Way. *MNRAS*, **291**, 219.

Genzel, R., Schödel, R., Ott, T., *et al.* (2003). Near-infared flares from accreting gas around the supermassive black hole at the Galactic centre. *Nature*, **425**, 934.

Ghez, A. M., Neugebauer, G., and Matthews, K. (1993). The multiplicity of T Tauri stars in the star forming regions Taurus–Auriga and Ophiuchus–Scoprius: a 2.2 micron speckle imaging survey. *AJ*, **106**, 2005.

Ghez, A. M., Weinberger, A. J., Neugebauer, G., Matthews, K., and McCarthy, D. W., Jr. (1995). Speckle imaging measurements of the relative tangential velocities of the components of T Tauri binary stars. *AJ*, **110**, 753.

Ghez, A. M., McCarthy, D. W., Patience, J. L., and Beck, T. L. (1997). The multiplicity of pre-main-sequence stars in southern star-forming regions. *ApJ*, **481**, 378.

Ghez, A. M., Klein, B. L., Morris, M., and Becklin, E. E. (1998). High proper-motion stars in the vicinity of Sagittarius A*: evidence for a supermassive black hole at the center of our Galaxy. *ApJ*, **509**, 678.

Ghez, A. M., Morris, M., Becklin, E. E., Tanner, A., and Kremenek, T. (2000). The accelerations of stars orbiting the Milky Way's central black hole. *Nature*, **407**, 349.

Ghez, A. M., Duchene, G., Matthews, K., *et al.* (2003). The first measurement of spectral lines in a short-period star bound to the Galaxy's central black hole: a paradox of youth. *ApJ Lett*, **586**, L127.

Ghez, A. M., Wright, S. A., Matthews, K., *et al.* (2004). Variable infrared emission from the supermassive black hole at the center of the Milky Way. *ApJ Lett*, **601**, L159.

Ghez, A. M., Salim, S., Hornstein, S. D., *et al.* (2005a). Stellar orbits around the Galactic center. *ApJ*, **620**, 744.

Ghez, A. M., Hornstein, S. D., Lu, J. R., *et al.* (2005b). The first laser guide star adaptive optics observations of the Galactic center: Sgr A*'s infrared color and the extended red emission in its vicinity. *ApJ*, **635**, 1087.

Ghez, A., Salim, S., Weinberg, N. N., *et al.* (2008). Measuring distance and properties of the Milky Way's central supermassive black hole with stellar orbits. *ApJ*, **689**, 1044.

Gillessen, S., Eisenhauer, F., Trippe, S., *et al.* (2009). Monitoring stellar orbits around the massive black hole in the Galactic center. *ApJ*, **692**, 1075.

Haller, J. W., Rieke, M. J., Rieke, G. H., *et al.* (1996). Stellar kinematics and the black hole in the Galactic center. *ApJ*, **456**, 194.

Hardy, J. W. (1991). Adaptive optics – a progress review. *Proc. SPIE*, **1542**, 2.

Hartkopf, W. I., McAlister, H. A., and Franz, O. G. (1989). Binary star orbits from speckle interferometry. II – Combined visual–speckle orbits of 28 close systems. *AJ*, **98**, 1014.

Henry, T. J. and McCarthy, D. W., Jr. (1990). A systematic search for brown dwarfs orbiting nearby stars. *ApJ*, **350**, 334.

Henry, T. J. and McCarthy, D. W., Jr. (1993). The mass-luminosity for stars of mass 1.0 to 0.08 solar mass. *AJ*, **106**, 773.

Hornstein, S. D., Matthews, K., Ghez, A. M., *et al.* (2007). A constant spectral index for Sagittarius A* during infrared/X-ray intensity variations. *ApJ*, **667**, 900.

Köhler, R., Kunkel, M., Leinert, C., and Zinnecker, H. (2000). Multiplicity of X-ray selected T Tauri stars in the Scorpius–Centaurus OB association. *A&A*, **356**, 541.

Konopacky, Q. M., Ghez, A. M., Barman, T. S., *et al.* (2010). High-precision dynamical masses of very low mass binaries. *ApJ*, **711**, 1087.

Labeyrie, A. (1970). Attainment of diffraction limited resolution in large telescopes by Fourier analysing speckle patterns in star images. *A&A*, **6**, 85.

Lacy, J. H., Townes, C. H., Geballe, T. R., and Hollenbach, D. J. (1980). Observations of the motion and distribution of the ionized gas in the central parsec of the Galaxy. II. *ApJ*, **241**, 132.

Leinert, C., Zinnecker, H., Weitzel, N., *et al.* (1993). A systematic approach for young binaries in Taurus. *A&A*, **278**, 129.

Leinert, C., Richichi, A., and Haas, M. (1997a). Binaries among Herbig Ae/Be stars. *A&A*, **318**, 472.

Leinert, C., Henry, T., Glindemann, A., and McCarthy, D. W., Jr. (1997b). A search for companions to nearby southern M dwarfs with near-infrared speckle interferometry. *A&A*, **325**, 159.

Liu, M. C., Dupuy, T. J., and Ireland, M. J. (2008). Keck laser guide star adaptive optics monitoring of 2MASS J15344984-2952274AB: first dynamical determination of a binary T dwarf. *ApJ*, **689**, 436.

Lohmann, A. W., Weigelt, G., and Wirnitzer, B. (1983). Speckle masking in astronomy – triple correlation theory and applications. *Appl. Opt.*, **22**, 4028.

Lu, J. R., Ghez, A. M., Hornstein, S. D., Morris, M. R., Becklin, E. E., and Matthews, K. (2009). A disk of young stars at the Galactic center as determined by individual stellar orbits. *ApJ*, **690**, 1463.

Maoz, D., Sternberg, A., and Ho, L. C. (1998). "Super star clusters" revealed in NICMOS images of circumnuclear rings. *A&AS*, **193**, 7604.

Marois, C., Macintosh, B., Barman, T., *et al.* (2008). Direct imaging of multiple planets orbiting the star HR 8799. *Science*, **322**, 1348.

Mason, B. D., Douglass, G. G., and Hartkopf, W. I. (1999). Binary star orbits from speckle interferometry. I. Improved orbital elements of 22 visual systems. *AJ*, **117**, 1023.

Mason, B. D., Gies, D. R., Hartkopf, W. I., *et al.* (1998). ICCD speckle observations of binary stars. XIX – An astrometric/spectroscopic survey of O stars. *AJ*, **115**, 821.

Mason, B. D., Hartkopf, W. I., and Wycoff, G. L. (2008). Speckle interferometry at the US Naval Obervatory. XIV. *AJ*, **136**, 2223.

Max, C. E., Avicola, K., Brase, J. M., *et al.* (1994). Design, layout, and early results of a feasibility experiment for sodium-layer laser-guide-star adaptive optics. *J. Opt. Soc. Am. A*, **11**, 813.

Max, C. E., Olivier, S. S., Friedman, H. W., *et al.* (1997). Image improvement from a sodium-layer laser guide star adaptive optics system. *Science*, **277**, 1649.

McAlister, H. A. (1977). Speckle interferometric measurements of binary stars. I. *ApJ*, **215**, 159.

McAlister, H. A., Hartkopf, W. I., Hutter, D. J., and Franz, O. G. (1987). ICCD speckle observations of binary stars. II – Measurements during 1982–1985 from the Kitt Peak 4 m telescope. *AJ*, **93**, 688.

McAlister, H. A., Mason, B. D., Hartkopf, W. I., Roberts, L. C., Jr., and Shara, M. M. (1996). ICCD speckle observations of binary stars. XIV. A brief survey for duplicity among white dwarf stars. *AJ*, **112**, 1169.

McCarthy, D. W., Jr., Henry, T. J., McLeod, B., and Christou, J. C. (1991). The low-mass companion of Gliese 22A – first results of the Steward Observatory infrared speckle camera. *AJ*, **101**, 214.

McGinn, M. T., Sellgren, K., Becklin, E. E., and Hall, D. N. B. (1989). Stellar kinematics in the Galactic center. *ApJ*, **338**, 824.

Meyer, L., Do, T., Ghez, A., *et al.* (2008). A 600 minute near-infrared light curve of Sagittarius A*. *ApJ Lett*, **688**, L17.

Moretti, A., Piotto, G., Arcidiacono, C., *et al.* (2009). MCAO near-IR photometry of the globular cluster NGC 7388: MAD observations in crowded fields. *A&A*, **493**, 539.

O'Dell, C. R. and Wen, Z. (1994). Postrefurbishment mission Hubble Space Telescope images of the core of the Orion Nebula: Proplyds, Herbig-Haro objects, and measurements of a circumstellar disk. *ApJ*, **436**, 194.

O'Dell, C. R., Wen, Z., and Hu, X. (1993). Discovery of new objects in the Orion nebula on HST images – shocks, compact sources, and protoplanetary disks. *ApJ*, **410**, 696.

Patience, J., Ghez, A. M., Reid, I. N., Weinberger, A. J., and Matthews, K. (1998). The multiplicity of the Hyades and its implications for binary star formation and evolution. *AJ*, **115**, 1972.

Patience, J., Ghez, A. M., Reid, I. N., and Matthews, K. (2002). A high angular resolution multiplicity survey of the open clusters α Persi and Praesepe. *AJ*, **123**, 1570.

Paumard, T., Genzel, R., Martins, F., *et al.* (2006). The two young star disks in the central parsec of the Galaxy: properties, dynamics, and formation. *ApJ*, **643**, 1011.

Peterson, D. M., and Solensky, R. (1988). 51 Tauri and the Hyades distance modulus. *ApJ*, **333**, 256.

Petr, M. G., Coude Du Foresto, V., Beckwith, S. V. W., Richichi, A., and McCaughrean, M. J. (1998). Binary stars in the Orion Trapezium cluster core. *ApJ*, **500**, 825.

Preibisch, T., Balega, Y., Hofmann, K.-H., Weigelt, G., and Zinnecker, H. (1999). Multiplicity of the massive stars in the Orion Nebula cluster. *New Astron.*, **4**, 531.

Roddier, F. J., Anuskiewicz, J., Graves, J. E., Northcott, M. J., and Roddier, C. A. (1994). Adaptive optics at the University of Hawaii I: current performance at the telescope. *Proc. SPIE*, **2201**, 2.

Schödel, R., Ott, T., Genzel, R., *et al.* (2002). A star in a 15.2-year orbit around the supermassive black hole at the centre of the Milky Way. *Nature*, **419**, 694.

Schödel, R., Ott, T., Genzel, R., *et al.* (2003). Stellar dynamics in the central arcsecond of our Galaxy. *ApJ*, **596**, 1015.

Sellgren, K., McGinn, M. T., Becklin, E. E., and Hall, D. N. B. (1990). Velocity dispersion and the stellar population in the central 1.2 parsecs of the Galaxy. *ApJ*, **359**, 112.

Stolte, A., Ghez, A. M., Morris, M., *et al.*, (2008). The proper motion of the Arches cluster with Keck laser-guide star adaptive optics. *ApJ*, **675**, 1278.

Torres, G., Stefanik, R. P., and Latham, D. W. (1997). The Hyades binaries Theta 1 Tauri and Theta 2 Tauri: the distance to the cluster and the mass-luminosity relation. *ApJ*, **485**, 167.

Weigelt, G. P. (1977). Modified astronomical speckle interferometry 'speckle masking'. *Opt. Commun.*, **21**, 55.

Weigelt, G., Balega, Y., Preibisch, T., *et al.* (1999). Bispectrum speckle interferometry of the Orion Trapezium stars: detection of a close (33 mas) companion to Theta (1) ORI C. *A&A*, **347**, L15.

11 Optical interferometry

ANDREAS GLINDEMANN

Introduction

Less than 300 years after Galilei's first telescope observations of celestial objects, Fizeau (1868) suggested a way to improve the measurement of stellar diameters by masking the telescope aperture with two small sub-apertures. Light passing through these sub-apertures would then interfere in the telescope focal plane. The first successful measurement using this principle was performed on Mt. Wilson in 1920 by Michelson and Pease (1921) who determined the diameter of α Orionis to be 0.047 arcsec. This was at a time when the smallest diameter that could be measured with a full aperture was about 1 arcsec, equivalent to the angular resolution of the telescope when observing through atmospheric turbulence.

Although the measurement of a stellar diameter is not the same as an image, the dramatic increase in angular resolution sparked enough interest in the new method that it was soon understood how such contrast measurements with different pairs of sub-apertures – different in separation and orientation – can be combined to form a high-resolution image not only of stars but of any type of object.

However, due to insurmountable technical problems with the mechanical stability at larger separations of the sub-apertures, optical[1] interferometry was abandoned in the late 1920s. It was not until 1974 that Labeyrie (1975) was able to combine the light from two independent telescopes at the Observatoire de la Côte d'Azur, demonstrating that optical interferometry was feasible.

While angular resolution increases linearly with the telescope diameter when eliminating atmospheric turbulence with adaptive optics, even today's largest telescopes cannot resolve features on the surface of individual stars. The diffraction limit is still so much larger than a star's disk that their images in the telescope focal plane are indistinguishable from point sources. For example, an angular resolution of 50 milliarcseconds (mas) on an 8-m telescope is only just about the angular size of Betelgeuse.

Combining individual telescopes to form an *optical interferometer*, the resolution is no longer determined by the individual telescope diameter but by the distance between the telescopes, called the *baseline B*. Combining two 8-m telescopes that are separated

[1] Throughout the text, *optical* refers to both visible and infrared wavelengths as long as the light is manipulated by optics.

Astrometry for Astrophysics: Methods, Models, and Applications, ed. William F. van Altena.
Published by Cambridge University Press. © Cambridge University Press 2013.

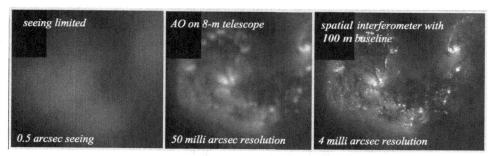

seeing limited

AO on 8-m telescope

spatial interferometer with
100 m baseline

0.5 arcsec seeing

50 milli arcsec resolution

4 milli arcsec resolution

Fig. 11.1   Comparison of the image quality on an 8-m telescope at an observing wavelength of 2 μm. On the left, the seeing-limited image is displayed for a seeing of 0.5 arcsec. In the middle, the diffraction-limited image of an 8-m telescope with adaptive optics is shown. On the right, the reconstructed image of an optical interferometer with a baseline of 100 m demonstrates the improvement over single-telescope observations. These images are only simulations for comparison of the achievable angular resolution. Especially for optical interferometers, an image with this amount of detail and this size has yet to be produced. From Glindemann (2011). With kind permission from Springer Science & Business Media.

by $B = 130$ m, like the telescopes of European Southern Observatory's (ESO's) Very Large Telescope Interferometer (VLTI), improves the angular resolution by a factor of *baseline/telescope diameter* $= 130/8 = 16$ to about 3 mas. In this way (Domiciano de Souza *et al.* 2003), a large number of stars can be resolved revealing shapes that are not necessarily circular. However, for good image quality, many observations must be made at different baselines, length, and orientation (see Section 11.5.1), to obtain an image as in Fig. 11.1.

Astrometry also benefits from the interferometric combination of telescopes. The accuracy of the interferometric distance measurement is improved, to first order, by the same factor as the angular resolution over single telescope measurements since the location of the (narrow) fringes is used instead of the (wide) Airy disk to determine the distance. Using a single baseline, the projection of the distance vector on the baseline vector is determined. Thus, at least two measurements are needed to determine the distance vector.

This improvement, however, comes at a price. In optical interferometry as in adaptive optics (see also Chapter 10) a bright guide star is needed to eliminate the effects of atmospheric turbulence (see also Chapter 9) to observe faint objects. Since the guide star has to be within 1 arcmin of the faint object, i.e. within the *isoplanatic patch* (or *region*), the number of faint objects that can be imaged is limited to the immediate neighbourhood of bright stars. In astrometry, one of the two objects needs to be bright enough to be used as a guide star.

In the following, we will introduce the theory of image formation and astrometry in optical interferometers and the influence of atmospheric turbulence on this process, following Glindemann (2011).

11.1 Preliminaries and definitions

Light as an electromagnetic wave can be represented by a dimensionless scalar $v(\mathbf{r}, t)$, called the *optical disturbance*, that is proportional to one component, e.g. E_x, of the electric-field vector **E** of the electromagnetic wave. The orthogonal component, E_y, can be treated independently. It is customary to write $v(\mathbf{r}, t)$ as a complex quantity. However, the physically relevant part representing the electromagnetic wave is the real part.

A monochromatic, plane wave propagating along the z-axis of a Cartesian coordinate system is then described by

$$v(z, t) = v_0 e^{-i(2\pi \nu t - kz)} \tag{11.1}$$

with ν the frequency and λ the wavelength of the monochromatic wave, and $k = 2\pi / \lambda$.

To discuss the propagation of light in space it is convenient to introduce the time-independent dimensionless *amplitude* $V(z)$ at frequency ν so that the monochromatic optical disturbance can be written as

$$v(z, t) = V(z) e^{-i2\pi \nu t} \tag{11.2}$$

In the optical, with frequencies between 10^{13} and 10^{15} Hz ($\lambda = 0.3 - 30 \, \mu m$), there are no detectors that are able to temporally resolve the signal and measure the optical disturbance. Only at wavelengths around $10 \, \mu m$, *heterodyne techniques*, familiar from radio astronomy, permit us to measure the optical disturbance (Johnson *et al.* 1974). The penalty is the low sensitivity due to the required narrow bandwidth. Usually, it is the intensity that we measure with optical detectors.

The intensity is related to the *energy flow density* given by the Poynting vector $\mathbf{S} = \mathbf{E} \times \mathbf{H}$, where **H** is the magnetic-field vector. The time average of the Poynting vector is called the *flux* (in optical astronomy) or the *power flux density* (in radio astronomy) of the electromagnetic wave in units of W/m^2. The measurable quantity with an optical detector is the integral of the flux over the detector area, i.e. the power in units of watt.

In the notation with the dimensionless optical disturbance $v(z, t)$ we define the dimensionless *intensity* as the time average of the product vv^*

$$I(z) = \langle v(z, t) v^*(z, t) \rangle \tag{11.3}$$

The intensity is proportional to the flux and, thus, proportional to the signal that is measured with optical detectors.

In terms of the amplitude $V(z)$, the intensity, using Eqs. (11.2) and (11.3), is written as

$$I(z) = |V(z)|^2 \tag{11.4}$$

In radio astronomy (see Chapter 12) with frequencies between 10^7 and 10^{11} Hz the optical disturbance $v(z, t)$ can be measured directly as a voltage proportional to the electric-field vector, and, if required, the intensity can be determined by computing the product vv^* and passing it through a low-pass filter, equivalent to temporal averaging in an optical detector. Using frequencies at least 10^3 lower than in the optical, sensitivity is not an issue when using heterodyne techniques at radio frequencies (Thomson *et al.* 1986).

In the case of polychromatic radiation, we use the *spectral intensity* $I(z, \nu)$ with dimension $1/Hz$. It is proportional to the *flux density* or *spectral power flux density* (in radio astronomy) in units of $W/m^2/Hz$. A common unit for the flux density is 1 jansky (Jy) $= 10^{-26} \, W/m^2/Hz$.

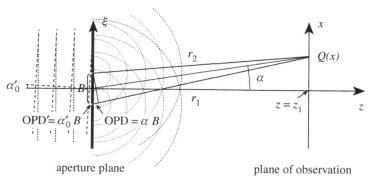

Fig. 11.2 The geometry of Young's experiment in a plane across the pinholes. A monochromatic plane wave from an on-axis point at a large distance illuminates the two pinholes separated by a distance B, called the *baseline*, in the aperture plane. For small angles, we have $\alpha = x/z_1$ with x the coordinate of Q, the point of observation. The difference $r_1 - r_2$ is called the optical path difference (OPD), with OPD $= \alpha B$ for small α. The OPD is related to the difference in arrival time τ between the light from the two pinholes by $\tau =$ OPD/c. A plane wave from a point source at α'_0 with external OPD, $\alpha'_0 B$, is also displayed. From Glindemann (2011). With kind permission from Springer Science & Business Media.

11.2 Principles of optical interferometry

Young's experiment in 1802 is the classical diffraction experiment that, in its simplicity, is perfectly suited to explain the concept of optical interferometry. It provides the experimental cornerstone for a demonstration of the wave nature of light.

Young illuminated a screen with two pinholes with light from a single point source at a large distance. On passing through the pinholes the light was diffracted and the waves from the two pinholes interfered. On a second screen, in the plane of observation, the diffraction pattern could be observed, displaying the characteristic fringe pattern. In Fig. 11.2, the experiment is depicted schematically for an illuminating point source at very large distance from the screen so that an approximately plane wave, with $V(\xi) = V_0$, illuminates the two pinholes at $z = 0$. A plane wave from a light source at an angle α'_0 is also displayed.

The amplitude V at point Q as a function of diffraction angle α in the plane of observation at distance z_1 is then the sum of the two spherical waves originating from the pinholes,

$$V(\alpha) = \frac{V_0}{r_1} e^{ikr_1} + \frac{V_0}{r_2} e^{ikr_2}$$
$$= \frac{V_0}{z_1} e^{ik(r_1+r_2)/2} \, 2\cos\big(k(r_1 - r_2)/2\big) \qquad (11.5)$$

where r_i is the distance between an individual pinhole and the point Q, with the approximation $r_1 = r_2 = z_1$ for the amplitudes V_0/r_i of the spherical waves. The difference between the optical path lengths, $r_1 - r_2$, is called the *optical path difference*, OPD.

The intensity is the squared modulus of the amplitude (11.4),

$$I(\alpha) = |V(\alpha)|^2 = \left(\frac{V_0}{z_1}\right)^2 2\big(1 + \cos\left(k(r_1 - r_2)\right)\big) = I_0\big(1 + \cos\left(k\alpha B\right)\big), \quad (11.6)$$

with $I_0 = 2(\frac{V_0}{z_1})^2$. This intensity distribution is called the *fringe pattern*, with a fringe spacing, the distance between the maxima of the cosine function, of λ/B (see Fig. 11.3a).

For a light source in direction α_0' we have to add the external OPD, $\alpha_0'B$, to $r_1 - r_2$ obtaining the total OPD, $(\alpha + \alpha_0')B$; see Fig. 11.2. This results in a fringe pattern that is shifted by α_0', yielding

$$I(\alpha) = I_0\left(1 + \cos\big(k(\alpha + \alpha_0')B\big)\right) \quad (11.7)$$

The OPD, $r_1 - r_2 = \alpha B$, is at the same time a difference in arrival time called the *time delay* τ between the light from the two pinholes, with $\tau = \alpha B/c$ and c the speed of light. We will see in the following that the diffraction pattern for increasing diffraction angles α – corresponding to increasing time delays τ – is constrained by the temporal coherence of the incoming light.

11.2.1 The visibility

The *visibility*, $\mathcal{V}$, is defined as the contrast of the fringe pattern. Monochromatic fringe patterns (11.6) have excellent contrast since the intensity oscillates between 0 and 1. This can be expressed more formally by writing the visibility as

$$\mathcal{V} = \frac{I_{max} - I_{min}}{I_{max} + I_{min}} \quad (11.8)$$

With $I_{min} = 0$ and $I_{max} = 1$, the contrast of the fringe pattern is $\mathcal{V} = 1$.

A contrast of 1 in a fringe pattern, i.e. perfect constructive and destructive interference in its maxima and minima, implies that the light waves from the two pinholes are perfectly coherent. In fact, the plane monochromatic wave illuminates the screen with the two pinholes and, thus, the light waves emerging from the pinholes are perfectly coherent.

If there is no coherence between the light from the pinholes, there is no fringe pattern but only a homogeneous illumination as a result of the diffraction of light at each individual aperture. Then, the visibility is zero and the light is called *incoherent*.

11.2.2 Temporal coherence

If we observe a source with a finite *spectral bandwidth* $\Delta\nu$ instead of a monochromatic source, the resulting fringe pattern is formed by adding up the interference patterns like (11.6) at different frequencies. This is displayed in Fig. 11.3b for the K-band, with $2.2 \pm 0.2\,\mu$m, $\Delta\lambda = 0.4\,\mu$m, and a baseline B of 100 m. For zero OPD, at $\alpha = 0$, all wavelengths have an intensity maximum. That is why this fringe is called the *white-light fringe*. The position of the first minimum $\alpha_{min} = \lambda/(2B)$ or $OPD = \lambda/2$ is then wavelength-dependent, as are the positions of the following maxima and minima.

This effect reduces the contrast of the resulting polychromatic fringe pattern (black curve in Fig. 11.3b) for increasing diffraction angles α. Since α is related to the difference in arrival time, the time delay τ, through $\tau = \alpha B/c$, this effect can be reformulated by stating that the contrast of the resulting fringe pattern is reduced with increasing τ. The time delay that is related to the quasi-loss of fringe contrast is called the *coherence time* τ_c, which is proportional to the reciprocal of the spectral bandwidth $\Delta\nu$. Consequently, the coherence length is defined as $\tau_c c$.

For the K-band, the coherence length is about 12 μm. Thus, if the OPD is larger than the coherence length, corresponding to a diffraction angle larger than 25 mas at a baseline of 100 m, there are no longer fringes (see Fig. 11.3b).

11.2.3 Spatial coherence

While the spectral bandwidth determines the temporal coherence, affecting the decrease of fringe contrast with increasing OPD, the size of the light source affects the fringe contrast of all fringes, which is determined by the spatial coherence of the light in the aperture plane.

First, we assume that the source is spatially incoherent, meaning that each point on its surface radiates independently, i.e. uncorrelated, from its neighbor. A thermal source like a star is a typical example of an incoherent source. We then compute the monochromatic fringe pattern for each source point using Eq. (11.6). Having an incoherent light source we add up the individual fringe patterns, i.e. the intensity distributions, that are slightly shifted with respect to each other. This operation is a convolution of the source intensity $I(\alpha')$ with the fringe pattern of an on-axis point source (11.6), which yields

$$
\begin{aligned}
I(\alpha) &= \int I(\alpha') \left(1 + \cos\left(k(\alpha + \alpha')B\right)\right) d\alpha' \\
&= \int I(\alpha')d\alpha' + \int I(\alpha') \cos\left(k(\alpha + \alpha')B\right) d\alpha' \\
&= I_0 + \mathrm{Re}\left(\int I(\alpha')e^{-ik\alpha'B} d\alpha' \, e^{-ik\alpha B}\right)
\end{aligned}
\tag{11.9}
$$

when I_0 is the integral over the source intensity. The second term is the real part of the complex Fourier transform of the source-intensity distribution $I(\alpha')$.

In Young's experiment, the baseline B is the distance between the pinholes. Discussing now a Fourier transform between the coordinate spaces α' and B, we have to interpret B as a coordinate that is the distance between any two points in the aperture plane.

In two-dimensional planes, we use the coordinate vector $\boldsymbol{\alpha}'$ in the source plane – the sky plane when observing celestial objects – and the difference coordinate $\mathbf{R} = \mathbf{B}/\lambda$ in the aperture plane. The baseline vector $\mathbf{R} = (u, v)$, calibrated by the wavelength, is formally the spatial frequency vector in the (u, v)-plane.

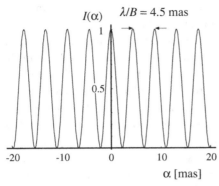

(a) point source, monochromatic

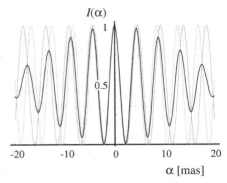

(b) point source, K-band

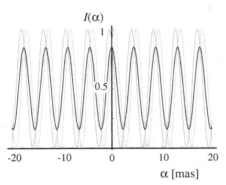

(c) extended source, monochromatic

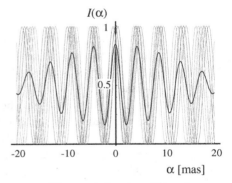

(d) extended source, K-band

Fig. 11.3 Summary of the influence of source size and spectral bandwidth on the fringe pattern. The pinhole separation, the baseline B, is 100 m in all figures. In (**a**) an individual fringe pattern for an observing wavelength of $\lambda = 2.2\ \mu m$ and a point source is displayed. In (**b**) the K-band fringe pattern is shown (black curve) when observing a point source. In (**c**) a monochromatic source with 2 mas diameter produces a fringe pattern (black curve) with reduced contrast. In (**d**) the resulting fringe pattern (black curve) for the K-band with a 2 mas source is displayed. The visibility is reduced around $\alpha = 0$ due to the source diameter and it is further reduced for increasing diffraction angles α due to the finite spectral bandwidth. From Glindemann (2011). With kind permission from Springer Science & Business Media.

Denoting the Fourier transform of the source intensity $I(\alpha')$ by $\mu(\mathbf{R})$, we obtain the monochromatic fringe pattern of an extended source, displayed in Fig. 11.3c, as

$$
\begin{aligned}
I(\alpha) &= I_0 \left(1 + \mathrm{Re}\left(\mu(\mathbf{R})e^{-i2\pi\,\alpha\cdot\mathbf{R}}\right)\right) \\
&= I_0\left(1 + |\mu(\mathbf{R})|\cos\left(\phi(\mathbf{R}) - 2\pi\,\alpha\cdot\mathbf{R}\right)\right)
\end{aligned} \tag{11.10}
$$

The contrast, i.e. the visibility, of the fringe pattern for a pair of pinholes separated by $\mathbf{B} = \lambda\mathbf{R}$ is determined by $|\mu(\mathbf{R})|$. Therefore, $\mu(\mathbf{R})$ is called the *visibility function* – in general a complex function with phase $\phi(\mathbf{R})$. It is straightforward to see that for $|\mu(\mathbf{R})| = 1$ the fringe contrast is 1, and that for a value of zero there is no fringe pattern, i.e. the contrast is zero. The phase $\phi(\mathbf{R})$ defines the position of the central fringe, called the *white-light fringe*.

When we observe with an optical interferometer, it is exactly the two quantities, fringe visibility $|\mu(\mathbf{R})|$ and its phase $\phi(\mathbf{R})$, that we are trying to determine.

The definition of the visibility function

$$\mu(\mathbf{R}) = \frac{\int I(\boldsymbol{\alpha}')e^{-i2\pi \boldsymbol{\alpha}' \cdot \mathbf{R}}d\boldsymbol{\alpha}'}{I_0} \tag{11.11}$$

is known as the *van Cittert-Zernike theorem*.

Due to the intensity distribution $I(\boldsymbol{\alpha}')$ being real and positive by definition, the modulus $|\mu(\mathbf{R})|$ of the visibility is symmetric and its phase $\phi(\mathbf{R})$ is anti-symmetric.

The two formulae (11.10) and (11.11) are the cornerstones of optical interferometry. They link the intensity distribution in the object $I(\boldsymbol{\alpha}')$ through a Fourier transform, (11.11), to the visibility of the fringe pattern (11.10). The inversion of this argument – measuring the visibilities for many different baseline vectors $\mathbf{R}$, i.e. for many points in the (u, v)-plane, and reconstructing the object intensity $I(\boldsymbol{\alpha}')$ through a Fourier back transform of $\mu(\mathbf{R})$ – was the important step to take optical interferometry from the mere determination of stellar diameters to an imaging tool of extremely high angular resolution.

11.2.4 Quasi-monochromatic approximation

In taking the step from monochromatic objects to a finite spectrum like the K-band, we obtain the fringe pattern displayed in Fig. 11.3d combining the fringe pattern for a star with a diameter of 2 mas in Fig. 11.3c with that of the K-band spectrum in Fig. 11.3b. The visibility is reduced at $\alpha = 0$ due to the finite size of the source and is further reduced for an increasing diffraction angle α, respectively for increasing time delay $\tau = \boldsymbol{\alpha} \cdot \mathbf{B}/c$.

The visibility of the fringes around $\alpha = 0$, i.e. around the white-light fringe, is approximately independent of the spectrum and only affected by the object shape so that the fringe pattern for small diffraction angles is described to a good approximation by the expression for the monochromatic fringe pattern (11.10). Therefore it is called the *quasi-monochromatic approximation*.

This approximation also has a consequence for the maximum object size. The object must be significantly smaller than the fringe package, i.e. smaller than 25 mas in the K-band with a 100 m baseline. Otherwise, fringes with low contrast would overlap with high-contrast fringes affecting the resulting visibility due to the spectral bandwidth rather than due to the object shape. This effect is called *bandwidth smearing* and has been studied extensively in radio astronomy (Bridle and Schwab 1999).

When a real object is observed in the sky, its coordinate vector $\boldsymbol{\alpha}'_0$ moves continuously due to the rotation of the Earth. We have first to make sure that we know all internal optical paths with micrometer precision so that the OPD is smaller than the coherence length $\tau_c c$, which is about 12 µm for the K-band, in order to see fringes. Additionally, we need to dynamically compensate the external OPD, $\boldsymbol{\alpha}'_0(t) \cdot \mathbf{B}$, so that we maintain the white-light fringe at a constant position requiring a precision well below one wavelength as we will see in Section 11.7.1.

11.3 Visibility and mutual coherence function

11.3.1 Amplitude interferometry

In the context of Young's experiment we introduced the coherence as a phenomenon related to the contrast of the fringe pattern. Here, we will use a more general definition based on statistical properties regarding the visibility as the correlation function of optical disturbances, hence the term amplitude interferometry. In Section 11.3.2, we will then discuss intensity interferometry as a higher-order correlation function.

The second-order correlation function of the optical disturbance is called the *mutual coherence function* (MCF), which is a function of time difference and of spatial coordinates. Remembering that the optical disturbance is proportional to one component of the electrical-field vector, we could say that we determine the correlation of two electrical-field vectors at two points in space (two telescopes) and two moments in time.

The MCF is defined as a time average:

$$\Gamma(\xi_1 - \xi_2, \tau) := \langle v(\xi_1, t + \tau) v^*(\xi_2, t) \rangle \tag{11.12}$$

where ξ_i are the coordinate vectors and τ is the time difference. For a very narrow spectrum, the MCF is a very wide function with respect to τ, indicating good coherence (correlation) for time differences up to the coherence time τ_c, whereas a very wide spectrum makes the MCF a very narrow function.

Regarding the spatial coordinates, the MCF in general depends on the absolute coordinates ξ_i. For our purpose, we can regard the MCF as a function of the coordinate difference only.

Using the generalized van Cittert-Zernike theorem for a narrow spectral band $\Delta \nu$ in quasi-monochromatic approximation (Goodman 1985), we regard the MCF at $\tau = 0$. We find that the MCF in the aperture plane is the Fourier transform of the object intensity $I(\alpha')$.

We can now relate the MCF in the aperture plane to the visibility by

$$\Gamma(\mathbf{B}, 0) = \mu \left(\frac{\mathbf{B}}{\lambda_0} \right) I_0 = \mathcal{F} \left(I(\alpha') \right) \tag{11.13}$$

with $\mathbf{B} = \xi_1 - \xi_2$, λ_0 the mean wavelength, and $\mathcal{F}$ denoting the Fourier transform.

Thus, by measuring the contrast of the *intensity* distribution of the fringe pattern we determine the correlation of *optical disturbances* at two points in the aperture plane that are separated by the baseline vector.

In analogy to the temporal coherence, the MCF is a very wide function with respect to the spatial coordinate difference $\mathbf{B}$ if the object size, given by $I(\alpha')$, is very small. Then, the light is coherent over a wide area of the wavefront. And the MCF is a very narrow function when observing a relatively large object.

11.3.2 Intensity interferometry

In intensity interferometry, we measure intensity fluctuations at two points in space (two telescopes) and two moments in time (Hanbury Brown and Twiss 1956a,b). We start by discussing instantaneous intensities $i(\xi, t) = v(\xi, t)v^*(\xi, t)$ applying the time average afterwards (Goodman 1985).

The product of two instantaneous intensities consists of a product of four optical disturbances $v(\xi, t)$, and we calculate the intensity correlation as the fourth-order moment of optical disturbances that can be reduced to second-order moments using the common assumption that optical disturbances follow a circular Gaussian random process (Goodman 1985).

Since we are interested in the intensity fluctuations, we write the intensity covariance by subtracting the average intensities $\langle i(\xi_i, t_i) \rangle$ from the intensity correlation, as

$$
\begin{aligned}
\langle i(\xi_1, t_1) i(\xi_2, t_2) \rangle_{\mathrm{cov}} &= \langle i(\xi_1, t_1) i(\xi_2, t_2) \rangle - \langle i(\xi_1, t_1) \rangle \langle i(\xi_2, t_2) \rangle \\
&= \langle v(\xi_1, t_1) v^*(\xi_1, t_1) v^*(\xi_2, t_2) v(\xi_2, t_2) \rangle \\
&\quad - \langle v(\xi_1, t_1) v^*(\xi_1, t_1) \rangle \langle v^*(\xi_2, t_2) v(\xi_2, t_2) \rangle \\
&= \langle v(\xi_1, t_1) v^*(\xi_2, t_2) \rangle \langle v^*(\xi_1, t_1) v(\xi_2, t_2) \rangle \\
&= |\Gamma(\xi_1 - \xi_2, \tau)|^2
\end{aligned}
\tag{11.14}
$$

with τ the time difference $t_1 - t_2$. We find that the intensity covariance is the squared modulus of the MCF $\Gamma(\xi_1 - \xi_2, \tau)$, which is the correlation function of the optical disturbance (11.12).

Since we determine the squared modulus of the MCF we lose its phase. However, determining the triple correlation of intensities at three telescopes and three moments in time, $\langle i_1(t + \tau_1) i_2(t + \tau_2) i_3(t) \rangle$, we obtain the sum of MCF phases $\phi_{12}(\tau_1 - \tau_2) + \phi_{23}(\tau_2) + \phi_{31}(\tau_1)$ that is called the *closure phase* because it is the sum of phases around a closed loop of three telescopes. By repeating this measurement for a large number of configurations the phases ϕ_{ij} of the individual baselines can be recovered from the closure phase.

The intensity covariance in (11.14) represents the result if the instantaneous intensities could be determined with infinite temporal resolution. Measuring only a time-averaged intensity we apply a temporal average over a period T to the instantaneous intensities on the left-hand side of (11.14) and to the MCF on the right-hand side.

The width of the MCF with respect to τ is determined by the coherence time τ_c. Assuming a spectral resolution $\Delta \nu / \nu$ of $1/1000$, τ_c is of the order of 10^{-12} s. Since the temporal resolution T of the available detectors is typically 10^{-9} s, the averaging process attenuates the signal by $\tau_c / T = \Delta f / \Delta \nu$ compared to an ideal detector (Mandel 1963), where $\Delta f = 1/T$ is the detector bandwidth. Reducing the spectral bandwidth $\Delta \nu$ reduces the intensity so that the overall situation does not change.

Therefore, the signal-to-noise ratio (SNR) is independent of the spectral bandwidth $\Delta \nu$ but it is proportional to the *spectral* intensity of the source – strictly speaking to the number of detected photoelectrons per unit optical bandwidth and per unit time – to the squared visibility, and to the square roots of the detector bandwidth $\sqrt{\Delta f}$ and of the integration time $\sqrt{t}$. This limits the sensitivity for a 5σ SNR to stars of approximately magnitude 5 for 10-m class telescopes and 10-min observations. This does not compare very favourably to amplitude interferometry reaching stars of magnitude 10 in a few tens of milliseconds.

The advantage compared to amplitude interferometry, on the other hand, is the requirement on OPD that is now determined by the temporal resolution of the detectors. With our assumption of $T = 10^{-9}$ s, we find that the optical paths between the telescopes should be equal to within $T \times c = 30$ cm, which is a very relaxed situation compared to the Young experiment when we need an OPD smaller than a few wavelengths if we want to see fringes. This was a big advantage in the 1950s when the precise control of the OPD was technically impossible. Since this problem has been resolved in the meantime, the disadvantage of a low sensitivity due to the limited detector resolution has prevented any new attempts in intensity interferometry.

11.4 Fizeau and Michelson interferometry

In combining individual telescopes into an interferometer array, it is the beam combination scheme that determines the characteristics of the interferometer. The two extreme cases are the Fizeau configuration when the image plane of the beam combining instrument shows fringes in every object in the field of view, and the Michelson configuration when the fringes are confined more or less to the Airy disk on-axis.

11.4.1 Fizeau configuration

In the Fizeau configuration, the output pupil, i.e. the image of the telescope apertures inside the beam-combining instrument, is an exact replica of the interferometer array so that the telescope apertures and their distance are reduced by the same factor. This is called *homothetic mapping* of the telescopes in the array. Then, the object–image relationship is the same as in a single telescope describing the image intensity as a convolution of the object intensity with the point-spread function (PSF). Since the imaging process can be computed like that of a large telescope with an aperture mask containing sub-apertures according to size and position of the individual telescopes, the PSF consists of an Airy disk of an individual sub-aperture with a narrow fringe pattern.

A true Fizeau configuration is difficult to build since homothetic mapping of the telescope apertures is a dynamic process. While observing a source, the effective baseline, i.e. the baseline as seen from the source, changes constantly due to the rotation of the Earth requiring the image of the telescopes in the exit pupil to be in permanent motion. The Large Binocular Telescope (LBT; Hill *et al.* 2006), with its two 8.4-m primary mirrors mounted on the same telescope structure, comes very close to a Fizeau configuration because the baseline is by design perpendicular to the line of sight so that only static aperture remapping is required. Most other interferometers, with individual telescopes on the ground, do not attempt to remap the telescope apertures homothetically.

11.4.2 Michelson configuration

The arbitrary arrangement of telescope apertures in the exit pupil is typical for the Michelson configuration named after the original experiment by Michelson and Pease (1921). In this case the object–image relationship can no longer be described by a convolution since the positions of the Airy disks in the field no longer coincide with equal optical path lengths in the interferometer so that, on-axis, the PSF is an Airy disk with fringes, while elsewhere in the field of view the PSF does not show fringes. Only small on-axis objects can be treated as in the Fizeau configuration, applying Eq. (11.10) to describe the fringe pattern.

In the limiting case of a Michelson configuration when the individual telescope apertures are imaged on top of each other – also called *co-axial combination* – the output pupil appears as a single aperture, and the PSF is an Airy disk without fringes. The PSF is either bright if the OPD between the two telescopes is zero, or it is dark if the OPD is $\lambda/2$. Thus, we do not have a spatial fringe pattern in the image plane any longer.

Measuring the visibility requires a modulation of the OPD, "scanning" through the fringe pattern that is formed by the varying brightness of the Airy disks. Usually, the OPD modulation is provided by an opto-mechanical unit in the beam-combining instrument.

Writing the integral over the Airy disk as a function of time delay $\tau = \text{OPD}/c$ between the two telescopes, we can rewrite the intensity distribution of the fringe pattern (11.7), using $\tau = \alpha \cdot \mathbf{B}/c$, as

$$I(\tau) = I_0 \left(1 + \cos\left(2\pi \nu \tau \right) \right) \tag{11.15}$$

The width of this fringe pattern, i.e. the maximum time delay τ before fringes disappear, is given by the coherence time τ_c of the observed spectral bandwidth. If the time delay is larger than the coherence time, the intensity has the constant value I_0. Scanning the complete fringe pattern requires an OPD modulation d_0 that is larger than the coherence length $\tau_c c$.

The angular coordinate α'_0 corresponding to the width d_0 of this fringe package can be computed using $d_0 = \alpha'_0 \cdot \mathbf{B}$. For the K-band and a baseline of 100 m, α'_0 is about 25 mas, as in Young's experiment. Writing the fringe pattern as a function of time delay does not change this property.

This situation is very similar to radio astronomy (Chapter 12). However, while we have assumed that the permanent external OPD variations due to the diurnal motion of the source are dynamically compensated for and the internal OPD is modulated to obtain a fringe pattern, in radio astronomy the internal OPD is kept constant and the diurnal motion of the source provides the required OPD variation.

In the following, we will discuss the situation for a small source in the Michelson configuration and we will then discuss astrometry of separated objects.

11.5 Small objects – imaging

In Michelson interferometry, the object has to be smaller than the Airy disk of an individual telescope so that the fringe pattern is given by Eq. (11.10). The visibility $\mu(\mathbf{R})$ of the fringe pattern is then determined by the Fourier transform of the object intensity, according to the van Cittert Zernike theorem (11.11).

In case of a finite spectral bandwidth, we apply the quasi-monochromatic approximation so that the permitted object size is additionally restricted by the width of the fringe package, and the visibility of the central fringes is given by $\mu(\mathbf{R})$.

Observing objects off-zenith with individual telescopes on the ground, the baseline vector is no longer perpendicular to the line of sight and we have to apply the effective baseline vector $\mathbf{R}_{\text{eff}}$ that is the projection of $\mathbf{R}$ onto the sky plane, which is perpendicular to the line of sight. Its length, R_{eff}, is reduced to $\cos(\zeta)R$ with ζ the off-zenith angle.

We can now write the intensity distribution of the fringe pattern in a Michelson configuration as

$$I(\tau) = I_0 \left(1 + |\mu(\mathbf{R}_{\text{eff}})| \cos\left(\phi(\mathbf{R}_{\text{eff}}) - 2\pi\nu\tau\right)\right) \tag{11.16}$$

with τ the time delay that is introduced through the modulation of the OPD. In the quasi-monochromatic approximation, this formula describes the fringe pattern for time delays much smaller than the coherence time τ_{c}.

The visibility modulus and phase can be determined using the *ABCD method* (Shao and Staelin 1977). Measuring four intensity values that are separated by $\Delta T/4$, scanning one fringe period ΔT, we obtain

$$|\mu(\mathbf{R}_{\text{eff}})| = \frac{\pi}{\sqrt{2}} \frac{\sqrt{(A-C)^2 + (B-D)^2}}{n}$$

$$\phi(\mathbf{R}_{\text{eff}}) = \tan^{-1}\left(\frac{A-C}{B-D}\right) \tag{11.17}$$

with A, B, C, and D the integrated intensity values in the intervals $[0, \Delta T/4], [\Delta T/4, \Delta T/2]$ etc., and $n = A+B+C+D$, the total number of detected photons during the measurement. The phase of the visibility can only be determined modulo 2π unless measurements at different wavelengths are combined so that the white-light fringe position, the position of $\tau = 0$, can be identified unambiguously.

The SNR of this measurement is proportional to $\sqrt{n}$ and to the visibility. That means that a partially resolved star with a visibility of 0.5 needs to be four times brighter than a point source to have the same SNR.

Combining visibility measurements for many baselines $\mathbf{R}_{\text{eff}}$ permits us to reconstruct the object intensity $I(\alpha')$, in principle as the Fourier back-transform of $\mu(\mathbf{R}_{\text{eff}})$; see Eq. (11.11). Then the smallest detail that can be resolved in the object is determined by the longest baseline of all measurements, which in turn is equivalent to the finest fringe pattern with fringe spacing $\lambda/B_{\text{eff}} = 1/R_{\text{eff}}$.

One should note that both modulus and phase of the visibility are needed for the object reconstruction. Attempting to reconstruct an object without the phase of the visibility results in a reconstruction that is lacking all the asymmetric parts. A binary for instance would be reconstructed as a central star with two symmetric companions so that the distance can be determined but its orientation has a 180° ambiguity.

11.5.1 Coverage of the (u, v)-plane

The number of independent parameters in the reconstructed object intensity is determined by the number of visibilities (modulus and phase) that are measured. Thus, a reconstruction of a 10×10 pixel image of e.g. a galaxy requires about 100 visibility measurements. The visibilities can be collected either by having a convenient distribution of the effective baselines $\mathbf{R}_{\text{eff}}$ by observing over several hours, or by moving the telescopes to have more physical baselines available, or both. Interferometry with more than two telescopes improves the (u, v)-plane coverage since the simultaneous combination of N telescopes provides $N(N-1)/2$ baselines.

Unlike radio interferometry where the electrical-field vector is measured directly before the correlations are computed, in optical interferometry the beams are combined before the detection. Combining N telescopes can either be done pairwise, distributing the light from each telescope into $N-1$ beams, or in any combination of telescopes up to combining the complete set simultaneously. In the latter case, the visibility measurement of a particular baseline is affected by light from the $N-2$ other telescopes contributing to the background noise. Thus, multi-telescope observations, although providing a more efficient (u, v)-plane coverage, always have a lower sensitivity than individual two-telescope observations.

Applying the Fourier transform of the measured visibility to reconstruct the object intensity provides an image quality that, depending on the distribution of baselines, is full of sidelobes and negative values. In radio interferometry (Chapter 12), this image is called the *dirty map*. The problem of *synthesis imaging* – systematically filling the gaps in the (u, v)-plane to improve the image quality – has been treated in radio interferometry over the last decades (Thomson *et al.* 1986).

Since it is not easy to collect a sufficient number of visibilities with a small number of telescopes, almost all interferometric observations in the optical attempt to fit model parameters rather than reconstructing an image. The measurement of stellar diameters is a good example for this technique. By assuming that the shape of the star can be modelled as a uniform disk, a single measurement can be sufficient to determine the one parameter, the diameter α'_0, that we are looking for. Different orientations of the baselines can reveal the elliptical form of a star as in the case of Achernar (Domiciano de Souza *et al.* 2003).

More measurements allow us to further refine the model, for instance by including the limb-darkening effect, i.e. a slight reduction of the intensity towards larger radii of the disk.

11.6 Widely separated objects – astrometry

Objects that are more widely separated than the fringe package form separate fringe packages so that one cannot determine a single visibility value based on the object morphology.

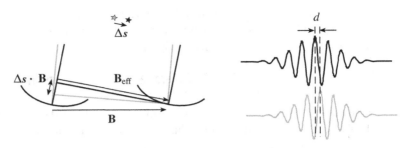

$$d = \Delta s \cdot \mathbf{B} + \phi(\mathbf{R}_{\text{eff}})\lambda/2\pi + \text{dOPD}_{\text{int}} + \text{dOPD}_{\text{turb}}$$

Fig. 11.4 Principle of astrometry with an interferometer. The differential delay d of the white-light fringe positions is determined by the differential external OPD $\Delta s \cdot \mathbf{B}$, by the phase of the visibility function $\phi(\mathbf{R}_{\text{eff}}) = \phi(\mathbf{B}_{\text{eff}}/\lambda)$, by the differential optical path fluctuations due to turbulence $\text{dOPD}_{\text{turb}}$, and by the differential internal optical path differences dOPD_{int}. From Glindemann (2011). With kind permission from Springer Science & Business Media.

Assuming a binary with one star on-axis and its companion at a distance Δs we have to compensate for an external OPD ($\Delta s \cdot \mathbf{B}$; see Eq. (11.7)) between them. Applying an internal OPD $d = -\Delta s \cdot \mathbf{B}$ we move from the fringe package of the on-axis star to the fringe package of the companion.

Thus, by precisely measuring the differential internal OPD, the differential delay d, we can determine the angular separation of the binary, or more precisely the projection of the separation vector Δs on the baseline vector $\mathbf{B}$. Repeating the measurement with a different baseline vector provides both the separation and the orientation of the binary.

In a real interferometer, the situation is not quite that simple. First, the objects are usually resolved by the individual telescopes forming separate Airy disks in the focal plane. One can use a *dual-feed system* to feed the light from each object through dedicated optical trains into individual beam-combining instruments. Each instrument then performs a measurement of the phase of the visibility as described in Eq. (11.17). The position of each white-light fringe is determined, providing the required internal OPD in each optical train to center the fringes at position $\tau = 0$.

Ideally, the differential delay, through $d = \Delta s \cdot \mathbf{B}$, provides a direct measure for the projection of the separation vector Δs on the baseline vector $\mathbf{B}$. However, the light, travelling through the atmosphere and through an extended optical system, suffers a number of delays, see Fig. 11.4, that need to be dealt with.

Four terms determine the differential delay d between the white-light fringe positions:

- the differential external OPD $\Delta s \cdot \mathbf{B}$,
- differential internal optical path differences between the two feeds dOPD_{int},
- the phase of the visibility $\phi(\mathbf{R}_{\text{eff}})$, and
- random differential fringe motion due to atmospheric turbulence, $\text{dOPD}_{\text{turb}}$.

The sum of the errors in each of these terms determines the achievable accuracy for the measurement of the angular separation. The largest error source is the optical path fluctuations due to atmospheric turbulence that limits the overall accuracy basically to

10 microarcseconds (μas; Shao and Colavita 1992), corresponding to an accuracy of $\sigma_d = 5$ nm for a 100-m baseline. This sets the limit for the error contributions of the individual terms.

We start by estimating the accuracy needed for the baseline length B. If we want to determine the separation vector with an accuracy of 10 μas for a 10 arcsec separation the length of the baseline vector, e.g. 100 m, needs to be known to about 100 μm, using $\partial \Delta s / \Delta s = \partial B / B$. This can be achieved by observing calibrator stars with known diameters.

The differential internal optical path difference, dOPD$_{\text{int}}$, needs to be monitored permanently with a laser metrology system with an accuracy of about 5 nm. The measured dOPD$_{\text{int}}$ between the stars and the telescopes is used as an offset for the differential delay compensating for internal path-length fluctuations due to imperfections in the optical paths. The required accuracy is achievable with available metrology systems.

The phase $\phi(\mathbf{R}_{\text{eff}})$ of the visibility of both objects is a-priori unknown. We have to make the assumption that they are both point-like so that the phases of their visibilities – even if they are slightly resolved – are zero. Only if their shape is asymmetric would there be a non-zero phase.

If one of the two objects does have a non-zero phase only a large number of measurements can be taken with different baselines to separate the distance term that is linear with the baseline, $\Delta s \cdot \mathbf{B}$, from the phase due to its shape.

On the other hand – moving from astrometry to imaging – these phase measurements provide the required information to reconstruct the object intensity. The remaining uncertainty in the exact separation Δs is irrelevant for the object reconstruction; the linear term moves the object laterally in the reconstruction without affecting its shape.

Thus, the dual-feed system provides not only an astrometric mode measuring the separation of the two objects but also an imaging mode, determining the shape of an object which, however, has to be smaller than the fringe package.

The fourth term of the differential delay, dOPD$_{\text{turb}}$, due to atmospheric turbulence, will be discussed in the following.

11.6.1 Atmospheric turbulence

The optical path difference fluctuations due to atmospheric turbulence (see also Chapter 9) have rather large values,

$$\sigma_{\text{OPD}} = 0.42 \, \lambda \, (B/r_0)^{5/6} [\text{m}] \tag{11.18}$$

amounting typically to 50 μm for a 100-m baseline and 0.7 arcsec seeing in the K-band. Here, r_0 is the Fried parameter, defined as the diameter of an atmospheric patch (or region) with 1 rad^2 variance of the wavefront aberrations. It is related to the seeing through λ/r_0. Observing a star with an interferometer means that the fringes in the K-band move about ± 23 fringes if we wait a sufficiently long time.

Due to the scaling of the Fried parameter with wavelength, $r_0 \propto \lambda^{6/5}$, the optical path length fluctuations are independent of wavelength while their impact on observations is determined by the phase fluctuations, $\sigma_\phi = \frac{2\pi}{\lambda} \sigma_{\text{OPD}}$. Thus, their impact is substantially larger at short wavelengths than at long wavelengths.

At radio frequencies, the path-length fluctuations are about the same as in the optical so that their impact on the phase fluctuations is rather benign.

When observing two objects, we are concerned about the differential motion of the fringes, i.e. about the astrometric position fluctuations $\sigma_{\Delta s}$, with $\Delta s_{\text{turb}} = \text{dOPD}_{\text{turb}}/B$. For large separations, about $1°$, $\sigma_{\Delta s}$ is independent of baseline and increases only slowly with angular separation since the optical paths are almost uncorrelated. For smaller separations, corresponding to a lateral separation of the beams in the upper atmosphere that is substantially smaller than the baseline, this behaviour changes substantially (Shao and Colavita 1992). For this *narrow-angle astrometry*, $\sigma_{\Delta s}$ increases linearly with the angular separation Δs and decreases with increasing baseline, according to $B^{-2/3}$.

For a 100-m baseline, the separation has to be much smaller than about 30 arcmin to enter into the narrow-angle regime. However, this assumes that both objects are sufficiently bright to be observed with short exposure times so that each exposure has fringes with good contrast. The separation will then be computed by averaging the fringe positions in all exposures of each object, obtaining the separation as the difference of the means.

If one of the two objects is faint requiring a long exposure time to integrate the fringe patterns on the detector, it is required that the fringe motion is sufficiently small not to wipe out the fringes. Then, the object separation has to be much smaller than 30 arcmin to ensure a sufficient correlation of the fringe motion as will be discussed in Section 11.7.1. This separation angle is called the isoplanatic angle, typically on the order of 1 arcmin.

In narrow-angle astrometry, the residual root-mean-square (rms) error of the angular separation due to atmospheric turbulence is

$$\sigma_{\Delta s} = C \Delta s \, B^{-2/3} \, t^{-1/2} \tag{11.19}$$

with the parameter C depending on the turbulence profile and on the seeing at the observing site, and with t the integration time (Shao and Colavita 1992). For a separation of $\Delta s = 10$ arcsec, a 100-m baseline and a seeing of about 1 arcsec, the rms error can be reduced to $10 \, \mu$as after about one hour of observations. See also Chapter 9 for an overall discussion of the precision limits for astrometry set by the atmosphere.

An accuracy of $100 \, \mu$as was achieved with the Palomar Testbed Interferometer (PTI) with a baseline of 110 m, measuring the angular separation of the two components in 61 Cygni with $\Delta s = 31$ arcsec (Lane *et al.* 2000). Also at the PTI, binaries with a separation of some hundred mas – within one Airy disk and not requiring a dual-feed system – were measured with an accuracy of $20 \, \mu$as (Muterspaugh *et al.* 2006).

11.7 Beating atmospheric turbulence

11.7.1 Fringe tracking

Atmospheric turbulence limits the sensitivity of an interferometer typically to magnitudes 10–12 on 10-m class telescopes since the exposure time has to be shorter than the atmospheric coherence time of some 10 ms. For that case, the fringes do not move more than about 1 rad, i.e. $1/2\pi$ of the fringe spacing, corresponding to a σ_{OPD} of 0.35 μm in the K-band, so that the visibility is reduced by less than 40%, permitting a reliable measurement.

Fainter objects can be observed if the fringe motion is reduced by measuring and stabilizing the fringes on a nearby bright guide star. This fringe tracking system limits the fringe motion of the faint object to the differential optical path fluctuations between the two objects. The differential fringe motion has to be much smaller than the fringe spacing to ensure good fringe contrast when integrating over several seconds. This defines the isoplanatic angle, typically 1 arcmin, that corresponds to a lateral separation of the beams in the upper atmosphere of about the Fried parameter r_0.

Using a dual-feed system with laser metrology to monitor all internal optical paths, as discussed in Section 11.6, the white-light fringe positions of the guide star and of the science object can be defined so that the phase $\phi(\mathbf{R}_{eff})$ of the visibility can be determined in addition to its modulus, providing an imaging mode for faint objects.

With a stable fringe pattern, the integration time on a faint science object can be increased to tens of seconds and the sensitivity is then improved by about seven magnitudes. However, a bright guide star of about magnitude 10–12 within 1 arcmin of the science object is required.

11.7.2 Closure phase

While the fringe position is completely random due to atmospheric turbulence when observing with two telescopes, adding a third telescope and measuring the visibilities on three baselines permits us to eliminate the influence of the turbulence by adding up the phases of all three baselines.

If we regard the random optical path difference fluctuations OPD_{turb} on each baseline $\mathbf{R}_i$ as the difference of the individual optical paths to each telescope, we find that the sum of the three $OPD_{turb}(\mathbf{R}_i)$ around a closed loop of three telescopes amounts to zero.

If the observed object has a non-zero phase, the measured phase on each baseline is the sum of the object phase $\phi(\mathbf{R}_i)$ plus a random term, $OPD_{turb}(\mathbf{R}_i)$. Adding up the measured phases of the three baselines then eliminates the random OPD fluctuations obtaining the closure phase as $\phi(\mathbf{R}_1) + \phi(\mathbf{R}_2) + \phi(\mathbf{R}_3)$.

It should be noted that the linear term of the phase, determining the source position, is also removed in the closure phase.

Combining closure phases of observations with different baselines permits us to reconstruct the phases of the individual baselines $\phi(\mathbf{R}_i)$. In intensity interferometry (Section 11.3.2) the closure phase is also employed to recover the object phase.

11.8 Optical interferometers and observatories

Following Labeyrie's successful demonstration of optical interferometry, there were several projects in the 1980s and 1990s, at the Observatoire de la Côte d'Azur in France (GI2T), in the USA (Mark I–III and CHARA at Mt. Wilson, NPOI in Flagstaff, PTI at Palomar, IOTA at Mt. Hopkins), SUSI in Narrabri, Australia, COAST in Cambridge, UK, and MIRA at the Mitaka Campus of the National Astronomical Observatory of Japan. These interferometers, with apertures between 0.4 and 1.5 m and baselines ranging from 2 to 640 m, have been operated in the visible or in the near-infrared.

The noticeable exception is the Infrared Spatial Interferometer (ISI) at Mt. Wilson, a three-telescope interferometer using heterodyne detection at an observing wavelength between 9 and 12 μm (N-band) with a maximum baseline of 56 m (Hale *et al.* 2000).

The Mark I interferometer was the first to demonstrate active fringe tracking (Shao and Staelin 1980), and the PTI had the first dual-feed system in operation (Colavita *et al.* 1999). The first closure-phase image was produced by the Cambridge Optical Aperture Synthesis Telescope (COAST; Baldwin *et al.* 1996).

In the meantime, the GI2T, the Mark I–III interferometers, the PTI, and the IOTA have been taken out of operation.

The scientific success of these interferometers was sufficient motivation to equip some of the world's largest telescopes with an interferometric mode. The Keck Observatory with two 10-m telescopes at Mauna Kea, Hawaii, and ESO's VLT observatory with four 8-m Unit Telescopes (UTs) at Paranal, Chile, combine the advantage of large apertures with that of long baselines. Since the telescope apertures are substantially larger than the Fried parameter r_0, they are all equipped with adaptive-optics systems.

The Keck Interferometer has a single baseline of 85 m, while the six baselines of the VLTI range from 47 to 130 m. In addition, the VLTI has four 1.8-m Auxiliary Telescopes (ATs) that can be moved on a grid providing baselines from 8 to 203 m.

Both the Keck Interferometer and the VLTI had first fringes with siderostats within 4 weeks early in 2001 (Colavita and Wizinowich 2002, Glindemann *et al.* 2002). Very quickly thereafter, a number of instruments were deployed in the near- and mid-infrared producing scientific results (Colavita *et al.* 2003, Kervella *et al.* 2003).

Currently, a dual-feed system is being installed at the VLTI, adding a Phase Referenced Imaging and a Micro-arcsecond Astrometry (PRIMA) mode (Delplancke *et al.* 2006). This two-telescope system provides the modulus and the phase of the visibility on any one of the six baselines, with either the 8-m telescopes or with the ATs, for the near- and mid-infrared instruments.

In the astrometric mode, two identical instruments measure the fringe position of each star, with a maximum separation of 1 arcmin (Launhardt *et al.* 2008). The goal for the accuracy is 10 μas on ATs – close to the atmospheric limit for 1-hour observations – and about 50 μas on UTs. The better-controlled environment on the ATs, basically without vibrations, ensures better performance.

The second generation of VLTI instruments will see four-beam instruments with six simultaneous baselines: MATISSE in the mid-infrared (Lopez *et al.* 2008), and GRAVITY in the near-infrared (Eisenhauer *et al.* 2008). The latter will provide narrow-angle astrometry and phase-referenced imaging with an accuracy of 10 μas on the UTs for objects separated up to 2 arcsec. The prime scientific goal is the Galactic center, studying motions to within a few times the radius of the event horizon and potentially testing general relativity in its strong field limit.

All of these interferometers are operated in Michelson configuration so that the field of view is reduced to less than an Airy disk. It is only the Large Binocular Telescope (LBT) that will use the Fizeau configuration providing a 20 arcsec field of view. The first Fizeau fringes were achieved in October 2010.

References

Baldwin, J. E., Beckett, M. G., Boysen, R. C., *et al.* (1996). The first images from an optical aperture synthesis array: mapping of Capella with COAST at two epochs. *A&A*, **306**, L13.

Bridle, A. H. and Schwab, F. R. (1999). Bandwidth and time-average smearing. *ASP Conf. Ser.*, **180**, 371.

Colavita, M. M. and Wizinowich, P. (2002). Keck Interferometer update. *Proc. SPIE*, **4838**, 79.

Colavita, M. M., Wallace, J. K., Hines, B. E., *et al.* (1999). The Palomar Testbed Interferometer. *AJ*, **510**, 505.

Colavita, M. M., Akeson, R., Wizinowich, P., *et al.* (2003). Observations of DG Tauri with the Keck Interferometer, *AJ*, **592**, L83.

Delplancke, F., Derie, F., Lévêque, S., *et al.* (2006). PRIMA for the VLTI: a status report. *Proc. SPIE*, **6268**, 62680U-1.

Domiciano de Souza, A., Kervella, P., Jankov, S., *et al.* (2003). The spinning-top Be star Achernar from VLTI-VINCI. *A&A*, **407**, L47.

Eisenhauer, F., Perrin, G., Brandner, W., *et al.* (2008). GRAVITY: getting to the event horizon of Sgr A*. *Proc. SPIE*, **7013**, 70132A-1.

Fizeau, H. (1868). Prix Borodin: rapport sur le concours de l'année 1867. *C. R. Acad. Sci.*, **66**, 932.

Glindemann, A. (2011). *Principles of Stellar Interferometry*. Berlin: Springer-Verlag.

Glindemann, A., Argomedo, J., Amestica, R., *et al.* (2002). The VLTI – a status report. *Proc. SPIE*, **4838**, 89.

Goodman, J. W. (1985). *Statistical Optics*. New York, NY: J. Wiley & Sons.

Hale, D. D. S., Bester, M., Danchi, W. C., *et al.* (2000). The Berkeley Infrared Spatial Interferometer: a heterodyne stellar interferometer for the mid-infrared. *AJ*, **537**, 998.

Hanbury Brown, R. and Twiss, R. Q. (1956a). Correlation between photons in two coherent beams of light. *Nature*, **177**, 27.

Hanbury Brown, R. and Twiss, R. Q. (1956b). A test of a new type of stellar interferometer on Sirius. *Nature*, **178**, 1046.

Hill, J. M., Green, R. F., and Slagle, J. H. (2006). The Large Binocular Telescope. *Proc. SPIE*, **6267**, 1.

Johnson, M. A., Betz, A. L., and Townes, C. H. (1974). 10-μm heterodyne stellar interferometer. *Phys. Rev. Lett.*, **33**, 1617.

Kervella, P., Thévenin, F., Ségransan, D., *et al.* (2003). The diameters of α Centauri A and B. *A&A*, **404**, 1087.

Labeyrie, A. (1975). Interference fringes obtained with Vega with two optical telescopes. *AJ*, **196**, L71.

Lane, B. F., Colavita, M. M., Boden, A. F., and Lawson, P. R. (2000). Palomar Testbed Interferometer – update. *Proc. SPIE*, **4006**, 453.

Launhardt, R., Henning, T., Queloz, D., *et al.* (2008). The ESPRI Project: astrometric exoplanet search with PRIMA. *Proc. SPIE*, **7013**, 70132I-1.

Lopez, B., Antonelli, P., Wolf, S., *et al.* (2008). MATISSE: perspective of imaging in the mid infrared at the VLTI. *Proc. SPIE*, **7013**, 70132B-1.

Mandel, L. (1963). Fluctuations of light beams. In *Progress in Optics II*, ed. E. Wolf. Amsterdam: North-Holland, p. 181.

Michelson, A. A. and Pease, F. G. (1921). Measurement of the diameter of α Orionis with the interferometer. *AJ*, **53**, 249.

Muterspaugh, M. W., Lane, B. F., and Konacki, M. (2006). Scientific results from high-precision astrometry at the Palomar Testbed Interferometer. *Proc. SPIE*, **6268**, 62680F-1.

Shao, M. and Colavita, M. M. (1992). Potential of long-baseline infrared interferometry for narrow-angle astrometry. *A&A*, **262**, 353.

Shao, M. and Staelin, D. H. (1977). Long-baseline optical interferometer for astrometry. *J. Opt. Soc. Am.*, **67**, 81.

Shao, M. and Staelin, D. H. (1980). First fringe measurements with a phase-tracking stellar interferometer. *Appl. Opt.*, **19**, 1519.

Thomson, A. R., Moran, J. M., and Swenson, G. W., Jr. (1986). *Interferometry and Synthesis in Radio Astronomy*. New York, NY: J. Wiley & Sons.

Radio astrometry

EDWARD B. FOMALONT

Introduction

Astrometry is the branch of astronomy that studies the position and motion of celestial objects in the Solar System, in the Milky Way Galaxy, and for galaxies that are near the limit of the observable Universe. In radio astronomy, the study can be separated into micro-astrometry and macro-astrometry. Micro-astrometry deals with the motion of individual or a small number of associated objects in order to determine their space motion, their distance from the Solar System, and their kinetic properties with respect to neighboring stars and planets. The observational techniques and reductions measure the separation of the target object from a nearby calibrator radio source with known and stable properties.

Macro-astrometry, on the other hand, deals with the absolute position of radio sources, which also requires the determination of the Earth's deformations, complex rotations, and space motion. This type of astrometric experiment observes many well-known compact radio sources over the sky within a 24-hour period. From the analysis of systematic residuals in the data, the absolute positions of the sources, as well as the astrometric and geodetic properties the Earth, are determined. From this 30+ year effort, the fundamental celestial inertial frame has been defined to an accuracy of about 0.01 milliarcsec (mas) using the position of nearly 300 radio sources.

For nearly 30 years, the highest astrometric precision has been obtained using radio-interferometric techniques because of several properties of radio waves. First, astronomers and engineers have been able to connect arrays of radio telescopes that span the Earth (even into Earth orbit) to achieve resolutions of a few mas and obtain positional accuracies well under 1 mas. This technology of radio interferometry is possible because radio waves can be combined coherently (essentially focused) between distant telescopes much more easily than electromagnetic waves at much higher frequency. Second, the media surrounding the Earth – the troposphere and ionosphere – produce relatively slow changes in the radio waves, so that the blurring and distortion of the high angular information can be ameliorated to a large extent. Finally, radio waves are emitted from many sources that are extremely distant and very small in angular diameter, making them excellent beacons for precise astrometric measurements. Additional information on the early development of radio astrometry is given by Walter and Sovers (2000, pp. 100–105) and Thompson, Moran and Swenson (2001; hereinafter TMS) (pp. 12–39).

Astrometry for Astrophysics: Methods, Models, and Applications, ed. William F. van Altena.
Published by Cambridge University Press. © Cambridge University Press 2013.

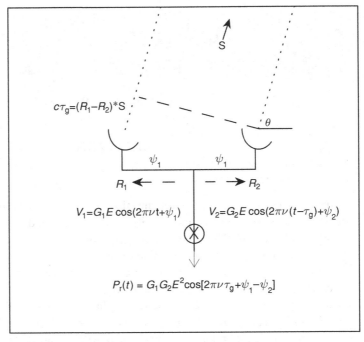

The plane wave intercepted figure contents:

$c\tau_g=(R_1-R_2)*S$

θ

S

ψ_1 ψ_1

$R_1 \leftarrow \quad - \quad - \rightarrow R_2$

$V_1=G_1E\cos(2\pi\nu t+\psi_1)$ $V_2=G_2E\cos(2\pi\nu(t-\tau_g)+\psi_2)$

$P_r(t) = G_1G_2E^2\cos[2\pi\nu\tau_g+\psi_1-\psi_2]$

Fig. 12.1   The basic interferometer: the response of an interferometer of baseline **B** from a source in direction **S**.

12.1 Fundamentals of radio astrometry

This section describes the radio-interferometric properties that lead to the fundamental observable, the visibility function. The geometric and Fourier properties of the visibility function are then described to illustrate the imaging and positional information contained in these data. Finally, the major instrumental imperfections, the effect of the media above the Earth and beyond, and the complexity of extracting the useful astrometric parameters from many other contaminants, are described.

12.1.1 The basic radio interferometer

A simple two-element radio interferometer is shown in Fig. 12.1. At first, the interferometer response for a point source in the direction **S**, emitting at a monochromatic frequency ν (wavelength $= \lambda$) and infinitely far away, will be considered. The plane wave intercepted by the two telescopes, separated by $\mathbf{B} = \mathbf{R}_1 - \mathbf{R}_2$, is identical except for the travel time difference between them, and this time is called the geometric delay $\tau_g = \mathbf{B} \cdot \mathbf{S}/c = B/c \cos(\theta)$, where c is the speed of light. Since most astrometric observations are made with Earth-bound telescopes that follow a source in its diurnal motion, the geometric delay changes quickly with time.

The radio waves from the source at each telescope focal point generate a voltage, proportional to the strength of the electric field E. The voltages are then amplified, often converted to a lower frequency, and transported to a location where the signals from all telescopes are combined in the correlator, as shown by the $\times$ in the diagram. The two voltage inputs to the correlator are

$$V_1(t) = G_1 E \cdot \cos(2\pi \nu t + \psi_1) \tag{12.1}$$
$$V_2(t) = G_2 E \cdot \cos(2\pi \nu (t - \tau_g) + \psi_2)$$

The instrumental gains G_1 and G_2 are the proportionality factors (assumed to be linear) between the electric-field strength and the voltages. The instrumental terms ψ_1 and ψ_2 are the additional phase change added by the telescope electronics, the media above the telescopes, and in the transportation of the signals to the correlator. Many of these changes can be determined or measured during the observations.

In most interferometers the two signals are multiplied and averaged in a device called a correlator, and then passed through a low-pass filter to remove the high-frequency term, producing the basic interferometric response

$$P_r(t) = G_1 G_2 E^2 \cos(2\pi \nu \tau_g + \psi_1 - \psi_2) \tag{12.2}$$

which is proportional to the intensity (power) in the incoming wave. The diurnal motion of the source through the fringe pattern in the sky produces a quasi-sinusoidal form, called natural fringes. For example, a telescope separation of 1000 km at an observing wavelength of 1 cm produces a maximum fringe rate of 7.3 kHz, and a minimum fringe separation in the sky of 2 mas. Generally, a source position can be determined to a small part of the fringe separation.

Most correlators form two independent products. The *real* correlator product is given above, while the *imaginary* correlator product is obtained after inserting a $(\pi/2)$ phase shift in one of the telescope paths with the response

$$P_i(t) = G_1 G_2 E^2 \sin(2\pi \nu \tau_g + \psi_1 - \psi_2) \tag{12.3}$$

At first glance these two correlation products appear redundant, but the noise fluctuations in each are independent, and they also form a convenient complex correlation pair. Further details concerning the optics of radio telescopes are given by Napier in TCP, Ch. 2 (see Taylor *et al.* 2003): the receiving electronics of a radio telescope are covered in Ch. 7 of TMS: the workings of a correlator can be found in ch. 8 of TMS.

12.1.2 The visibility function and Fourier relationships

The complex correlator outputs for two point sources, at positions $\mathbf{S} + \sigma_1$ and $\mathbf{S} + \sigma_2$ with intensities I_1 and I_2, neglecting the instrumental terms, are

$$P_r(t) = I_1 \cos(2\pi \nu \tau_{g1}) + I_2 \cos(2\pi \nu \tau_{g2})$$
$$P_i(t) = I_1 \sin(2\pi \nu \tau_{g1}) + I_2 \sin(2\pi \nu \tau_{g2}) \tag{12.4}$$

where τ_{g1} and τ_{g2} are the geometric delays associated with each source. When dealing with the combination of sinusoids at nearly the same frequency, it is more convenient to use the phasor representation of the sinusoids,[1] and the complex sum of the two correlation products $P(t) = P_r(t) + iP_i(t)$ can be written as $(i = \sqrt{-1})$

$$P(t) = e^{i2\pi v \tau_g}\left[I_1 e^{i2\pi \frac{v}{c}(\mathbf{B} \cdot \sigma_1)} + I_2 e^{i2\pi \frac{v}{c}(\mathbf{B} \cdot \sigma_2)}\right] \tag{12.5}$$

The leading exponential term gives the natural fringe phase change associated with a source at position $\mathbf{S}$, called the *phase center*, and produces the high-frequency part of the response as the source moves through the fringe pattern. The additional term is a complex quantity called the *visibility function* V, with an amplitude proportional to the fringe amplitude, and a phase equal to the phase shift of the response compared with that of a point-source response at the phase center.

The two-point-source model is easily generalized into a continuous distribution of radio emission in the sky, the radio brightness $I(x, y)$, where the (x, y)-plane is the sky plane with its tangent point at the phase center, and with $+x$ to the east ($+\alpha \cos \delta$) and $+y$ to the north ($+\delta$). The projection of the telescope separation on the sky, $\mathbf{B} \cdot \mathbf{S}/c$, can be expressed in the rectangular coordinates (u, v), denoted as the spatial frequencies toward the east and north directions, respectively. Then, the visibility function corresponding to this brightness distribution when observed with (u, v) spatial frequencies is $\mathcal{V}(u, v)$, with

$$\mathcal{V}(u, v) = \int_{sky} I(x, y)e^{i2\pi v(ux+vy)}dx\, dy \tag{12.6}$$

This result is known as the van Cittert-Zernike theorem (Born and Wolfe 1999, TMS pp. 594–612), relating the spatial coherence function (visibility function) with the sky brightness by a two-dimension Fourier transform. The units of (u, v) in this formulation are in time (delay) by dividing through by c. Dividing through by the frequency will produce (u, v) in units of wavelengths. The units of the spatial frequency (meters, wavelengths, and delay) are used interchangeably.

The inverse relationship is

$$I(x, y) = \int_{uv\text{-}plane} \mathcal{V}(u, v)e^{-i2\pi v(ux+vy)}du\, dv \tag{12.7}$$

The visibility function is generally sparsely sampled in a contiguous region of the (u, v)-plane, so the determination of $I(x, y)$ is problematical. However, a first approximation of the source brightness (principal solution, but euphemistically called the dirty image), $I_d(x, y)$, is the Fourier sum of the m measured visibility functions

$$I_d(x, y) = \sum_{i=1}^{m} \mathcal{V}(x, y)e^{-i2\pi v(ux+vy)} \tag{12.8}$$

A detailed discussion of methods for obtaining a more accurate representation of the true source brightness from the dirty image is given by Cornwell *et al.* in TCP (pp. 151–186).

The two-dimensional Fourier relation between I and $\mathcal{V}$ is an approximation. If the source emission extends over many arcminutes and the antenna array is non-planar, as for most very long baseline interferometry (VLBI) arrays, there is an additional term in the exponential that increases quadratically with distance from the phase center. This complication is rarely

---

[1] See: http://en.wikipedia.org/wiki/Phasor_(electronics).

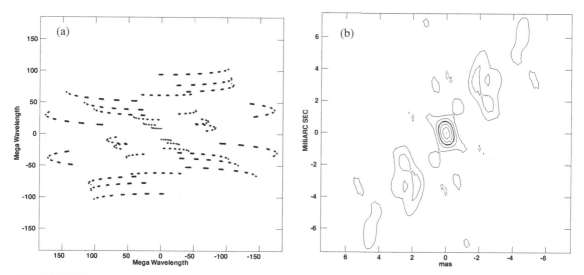

Fig. 12.2   (a): The (u, v) coverage (v is the ordinate and u the abscissa) for six VLBA telescopes for a source at $\delta = -5°$. The tracks are not continuous because the source observations were alternated with another source. (b) The derived point-source response on the sky (y is the ordinate and x the abscissa) using this (u, v) coverage. The contour levels are at 1/6 of the peak intensity with the heavy contour at the 50% level.

important for astrometric observations since most objects are small in angular size, but the Fourier extension is discussed by Perley in TCP (pp. 383–400).

An important property of the Fourier transform is its coherence and linearity; that is, an image can be made from the summation of visibility-function samples over a long period of time using many telescope baselines – so long as the source brightness does not vary significantly over the observing period. For example, integration times of days on large arrays are often made in order to detect and image weak radio sources. For such objects, Fourier techniques are the reliable method for determining the source position and structure. It is often the method used to determine micro-astrometric results. Most macro-astrometric observations cannot use the Fourier inversion directly, and the reason for this differentiation will be discussed below.

12.1.3 Resolution and (u, v) coverage

The sparse sampling of the visibility function in the (u, v)-plane is alleviated in two ways: First, the use of an n-element array produces $n \times (n-1)/2$ independent measures of the visibility function. Second, as the source moves in its diurnal path, the (u, v) coordinates change orientation and spacing.

An example of a (u, v) coverage is shown in Fig. 12.2a. The resultant image at 22 GHz of a point source, using Eq. 12.8 with this sampling, is shown in Fig. 12.2b. Large-scale image distortions, called side-lobes, are present at the 30% level. Additional examples of the (u, v) coverage geometry can be found in TMS (pp. 86–93). For instruments that will image large sources with complex structure, the configuration of the array must be designed to fill the (u, v)-plane relatively uniformly, especially for observations as short as a few minutes (TMS, p. 150–155).

12.1.4 The correlator output

The complex correlator output, Eq. (12.5), in more generic form, including the error terms, is

$$P(t) = e^{i[2\pi \frac{v}{c}(\mathbf{R}_i - \mathbf{R}_j) \cdot \mathbf{S} + \psi_M(t, v)]} G_i G_j \, \mathcal{V}_{i,j}(t, v) \; e^{i(\psi_i - \psi_j)} + \eta_{i,j} \tag{12.9}$$

The first term (exponential) is called the *correlator model* and consists two terms; the natural fringe response associated with the geometric delay of a point source at the phase center, plus any additional terms ψ_M that are known during the correlation of the data. These might include the estimate of the tropospheric delay (from ground measurements of GPS measures) or the clock delay offset. The correlator model is usually given as a delay. Hence, the phase for any baseline (i, j) and frequency can be determined from the appropriate telescope pair delay difference multiplied by the frequency of the data stream.

The remaining terms represent the more slowly varying part of the response called the observed visibility function $\mathcal{V}^\circ$

$$\mathcal{V}_{i,j}^\circ(t, v) = G_i G_j \mathcal{V}_{i,j}(t, v) \; e^{i(\psi_i - \psi_j)} + \eta_{i,j} \tag{12.10}$$

It contains information about the source structure, its position and all of the instrumental gain and phase variations that are not included in the correlator model.

The correlator model and observed visibility function are generally separated in the output data base. The model, although rapidly changing, is a smooth function of time and can be accurately represented by a polynomial expansion with slowly changing coefficients updated every minute. The observed visibility function, on the other hand, varies sufficiently slowly so that many seconds of interferometric data can be averaged before it is placed in the correlator output.

In addition to the observed visibility (real and imaginary value) at each time stamp, a data weight (usually proportional to the inverse square of the a-priori root-mean-square (rms) noise) and the (u, v) spatial frequencies are tabulated. Although only monochromatic emission from a source has been considered above, most observations cover a wide frequency range, and this wide-bandwidth signal is usually electronically separated into many smaller, equally spaced frequency channels that are correlated in parallel and then placed in the output data base. (This is an over-simplification. See ch. 8 in TMS which covers the two methods in which a wide-band signal is processed in a correlator.) For reasons that will become clear in the following sections, most experiments consist of a collection of many short observations of a given source, called *scans*, each for about 30 seconds to 10 minutes with the observed visibility sampled every 1 to 10 seconds over all of the relevant frequency data streams. The source and scan selection depends on the goals of the experiment.

The observed visibility function is corrupted by many sources of error, but most of these corruptions modify the amplitude and phase independently. For this reason, the visibility function is generally separated as

$$\mathcal{V} = \mathcal{A} \, e^{i\phi} \tag{12.11}$$

where $\mathcal{A}$ is the visibility amplitude and ϕ is the visibility phase in radians. The noise term, η, however, is distributed randomly equally in the real and imaginary part of the visibility function, and is not conveniently separated into amplitude and phase.

12.1.5 Telescope-based imperfections, closure and self-calibration

Except for the source-structure effects, the observed visibility functions for each baseline, time sample, and frequency channel are nearly always affected by instrumental and astrometric errors that are telescope-based. Hence, any change with baseline (i, j) is associated with a change in telescope i and/or telescope j. This condition of *baseline closure* has already been assumed implicitly in Eq. (12.2) by using $G_1 G_2$ as the gain error for baseline $(1, 2)$, rather than $G_{1,2}$. The closure condition provides a powerful constraint when determining calibration and astrometric parameters. For example, if observations are made with an n-element array, $n(n-1)/2$ baseline visibility functions are observed but only n telescope amplitudes and $(n-1)$ telescope phases (one telescope is used as the reference for the phase) need be determined. This redundancy between the number of data points and the number of unknown parameters enables robust solutions to be obtained for the telescope-based errors. A few telescope errors do not have baseline closure: correlator malfunctions, the effect of interference, second-order effects of large telescope-based errors, polarization impurity of the feed/telescope system, and the gross effects of noise. These effects can usually be removed to sufficient accuracy.

Once all non-closure instrumental effects are removed, the closure conditions permit the determination of the source structure in a simple manner. For the observed visibility phase data, the telescope-based phase errors ψ_j summed around any triangle of baselines will cancel, leaving only the sum of the true visibility phases remaining,

$$\phi^{\circ}_{i,j} + \phi^{\circ}_{j,k} + \phi^{\circ}_{k,i} = \phi_{i,j} + \phi_{j,k} + \phi_{k,i} + \eta_{i,j} + \eta_{j,k} + \eta_{k,i} \qquad (12.12)$$

For the visibility amplitude, four baselines are needed with the closure ratios (noise contribution not included),

$$\frac{A_{i,j} A_{k,l}}{A_{i,k} A_{j,l}} = \frac{A^{\circ}_{i,j} A^{\circ}_{k,l}}{A^{\circ}_{i,k} A^{\circ}_{j,l}} \qquad (12.13)$$

Assuming that the noise contribution is sufficiently small, there are algebraic techniques (called hybrid mapping) that can be used to determine the source brightness (Readhead *et al.* 1980, TMS pp. 438–440). This method requires absolutely no knowledge of the instrumental errors to obtain the source structure, but certain source-structure symmetries are ambiguous. It is difficult to use this method for sources with complicated structures.

The more common method of incorporating hybrid mapping with more conventional fast-Fourier-transform (FFT) mapping is: (1) Determine a radio-source structure that approximates the observed closure errors. Often a point-source model is a good-enough first guess (i.e. closure phase = 0). (2) Determine the telescope-based calibration errors needed to produce the assumed radio structure from the visibility data. (3) Image the data with these calibrations in order to obtain an improved source model. (4) Compare the closure quantities of the improved source model with that observed. (5) If converging, go to step

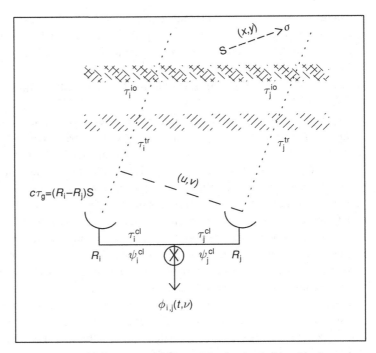

Fig. 12.3 Error terms of an interferometer: the major delay and phase errors associated with an interferometer.

(2) and iterate several times. This iterative method is called *self-calibration*. Even if the original data are sufficiently well calibrated so that an accurate image can be obtained, the self-calibration technique is used to remove small amplitude and phase errors that remain after normal calibration, and image accuracies can be increased from 1% to <0.1%. The dangers and assumptions in using self-calibration techniques are described by Cornwell and Fomalont in TCP, ch. 10.

12.1.6 The astrometric phase equation

Most radio astrometric programs use strong, compact radio sources in order to obtain the highest-accuracy astrometry. Hence, the analysis relies heavily on the visibility-phase measurements. Figure 12.3 shows the major error terms that form the basis for much of the remainder of this chapter. For a point radio source located at a position offset σ with Cartesian offsets (x, y) from the phase center $\mathbf{S}$ the observed visibility phase over a scan of about one minute, $\phi_{i,j}^o(t, \nu)$, measured on baseline (i, j) at time t and frequency ν, is

$$\phi_{i,j}^o(t, \nu) = \phi_{i,j} + 2\pi\nu\big[(ux + vy)$$
$$+ \Delta(\mathbf{R_i} - \mathbf{R_j}) \cdot \mathbf{S}/c + \big(\tau_i^{cl}(t) - \tau_j^{cl}(t)\big) + \big(\tau_i^{tr}(t, \sigma) - \tau_j^{tr}(t, \sigma)\big)\big]$$
$$+ \frac{1}{\nu}\big[\tau_i^{io}(t, \sigma) - \tau_j^{io}(t, \sigma)\big] + (\psi_i(t, \nu) - \psi_j(t, \nu)) + \eta_{i,j} + 2\pi N \quad (12.14)$$

Table 12.1 Uncertainties in the VLBI delay model		
Model component	Delay maximum (mm)	Delay error (mm)
Pure geometric delay	6×10^9	–
Earth orbital motion	6×10^5	1
Gravitational delay	2×10^3	2
UT and polar motion	2×10^4	2
Nutation/precession	3×10^5	5
Tectonic motion	100	1
Tidal motion	500	3
Non-tidal motion	50	5
Antenna structure	1000	4
Instrumentation	30 000	5
Source structure	50	10
Ionosphere (8 GHz)	100	10
Troposphere at zenith	2000	20

The first line contains the two parts of the source visibility phase that describe the source-emission properties: the structure phase, and the phase associated with the source position offset from the phase center. All of the remaining *telescope* error terms must be determined before an accurate position and structure can be estimated.

On the second line, the first term is associated with an error in the separation of the two telescopes. This innocuous term is extremely complex because the telescopes, while solidly fixed on the ground, partake in the complicated motion of the Earth in space, spin-axis changes, and crustal motions. Hence, the telescope baseline error with respect to an inertial frame defined by the distant sources has a complicated trajectory and can be in error by many centimeters. The next two terms of this line represent the clock/path delay error τ^{cl} and the troposphere model error τ^{tr}. These terms, as well as the position offset term, are pure delays (non-dispersive) and the phase errors scale linearly with frequency.

The next line contains two dispersive terms: the ionosphere phase τ^{io} that varies with $1/\nu$, and an additional instrumental phase term ψ not specifically included in any of the previous terms. The stochastic phase noise error is denoted by η, and its magnitude is roughly equal to the signal-to-noise ratio (SNR) in the scan. For example, if the signal is 100 times that of the noise, then this noise phase error is about 0.01 rad, or about 0.5°. If the SNR is near unity, then the resultant phase errors make most astrometric analysis difficult or impossible. Finally, the last line emphasizes that the observed visibility phase is only defined through one cycle; hence the true phase residual may contain an arbitrary number of cycles.

For many VLBI experiments, the errors terms are larger than one turn (equivalently, the delay errors are larger than the observing wavelength) so that many analytical techniques in determining parameters directly from the phase measurements are complicated by possible phase-lobe ambiguities. A list of the residual delays that remain even after using the most accurate a-priori VLBI correlator models are given in Table 12.1, and described in detail by Sovers *et al.* (1998). The second column shows the magnitude of the delay component and the third column gives the typical residual error after removing the best model. The troposphere error of 20 mm is the largest error and is among the most variable of the terms (it also increases with lower source elevation). For example, an error of 30 mm corresponds to a delay error of 0.1×10^{-9} s and about one revolution in phase at 8 GHz.

12.2 Macro-astrometry

In the early 1970s it was realized that by using radio astrometric techniques it would be possible to define a quasi-inertial frame more accurately than the existing optically-based FK5 astrometric grid that was adopted in 1976 (e.g. Walter and Sovers 2000, pp. 66–68). Three reasons for radio's advantage were: the higher intrinsic accuracy of radio interferometry; the relative ease in connecting distant telescopes; and the availability of targets (quasars) that were compact and at enormous distances. In this section, the radio astrometric observing strategies and analysis methods that were used to develop the International Celestial Reference Frame (ICRF), the major macro-astrometric program over the last 30 years, are outlined. A large part of this endeavor requires the determination of the precise locations of the telescopes, which in turn require the simultaneous determination of Earth's orientation parameters (EOPs), nutation and polar motion, and telescope motions associated with plate tectonics and crustal deformations. The motion of the Earth's geocenter with respect to the Solar System barycenter and the rotation of the Sun around the Galactic center can be ascertained in conjunction with spacecraft navigation (see also Chapters 5 and 7).

Other technologies also impact and support the VLBI astrometric development: the Global Positioning System (GPS) ranging system of Earth-orbiting satellites; the satellite laser ranging (SLR) system; the lunar laser ranging (LLR) system; and Doppler orbitograph (DORIS) system. A description of these technologies and their relationship to VLBI is available.[2] More information on these technologies and their astrometric importance can be obtained from Kovelevsky and Seidelmann (2004, pp. 23–28). Additional history of the development of the ICRF is given by Walter and Sovers (2000, pp. 164–169) and Ma *et al.* (1998).

12.2.1 The group delay

Although correlator models have become more accurate, parameter uncertainties lead to visibility phase errors that are significantly larger than one wavelength (see Table 12.1). Rogers (1970) suggested that the phase-ambiguity problem could be removed by observing simultaneously across a large spanned bandwidth. With proper calibration of the relative phase response among the frequency channels, the phase slope with frequency, called the *observed group delay*, $D_{i,j}^o(t, \nu) = [d/d\nu]\phi_{i,j}^o(t, \nu)$, can be determined for each scan and baseline, with an accuracy proportional to the spanned bandwidth. The group-delay

---

[2] http://hpiers.obspm.fr/eop-pc

equation derived from Eq. (12.14) is

$$D_{i,j}^{o}(t, \nu) = 2\pi \left[(ux + vy) + \frac{\mathrm{d}}{\mathrm{d}\nu}[\phi_{i,j}(t, \nu) - \phi_{i,j}(t, \nu)] \right.$$

$$+ \Delta(\mathbf{R}_1 - \mathbf{R}_2) \cdot \mathbf{S}/c + \left(\tau_i^{\mathrm{cl}}(t) - \tau_j^{\mathrm{cl}}(t) \right) + \left(\tau_i^{\mathrm{tr}}(t, \sigma) - \tau_j^{\mathrm{tr}}(t, \sigma) \right) \Big]$$

$$- \frac{1}{\nu^2} \left[\tau_i^{\mathrm{io}}(t, \sigma) - \tau_j^{\mathrm{io}}(t, \sigma) \right] + \frac{\mathrm{d}}{\mathrm{d}\nu}[\psi_i(t, \nu) - \psi_j(t, \nu)] \qquad (12.15)$$

The major advantages of using and analyzing the group delay is the lack of phase ambiguities of this observable. Two complications to using the group delay are: (1) the instrumental phase derivative term with frequency (last term on line 3) depends on the frequency stability of the instrumental electronics, but most telescope systems are closely monitored to significantly lessen its effect; (2) the structure phase ϕ variation with frequency is small if compact radio sources are used.

The most important astrometric parameter is the total delay measured between any two telescopes for the radio signals from a source. This total delay is the sum of the model delay used in the correlator, $D_{i,j}^{\mathrm{M}} = 2\pi(\mathbf{R}_i - \mathbf{R}_j) \cdot \mathbf{S}/c$, plus the residual delay $D_{i,j}^{o}(t, \nu)$ that is shown in Eq. (12.15). Although it is still important to use an accurate correlator model to keep the phase/delay residuals small and slowly variable, astrometric analysis of the total delay data is independent of the model delay used for the observations. The total phase can be defined in the same manner, but it is not as useful for analysis unless the model is so accurate that there is little chance for phase-lobe ambiguities.

The Fourier inversion of the visibility function in Eq. (12.7) to image a radio source has no group-delay analog. In fact, each scan must have a signal well above the noise in order to determine sufficiently accurate group delays to be useful. With the present sensitivity limits of the VLBA at 8 GHz, sources must be stronger than about 0.05 Jy for group-delay determinations, whereas there is no limit of source weakness if the Fourier phase inversion is used. In practice, algebraic (least-squares) methods are used to determine the source position and many astrometric quantities from Eq. (12.15).

12.2.2 Use of simultaneous dual frequencies

It was realized in the 1970s that the ionosphere term τ^{io}, which is often larger than the tropospheric term below 5 GHz, can be estimated by observing the group delay simultaneously at two widely separated frequencies, ν_1 and ν_2 (e.g. Clark et al. 1985). This technique was found to be so effective that it led to adding dual-frequency capability to older telescopes and adding this capability to most new telescopes and arrays, such as the Very Long Baseline Array (VLBA) and the VLBI Exploration of Radio Astrometry (VERA) array.

The ionospheric component is obtained in a straightforward manner. If D_{ν_1} and D_{ν_2} are the two measured group delays for any baseline and time, then the ionosphere-corrected group delay at $D_{\nu_1}^c$ is

$$D_{\nu_1}^c = \left(\nu_2^2 D_{\nu_1} - \nu_1^2 D_{\nu_2} \right) / \left(\nu_1^2 - \nu_2^2 \right) \qquad (12.16)$$

12.2.3 Observing schedules

The observing strategy for a typical astrometric session follows (additional motivation and details are given by Ma *et al.* 1998). A session usually spans 24 hours because the delay variations associated with a source position offset and the five EOPs vary with the $\Delta(\mathbf{R}_i - \mathbf{R}_j) \cdot \mathbf{S}$ terms that have a sidereal-day period. The full day need not be covered entirely, but shorter observing spans can produce large correlations among the parameter solutions. For the most accurate sub-mas astrometric results, the baseline dimensions of the telescope array should extend to 5000 km. Some experiments consists of more than 10 telescopes used over the 24-hour period and these can determine the full range of astrometric and geodetic parameters as well as improved source positions and structures. However, useful astrometry, limited to a few parameters like the daily monitoring of EOPs, can be obtained with one sufficiently long baseline, e.g. Australia to South Africa, using strong and unresolved radio sources or using a subset of the VLBA over a few hours.

Superposed on the diurnal-delay error terms are additional changes caused by variable tropospheric refraction over the sky (we are assuming that the ionospheric refraction has been successfully removed – if not, its variability characteristics are similar to that of the troposphere). Although accurate refraction models at each telescope can be obtained from ground-based and GPS-based measurements, these measurements do not include refraction components that produce path-length changes of 2 cm or more over angular scales of 5 to 10 degrees and timescales of about 10 minutes to several hours. In order to sample the changing tropospheric screen, the typical schedule cycles quickly among many sources over a large part of the sky within about a 30-minute to 60-minute period, and the observing sequence used is carefully chosen to maximize the solution robustness. Observations at elevations as low as 5° where the tropospheric error dominates are crucial. Adequate sky coverage requires fast slewing speeds on the telescopes; hence the use of large telescopes in these arrays can be a limitation. About 250 scans are observed over a 24-hour period. Unless particular sources are under investigation to determine their more accurate position or structure, the necessary scheduling properties can be obtained by including about 60 ICRF sources, and the best-quality (strongest and most compact) sources are usually chosen.

Most source scans have an integration time of a few minutes, long enough for good SNR to obtain an accurate group delay, but shorter than the coherence time (see Section 12.3.2). The observations are made simultaneously at 2.3 GHz and 8.4 GHz with many frequency channels that span a frequency range of about 100 MHz and 800 MHz, respectively, in order to remove the ionospheric component. The correlator models use the most accurate source positions, telescope locations EOPs, and weather data, so the delay errors in Eq. (12.15) are as small as possible. Sessions with large arrays that include the VLBA are scheduled every few months, and sessions with a smaller number of telescopes are scheduled twice a week to determine the short-term changes of the (EOPs). There have been several continuous VLBI campaigns lasting over 2 weeks to investigate the short-term fluctuations (Haas 2006) and tests of various solution methods.

12.2.4 Session astrometric solutions

There are several software packages that analyze either one session or many sessions. The software developed at NASA/Goddard is called Calc/Solve.[3] The Calc part contains the a-priori values for all astrometric/geodetic parameters and calculates the model delay for each telescope as well as its partial derivative with respect to many parameters. (This software package is used on-line in many arrays for generating their correlator models.) The software system that does the least-squares fitting of the observed total delay to obtain many hundreds of relevant parameters is called Solve. A similar package developed by the Jet Propulsion Laboratory is called MODEST. The details of the VLBI astrometric modeling is given by Sovers *et al.* (1998) and a general description of the MODEST package is given by Sovers *et al.* (2004). Other packages are OCCAM (Titov 2004), and SteelBreeze from Kiev (Boloton, personal communication).

The input data for all packages include: the measured total delay for each scan and baseline, which are used to determine the relevant accurate astrometric and geodetic parameters; updated positions of some sources and/or improved antenna locations (Ma *et al.* 1998). The two major time-variable parameters are the antenna-clock delay and tropospheric refraction delay. The clock delay for each antenna is a function of time and not source location on the sky, and its dependence is parametrized by an offset, rate, and acceleration over the experiment period to characterize the long-term delay behavior of the maser reference clocks at each antenna. Superposed on this long-term behavior are additional hour-to-hour clock variations that must be determined within certain statistical limitations.

The optimum strategy for the parametrization of the troposphere refraction at each telescope has been a subject of intense study and trial, and has impacted the scheduling of the observations as described above. The methods vary, but generally the zenith-path delay is determined every 30 to 60 minutes; and the gradient term in the E/W and the N/S directions every 4 to 6 hours. The dependence of these parameters with elevation is called the mapping function and depends on the scale height of the refractive media (wet and dry components differ) and is a function of the local weather conditions. The combination of least-squares analysis and Kalman filtering algorithms are used to determine robust solutions (Herring *et al.* 1990).

The flavor of the effort to determine the tropospheric refraction can be obtained from Lanyi (1984) and Niell (2006). A new initiative called VLBI2010 is constructing a global array of small telescopes and wide bandwidth to cover the sky more quickly in order to decrease the residual tropospheric delay error, and to determine much more accurate group/phase delays.[4]

12.2.5 Global astrometric solutions and the ICRF2

An analysis of data from many sessions produces *global* solutions that determine the value of relevant astrometric and geodetic parameters that do not vary during any one

---

[3] http://gemini.gsfc.nasa.gov/solve
[4] http://www.haystack.mit.edu/geo/pubs/IAG2005_VLBI2010_rev1.pdf

session – or not at all. Such parameters are the position of radio sources (at least the ones that are compact with little extended structure), the telescope locations and velocities, the secular changes in the Earth rotation, orientation, precession, and nutation. In contrast, the parameters that vary from session to session are called *arc* parameters and include the clock and troposphere, and daily solutions for the polar-motion and short-term nutation offsets from the nominal long-term model. Some parameters can switch between global or arc, depending on the intent of the analysis. For example, to determine a more accurate source position (or to evaluate source-position errors), its position would become arc parameters.

From global VLBI solutions for over 1000 sessions between about 1979 to 1995, the position of over 600 strong and compact sources have been determined. In 1997, the International Astronomical Union (IAU) voted to adopt a set of observing and analysis strategies, called the ICRF, on January 1, 1998. Two hundred and twelve high-quality radio source positions around the sky were the basis of the ICRF to define the celestial frame. The position accuracy for a typical source is about 0.2 mas and this leads to an ICRF frame that is stationary and non-rotating to about 0.02 mas. The ICRF list of sources is given by Ma *et al.* (1998) and has been updated to the ICRF-ext2 list by Fey *et al.* (2004).

The second realization of the ICRF was adopted in 2010 by the International Earth Rotation and Reference Systems Service (IERS), the IVS (International VLBI Service for Geodesy and Astrometry), and the IAU. Compared with the ICRF, this new frame, designated as ICRF2, contains 14 years of additional observations, numerous advances in analysis and modeling techniques, and over 3000 source positions. The noise floor of the ICRF2, to accommodate unknown systematic errors, is five times lower than that of the ICRF. A description of the ICRF2 is given by Fey *et al.* (2009). The reference frame is defined by the positions of 295 compact radio sources over the sky, many with positions accurate to 0.05 mas.

Many of the current sessions are also designed to monitor the EOPs in a consistent manner under the auspices of the IERS and the results are available at www.iers.org. These parameters are important inputs to the correlator models that are used for VLBI observations, as well as valuable information for geodetic research. Updated parameter values are obtained about 10 days after the observations.

12.2.6 Radio-source calibrators over the sky

It is important to compile a catalog with a dense grid of calibrators (compact sources with accurate positions) so that any target source is likely to be within a few degrees of a calibrator. These calibrator/target pairs are used to determine the micro-astrometric results that are discussed next. For this reason, a series of about ten 24-hour sessions were observed with the VLBA between 1997 and the present, called the VLBA Calibrator Surveys (VCS) (last one reported in Petrov *et al.* 2008). In these sessions about 15% of the observations use ICRF sources, but the remainder of the schedule includes candidate radio sources which might be sufficiently strong and compact for use as calibrators for differential astrometry. The analysis techniques are similar to those described above, and position estimates were obtained for the new sources. Because of the limited number of VLBI facilities in the southern hemisphere, the density of calibrators decreases significantly south of declination $-30°$. However, this dearth is being rectified with recent observations of the Australian Long Baseline Array (LBA; Ohja 2005, Petrov *et al.* 2011). A description and listing of the ICRF2 Catalog, and the most complete catalog that contains the results of the VLBI observations over the last 30 years are available on-line.[5,6]

12.3 Micro-astrometry

The position and motion of radio sources in the sky give insight into a wide range of astrophysical problems, such as: obtaining the distance by measuring the trigonometric parallax; measuring an accurate position for identification with objects detected at other wavebands; following the motion of binaries to determine their relative masses; measuring the space motion to determine group associations; and testing the validity of general relativity.

12.3.1 Phase-referencing technique

Obtaining the astrometric/structure information for an individual or small set of sources in close proximity in the sky requires somewhat different observing strategies and analysis techniques compared with that described for the ICRF-type sessions. For the macro-astrometric projects, many sources were observed over the sky in order to sample the group delay over a wide range of sky positions, time, and baseline so that the many astrometric parameters could be accurately determined from the set of measured delays. For the micro-astrometric projects described below, quickly alternating observations are made between the target object and a suitable calibrator source. The calibrator data can be regarded as a test signal that, with proper analysis, provides the calibration state of the array. If the target is sufficiently close to the calibrator in the sky and the observations are quickly alternated, then the state of the array determined from the calibrator can be transferred accurately to the target; hence its structure and position can be obtained. This nodding technique is called *phase referencing* and has been used for decades for many interferometric observations, including kilometer-sized arrays such as the Very Large Array (VLA), the VLBA (Beasley and Conway 1995), and the Australia Telescope Compact Array (ATCA).

The basic phase-referencing experiment consists of the following: Scans of the calibrator and target are alternated every few minutes, the time sequence depending on the strength of the sources, the frequency of the observations, and the weather conditions. For each calibrator scan the residual phases for each telescope and frequency data stream are determined from the visibility data. Usually the average value of the phase over the length of a scan is sufficient. These values are then interpolated between successive scans and removed from the observed phase of the target data. The basic assumption of phase referencing is that the residual amplitude and phase changes that are needed to produce a calibrator image with a

---

[5] http://www.iers.org/IERS/EN/Publications/TechnicalNotes/tn35.html
[6] http://astrogeo.org/vlbi/solutions/rfc_2012a/rfc_2012a_cat.tt

known flux density and position also applies to the target. This "calibrated" target data can then be imaged using FFT and deconvolution methods described previously to obtain its position in the sky. For strong and compact targets, algebraic methods are also often used on the visibility data directly to determine its position.

The phase-referencing technique measures the relative position between the calibrator and target. Thus, the target position is tied to the assumed position of the calibrator. Although there is always an uncertainty in the calibrator position, its offset should generally be less than a few milliarcseconds in order to avoid second-order phase errors that will distort the target (Reid *et al.* 2009). If the calibrator is not a point source, its structure phase can often be determined by self-calibration techniques (Section 12.1.5) and removed before phase referencing. However, if the calibrator structure varies with time, then the position of the calibrator is not well defined and may vary in position as different part of the calibrator change their relative strength. For long-term astrometric projects (e.g. measuring a target parallax), this calibrator effective position change can add significant error to the results.

12.3.2 Temporal and angular coherence

The phase-referencing technique will be successful if the phase errors associated with the calibrator are nearly the same as that for the target. Because calibrator and target are observed at slightly different times and are in different directions, the phase errors do slightly differ between them and introduce errors in the image and position of the target.[7] In the following analysis of the phase-referencing technique, the calibrator and target separation $\Delta\sigma$ in the sky, and the time Δt between consecutive calibrator scans will be used. The measure of the phase change as a function of time is called the *temporal coherence* and a measure of the phase change as a function of sky position is called the *spatial coherence*. Detailed modeling of the effects of the angular and temporal correlation of tropospheric and ionospheric refraction are given by Pradel *et al.* (2006) and Asaki *et al.* (2007).

The temporal coherence is generally associated with the small-scale fluctuations in the troposphere (usually the wet component) and ionosphere refraction moving above each telescope. These produce delay changes over seconds to minutes of time that depend on weather conditions. For telescopes separated by more than about 100 km, the tropospheric-delay difference over each telescope is largely independent and can be as little as 0.5 mm at a dry high location to 100 mm at a humid coastal location. The ionospheric delay is comparable to the tropospheric phase delay at 5 GHz, and becomes dominant at lower frequencies. The variability in temporal coherence is evident from the example of the phase stability/coherence time for a VLBA experiment at 23 GHz as shown in Fig. 12.4. For the BR–LA baselines, the phase is remarkably constant with a change over 15 m of about 20° of phase, or <1 mm delay. For the SC–LA or PT–LA baselines, there are periods when the phase changes considerably in less than 1 minute. For example, the interpolation of

---

[7] If the two sources are sufficiently close in the sky and are within the telescope reception area, or a dual-beam system like VERA is used, then the objects can be observed simultaneously.

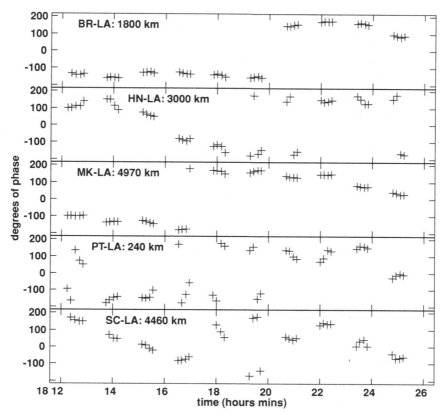

Fig. 12.4   Phase coherence. The plot shows the phase stability at 23 GHz for 16-min period during a much longer phase-referencing experiment. The crosses show the calibrator phase every 10 s over each 40-s scan. Between each scan is a target source. The coherence times vary from more than 10 min to 1 min or less. The telescopes are BR (Brewster, WA), HN (Hancock, NH), MK (Mauna Kea, HI), PT (Pie Town, NM), SC (Saint Croix, VI) and LA (Los Alamos, NM).

the phase for the SC–LA baseline between the two calibrator scans at 18 h 18.3 m to 18 h 19.5 m is uncertain. The term coherence time is the period at which the interpolation of the phase between two successive measurements cannot be interpolated to an accuracy better than about 1 rad. In many experiments, especially at high frequencies, a significant fraction of an experiment may have to be edited because of large phase scatter. These periods are associated with known poor weather at some telescopes, and often occur in windy conditions. Including data where the temporal coherence is marginal usually decreases the astrometric accuracy.

In an analogous manner, the angular separation between the target and calibrator introduce directionally dependent phase-error terms (see Eq. 12.15) between them even with perfect calibrator phase temporal interpolation. The source separation for which the phase difference becomes larger than than about 1 rad is called the *coherence angle*, and this limit is usually produced by unmodeled troposphere and ionosphere refraction and by antenna-separation errors in the correlator model. The statistical nature of the the temporal-scale and the angular-scale errors are different. In poor weather conditions, the temporal-phase scatter is large but somewhat random over periods of minutes. Its main effect is to decrease the coherence of the data and the average target visibility is decreased to the point of being useless. On the other hand, the errors associated with significant angular-coherence problems are often constant over periods of many minutes and hours. Even a small phase offset of $15°$, if persistent, will produce relatively large imaging and astrometric errors. For this reason, the correlator model used for phase-referencing experiments should be as accurate as possible, and specialized observations to further decrease the correlation angle are discussed in Section 12.3.4. For VLBI observations the coherence-angle limit is about $5°$ at 8 GHz, and $1°$ at 43 GHz, and that is the reason why the dense grid of calibrators is of crucial importance for phase-referencing success.

12.3.3 Phase-referencing equation

The observed phase ϕ° of the target (for any baseline), after interpolating the phase for the calibrator observations, can be derived from Eq. (12.14) and, to first order in the expansion of the phase with time and angle, is

$$\phi^\circ(t, \Delta\sigma) = \phi(t) + 2\pi v(ux + vy) + \ddot{\psi}^{\text{cal}}(t)\Delta t + \nabla\psi^{\text{cal}}(t) \cdot \Delta\sigma \qquad (12.17)$$

where (u, v) is the spatial frequency of the target, and (x_t, y_t) is the target offset position from the phase center. The structure phase of the target, $\phi(t)$, is associated with the source structure, and its offset from the phase center is given by the $2\pi v(ux + vy)$ term. If the phase referencing errors for the target are believed to be less than ≈ 0.2 rad, then the target visibility is sufficiently well-calibrated and images can be obtained using the techniques of Eq. (12.14) with the clean deconvolution algorithm. The amplitude calibration (ignored so far) is generally straightforward. For sources that are relatively simple in structure, fitting the visibility data directly with simple brightness models is useful. The software package DIFMAP (Shepherd *et al.* 1994) is the most convenient method for this type of model fitting. If the phase error is significantly larger than 1 rad (it can be many cycles), then analysis of the group delay, as for macro-astrometric reductions, must be used.

The quality of the image will be associated with the size of the phase-error terms that are shown in the last two terms in Eq. (12.17). First, the temporal phase error can be approximated by the product of the non-linearity of the observed calibrator phase over the switching cycle time, $\ddot{\psi}^{\text{cal}}(t)\Delta t$. As discussed above, it is dominated by short-term changes in τ^{cl}, τ^{tr} and τ^{io}. If the phase difference between two calibrator scans differ by more than about $60°$, a significant phase error will be associated with the calculated target phase.

The coherence angle limit is shown by the $\nabla\psi^{\text{cal}}(t) \cdot \Delta\sigma$ term. It is produced by many astrometric errors and by persistent troposphere and ionosphere refraction-model errors that produce a phase screen above each telescope. Over an area as large as radius $10°$, the phase screen is sufficiently linear so that the first-order term in $\Delta\sigma$ is sufficiently accurate. Clearly, the closer the calibrator is to the target, the smaller this spatial error is. The above separation of the phase errors into short-term temporal- and long-term angular-phase changes is an approximation. Of course, before any long-period significant angular-phase dependence can be determined, the short-term temporal-phase changes must be sufficiently small.

Most of the errors described above are delay errors and the resultant phase changes scale with frequency. For bandwidths that are less than about 10% of the observing frequency, this scaling can be ignored and the phase change with time assumed equal over the entire observing band. This allows wide bandwidths to be averaged in order to determine the accurate temporal-phase change. If the observing bandwidth is larger than 10% of the observing frequency, phase scaling should be done. It is also possible to distinguish between ionospheric refraction (varies as inverse frequency) from the tropospheric refraction (varies as frequency) in a way similar to that for the group delay, shown in Eq. (12.15). This phase-separation method used in phase referencing is described by Brisken *et al.* (2000).

The location of the target depends on two fiducial positions used in the correlator. First, the assumed position of the calibrator defines the reference frame of the experiment. If the assumed calibrator position is displaced in a certain direction, then the derived position of the target will move by the same angle, to first order. For this reason, it is not advised to change calibrators in different sessions, at or at different frequencies, in order to determine consistent positions for a source. Second, the assumed target position defines the location of the central pixel of the image, or the offset position that is determined with any astrometric analysis. Hence, if the calibrator position is accurate, then the displacement of the target on the image gives its true offset position.

12.3.4 Improved phase-referencing images and positions

There are several methods in which the phase-referencing accuracy can be improved, either by measurements of relevant properties during the observations (measurement of the tropospheric delay at each telescope in the direction to the GPS satellites) or by including specialized observations during an experiment. If the coherence time is short so that phase fluctuations over any calibrator or target scan are larger than about 1 rad over tens of seconds, then the observations are nearly useless even if extremely fast calibrator–target switching is possible. Recent advances in the measurement and analysis of the water emission in the troposphere along the path of the radio wave are now being used to estimate the variable refraction associated with water vapor along the path from each telescope toward the radio source. This technique, using the water line at 181 GHz, is being successful used for the high-frequency ALMA array, now being built in Chile (Nikolic *et al.* 2011). Application of the correction can reduce the residual path length of the water-vapor component to about 0.2 mm delay.

The coherence angle can be increased with supporting observations during the phase-referencing experiment by measuring some of the astrometric parameters that can limit the coherence. For example, if two calibrators are located on either side of the target, then a suitable average of the telescope phases from the two sources can be combined to estimate the instrumental phase in the direction of the target. This scheme is described by Fomalont and Kopeikin (2003). In many astrometric experiments, a second calibrator is often added to the observation schedule and treated as a second target source. The derived position of this second calibrator, after using the prime calibrator, is a good "sanity" check on the astrometric quality of the observations. Other reasons for using more than one calibrator are discussed in the following section.

An observation method to estimate the residual tropospheric (and ionospheric) zenith-path delay during a phase-referencing experiment utilizes about 30 to 60 minutes of observations that are similar in style to the macro-astrometric-style observation: short scans of ICRF sources distributed over the sky with a sufficiently wide frequency range to determine accurate group delays. These data are then fit to a simple model that produces the largest systematic phase-referencing errors: an antenna-based clock delay and Zenith path delay offset in the assumed correlator delay model. These corrections are then applied to the visibility data before the phase-referencing analysis (Reid *et al.* 2009). For VLBA observations, this technique reduces the residual tropospheric zenith-path-delay error from about 5 cm to 1 cm. For a calibrator–target separation of 1° at an observing frequency of 23 GHz, this reduces the systematic phase offset between the two sources from 0.8 rad to about 0.16 rad.

12.3.5 Astrometric examples using phase referencing

Four arrays, the VERA, the VLBA, the European VLBI Network (EVN), and the LBA, are the main contributors to accurate differential radio astrometric observations. The unique feature of the Japanese VERA array is that each of the four telescopes contains two independent feed/receiver systems at 23 GHz or 43 GHz that permit simultaneous observations of two objects with an angular separation between 0.3° and 2.3°. This configuration reduces the coherence-time limitation since there are no observation gaps between the calibrator and target. Significant VERA results are reported in a 2008 Publication of the Astronomical Society of Japan (volume 60), and many are associated with proper-motion and parallax measurements of Galactic objects using the emission from narrow-frequency spectral features. When these astrometric results are combined with the Doppler information, the three-dimensional kinematics from maser emission associated with the hot ionized medium in nebulae or with the halo of massive stars, lead to astrophysical insights into the formation and evolution of these regions. A further example of the astrometric power of VERA is displayed in the SiO maser results from Orion-KL (Kim *et al.* 2008).

A range of astrometric results from the VLBA and the EVN can be seen in the following references: pulsar proper-motion and parallax determinations (Chatterjee *et al.* 2009) emphasize the use of in-beam-phase referencing and imaging; parallax determination of methanol masers at 6.7 and 12 GHz shows accurate results and have an excellent description of the phase-referencing technique (Reid *et al.* 2009); a test of general relativity by measuring light-ray bending by the Sun among four sources describes the details of phase referencing and coronal (ionospheric) refraction (Fomalont *et al.* 2009); the summary of recent determinations of the Hubble constant from the maser emission motions in NGC 4258 (Argon *et al.* 2007); the residual motion of SGR A*, associated with a black hole at the center of the Milky Way Galaxy, from which its mass can be determined (Reid and Brunthaler 2004); the alignment of the compact components in the radio galaxy M81 at several frequencies using phase referencing with the VLBA plus the EVN describes the complex nature in the emission of a typical active galactic nuclei (Bietenholz *et al.* 2004).

The LBA has determined the kinematics of many pulsars. For example, recent phase-referencing results for PSRJ0437-4715 have measured its proper motion and parallax with respect to the ICRF to an accuracy of about 0.05 mas. The position of pulsars can also be obtained by measuring the change of the pulsar arrival time as the Earth orbits the Sun. If the pulse period is less than 0.01 s and extremely stable, then its position can be obtained even more accurately than from VLBI techniques (see Verbiest *et al.* 2008). Since the pulsar timing analysis is tied to the Solar System ephemerides (most critically to the position of the Solar System barycenter) and is affected by the gravitational influence of Solar System bodies (hence their assumed masses), the comparison of ICRF and timing positions can determine differences in the two reference frames. For example, the current excellent agreement between the two frames limits the amount of unknown mass in the Solar System, and the possibility of a large Jupiter-sized planet beyond the orbit of Pluto can be eliminated.

12.3.6 Spacecraft navigation

Radio arrays are now used to determine the position of spacecraft to support their navigation to planets and satellites. This sky-plane position is complementary to the Doppler-delay navigation results that measure the spacecraft distance. The correlator models must include relatively accurate predicted orbits of the spacecraft to keep the residual phase and delay relatively small. Other corrections not normally encountered with observations of distant quasars are the spacecraft parallax correlation, the wavefront near-field corrections (the wavefront from the spacecraft has a significant curvature between telescopes that are thousands of kilometers apart) and the correct solar and planetary gravitational delay of the spacecraft.

The VLBA, with its good phase stability and accurate correlator model, uses conventional phase-referencing techniques to determine the spacecraft position with respect to a nearby calibrator source. The telemetry signal from the spacecraft is sufficiently strong and wide-band, and its interferometric properties are identical to that of a celestial source. Other arrays, with somewhat less phase stability, use delay-referencing techniques. In this case, the spacecraft must be also transmit two beacons that are sufficiently separated in frequency (≈ 100 MHz) in order for an accurate group delay to be measured.[8]

12.3.7 Position precision and ICRF source stability

The determination of the accurate parallax and proper motion of targets and the accuracy of the ICRF2 frame depends on the stability and the compactness of the emission from the calibrators. Based on studies of many radio sources, the most compact radio component (called the radio core) in a radio source is associated with the inner jet region of a galaxy, perhaps displaced from 0.001 to 0.01 mas from the galactic nucleus that may contain a black

---

[8] http://ieeexplore.ieee.org/stamp/stamp.jsp?arnumber=04390042

hole. This core is optically thick and is often faint at frequencies below about 22 GHz. The structures of most radio sources undergo changes in intensity and structure over periods of time from a few hours (intra-day variables, Lovell *et al.* 2008) to years. The most significant changes occur when material is ejected from the nuclear regions and then moves away from the core at a velocity of up to 1 mas per year (Charlot 2004).

A recent analysis of four compact ICRF2 radio sources over a period of a year, between 8 GHz and 43 GHz, has determined the uncertainties in determining the position of the radio core of even the most "compact" radio sources (Fomalont *et al.* 2011). First, observations of intercontinental baselines at 43 GHz over a period of a year are often needed to resolve the radio-core component from the other emission within a 1 mas-sized region. Once identified, these four radio cores were stationary in the sky for one year at the 0.02-mas level. Second, at lower frequencies, the location of the brightest emission peak can be up to 0.5 mas displaced from the core. Third, since the peak brightness at the lower frequencies can be associated with moving material, its position will change with time, and change the apparent position of the ICRF source at the lower resolutions where many radio components near the core are averaged. These position ambiguities are similar to the frequency-dependent position of radio sources found by Kovalev *et al.* (2008). These changes in the calibrator structures also add uncertainty in determining the parallax and proper motion of targets since some of the target motion will reflect the apparent calibrator position changes.

These small position changes do not impact the present accuracy of the ICRF frame since they will average out over the nearly 300 calibrators that define the frame. In the future, when the main limitation of position accuracy, the residual tropospheric refraction, is decreased by a factor of three, the structure changes in calibrator sources may produce the dominant astrometric uncertainty at the tens of microarcsec level. To guard against this possibility, including several calibrators in a phase-reference experiment can, at least, suggest when one of the calibrators is particularly variable in its structure and effective positions.

In the next decade when optical interferometers determine the positions of stellar objects with an accuracy of 0.01 mas, a comparison with the radio position will require accurate knowledge of the radio (and perhaps optical) emission structure and changes with time.

References

Argon, A. L., Greenhill, L. J., Reid, M. J., Moran, J. M., and Humphreys, E. M. L. (2007). *ApJ*, **659**, 1040.

Asaki, Y., Sudou, H., Kono, Y., *et al.* (2007). *PASJ*, **59**, 397.

Beasley, A. J. and Conway, J. E. (1995). *ASP conf. Ser.*, **82**, 327.

Bietenholz, M. F., Bartel, N., and Rupen, M. P. (2004). *ApJ*, **615**, 173–180.

Born, M. and Wolf, E. (1999). *Principles of Optics*, 7th edn. Cambridge: Cambridge University.

Brisken, W. F., Benson, J. M., Beasley, A. J., *et al.* (2000). *ApJ*, **541**, 959.

Charlot, P. (2004). IVS 2004 General Meeting Proceedings, Ottawa, Canada, February 9–11, pp. 12–21. See http://ivscc.gsfc.nasa.gov/publications/gm2004/charlot/

Clark, T. A., Corey, B. E., Davis, J. L., *et al.* (1985). *IEEE Trans. Geosci. Remote Sensing*, **GE-23**, 391–397.

Chatterjee, S., Ma, C., Arias, E. F., *et al.* (2009). *ApJ*, **698**, 250.

Fey, A. L., Briskin, W. F., Vlemmings, W. H. T., *et al.* (2004). *AJ*, **127**, 3587.

Fey, A. L., Gordon, D., and Jacobs, C. S., eds. (2009). IERS Technical Note 35, Frankfurt am Main: Verlag des Bundesamts für Kartographie und Geodäsie. See http://www.iers.org/MainDisp.csl?pid=46-1100252.

Fomalont, E. B. and Kopeikin, S. M. (2003). *ApJ*, **598**, 704.

Fomalont, E. B., Kopeikin, S. M., Lanyi, G., and Benson, J. M. (2009). *ApJ*, **699**, 1395.

Fomalont, E. B., Johnston, K. J., Fey, A., *et al.* (2011). *AJ*, **141**, 91.

Haas, R. (2006). IVS 2006 General Meeting Proceedings, Concepción, Chile, January 9–11. See http://ivscc.gsfc.nasa.gov/publications/gm2006/haas.

Herring, T. A., Davis, J. L., and Shapiro, I. I. (1990). *J. Geophys. Res.*, **95**, 12561.

Kim, M. K., Hirota, T., Honma, M., *et al.* (2008). *PASJ*, **60**, 991–999.

Kovalev, Y. Y., Lobanov, A. P., Pushkarev, A. B., and Zensus, J. A. (2008). *A&A*, **493**, 759.

Kovalevsky, J. and Seidelmann, P. K. (2004). *Fundamentals of Astrometry*, Cambridge: Cambridge University Press.

Lanyi, G. (1984). *TDAPR*, **78**, 152.

Lovell, J. E. J., Rickett, B. J., Macquart, J. P., *et al.* (2008). *ApJ*, **689**, 108.

Ma, C., Arias, E. F., Eubanks, T. M., *et al.* (1998). *AJ*, **116**, 516.

Niell, A. (2006). IVS 2006 General Meeting Proceedings, Concepción, Chile, January 9–11. See http://ivscc.gsfc.nasa.gov/publications/gm2006/niell.

Nikolic, B., Richer, J., Bolton, R., and Hills, R. (2011). *The Messenger*, **143**, 11.

Ohja, R., Fey, A. L., Charlot, P., *et al.* (2005). *AJ*, **130**, 2529.

PASJ (2008). *Publ Astr. Soc. Jn.*, **60**, No. 5.

Petrov, L., Kovalev, Y. Y., Fomalont, E., and Gordon, D. (2008). *AJ*, **136**, 580.

Petrov, L., Phillips, C., Bertarini, A., Murphy, T., and Sadler, E. M. (2011). *MNRAS*, **414**, 2528.

Pradel, N., Charlot, P., and Lestrade, J.-F. (2006). *A&A*, **452**, 1099.

Readhead, A. C. S., Walker, R. C., Pearson, T. J., and Cohen, M. H. (1980). *Nature*, **285**, 137.

Reid, M. J. and Brunthaler, A. (2004). *ApJ*, **616**, 872.

Reid, M. J., Menton, K. M., Brunthaler, A., *et al.* (2009). *ApJ*, **693**, 397.

Rogers, A. R. R. (1970). *Radio Science*, **5**, 1289.

Shepherd, M. C., Pearson, J. J., and Taylor, G. B. (1994). *BAAS*, **26**, 987.

Sovers, J., Fanselow, J. L., and Jacobs, C. S. (1998). *Reviews of Modern Physics*, **70**, 1393.

Sovers, J., Jacobs, S., and Lanyi, G. E. (2004). IVS 2004 General Meeting Proceedings, Ottawa, Canada, February 9–11. See http://ivscc.gsfc.nasa.gov/publications/gm2004/sovers.

Taylor, G. B., Carilli, C. L., and Perley, R. A. (2003). Synthesis imaging in radio astronomy II. *ASP Conf. Ser.*, **180**. (TCP)

Thompson, A. R., Moran, J. M., and Swenson, G. W. Jr. (2001). *Interferometry and Synthesis in Radio Astronomy*, 2nd edn. New York, NY: John Wiley & Sons. (TMS)

Titov, O. (2006). IVS 2006 General Meeting Proceedings, Concepción, Chile, January 9–11. See http://ivscc.gsfc.nasa.gov/publications/gm2006/titov.

Verbiest, J. P. W., Bailes, M., van Straten, W., *et al.* (2008). *ApJ*, **679**, 675–680.

Walter, Hans G. and Sovers, O. (2000). *Astrometry of Fundamental Catalogs*. Heidelberg: Springer.

PART IV

FROM DETECTED PHOTONS TO THE CELESTIAL SPHERE

13 Geometrical optics and astrometry

DANIEL J. SCHROEDER

Introduction

As is evident from the preceding chapters, the field of astrometry requires the precise positions of stars and other celestial objects relative to one another within a chosen reference frame. In the old days, images of celestial objects were recorded on photographic plates affixed in the focal planes of telescopes, often long-focus refractors, with image positions subsequently determined with plate-measuring machines. In recent years, practically all image recording has been done with CCD and CMOS electronic detectors (see Chapter 14) instead of photographic plates, with positions determined by computer analysis and sophisticated software.

Whatever the recording medium, the telescope is an instrument to transform the positions of objects on the sky to recorded images on the detector. That is, the telescope is simply a device to gather light and image distant object space on to the detector in the telescope focal surface. This re-imaging process is usually not perfect because of aberrations associated with the telescope. In this chapter we consider the types of optical aberrations and their effects on astrometric analyses.

For the purposes of astrometry the transformation from object space to image space should be done without distortion, though in practice this is rarely the case. By definition, transformation without distortion means that the pattern of multiple objects recorded in image space is identical and in one-to-one correspondence with the pattern of these objects on the sky. Throughout this chapter the term distortion will refer to optical field-angle distortion, or OFAD.

A distortion-free optical system is the pinhole camera shown in Fig. 13.1. Points Q_1 and Q_2 are on light rays from distant stars far to the left and Q_1' and Q_2' are "images" of these stars. The line MM' is a reference line from which angular distances in object space to the left and in image space to the right are measured. If corresponding angles in image and object space satisfy the relation

$$\tan u' / \tan u = \text{constant} \tag{13.1}$$

and all of the points in Fig. 13.1 are coplanar, then distortion is zero. A pinhole camera, unfortunately, does not gather much light and, except for showing rather good images of a partly eclipsed Sun, is not especially useful.

Astrometry for Astrophysics: Methods, Models, and Applications, ed. William F. van Altena.
Published by Cambridge University Press. © Cambridge University Press 2013.

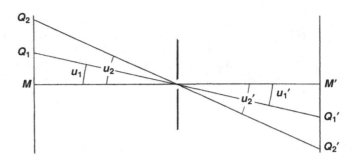

Fig. 13.1   Schematic of the pinhole camera.

There is one seemingly more practical optical system which comes close to being free of distortion; it is a spherical mirror with an aperture stop centered at the center of curvature of the mirror. Let the ray from any distant star which passes through the center of the stop be called its chief ray. From a point on any chief ray the optical system looks the same to the incident light. That is, there is no preferred axis and light from any star is imaged in exactly the same way on the focal surface. Note that any reflected chief ray retraces the path of its incident ray, hence $u' = u$ and Eq. (13.1) is satisfied. If the images were of high quality, this would be an ideal distortion-free system. But, alas, the images suffer from spherical aberration (defined and discussed below) and the system as shown is not an especially useful one

As the reader knows, a perfect image of a single star is obtained if the sphere is replaced by a paraboloid. But now the system has a preferred axis and images of stars not on this axis are no longer perfect. Because of these off-axis aberrations, the paraboloid is not distortion-free, as defined above. It is probably safe to say that any practical telescope has aberrations associated with it and account of these must be made for precise astrometric results.

We now proceed to discuss paraxial optics, the various types of aberrations, and their presence in selected types of telescopes and components of optical systems.

13.1 Paraxial optics

Before beginning our discussion of geometrical optics, it is necessary to point out that much of the material in this and following sections is taken from the book *Astronomical Optics*, 2nd edition (Schroeder 2000; hereafter referred to as AO2). This is done with the kind permission of the author of that text, also the author of this chapter. Our discussion here is a much abbreviated version of that found in AO2 and the reader of this chapter is strongly encouraged to have a copy of AO2 close at hand. Specific references to chapter sections and figure numbers are given in the text that follows. The reader may also find it useful to refer to the textbooks on aspects of optics by Longhurst (1967), Jenkins and White (1976), Born and Wolf (1980), Wilson (1996), and Mahajan (1998), as well as various SPIE publications.

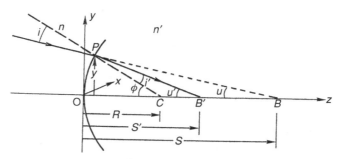

Fig. 13.2 Refraction at a spherical interface. All angles and distances are positive in the diagram; see text for discussion.

In this section we give some of the basic relations needed for a first-order or *paraxial* analysis of an optical system. (This is the starting point upon which our discussion of aberrations, a third-order theory, is necessarily based.) By first-order we mean that point P in Fig. 13.2 (also Fig. 2.1 in AO2) is close enough to the optical axis so that sines and tangents of angles can be replaced with the angles themselves. In this approximation any ray is close to the axis and nearly parallel to it, hence the term *paraxial approximation*.

13.1.1 Sign conventions

The coordinate system within which surface locations and ray directions are defined is the standard right-hand Cartesian frame shown in Fig. 13.2. For a single refracting or reflecting surface the z-axis coincides with the optical axis, the line of symmetry of the elements in an optical system. Fig. 13.2 illustrates refraction at a spherical surface with an incident ray directed from left to right. Rays from an initial object are always assumed to travel in this direction. The indices of refraction are n and n' to the left and right of the surface, respectively.

Unprimed symbols in Fig. 13.2 refer to the ray before refraction, while primed symbols refer to the ray after refraction. The symbols s and s' denote the object and image distances, respectively, and R represents the radius of curvature of the surface, measured at the vertex. The sign convention for distances is the same as for Cartesian geometry. Hence, distances s, s', and R are positive when the points B, B', and C are to the right of the vertex, and distances from the optical axis are positive if measured upward. (See Section 2.1 in AO2 for more details.) The advantage of these conventions is that both refracting and reflecting surfaces can be treated with the same relations. All ray-tracing programs use these same conventions. It is also worth noting that formulas for reflecting surfaces are obtained directly by letting $n' = -n$ in the formulas given for refracting surfaces.

13.1.2 Paraxial equations for refraction

With the help of Fig. 13.2 we can easily determine the relation between s and s' when the distance y and all angles are small, that is, in the paraxial approximation. The exact form of Snell's law of refraction is $n \sin i = n' \sin i'$, or $ni = n'i'$ in the paraxial approximation.

Following the discussion in Section 2.2 in AO2 we find

$$\frac{n'}{s'} - \frac{n}{s} = \frac{n'-n}{R} = \frac{n'}{f'} = -\frac{n}{f} \qquad (13.2)$$

The points at distances s and s' from the vertex are called conjugate points, that is, the image is conjugate to the object and vice versa. If either s or $s' = \infty$, then the conjugate distance is the focal length, that is, $s = f$ when $s' = \infty$ and $s' = f'$ when $s = \infty$. This is the first-order or Gaussian equation for a single refracting surface and is the starting point for analyzing systems that have several surfaces. For multi-surface systems the image formed by the first surface serves as the object for the second surface, and so on. Equation (13.2) does not contain height y and hence applies to any ray passing through B before refraction, provided of course the paraxial approximation is valid. This equation also applies to object and image points that are not on the optical axis, provided these points are close to B and B' in Fig. 13.2 and lie on a line passing through point C. (See Fig. 2.2 in AO2 for an illustration.)

The geometry in Fig. 2.2 in AO2 can be used to determine the *transverse* or *lateral magnification m*, defined as the ratio of image height to object height, with the result

$$m = \frac{h'}{h} = \frac{s'-R}{s-R} = \frac{ns'}{n's} \qquad (13.3)$$

If $m < 0$, the image is *inverted* relative to the object; in the case where $m > 0$ the image is said to be *erect*. (See Section 2.2 in AO2 for details.)

13.1.3 Paraxial equations for reflection

The Gaussian equations for a reflecting surface in the paraxial approximation are found by letting $n' = -n$ in Eq. (13.2) and (13.3). The results are

$$\frac{1}{s'} + \frac{1}{s} = \frac{2}{R} = \frac{1}{f'} = \frac{1}{f} \quad , \quad m = -\frac{s'}{s} \qquad (13.4)$$

A negative index of refraction simply means that the light is traveling in the direction of the $-z$-axis, or from right to left. Consistent use of this convention, together with the other sign conventions, allows us to work with any set of refracting and/or reflecting surfaces in combination. In many situations it is convenient to take $f > 0$ for a concave mirror and $f < 0$ for a convex mirror, independent of the direction of the incident light. We will adopt this convention for convenience, keeping in mind that it violates the strict sign convention. The sign convention for s, s', and R will always be observed. (See Section 2.3, especially Fig. 2.4, in AO2 for further details.)

13.1.4 Two-surface refracting elements

We now take the results from Section 2.4 in AO2 for several systems with two refracting surfaces: thick lens, thin lens, and thick plane-parallel plate.

Thick lens

A schematic cross-section of a thick lens is shown in Fig. 2.5 in AO2. Applying Eq. (13.2) to each surface, with radii of curvature R_1 and R_2, respectively, we find the effective focal length f'

$$\frac{1}{f'} = \frac{n-1}{R_1} - \frac{n-1}{R_2} + \left(\frac{d}{n}\right)\frac{(n-1)^2}{R_1 R_2} \tag{13.5}$$

The effective focal length f' in Eq. (13.5) is measured from the intersection of two extended rays, the incident ray to the right and the refracted ray to the left.

Thin lens

A thin lens is defined as one in which the separation of the two surfaces is negligible compared to other axial distances. Applying Eq. (13.2) in succession or setting $d = 0$ in Eq. (13.5) we find

$$\frac{1}{s'} - \frac{1}{s} = (n-1)\left(\frac{1}{R_1} - \frac{1}{R_2}\right) = \frac{1}{f'} \tag{13.6}$$

Note that all axial distances, s, s', and f', are measured from the thin lens. The net magnification of the thin lens is the product of the surface magnifications, hence $m = m_1 m_2 = s'/s$.

As a final item for thin lenses, we note that two separated thin lenses in air have a focal length for the combination given by a relation similar to that of Eq. (13.5). Two thin lenses with focal lengths f_1' and f_2', separated by distance d, have a focal length given by

$$\frac{1}{f'} = \frac{1}{f_1'} + \frac{1}{f_2'} - \frac{d}{f_1' f_2'} \tag{13.7}$$

Although thin lenses are only an approximation to real lenses, combinations of lenses described by Eq. (13.7) are a good starting point for a wide variety of practical systems. Examples include telephoto lenses, simple microscopes and refracting telescopes, eyepieces, and achromatic doublets.

Thick plane-parallel plate

A thick plane-parallel plate, as shown in Fig. 2.6 in AO2, has an image that is displaced along the optical axis relative to the object. As shown in Section 2.4 in AO2, the axial shift is $\Delta = d[1 - (1/n)]$. Note that the displacement Δ is independent of the object distance and, as is true in all cases in the paraxial approximation, independent of height y. For a typical glass with $n \cong 1.5$, we see that $\Delta \cong d/3$.

In the paraxial approximation, an optical system is free of any aberrations, that is, an object point is imaged precisely into an image point. When the exact form of Snell's law is used, most systems will have some form of aberration. A thick plate is a good example of a simple system with aberration, that is, it fails to take all rays from a single object point into a single image point. Applying Snell's law in its exact form at each surface, the geometry

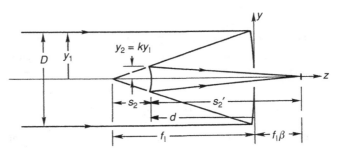

Fig. 13.3 Schematic diagram of a two-mirror Cassegrain telescope. Designated parameters are y_1 and y_2, height of ray at margin of primary and secondary, respectively; D, telescope diameter; R_1 and R_2, vertex radius of curvature of primary and secondary mirror, respectively; f_1, focal length of primary mirror; and d, distance from primary to secondary, $d < 0$. See Table 13.1 for definitions of normalized parameters.

of Fig. 2.6 in AO2 leads to

$$\Delta = d \left(1 - \frac{\cos u}{n \cos u'} \right) \tag{13.8}$$

Combining Eq. (13.8) with the paraxial relation for Δ gives

$$\Delta_{\text{exact}} - \Delta_{\text{par}} = \frac{d}{n} \left(1 - \frac{\cos u}{\cos u'} \right) \cong \frac{y_1^2 d \left(n^2 - 1 \right)}{2 s_1^2 n^3} \tag{13.9}$$

hence the image position depends on the ray height at the first surface. We note that flat plates are parts of nearly all optical systems, either as filters or windows, and thus aberrations of a thick plate must be considered in detail. We do so after a discussion of the various types of aberrations.

13.1.5 Two-mirror telescopes

We now extend the results of the preceding sections to the most common type of two-mirror telescope, the Cassegrain shown in Fig. 13.3. Here we are concerned only with the paraxial properties of this system, given in terms of parameters as defined in the legend. For a more inclusive discussion of two-mirror telescopes see Chapter 2 in AO2, especially Section 2.5.

It is helpful to describe any two-mirror system in terms of a set of dimensionless or *normalized* parameters, defined as given in Table 13.1. For the entries in Table 13.1 note that focal ratios and focal lengths are positive quantities, while the radii of the two mirrors in a Cassegrain are negative. In Fig. 13.3 we have $\beta > 0$ when the focal point lies to the right of the primary vertex. We also note that k and m are each positive for a Cassegrain.

The relationships between these parameters are obtained with the aid of Eq. (13.4) applied to the secondary. The results are

$$m = \frac{\rho}{\rho - k}; \quad \rho = \frac{mk}{m - 1}; \quad k = \frac{\rho(m - 1)}{m}; \quad 1 + \beta = k(m + 1) \tag{13.10}$$

The focal length of a two-mirror telescope is found directly from Eq. (13.5). Note that $n = -1$ because the light between the two mirrors is traveling from right to left, and that $d < 0$

Table 13.1 Normalized parameters for two-mirror Cassegrain telescopes

Parameter	Definition
$k = y_2/y_1$	Ratio of ray heights at mirror margins
$\rho = R_2/R_1$	Ratio of mirror radii of curvature
$m = -s_2'/s_2$	Transverse magnification of secondary
$f_1\beta$	Back focal distance
$F_1 = f_1/D$	Primary mirror focal ratio
$W = (1-k)f_1$	Distance from secondary to primary mirror
$F = f/D$	System focal ratio, where f is the telescope focal length

because the secondary is to the left of the primary mirror. With $d = -W = (1-k)/(R_1/2)$ from Table 13.1, Eq. (13.4), and $n = -1$ put into Eq. (13.5) we get

$$\frac{1}{f'} = -\frac{2}{R_1}\left[1 - \frac{1}{\rho} + \frac{(1-k)}{\rho}\right] = -\frac{2}{R_1}\cdot\frac{1}{m}; \quad f' = f = mf_1 \qquad (13.11)$$

Consider the rays reflected from the secondary in Fig. 13.3. If these rays are extended to the left until they intersect their corresponding incident rays, the distance between the intersection plane and the focal point, measured along the axis, is the focal length. The focal point lies to the right of the intersecting rays, hence the focal length is positive.

As for the magnification, its sign according to our convention is positive if the image made by the secondary has the same orientation as the object for the secondary. This is the case for the secondary mirror in a Cassegrain telescope. But this telescope has a final image that is inverted with respect to the original object on the sky. This is because the image given by the primary is inverted, hence the final image is also inverted. Thus the Cassegrain telescope is simply characterized as one with $f > 0$, $m > 0$, and final image inverted.

For a given pair of primary and secondary mirrors the location of the telescope focus depends on the location of the secondary. If the secondary is moved along the optical axis, then both m and k are changed, and so also is the focal position. From the discussion in Section 2.5.b in AO2 we find $d\beta = (m^2 + 1)dk$, where $f_1 d\beta$ is the shift of focus relative to the primary-mirror vertex when the secondary mirror is moved by $f_1 dk$.

Although this shift does not set any apparent limit on how far the secondary can be moved, there is a limit set by the onset of aberrations. Two-mirror telescopes generally have a mirror separation set to make the on-axis aberration zero. For a different secondary position the on-axis aberration is no longer zero, and its size sets a practical limit to the amount of secondary displacement

A final, and very significant, advantage of two-mirror systems is the additional freedom provided for controlling image quality, a topic considered below.

13.1.6 Object space to image space revisited

With selected concepts of paraxial optics now in place, we return to Eq. (13.1) and see if it is satisfied by any of the two-surface optical systems we have considered thus far. If Eq. (13.1) is satisfied, then we can state that such a system is distortion-free.

Consider first the simplest of systems, a thin lens. Let the ray from a distant star hitting the lens at its center at angle θ with the z-axis be called the *chief ray*. A thin lens is defined as one with zero thickness, hence the chief ray emerges into image space undeviated and Eq. (13.1) is satisfied. But, of course, a thin lens is an idealization and does not exist in practice.

Consider now a real lens with axial thickness d and the chief ray from a distant star hitting the lens at its center at angle θ with the z-axis. Near its center the lens can be considered nearly a plane-parallel plate. With the aid of Fig. 2.6 in AO2, and following the same procedure used in arriving at Eqs. (13.8) and (13.9) above, it turns out that there is a lateral shift of the chief ray at the detector $\propto d\theta^3$. Thus, the pattern in image space at the detector is no longer identical with the pattern of stars on the sky and Eq. (13.1) is not satisfied.

In the case of a Cassegrain telescope we trace the chief ray as if there is no hole in the center of the primary and no obscuration due to the secondary. This ray, after reflection from the secondary, appears to come from a point on the axis of the telescope. This point is at the center of a virtual surface called the *exit pupil*. If the angle of the chief ray entering the telescope is θ relative to its axis, the angle with respect to the same axis is ψ after reflection from the secondary. As shown in Section 2.6.b in AO2, the angle ψ is of order m larger than the incident chief-ray angle. Although $\psi/\theta = $ constant in the paraxial approximation, it does not follow that $\tan\psi/\tan\theta$ in Eq. (13.1) is also constant. It turns out that the lateral shift at the detector is also $\propto \theta^3$, as in the case of a thick lens.

(For two-mirror telescopes the exit pupil is the image of the aperture stop, usually the boundary of the primary, produced by the secondary; for more details on stops and pupils, including examples, read Sections 2.6.b and 2.6.c in AO2.)

13.2 Aberrations

Different approaches to aberrations of optical systems are possible, but the one that is especially illuminating proceeds via Fermat's principle. Briefly stated, this principle states that the path followed by an actual light ray in going from a point in object space to its conjugate point in image space is such that the time of travel required has a stationary value with respect to small changes of that path. For a single-plane refracting or reflecting surface, the time of travel is a minimum. For an aberration-free optical system with several surfaces, such as a telescope or microscope, the time of travel or optical path length is the same for any two rays between a given pair of conjugate points. For a system with aberrations there will be deviations from this stated principle. In this section we use reflecting surfaces to illustrate how aberrations arise.

13.2.1 Conic sections

In our discussion of paraxial optics we assumed all curved surfaces were spherical in cross-section. But a spherical reflecting surface is perfect only if the object and image point are

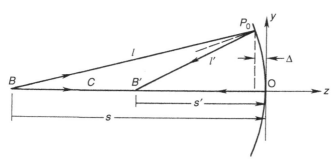

Fig. 13.4   Rays between conjugates at finite distances via a concave reflector, where l (l') is the distance from P_0 to B (B').

coincident and located at the mirror's center of curvature. As we show directly, curved reflecting surfaces will satisfy Fermat's principle for other pairs of conjugate points if we replace spherical surfaces by *conic sections*.

Consider first a concave mirror, as shown in Fig. 13.4 with an object point at B and the corresponding image point at B', both on the z-axis. We follow the sign convention for s and s', while choosing l and l' positive. Given s and $s' < 0$ in Fig. 13.4, the application of Fermat's principle to the two rays leaving B gives $l + l' = -(s + s')$. The details leading to the equation of the curve in Fig. 13.4 are found in Sec. 3.4.b in AO2; the result is an ellipse.

It is not surprising that Fermat's principle leads to an ellipse as the appropriate curve with the two conjugate points at the foci of the ellipse, considering the standard technique for drawing an ellipse with pencil, string, and two pins. A rotation of the ellipse about the z-axis gives an *ellipsoid*, with the surface equation replacing y^2 by $x^2 + y^2$. Note that the sphere is a special case of an ellipsoid in which $s = s'$. A paraboloid is a second special case; in this case $s = \infty$ and $s' = -f$.

Each of the surface cross-sections given above is a conic section and it is appropriate to find a single equation describing the family of such curves with the vertex at the origin. The result, with steps given in Section 3.5 in AO2, is

$$r^2 - 2Rz + (1 - e^2)z^2 = 0 \quad \text{or} \quad r^2 - 2Rz + (1 + K)z^2 = 0 \qquad (13.12)$$

where $r^2 = x^2 + y^2$ and $K = -e^2$.

Although derived from the ellipse equation, the relation in Eq. (13.12) describes the family of conic sections, provided we choose the *eccentricity e* appropriately. In the literature conic section is often described in terms of a *conic constant K*. In terms of e and K the various conic sections are as follows:

$$
\begin{array}{lll}
\text{oblate ellipsoid:} & e^2 < 0 & K > 0 \\
\text{sphere:} & e = 0 & K = 0 \\
\text{prolate ellipsoid:} & 0 < e < 1 & -1 < K < 0 \\
\text{paraboloid:} & e = 1 & K = -1 \\
\text{hyperboloid:} & e > 1 & K < -1
\end{array}
$$

In all of the discussion to follow, we use K to describe the conic sections.

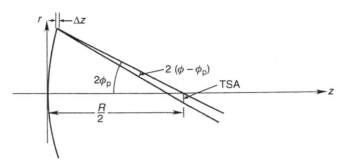

Fig. 13.5 Relation between TSA, angular difference between ray paths after reflection, and path difference between rays reflected from paraboloid and conic surfaces. The size of Δz is too small to be shown in diagram.

In summary, then, conic surfaces used as mirrors can provide perfect imagery for a single pair of conjugates. A given conic mirror, however, will not strictly satisfy Fermat's principle at any other pair of conjugates. This failure to image a point into a point indicates the presence of aberrations. In spite of this apparent limitation, the family of conic surfaces is the basis for most multi-mirror systems.

13.2.2 Aberration example

To illustrate the onset of aberrations we consider a very simple optical system, the single conic mirror with conic constant K shown in Fig. 13.4. Let the incident collimated light from a distant star be parallel to the mirror axis, hence $s = \infty$ and $s' = f$. The desired quantity is the distance from the mirror vertex to the point where a reflected ray intersects the optical axis. The analysis is given in Section 4.1 of AO2, with the result

$$s' = f = \frac{R}{2} - \frac{(1+K)r^2}{4R} - \frac{(1+K)(3+K)r^4}{16R^3} - \cdots \qquad (13.13)$$

As expected, Fermat's principle is strictly satisfied, hence f is constant for any r, only for a paraboloid with $K = -1$ when the object is at infinity. For a sphere or ellipsoid, the conic constant $K > -1$ and $f < R/2$, while for a hyperboloid $f > R/2$. Thus, for any conic surface other than a paraboloid the image of a distant object on the optical axis is blurred. This image defect is called *spherical aberration* (SA). The character of the blurred image can be seen in a ray fan and set of spot diagrams in Figs. 4.5 and 4.6, respectively, in AO2.

The size of this aberrated image can be specified in several ways: *longitudinal* (LSA), *transverse* (TSA), or *angular* (ASA). The connection between these types of descriptions is shown in Fig. 13.5. LSA is defined as the distance from the paraxial focal plane to the point where a ray from height r crosses the z-axis. In this example the paraxial focal plane is $R/2$ from the vertex and LSA is $\Delta f = f - R/2$ from Eq. (13.13). TSA is defined as the intersection of a ray from height r on the mirror with the paraxial focal plane, and is measured perpendicular to the z-axis. The relation needed to get TSA from LSA is given in Section 4.2 in AO2.

ASA is defined as the difference in direction between a reflected ray and a pseudo-ray passing through the paraxial focal point. From the geometry in Fig. 13.5 we see directly the relation between transverse and angular aberration in this example, namely, TSA $= (R/2)$ (ASA), provided the angle φ is not too large. ASA, in turn, is related to the difference in slope of the actual mirror from that of the paraboloid at the point of reflection. As shown in Section 4.2 in AO2, the difference in slope can be expressed in terms of the optical path difference between the actual and reference mirrors.

The procedure outlined here for collimated light incident on a conic mirror can be generalized to any mirror and any pair of object and image conjugates. All that is needed is Δz, the difference between the reflecting surface that images without aberration and the actual surface. This procedure can be further extended and applied to optical systems with more than one surface; once Δz is found, it is then a straightforward matter to calculate angular and transverse aberrations.

13.2.3 Off-axis aberrations

We now turn our attention briefly to off-axis aberrations, those aberrations present when the object point does not lie on the optical axis. Here we want to indicate the nature of these aberrations and their dependence on aperture radius and field angle. The analysis in Section 5.1 of AO2 gives Δz, the optical path difference between two rays from an off-axis object point in the yz-plane to its conjugate point. These rays are the chief ray, that ray intersecting the surface at its vertex, and an arbitrary ray meeting the surface at (x, y). The result that follows is the angular aberration to third-order, derived from Δz in the special case $x = 0$:

$$\text{AA3} = a_3 \frac{y^3}{R^3} + a_2 \frac{y^2 \theta}{R^2} + a_1 \frac{y \theta^2}{R} + a_0 \theta^3 \tag{13.14}$$

The terms in Eq. (13.14) represent different aberrations: from left to right they are *spherical aberration*, *coma*, *astigmatism*, and *distortion*. The character of each aberration is quite different because of the different way in which each depends on y and θ. Equation (13.14) applies to both refracting and reflecting surfaces, with different coefficients for corresponding terms. Our following description of each aberration is based partly on Eq. (13.14), along with additional insight gained from spot diagrams near the image surface.

Spherical aberration does not depend on θ, hence is constant over the object field. A change in the sign of y changes the sign of this aberration and thus rays from opposite sides of the aperture are on opposite sides of the chief ray at the image. When rays over the entire aperture are included, the image is seen to be circularly symmetric about the chief ray, as shown in Fig. 4.6 in AO2.

Coma is proportional to $y^2 \theta$ and hence is changed in sign when θ changes sign. Coma is invariant to the sign of y and therefore rays from opposite sides of the surface are on the same side of the chief ray at the conjugate image. The result is an asymmetric image that looks like a comet, hence the name coma. A set of through-focus spot diagrams is shown in Fig. 13.6(a) (Fig. 5.9 in AO2.).

Astigmatism is proportional to $y \theta^2$ and hence is unchanged by a sign change in θ. Comments about the sign of y for spherical aberration are true for astigmatism as well, but

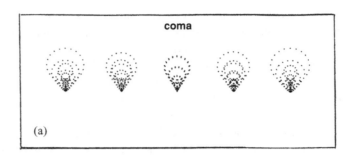

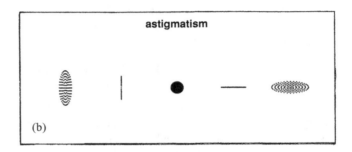

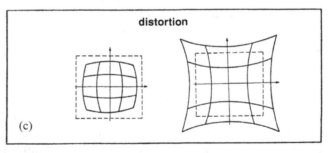

Fig. 13.6 (a) Through-focus spot diagrams of an image with pure coma; (b) Spot diagrams for an image with astigmatism. The tangential (T) and sagittal (S) images are to the right and left, respectively, of the astigmatic blur circle; (c) Image with barrel distortion (left), and pincushion distortion (right). Object is a dashed square.

the character of the image is quite different. This is easily seen in Fig. 13.6(b) (Fig. 5.3 in AO2). In going through focus of an astigmatic image we find two orthogonal line images with a circular blur located midway between the line images. Each of the line images lies on its own curved focal surface, as does the circular blur. Each of these surfaces has its own radius of curvature and is related to an aberration that does not appear in Eq. (13.14): *curvature of field*. We consider this aberration when we give characteristics of specific telescopes.

Distortion is proportional to θ^3 and does not depend on y. Thus, this aberration, if it is the only one present, does not affect the image quality, only its position. For a set of point objects equally spaced perpendicular to the optical axis, the set of images would not be equally spaced if distortion is present. In Fig. 13.6(c) are shown the patterns on a flat detector of a square object field by systems with distortion; one is called barrel, the other pincushion.

At this stage, five third-order monochromatic aberrations have been identified: spherical aberration, coma, astigmatism, distortion, and curvature of field. With the exception of spherical aberration, all depend on some power of θ. The first three aberrations affect image quality; the last two affect only image position.

It is important at this point to indicate how each of these aberrations can affect astrometric measurements. Our reference is that of the perfect image, one whose full width at half maximum (FWHM) is set by diffraction. (This topic is covered in Chapter 10 in this book.) An image with spherical aberration is less sharply peaked and has a larger FWHM than a perfect image, thus a larger uncertainty in its central location. Because an image with this defect is circularly symmetric about the optical axis, the center of gravity (CG) of the image is unchanged and image positions do not need adjustment.

Astigmatic images, by contrast, are circularly symmetric only midway between the line images. If astigmatism is the dominant aberration, as it is for one type of telescope we discuss below, then the focal surface of interest is the one for these circularly symmetric images. This focal surface is generally curved and this, too, complicates any adjustment needed to accurately locate image centers.

An image with coma is obviously one in which the peak of the measured intensity is not coincident with the chief ray, hence some kind of adjustment must be made to more accurately locate the proper image positions. Detailed modeling is needed to determine the size of this field-dependent effect and make adjustments.

The final monochromatic aberration, distortion, obviously affects image positions and must be "backed out" for accurate relative positions as discussed in Chapter 19. For purely reflecting telescopes, the magnitude and sign of distortion can be calculated; the situation becomes more complicated when refracting elements are added. Examples to follow show the significance of this aberration.

We have so far limited our discussion to monochromatic aberrations, but must now note that chromatic aberrations arise for any refracting surface when a range of wavelengths is considered. All refracting materials have an index of refraction that changes with changing wavelength and *longitudinal chromatic aberration* and *lateral chromatic aberration* are both present for a single surface. The former is analogous to the change in focus for a conic section discussed above, while the latter is a transverse shift along a focal surface. One or both of these aberrations can be reduced to negligible levels with careful design of refracting systems.

13.2.4 Telescopes and aberrations

With an overview of the nature of aberrations and their functional relation to field angle θ and aperture radius y in hand, we now consider several specific types of telescopes and their aberrations. For reflecting telescopes and simple refracting elements we give explicit expressions for the aberrations; for more complicated combinations we depend on results from ray-tracing. Results are given without derivation and are, for the most part, taken from Chapters 5, 6, and 9 in AO2.

We begin by giving a two-letter designations for each aberration as follows: SA, spherical aberration; TC, tangential coma; AS, astigmatism; and DI, distortion. If the aberration is transverse, a prefix T is attached; if the aberration is angular, a prefix A is attached. Thus TTC is transverse tangential coma. For a telescope, the relation between transverse and angular aberrations is $Axx = Txx/f$, where f is the telescope focal length. A useful conversion factor from one to the other is the telescope scale S expressed in arcsec/mm or its inverse given in µm/arcsec.

Without regard for sign, each of the transverse aberrations has the following interpretation: TSA = blur radius at paraxial focus; TTC = length of comatic flare; TAS = astigmatic blur circle diameter; TDI = transverse displacement of chief ray. We use the symbol κ to denote surface curvature and u to represent image surface sag from a plane surface.

For each type of telescope to follow we choose an existing one, give the necessary optical parameters and aberration relations for that telescope type, and discuss the nature of the focal surface and character of the images. We start with a simple paraboloid, then on to more complex systems, including the addition of refracting plates to simulate filters and/or detector windows and correcting lenses. For each of these telescopes we assume no misalignments, hence all optical axes coincide.

Paraboloid

The simplest telescope that is free from spherical aberration is a paraboloid. Our example of this type is the US Naval Observatory astrometric reflector at its Flagstaff station. It is a folded Newtonian with a plane mirror redirecting the converging beam from the paraboloid back through a hole in its center. The relevant optical parameters are $D = 1.55$ m, $F = 9.8$, $f = 15.2$ m, and scale $S = 13.6$ arcsec/mm or 74 µm/arcsec.

The aberrations of a parabolic telescope and its image surface characteristics are given in Table 6.1 in AO2, with a summary as follows:

$$ATC = \frac{3\theta}{16F^2}; \quad AAS = \frac{\theta^2}{2F}; \quad ADI = 0; \quad \kappa_m = \frac{1}{f}; \quad u_m = \frac{f\theta^2}{2} \tag{13.15}$$

Note (1) that distortion is zero and (2) that the image surface is concave as seen from the direction of the light incident upon it.

An analysis of the aberrations in Eq. (13.15), where θ is measured in radians, shows that coma is dominant over astigmatism with, for example, ATC = 0.59 arcsec and AAS = 0.022 arcsec for $\theta = 5$ arcmin. We assume this field angle as the limiting one for which astrometric plates are taken, with comatic images beginning to degrade image quality under excellent seeing conditions. At this limiting angle the sag u_m of the curved focal surface is about 16 µm, hence the focal surface can be considered as flat within the depth of focus of the $f/9.8$ beam.

Given the asymmetric character of a comatic image, as seen in Fig. 13.6(a), the location of the true center of an image near the edge of the field will likely differ from the CG of the recorded image. In this case careful modeling of a comatic image, with central obscuration of the folding flat included, and its convolution with a seeing disk would be required to map the recorded images into a true map.

Classical Cassegrain

This type of two-mirror telescope is one for which the primary is a paraboloid and the secondary is a hyperboloid with its conic constant chosen to give zero spherical aberration. As noted in Section 6.2.a in AO2, coma for this type of telescope is identical with that of the paraboloid of the same focal ratio F. Although astigmatism is larger for a classical Cassegrain, 0.07 arcsec compared to 0.022 arcsec above, coma still dominates at $\theta = 5$ arcmin. Distortion is negligible at this field angle, with a value of about 2 milliarcsec.

Another major difference between the paraboloid and the classical Cassegrain is the curvature of the focal surface. The reciprocal of the curvature, or radius of curvature, is 15.2 m for the paraboloid and about 1.1 m for a classical Cass of the same focal ratio and diameter. Thus the sag of the focal surface at the same field angle is some 14× larger and its curvature must be taken into account in an astrometric analysis.

Aplanatic Cassegrain

Any optical system that is free from both coma and spherical aberration is called an *aplanat*. Because image quality is significantly better over a larger field, this type of two-mirror telescope, commonly called a *Ritchey–Chrétien*, has been the overwhelming choice for large telescopes in recent years. Both mirrors for this type are hyperbolic in cross-section, with relations for the conic constants given in Section 6.2.b in AO2. Our example of an aplanatic telescope is the WIYN telescope located on Kitt Peak in Arizona. The relevant optical parameters are $D = 3.5$ m, $F_1 = 1.75$, $F = 6.28$, $m = 3.59$, $k = 0.314$, $\beta = 0.442$, and scale $S = 9.38$ arcsec/mm or 106.6 μm/arcsec.

Expressions for the aberrations of the class of aplanatic two-mirror telescopes, taken from Table 6.9 in AO2 (with the corrected relation for ADI), are as follows:

$$\text{AAS} = \frac{\theta^2}{2F}\left[\frac{m(2m+1)+\beta}{2m(1+\beta)}\right]; \quad \kappa_m = \frac{2}{R_1}\left[\frac{(m+1)}{m^2(1+\beta)}\left(m^2 - \beta(m-1)\right)\right] = \frac{1}{R_m}$$

$$\text{ADI} = \frac{\theta^3(m-\beta)}{4m^2(1+\beta)^2}[m(m^2-2)+\beta(3m^2-2)] \tag{13.16}$$

Using these relations we find the following aberrations at a field angle $\theta = 12$ arcmin: AAS = 0.576 arcsec, ADI = 0.014 arcsec, $R_m = -2115$ mm, and focal surface sag $u_m = 1.4$ mm. Compared to the field size available with either of the telescope types discussed above, the gain of about an order of magnitude in area is significant. Note also that the image blur is now symmetric and the image centers can be located more accurately.

Although the gain in field coverage makes it obvious why this type of telescope is preferred over other two-mirror types, there still remains the question of how best to either eliminate or reduce one or both of the remaining aberrations, astigmatism and curvature of field. Before exploring answers to this question we first consider the aberrations of refracting plates and lenses. This is necessary because answers to the question just posed usually require the introduction of refracting elements.

Refracting plates

In Section 13.1.4 above we noted that a flat plate of thickness d and index n causes an axial or longitudinal shift of an incident light ray by an amount $\Delta = d[1 - (1/n)]$ in the paraxial approximation. Thus a flat plate placed ahead of the focal surface in a telescope will require a refocus of the telescope by this amount. There is also a transverse shift of $u\Delta$ in this same approximation. The net result of these two shifts is that all rays intersect the image surfaces in exactly the same places in the two situations, at least in the paraxial case.

As also noted in Section 13.1.4, aberrations do arise when light passes through a flat plate. Aberrations that are present, without regard for sign, are as follows:

$$\text{TSA} = d\left(\frac{1}{16F^3}\right) \cdot \text{f}(n); \qquad \text{TTC} = d\left(\frac{3\psi}{8F^2}\right) \cdot \text{f}(n); \qquad \text{TAS} = d\left(\frac{\psi^2}{2F}\right) \cdot \text{f}(n)$$

$$\text{TDI} = d\left(\frac{\psi^3}{2}\right) \cdot \text{f}(n); \quad \text{where} \quad \text{f}(n) = \frac{(n^2 - 1)}{n^3}; \quad \psi = \theta\left(\frac{m^2 + \beta}{m(1 + \beta)}\right) \quad (13.17)$$

$$\Delta = d\left(1 - \frac{1}{n}\right); \qquad \text{LCA} = \frac{d\Delta}{dn}dn = d\frac{dn}{n^2}; \qquad \text{TCA} = \text{LCA} \cdot \psi \quad (13.18)$$

Monochromatic and chromatic aberrations are given in Eqs. (13.17) and (13.18), respectively. Included in Eq. (13.18) is the axial shift Δ first introduced in Section 13.1.4, along with LCA, longitudinal chromatic aberration and TCA, transverse (or lateral) chromatic aberration.

The angle ψ in both of these equations was first introduced in Section 13.1.6. It is the angle of a chief ray leaving the telescope exit pupil that first entered the telescope at angle θ, with both angles measured from the telescope axis. In a paraboloidal telescope $\psi = \theta$, with this same equality nearly true for a refracting telescope. In a two-mirror telescope, however, the exit pupil is located near the focal point of the primary and the relation between ψ and θ is that given in Eq. (13.17). The result for the WIYN telescope is $\psi = 2.575\theta$.

Note that the aberrations of a flat plate in Eqs. (13.17) and (13.18) do not depend on where the plate is located in the beam following the exit pupil. (Except of course if the filter is not flat and then the distance from the focal plane becomes critical. See for example van Altena and Monnier 1968); all that matters are the parameters of the plate and the angle and focal ratio of the beam passing through it. As an example we put a 20-mm thick plate of BK7 glass in the WIYN telescope with $\theta = 12$ arcmin. The aberrations, with units of μm, are as follows: TSA = 1.88, TTC = 0.64, TAS = 0.05, TDI = 0.003, LCA = 117, and TCA = 1.0. Given the scale of 106.6 μm/arcsec, it is clear that all of the aberrations introduced by the plate, except possibly LCA, are negligible. The wavelength span for the chromatic aberrations is 0.45–0.75 μm, an unlikely range in practice. The color blur for beams spanning this wavelength range is about 20 μm or 0.2 arcsec.

Lenses

There are no simple relations for lenses comparable to those in Eqs. (13.17) and (13.18), but we can still make some general statements about their aberrations. It turns out that lens aberrations are largely determined by the shape of the lens and the location of the aperture stop.

Consider first a simple lens with paraxial relations given in Eqs. (13.5) and (13.6). As seen from these equations, there are an unlimited number of combinations of R_1 and R_2 for any given focal length. Each such combination gives a lens with a different shape and it is the shape of the lens which determine the sizes of both the spherical aberration and coma. For a single lens there is one shape for which coma is zero; the same cannot be said for spherical aberration.

Astigmatism of a single lens, on the other hand, is not influenced by lens shape and is determined solely by the focal length; the longer is the focal length the less is the astigmatism for a given lens diameter. Because astigmatism is not zero, field curvature is also present. For separated lenses, spacing between lenses and locations of stops also influences the astigmatism and field curvature.

The amount and type of distortion introduced by a lens depends largely on the location of the aperture stop. If the stop is at the edge of a thin lens, distortion is zero. Distortion is present if the stop is separated from the lens, with the amount of the distortion proportional to the cube of this distance. The telescope exit pupil is the effective stop for a lens placed in a telescope following the secondary.

Aplanat with correcting optics

A two-mirror aplanat, such as the WIYN telescope, provides a moderate-sized field with good image quality. Increasing the field size by reducing astigmatism to a negligible level is the next logical step for an aplanat and several ways of doing this are possible. It is possible to derive the conditions for an astigmatism-free two mirror telescope. Two such telescopes, so-called anastigmatic aplanats, are discussed in Section 6.2.c in AO2. As pointed out there, both have significant drawbacks and neither is a viable option.

The other options are basically two-fold: (1) for an existing aplanat such as WIYN add additional optics, usually refracting elements, to eliminate astigmatism and reduce field curvature, or (2) start from the ground and design the telescope and additional correcting optics as a single system with the desired optical characteristics. We first consider briefly possible refracting optics to implement option (1); we then discuss briefly a telescope built according to (2).

For an aplanat telescope we can nearly eliminate astigmatism with a so-called Gascoigne corrector. The approach to designing such an aspheric corrector is described in Section 9.3.a in AO2 for an $f/10$ aplanat. As described there, this corrector indeed does its job, flattening the field somewhat and reducing AAS from 2.86 arcsec at $\theta = 30$ arcmin to 0.17 arcsec while introducing only a modest amount of coma, ATC $= 0.16$ arcsec. Although there is very good image quality over a $1°$ diameter field, computer ray-tracing reveals that ADI $= 12$ arcsec at the edge of this field, a distortion that is 40 times larger than without the corrector. We also find that lateral color is now significant with a shift on the outer parts of the focal surface of about 0.5 arcsec over a wavelength span of 0.4–0.7 μm. This color shift is evident in Fig. 9.8 in AO2.

One alternative to a Gascoigne corrector is a closely spaced zero-power doublet placed in about the same location in the telescope beam. The author has done this as an exercise using the WIYN telescope and an optimizing ray-trace program. The result is a focal surface some 2.5 times flatter than that of the bare telescope, with image blurs $\cong$ 0.25 arcsec at the edge of a 1° diameter field. Once again, large distortion is introduced, about 50 times larger than that of the bare telescope. A small amount of color aberration is also present.

The upshot of the discussion in this subsection is to demonstrate that the introduction of refracting elements may solve one problem but at the expense of introducing others. It is evident that careful modeling is required to determine the aberrations present before beginning an astrometric analysis.

SDSS telescope

The other option noted above, that of starting from the ground up, was taken by the Astrophysical Research Consortium (ARC). An excellent and exhaustive description of the Sloan Digital Sky Survey telescope and how it came into being is given by Gunn *et al.* (2006).

The gist of the optical side of this telescope is best given by quoting from the abstract of the Gunn paper, as follows:

> The telescope is a modified two-corrector Ritchey–Chrétien design with a 2.5 m $f/2.25$ primary, a 1.08 m secondary, a Gascoigne astigmatism corrector, and one of a pair of interchangeable highly aspheric correctors near the focal plane, one for imaging and the other for spectroscopy. The final focal ratio is $f/5$.... Novel features of the telescope include ... a 3° diameter (0.65 m) focal plane that has excellent image quality and small geometric distortions over a wide wavelength range (3000−10,600 Å) in the imaging mode....

The complete telescope is a modified Ritchey–Chrétien design, meaning that the bare telescope is no longer an aplanat. For a description of telescopes of this type see Section 9.3.c in AO2 and the article by Bowen and Vaughan (1973). Many more details are given in Gunn *et al.* (2006). The astrometric calibration is described by Pier *et al.* (2003).

13.2.5 Misalignments and aberrations

We have considered the aberrations of a variety of reflecting telescopes and/or refracting optics, with the aim of showing the nature of the aberrations and steps that can be taken to eliminate one or more of them. An implicit assumption in this discussion was that all optical elements were aligned to a common axis. We now discuss aberrations that arise when one or more elements is misaligned, with our remarks based in part on Section 6.3 in AO2 and in part on ray-traces.

Misalignments in two-mirror telescopes

Relative to the primary mirror in a telescope, the secondary mirror in a telescope can have one of three position errors: *decenter, tilt,* or *despace.* Decenter is a transverse shift of the secondary axis relative to that of the primary, tilt is a rotation of the secondary axis, and despace is an axial translation of the secondary toward or away from the primary. The first two are true misalignments, while despace is a shift away from the design location that gives zero spherical aberration. We limit our discussion to misalignments of aplanatic Cassegrain telescopes, with or without accompanying refracting elements.

Assume a secondary decentered by l and tilted by angle α with respect to the axis of the primary, as shown in Fig. 6.6 in AO2. As shown there, the decenter and tilt angle are in the same plane to allow for the possibility of one cancelling the other. For an aplanatic telescope this misalignment gives rise to coma according to:

$$\text{ATC} = -\frac{3(1+\beta)(m-1)}{16F^2}\left[\alpha - \frac{l}{f}\frac{m}{k}\left\{1 + \frac{1}{(m-\beta)(m-1)}\right\}\right] \qquad (13.19)$$

The principal feature of this expression is no dependence on field angle θ, hence coma due to mirror misalignment is *constant* over the field.

To illustrate the size of this coma we take the parameters for the WIYN telescope given above and find

$$\text{ATC(arcsec)} = 1.065\alpha(\text{arcmin}) - 2.14\,l(\text{mm}) \qquad (13.20)$$

Thus, either a tilt of 2 arcmin or a decenter of 1 mm gives coma of about 2 arcsec over the entire field, with each accompanied by image shifts across the field. This coma would be easily noticed in any image and steps would be taken to realign the secondary. Note also, however, that a combined tilt and decenter in the same plane can give zero coma, hence an apparently aligned telescope.

In this case there remains an astigmatism introduced by the misalignment; this adds to that already there in an aplanat. This astigmatism consists of a constant term (almost always negligible) and a term that is *linear* in the field angle. For the WIYN telescope the linear AST(arcsec) $\cong 0.01\theta_y$ (arcmin) l(mm).

We now give results from ray traces using the WIYN telescope as our reference. We consider a misaligned WIYN with (1) no additional optics, and (2) with a refracting corrector (closely spaced doublet discussed briefly above).

(1) Image quality is similar to that of an aligned WIYN over the field of good images ($\theta \approx 0.2°$), with linear astigmatism causing a slight asymmetry of images on opposite sides of the field. (Adding 20 mm of flat glass, either normal to the primary axis or tilted, has no noticeable effect on image quality.)

(2) When a refracting corrector is added to a misaligned, but coma-free, WIYN telescope, the effect is more dramatic. Linear astigmatism is seen at the limits of a 1° diameter field for l as small as 0.5 mm, and through-focus images show focal surface tilt. More troubling is the left–right difference between two images at $\theta = \pm0.5°$ and one at $\theta = 0$. For $l = 0.5$ mm this difference is about 100 μm or nearly 1 arcsec projected on the sky. Such a difference would give rise to significant errors in astrometric measurements.

Other interesting effects are also evident in ray-traces of this system. It is possible to decenter the refracting corrector and change the left–right difference, even improving the image quality in the process. A decenter of one of the lenses within the doublet has similar effects. Likewise, an aligned telescope with tilts and/or decenters of one or more of the lenses within the corrector can show similar kinds of image asymmetries and shifts.

As another example let the final lens in the SDSS telescope two-lens corrector be decentered by 1 mm. Because this lens is fairly close to the focal plane, changes in the image quality are hardly noticeable. But the left–right difference of the type noted above is again present with a value of 35 μm or 0.6 arcsec at $\theta = \pm 1°$.

Possible combinations of tilts and decenter are almost endless and an astrometrist should be ever conscious of the possibility of misalignments, should develop observing strategies to uncover their presence, and be ready to model fully any optical system used to acquire images (see Chapter 19).

13.3 Concluding comments

In this chapter we have tried to highlight some of the important relations for both paraxial or first-order optics, where there are no aberrations, and the real world of telescopes and refracting optics where aberrations abound. Much of our discussion of aberrations is given in terms of third-order formulas; these relations, though not exact, give a good picture of the kinds of effects optical systems have on bundles of light rays. We have not described the methods used to correct recorded images for any of these aberrations, but simply pointed out that coma, distortion, and chromatic aberration are especially in need of correction as is discussed in detail in Chapter 19.

There is no system that is ideal in the sense of the pinhole camera, but awareness of the potential errors in images taken with a real telescope, either aligned or misaligned, is necessary to fully and correctly analyze these images.

References

Born, M. and Wolf, E. (1980). *Principles of Optics*, 6th edn. Oxford: Pergamon.

Bowen, I. and Vaughan, A. (1973). The optical design of the 40-in. telescope and of the Irenee DuPont telescope at Las Campanas Observatory, Chile. *Appl. Opt.*, **12**, 1430.

Gunn, J. Siegmund, W. A., Mannery, E. J., *et al.* (2006). The 2.5 m telescope of the Sloan Digital Sky Survey. *AJ*, **131**, 2332.

Jenkins, F. and White, H. (1976). *Fundamentals of Optics*, 4th edn. New York, NY: McGraw-Hill.

Longhurst, R. (1967). *Geometrical and Physical Optics*, 2nd edn. New York, NY: John Wiley.

Mahajan, V. (1998). *Optical Imaging and Aberrations, Part I, Ray Geometrical Optics*, SPIE (Bellingham, WA: SPIE Optical Engineering Press).

Pier, J., Munn, J. A., Hindsley, R. B. *et al.* (2003). Astrometric calibration of the Sloan Digital Sky Survey. *AJ*, **125**, 1559.

Schroeder, D. (2000). *Astronomical Optics*, 2nd edn. San Diego, CA: Academic Press.

van Altena, W. F. and Monnier, R. C. (1968). Astrometric accuracy and the flatness of glass filters. *AJ*, **73**, 649.

Wilson, R. (1996). *Reflecting Telescope Optics I*. Berlin: Springer.

CCD imaging detectors

STEVE B. HOWELL

Introduction

Charge-coupled devices (CCDs), the standard imagers at all observatories today, consist of integrated circuits made through the same process as computer memory or the chips in cell phones. Complementary metal oxide semiconductors (CMOS) are an alternative image-sensor technology with high noise immunity and low static power consumption; however, CCDs are the dominant imagers today, so we will concentrate our discussion on their use.

Silicon crystals are sensitive to light through the process by which incident photons of sufficient energy can excite electrons into the valence levels of the atom. Photons with energies less than the valence levels fail to create photoelectrons and are therefore not detected, while the higher-energy photons are absorbed near the surface of the silicon layer before creating usable photoelectrons. If one applies a voltage to the silicon in a controlled manner, these photoelectrons can be either held in place (during the integration) or moved through the silicon lattice (during readout) and collected.

14.1 What is a charge-coupled device?

When a CCD is constructed, a square grid of microscopic electrodes called gates is fabricated on the surface of a silicon wafer (see Fig. 14.1). The orthogonal axes of the grid are called columns and rows and the grid elements are the pixels, which have typical sizes of 10–20 μm. When exposed to optical light, the silicon substrate reacts to each absorbed photon by creating one photoelectron–hole pair. The gate voltages control the movement and position of these photoelectrons. During a stare exposure, e.g. when the telescope is tracking the target and the shutter is open, the voltages are fixed so that the photoelectrons generated within each pixel are collected and held in place. In drift-scan mode (see Chapter 15), the sky field is allowed to drift across the array in the vertical direction during the exposure, and readout is continuous with the vertical shift rate matched to the drift rate of the sky. The number of photoelectrons within each pixel is directly proportional to the

Astrometry for Astrophysics: Methods, Models, and Applications, ed. William F. van Altena.
Published by Cambridge University Press. © Cambridge University Press 2013.

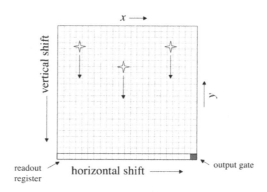

Fig. 14.1 A diagram representing the readout of a CCD. The pixels form a grid defined by columns (vertical) and rows (horizontal). Photoelectrons collected within each pixel are shifted, row by row, into the readout register. Between each vertical shift, electrons in the readout register are shifted horizontally to the output gate where they are sampled. In stare mode, the entire array is read out between exposures. In drift-scan mode (Chapter 15), the sky field is allowed to drift across the array in the vertical direction during the exposure, and readout is continuous with the vertical shift rate matched to the drift rate of the sky.

number of object photons incident on the pixel up to a saturation value, either the well depth or the analog-to-digital (A/D) limit. Hence a CCD maps light intensity versus position on its surface into electron number versus pixel position within its substrate.

To read out the electronic image, the gate voltages are clocked in such a way that the photoelectrons shift from one pixel to the next within each column. On each clock cycle, all the pixels in all the columns are shifted at the same time. The effect is to shift the whole electronic image in the column direction, one row at a time. This is called a vertical shift. The shift direction is towards a special row at the end position of the columns known as the serial readout register. On each cycle of a vertical shift, a new row of electrons is shifted into the serial readout register.

The gates of the serial register are connected so that the electrons can be shifted in the horizontal or row direction, a pixel at a time, towards one special pixel at the end of the readout register called the output gate, which is connected to an on-chip amplifier. Between horizontal shifts, the total charge in the output gate is converted into a voltage, amplified, digitized, and recorded by a computer. Each time there is a vertical shift by one row, all the electrons in the serial register are shifted horizontally and sampled at the output gate. For a CCD with N_r rows and N_c columns, a complete readout in stare mode consists of $N_r \times N_c$ samples at the output gate. This occurs while the camera shutter is closed so that no new photoelectrons are generated during the readout. Upon completion, all the CCD pixels have had their collection wells emptied, and the array is ready for a new exposure. Additional false pixels are often sampled (usually 32 pixels at the end of each row) during readout to provide a light-free "zero level" yielding a row-by-row measurement of the electronic noise. These extra pixels are called the overscan.

The output number, analog-to-digital unit (ADU) or data number (DN), is a representation of the number of collected photons in each pixel. How this conversion takes place, the accuracy of the conversion, and other details related to A/D converters is covered in detail in Janesick (2001) and Howell (2006).

The voltage manipulation of the photoelectrons involved in reading out the CCD is termed clocking the device. For most modern CCDs a three-phase clocking is used where the voltages generally change over the range of 0 to +15 volts, in 5-volt increments (see Howell 2006, Fig. 2.2). Special CCDs and CCD controllers use variants on this shifting process but the general principles are the same. The CCDs are referred to by their manufacturer and/or (x, y) pixel size. For example, a Tek2048 CCD would mean it was manufactured by Tektronics and is 2048 × 2048 pixels in size. Not all CCDs are square and spectroscopic applications often call for a rectangular CCD, possibly of size 1024 × 4096.

Charge-coupled devices also come in a variety of types, of which front-side-illuminated and back-side-illuminated are the most common, the difference being which side of the device the incident light falls upon. In front-side-illuminated devices the light is obstructed in part by the gate structures that provide the readout capability of the CCD; hence their effective sensitivity to light is lowered. Back-side devices are thinned, that is physically made thinner after manufacturing by chemical etching processes. It was experimentally found that thinning and using CCDs in a back-side-illuminated manner greatly increases their sensitivity to light, especially to the blue light that is otherwise absorbed in the surface layers of the silicon before creating usable photoelectrons. The sensitivity is also generally lower to red light since the lower photon energies cannot generate photoelectrons or carriers. We will discuss these issues further below.

There are many other types of CCDs, such as surface-channel or buried-channel devices, interline or frame-transfer devices, anti-blooming- and multi-pinned-phase CCDs, orthogonal-transfer CCDs, and low-light-level or electron-multiplying CCDs, and a wealth of other types of imagers. The general references listed at the end of the chapter, especially Janesick (2001) and Mackay (1986), discuss this cornucopia of digital detectors in far more detail than presented here.

14.2 CCD characterization parameters

14.2.1 Quantum efficiency

Responsive quantum efficiency or RQE of a CCD is a prime metric as it tells the user at a glance how effective a specific device will be for observations. If a CCD has 80% RQE at a specific wavelength, it should collect 80% of the light at this wavelength incident on the detector and convert it into photoelectrons within the pixel.

$$RQE = N(\text{detected})/N(\text{incident}) \tag{14.1}$$

Quantum efficiency has made great gains in the past 25 years since the introduction of CCDs into astronomy. In the 1980s, CCD RQE was near 20–40% at best and today it can be as high as 90–95% over some regions of the optical bandpass. Modern back-side-illuminated

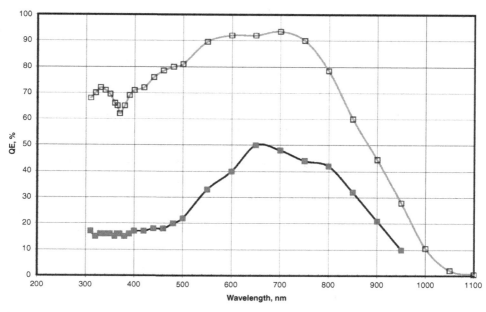

Fig. 14.2   A modern thinned back-side-illuminated CCD (open squares) and a Lumigen-coated thick front-side-illuminated CCD (filled squares). The thinned device has far better quantum efficiency over the entire optical range while the Lumigen coating extends the thick device RQE to 300 nm. These two RQE curves are from new-generation CCDs developed and being tested at the WIYN Observatory on Kitt Peak in southern Arizona.

CCDs usually have RQE values over 70% from about 400 nm to 800 nm with sensitivity to 1 μm. Thick, front-side-illuminated devices peak in RQE near 50% at 650 nm and fall off in RQE to both the red and blue often with essentially no sensitivity blue-ward of about 400 nm. Adding a coating, such as Lumigen (Howell, 2006), to a thick CCD can enhance and extend its blue response to about 300 nm. This general discussion is illustrated in Fig. 14.2, which shows a modern thinned back-side-illuminated CCD and a Lumigen-coated thick front-side device. The detective quantum efficiency or DQE defined in Eq. (14.2) is a measure of the detection efficiency of the CCD in terms of how well it preserves the input signal-to-noise ratio (SNR).

$$DQE = (SNR)_{\text{in}}^2 / (SNR)_{\text{out}}^2 \qquad (14.2)$$

CCDs are usually characterized by their RQE, but when discussing the precision of data obtained with a CCD, DQE is the important parameter.

Finally, we take another look at the quantum efficiency of a CCD. Many CCD observations and current and planned surveys focus on the red portion of the optical bandpass. For a CCD, the exact RQE in the red bandpasses depends on the temperature of the CCD. For wavelengths from 800 nm to 1100 nm, the RQE for incident photons can vary by a factor of two with changes of 60 °C in the CCD temperature. Most CCD systems regulate the temperature of the detector to within 1 °C, so the change in RQE is essentially zero. Some new systems are planning to exploit this change in RQE and run the detector slightly hotter or colder by a few degrees to optimize the RQE for bluer or redder observations.

14.2.2 Readout noise

Readout noise, N_r, or just read noise, is another important characterization factor in the CCD lexicon. CCDs are analog detectors and each time the integrated charge is read, Johnson noise in the switch introduces an uncertainty in the bias voltage actually applied. Since this noise is a function of how rapidly the charge is read, many CCDs offer the option to read the frame rapidly and accept a larger read noise. This feature is often of use when many sequential exposures are taken at different focus settings to determine the best focus. Read noise is added in quadrature to the object frame signal and the DQE is lowered as a consequence, so the goal is normally to have very low read noise. Modern CCDs can have read-noise values that are amazingly low, near 2–4 electrons per second.

14.2.3 Dark current

Dark current is generally caused by thermally generated electrons in the CCD that increase the signal in a pixel with time. At room temperature many CCDs saturate in about 1 second. However, since it is a thermal effect with a typical doubling of the dark current rate every $\sim$8 K, cooling the CCD generally results in very low dark current. In general the dark current rate, R_d, is given by

$$R_d = R_{d_0} 2^{(T-T_o)/dT} \tag{14.3}$$

where R_{d0} is the dark current per second at reference temperature T_o and dT is its doubling temperature. Some pixels will be found to be *hot* and have substantially higher dark-current rates than the majority of pixels. They must be identified and dealt with separately by creating a mask of their (x, y) locations and deleting them from the analysis. Dark current is the major reason for placing CCDs in vacuum dewars and operating them at cold temperatures such as $-120\,^\circ$C. As the operating temperature drops, a CCD's dark-current rate will drop until it reaches an acceptable level, generally near 0.001 thermally generated electrons per pixel per second. It is often necessary to find an optimum operating temperature to balance resources using liquid nitrogen and a vacuum dewar or a thermoelectric cooler as well as optimizing the CCD performance. For example, operating a CCD at a temperature to minimize the dark current may significantly impair the charge-transfer efficiency (CTE) and result in image shape distortions and limit the astrometric and photometric capabilities of the CCD. Zacharias *et al.* (2000) discuss in detail the effect of CCD temperature on CTE as well as its calibration.

14.2.4 Linearity of response

CCDs are linear devices meaning that the output DN value is proportional to the input signal (incident photons) at all signal levels until the well is near saturation, which is often 100 000 electrons or more. For example, if a CCD collects 1000 DN per pixel in a 30-second exposure, that same CCD will collect 10 000 DN per pixel for a 300-second exposure, keeping all other factors equal. CCDs are nearly linear but do show small deviations, often of less than 1% over the range of collected photoelectrons, which often reaches six magnitudes. The degree of a CCD's linearity is highly related to the type and method of A/D conversion used in the output. The linearity of a CCD is an important part of the data-reduction process since most photometric reduction schemes assume that the device is linear. While CCD linearity is highly desirable for accurate photometry, it is not needed for astrometry; however, the best photometry requires precise image centering so it makes sense to require linearity in CCDs used for astrometry. Rough calibration of linearity is relatively simple since it only requires taking several exposures, doubling the exposure each time.

14.2.5 Gain

The gain of a CCD is set by the control electronics and provides a mechanism by which collected photoelectrons are converted to output ADUs. For example, if the manufacturer says that the full well is 250 000 electrons and the CCD uses a 16-bit A/D converter, then the number of measurable levels will be 65 535 ADUs and the gain could be set to 250 000/65 535 = 3.81 electrons/ADU. A CCD gain of 3.81 electrons/DN means that for every 3.81 photoelectrons collected by a pixel, 1 DN will be produced by the A/D converter and sent to the computer. For a typical CCD exposure, many photoelectrons are generated and, on average, N collected photoelectrons will be converted to $N/3.81$ DN of ADU signal.

14.2.6 Charge-transfer efficiency (charge diffusion)

As we noted earlier, photoelectrons collected in pixels of a CCD are shifted many times before being A/D converted and stored as a CCD "frame." Each shift of the pixel's collected charge has some probability of losing or leaving behind some of its photoelectrons. The level at which this happens is called the charge-transfer efficiency or CTE. CTE values are given as the probability of each photoelectron actually being shifted and current values are often near 0.999999. Thus, the charge collected in a given pixel is highly likely to be essentially the same as when it finally reaches the output register and A/D converter. However, when there is significant charge-transfer inefficiency or CTI, then the measurable effects include loss of flux and systematic centroid shifts (Stetson 1998, Zacharias et al. 2000, Goudfrooij et al. 2006, Kozhurina-Platais et al. 2007, Lindegren 2010) that depend on (among other things) the temperature of the chip (Biretta 2005), position on the chip, the size of the signal, and the background level (Neeser et al. 2002).

Within each pixel, the collected photoelectrons are held in place during the active integration by an applied voltage. This voltage sets up a potential well within each pixel that attempts to keep each photoelectron held in the pixel in which it was generated. While this is a robust operation, some of the photoelectrons can find their way to other pixels or to local deviations within the potential wells where they may remain for long times compared to the CCD readout. These diffused electrons are more problematic in some CCDs than others (for example thick or deep depletion devices) and can affect measurements either by removal of counts within a source or broadening the profile of a source.

A number of other less-important CCD characteristics are not discussed here as well as the underlying derivation of the CCD equation and some associated subtleties; these are dealt with in Merline and Howell (1995).

14.3   Types of CCD exposures

A variety of CCD exposure types exist for the purpose of calibrating the "target" exposure taken for some specific study. Among these we include: bias frames, flats, and dark exposures. We will discuss them in the order in which they are generally obtained at a telescope.

14.3.1   Bias frame, pedestal or zero exposure

This type of CCD exposure is used for calibration and consists of a zero-second exposure. That is, the shutter is not opened, the CCD is not exposed to light, and the data obtained yield a measurement of the CCD electronic noise level due solely to the readout electronics and A/D conversion. Within the bias frame the noise can be monitored on a row-by-row basis to search for readout problems in the electronics as well as any two-dimensional structure in the bias. A master bias frame is usually created from the median of many bias frames, since an individual bias frame may have spurious data points due to cosmic rays, hot pixels, and general sampling noise.

14.3.2   Flat-field exposure

This is a calibration exposure and provides an (x, y) mapping of the individual pixel's response to light. While each pixel is nearly identical, small manufacturing defects, variations in the silicon from which the CCD was made, and optical-system imperfections such as dust on the filter and dewar window, make the actual response vary across the CCD. Flat-field exposures are often obtained in the day time by pointing the telescope to a large white spot hanging on the inside of the dome that is illuminated by high-intensity lamps. This type of flat-field exposure is called a dome flat and each exposure is usually only a few to tens of seconds exposure. Observations of the twilight sky (sky flats – see for example Chromey and Hasselbacker 1996), diffusers over the telescope objective (see for example Zhou *et al.* 2004), and the dithered night-time sky itself (often called a super-sky flat) can also be used as flat-field exposures. In the latter case care must be taken since the median averaging of many frames assumes that the star positions are random and if the fields are dense or the images have some characteristic spacing, e.g. when an objective wire grating is used, then a poor flat will be derived. In all cases care must be taken that the SNR of the derived master flat field is very high, since that will limit the SNR of the final processed object frame. Finally, fringing in narrow-band observations and nearly monochromatic light, e.g.

emission-line nebulae and night-sky lines, can cause severe problems in the redder wavelengths due to the interference from reflected light within the CCD. Fringing is often very difficult or nearly impossible to remove.

14.3.3 Dark frame

Dark exposures are obtained with the shutter closed but are longer than zero seconds. They are calibration exposures whose aim is to discover and correct for any dark current present in the CCD. The dark frame may also contain hot pixels, which are pixels that have an unusually high dark current often due to a defect in structure of the CCD. These hot pixels are generally removed by creating a mask of their (x, y) positions and removing them from the analysis. A master dark frame is usually created from the median of many dark frames, since an individual frame will have spurious data points due to cosmic rays and other bad data.

14.3.4 Object frame

Object frames are exposures taken with the shutter open to the night sky of the objects of interest to the astronomer. They are exposures that can range in exposure time but are typically a few minutes to 20 minutes or longer depending on the science goals, telescope, filter, and type of CCD used. These exposures collect sky light as well as light from astronomical targets. They also include the zero or bias level, pixel-to-pixel variations, and any dark current present in the CCD. The calibration images described above will be used in the reduction process to correct for these non-astronomical sources of uncertainty and error.

A few other similar CCD frame types occur but they are related to spectroscopic observations and therefore beyond the scope of this chapter.

14.4 Components of the CCD signal

The components of the CCD signal include photoelectrons released by photons from the target of interest as well as the sky and detector contributions such as dark current and read noise. In addition, the photons from the target and sky pass through the telescope and camera optics and are then detected by pixels with varying characteristics. Equation (14.4) expresses the number of detected photoelectrons S_{pix} in a pixel as a function of the vignetting V in the telescope and camera optics, the shadowing s by dust and defects in the filters and CCD window, the RQE of the pixel, the photon rate R_* of the target object, the photon rate of the sky R_s, the exposure time t, the dark current rate R_d, the time since the CCD was last read t_d and the bias level B. Note that t_d may differ from t if the CCD was not read or cleared immediately prior to the beginning of the target exposure or if the exposure was paused. Once the sky rate R_s has been determined, the desired target rate R_* in a pixel can be obtained by subtracting the dark count $R_d t_d$ and bias level B from the signal S_{pix} and dividing by the flat field $[V s RQE]$. To obtain the total count rate from the target it is necessary to add up all the counts within the whole target image and of course accurately determine the contribution of the sky to subtract off. The latter two operations can be simple or complex functions used in determining the photometry and astrometry of the targets. The detected photo-electrons can be expressed by the following expression, which is strictly a function of wavelength as well.

$$S_{\text{pix}} = [V s\, RQE]\, (R_* + R_s)\, t + R_d t_d + B \qquad (14.4)$$

14.4.1 Signal-to-noise ratio (SNR)

Finally, in this section, we need to discuss the signal-to-noise ratio expected for a CCD observation. The signal is easy to assess as it is merely the total counts from the target of interest. If a CCD observation is made of a star and within the extracted stellar profile we measure 60 000 photoelectrons collected by the CCD then that number is the signal, often listed as N. The noise of a CCD is a bit more complex as there are many factors that contribute to it. Excluding some specific noise sources which are generally small contributors for most typical CCD observations, we list the five major noise sources in a CCD observation.

(1) The object itself provides Poisson noise (or photon noise) that enters as the square root of the signal.
(2) The sky provides an additional Poisson noise source as photons from it fall onto the CCD and cannot be distinguished from object photons. This noise source is multiplied by the number of pixels over which the object is measured.
(3) Dark current is usually a zero contributor to modern, well-cooled CCDs. It can become a noise source for CCD systems that are thermoelectrically cooled (see Zacharias *et al.* 2000) or for which the CCD temperature varies during exposures.
(4) Bias and flat field are usually small noise sources, but their calibration is necessary for nearly all types of CCD astrometry.
(5) Read, or Johnson, noise is a shot type of noise and as such enters as the square of the signal. Thus, low read noise is very important for a CCD, especially those desiring to determine image centers for faint objects, low-surface-brightness sources, and/or extended objects.

When combined into the "CCD equation," the SNR for a CCD measurement is given by

$$SNR = \frac{R_* t}{\sqrt{R_* t + n_{\text{pix}} \left(R_s t + R_d t_d + N_b + N_r^2\right)}} \qquad (14.5)$$

where R is the rate of the target, sky, and dark respectively and the subscripts refer to the source of the noise: * = object, s = sky, d = dark, b = bias, and r = read. The lower-case n is the number of pixels used to measure the SNR. We can see from the equation above that for a bright source, one for which the object counts are large, i.e. far greater than any of the other noise sources, the SNR of the observation is simply given by the square root of the object counts themselves. Note that all values used in the CCD equation should be in electrons, where the CCD gain is used to convert output DN to (photo)electrons. Additional information on the characterization of SNR is given in Howell (1992, 2006) and Newberry

(1991, 1994), while the specific details of the SNR in two-dimensional image extraction is given in Howell (1989). A discussion of how the SNR can be correctly determined when using ensembles of stars is discussed in Gilliland *et al.* (1993) and Everett and Howell (2001).

14.5 Photometry and astrometry with CCDs

We will examine photometric observations with CCDs for the specific case of point sources, as they simplify the visualization and understanding of the techniques we will be discussing. In addition, point sources are the cornerstones of photometric and astrometric astronomy. Detailed work on the shape and extent of a stellar image (i.e. its point-spread function) are given by King (1983) and Diego (1985). The techniques discussed are fully applicable to extended sources, albeit with slight modifications.

14.5.1 Magnitude and color systems

Photometry, in its simplest form, involves the measurement of the received flux from a source. This may be determined in a specific filter such as the V bandpass or it may be over a specific wavelength range (e.g. 440–445 nm). The Johnson broad band UBVRI system, the Stromgren system, and the SDSS ugriz system are specific examples of a few major photometric systems in common use today. The different photometric systems have different zero points and intricacies, but in general they all work in the same way. A full discussion of these photometric systems is far beyond the scope of this volume but a good knowledge of them is crucial for photometric observations. A good reference for this topic is the conference proceedings edited by Sterkin (2007).

14.5.2 Measuring the flux

After proper data reduction, the flux of a point source has to be determined from the CCD exposure containing it. One must first be able to find the source, center on it, and derive a method by which the counts in the star can be isolated from other sources (e.g. sky photons) and turned into a proper magnitude or flux. A number of software applications have been developed to find sources (point and extended) on a CCD image. Examples are DAOPHOT (Stetson 1987) and SExtractor (Bertin and Arnouts 1996) but others exist as well. The reference list at the end of this chapter provides some starting points.

14.5.3 Image centers

All routines that find sources start with an estimation of their center. A simple centering idea is to take a small square box of say 30 by 30 pixels containing the point source. Next, sum the counts in each row and column in both the x and y directions to form two distributions. These are called marginal sums and their peaks, one in x and one in y, are simple yet very robust estimates of the centroid of the source. More sophisticated centering routines based on fitting functions are discussed in Auer and van Altena (1978) and Lee and van Altena (1983), and the user must decide which is appropriate for the desired output result, e.g. speed of reduction or the centering precision.

The quality of the image is determined by the optical system of the telescope and the turbulence due to the atmosphere. The distribution of light in the image resulting from a point source at infinity is termed the point-spread function or PSF. This distribution is often approximated by a Gaussian, or more complicated, function. The width of the PSF is often parameterized by measures such as the full width at half maximum (FWHM), the Strehl ratio, or the encircled energy. These metrics give an indication of the sharpness of the source profile and as such tell the user immediately how concentrated the source is and how good the image will be. For ground-based CCD imagery FWHM values produced with typical telescopes are near 1.0 arcsecond in size while those from space-based telescopes can be as small as 0.01 arcsecond. More details on the determination of precision image centers are discussed in Section 19.5.

14.5.4 Sky background

Once a source has its center determined and some method of profile fitting has provided a good estimate of its extent, the pixels involved in the source can be summed to estimate the source flux. This summing can be simple (place a circle around the source and count the pixel values inside) or more complicated (fit the source with a mathematical function and use it to estimate the counts) and other choices exist as well. We next need to estimate the extra counts in these pixels due to extraneous photons from the sky (and/or other sources) that were collected within the same pixels as those containing the source. To make such as estimate, we cannot "look under" the source and see the extra photons, so we need to estimate them in another way. Typically an area of pixels near the source and as devoid as possible of other sources is used to estimate the background sky value on a per-pixel basis. Often, this area is an annulus surrounding the source with an inner radius just outside the outer radius of the source and a width of a few pixels. The pixel values from this ring are summed and divided by the number of pixels in the annulus to provide a sky or background estimate.

The source flux is then simply determined as the total counts within the source minus the total background counts estimated to be "under" or contained within the source itself. Additional details about the source and "sky" value determinations and more intricate methods to estimate them are discussed by Stetson (1987), Bertin and Arnouts (1996) and others.

14.5.5 Sampling and pixel size

We now add to the discussion of this section a few additional topics that are directly related to the ability to photometrically and astrometrically measure sources imaged on CCDs. Two basic parameters that effect both photometry and astrometry are the pixel sampling and the plate scale. Typical plate scales are 0.2 to 0.4 arcsecond/pixel although systems developed for astrometry, high-precision photometry, or speckle observations often have plate scales of 0.1 or less arcsecond/pixel.

Images containing well-sampled sources, that is, sources with many pixels in them, are far better and easier to measure and they provide more robust results. The measure of image sampling is based on Nyquist's theorem and is generally expressed as the sampling parameter $r = $ FWHM/pixel scale. For r less than approximately 1.5, the image is considered undersampled. For such images, a typical PSF will only be collected on one or two pixels making it difficult to accurately determine the center. A rule of thumb for high-precision centering routines is that the best results are obtained when there are more than five pixels per FWHM, i.e. considerably more than the Nyquist theorem would dictate. Lindegren (2010) gives a more detailed discussion of this sampling problem and its relation to the design of imaging systems for astrometry. Suffice it to say that when attempting to determine image centers for images near the undersampled limit, the shape of the actual image becomes extremely important. Lindegren estimates that the undersampled nature of the CCDs in the Hubble Space Telescope ACS/WFC camera has degraded its image-centering precision by about 50%. The analysis of poorly sampled images is discussed in Chapter 17, while the deconvolution of images with imperfect PSFs is described in Chapter 18.

14.6 Summary

Charge-coupled devices are wonderful devices that have revolutionized astronomy. They have allowed photometric and astrometric observations to reach new levels of precision as well as providing these levels over large fields of view and for large-area surveys. Current software and techniques are at a high level of sophistication but will be put to the test with the upcoming very large sky surveys such as Pan-STARRS, the LSST, and the Dark-Energy survey. Those reading this volume will be the ones to take on these challenges.

References

Auer, L. H. and van Altena, W. F. (1978). Digital image centering. II. *AJ*, **83**, 640.

Bertin, E. and Arnouts, S. (1996). SExtractor: software for source extraction. *A&AS*, **117**, 393.

Biretta, J. (2005). WFPC2 status and calibration. In *The 2005 HST Calibration Workshop*, ed. A. M. Koekemoer, P. Goudfrooij, and L. L. Dressel. Baltimore, MD: Space Telescope Science Institute.

Chromey, F. R. and Hasselbacker, D. A. (1996). The flat sky: calibration and background uniformity in wide field astronomical images. *PASP*, **108**, 944.

Diego, F. (1985). Stellar image profiles from linear detectors and the throughput of astronomical instruments. *PASP*, **97**, 1209.

Everett, M. and Howell, S. B. (2001). A technique for ultrahigh-precision CCD photometry. *PASP*, **113**, 1428.

Gilliland, R. L., Brown, T. M., Kjeldsen, H., *et al.* (1993). A search for solar-like oscillations in the stars of M67 with CCD ensemble photometry on a network of 4 M telescopes. *AJ*, **106**, 2441.

Goudfrooij, P., Bohlin, R., Maíz-Apellániz, J., and Kimble, R. (2006). Empirical corrections for charge transfer inefficiency and associated centroid shifts for STIS CCD observations. *PASP*, **118**, 1455.

Howell, S. B. (1989). Two-dimensional aperture photometry – signal-to-noise ratio of point-source observations and optimal data-extraction techniques. *PASP*, **101**, 616.

Howell, S. B. (1992). Astronomical CCD observing and reduction techniques. *ASP Conf. Ser.*, **23**, 345.

Howell, S. B. (2006). *Handbook of CCD Astronomy*, 2nd edn. Cambridge: Cambridge University Press.

Janesick, J. (2001). *Scientific Charge-Coupled Devices*. Bellingham, WA: SPIE Press.

King, I. (1983). Accuracy of measurement of star images on a pixel array. *PASP*, **95**, 163.

Kozhurina-Platais, V., Goudfrooij, P. and Puzia, T. H. (2007). *ACS/WFC: Differential CTE Corrections for Photometry and Astronomy from Non-Drizzled Images*. Space Telescope Science Institute, Instrument Science Report ACS 2007-04d.

Lee, J.-F. and van Altena, W. F. (1983). Theoretical studies of the effects of grain noise on photographic stellar astrometry and photometry. *AJ*, **88**, 1683.

Lindegren, L. (2010). High-accuracy positioning: astrometry. In *Observing Photons in Space*, ed. M. C. E. Huber, A. Pauluhn, J. L. Culhane, J. G. Timothy, K. Wilhelm, and A. Zehnder. ESA/ISSI, ISSI Scientific Reports Series, 279.

Mackay, C. (1986). Charge-coupled devices in astronomy. *Ann. Rev. Astron. Astr.*, **24**, 255.

Merline, W. and Howell, S. B. (1995). A realistic model for point-sources imaged on array detectors: the model and initial results. *Exp. Ast.*, **6**, 163.

Neeser, M. J., Sackett, P. D., De Marchi, G., and Paresce, F. (2002). Detection of a thick disk in the edge-on low surface brightness galaxy ESO 342-G017. I. VLT photometry in V and R bands. *A&A*, **383**, 472.

Newberry, M. V. (1991). Signal-to-noise considerations for sky-subtracted CCD data. *PASP*, **103**, 122.

Newberry, M. V. (1994). The signal-to connection. *CCD Astron.*, **1**, No. 2/Summer, 34.

Sterken, C. (2007). The future of photometric, spectrophotometric, and polarimetric standardization. *ASP Conf. Series*, **364**.

Stetson, P. (1987). DAOPHOT – A computer program for crowded-field stellar photometry. *PASP*, **99**, 191.

Stetson, P. (1998). On the photometric consequences of charge-transfer inefficiency in WFPC2. *PASP*, **110**, 1448.

Zacharias, N., Urban, S. E., Zacharias, M. I., *et al.* (2000). The First US Naval Observatory CCD Astrograph Catalog. *AJ*, **120**, 2131.

Zhou, X., Burstein, D., Byun, Y. I., *et al.* (2004). Dome-diffuser flat-fielding for Schmidt telescopes. *AJ*, **127**, 3642.

DAVID L. RABINOWITZ

Introduction

Time-delay integration (TDI), also known as drift scanning, is a mode of reading out a charge-coupled device (CCD) camera that allows a continuous image or scan of the sky to be recorded. Normally, most astronomers use CCDs in the point-and-shoot or stare mode. A telescope is pointed to a particular position of interest on the sky and made to track at that position. A shutter is opened to expose the CCD, and then closed while the electronic exposure recorded by the CCD is read out. In drift-scan mode, the telescope is parked, tracking is turned off, and the camera shutter is held open. As the sky drifts across the field, the electronic exposure recorded by the CCD is shifted across the pixel array, row by row, to match the drift rate of the sky. The time it takes a source in the field to drift across the whole array is the exposure time of the scan. Since the readout is continuous, this is the most time-efficient way to survey large areas of sky. There is no pause between exposures to wait for the readout of the camera. The largest-area photometric surveys to date have been made with drift-scanning cameras. Smaller-scale surveys have used drift scans for astrometry of faint standards, suitable for the re-calibration of the relatively imprecise positions of the large photometric and Schmidt-plate catalogs.

Charge-coupled device cameras were first used in drift-scan mode on ground-based telescopes beginning in the 1980s with the CCD Transit Instrument (McGraw *et al.* 1980) and the Spacewatch Telescope (Gehrels *et al.* 1986). These early instruments consisted of CCDs with field sizes of ∼10 arcmin, capable of covering 10–20 square degrees in a single scan. The Spacewatch camera was the first to use a CCD for automated detection of near-Earth asteroids (Rabinowitz 1991). The CCD Transit Instrument had two CCDs aligned east–west allowing simultaneous scans of the same field in two different passbands. Advancements in computer speed and capacity have since allowed the construction of much larger scanning cameras made up of CCD mosaics. The photometric camera used by the Sloan Digital Sky Survey was a 6 × 5 mosaic of 30 CCDs with a total field size of 2.5 degrees (Gunn *et al.* 1998). It produced the largest multi-color photometric catalog of stars and galaxies to date, covering more than 11 000 square degrees and 350 million objects (Adelman-McCarthy *et al.* 2007). The Palomar Quest camera is a 4 × 28 mosaic of 112 CCDs spanning 4.5 degrees (Baltay *et al.* 2007). Covering 15 000 square degrees

Astrometry for Astrophysics: Methods, Models, and Applications, ed. William F. van Altena.
Published by Cambridge University Press. © Cambridge University Press 2013.

at multiple epochs and in multiple passbands over a time span of 5 years, the instrument has been used for the detection of nearby supernovae, optical transients and variables, and bright Kuiper-belt objects (Andrews *et al.* 2007).

Beyond their use for photometry and the detection of transient phenomena, drift-scanning cameras have also been used for precision astrometry. The Flagstaff Astrometric Scanning Transit Telescope (FASTT) was one of the first to use a CCD in scanning mode for this purpose (Stone 2000). The Carlsberg Meridian Telescope has been producing accurate astrometric catalogs from drift scans since 1998 (Evans *et al.* 2002). These systems rely on accurate catalog standards (e.g Tycho 2) to calibrate the positions of fainter field stars, which can then be used as secondary standards to improve the accuracy of Schmidt plate catalogs. The most precise implementation of drift-scanning technology yet contemplated is the astrometric camera of the Gaia spacecraft, expected to be launched in 2013. The camera will be a mosaic of 170 CCDs, with the rotation of the spacecraft providing the motion of the sky across the focal plane (Perryman *et al.* 2001, Lindegren *et al.* 2008, and Chapter 2). The goal of the survey is to produce a complete astrometric catalog to $V = 20$ mag and accurate to 15 microarcseconds (μas) at $V = 15$ mag and 300 μas.

This chapter will review some of the details of CCD operation, and provide a basic description of how the drift scan mode is implemented. This will lead to a discussion of the image distortions intrinsic to drift scans, and their consequences for precision astrometry. A description follows of the general techniques used to detect point sources in the drift scans and to obtain the transformations, derived from catalog standards, relating pixel coordinates in the scan to astrometric positions in a standard reference frame. This chapter ends with a discussion of the Palomar Quest camera, which has been designed to cover a larger area than any other drift-scanning instrument while minimizing the effects of drift-scan distortions.

15.1 The drift-scan method

To understand how drift scanning is implemented, we need to understand the basic principles of CCD operation (see Chapter 14 and Howell 2006, and for an in-depth description Janesick 2001). Unlike stare mode, the readout of the CCD in drift-scan mode is continuous throughout the duration of the scan and the shutter is never closed. New photoelectrons are intentionally collected as the electronic scan is shifted in the vertical direction. The columns of the CCD are carefully aligned to the drift direction of sky and the electronic scan is shifted at the same pixel rate w at which the sky is drifting. Hence, the newly collected photoelectrons simply add to the electronic scan as it is shifted across the array. The exposure time for each pixel is N_r/w, which is the time it takes a row to be shifted across the entire array. A new row begins its exposure each time a fully exposed row is shifted into the readout register. The duration DT of the scan can last an entire night, yielding a long scan with pixel dimensions N_c in the row direction and DT/w in the column direction.

15.2 Field distortions in drift scanning

The drift-scan mode has both advantages and disadvantages compared to stare mode with regard to astrometry. The main advantage is the increase in survey efficiency owing to the continuous readout. However, it is also advantageous that the signal in each pixel of the image is the sum of the response from each pixel in the array along the column or drift direction. Any optical distortions in the focal plane, as discussed in Chapters 13 and 19, are averaged in this direction during the image readout. Provided that the readout rate is held constant throughout the exposure, and that the motion of the sky image remains uniform, there are no distortions that are row-dependent in the recorded image. For precise astrometry, these conditions can never be perfectly met owing to variable refraction in the atmosphere (see Chapter 9) during the drift scan, to small telescope motions caused by the wind, and to minute flexures of the telescope caused by temperature variations. These variations are the factors that limit the precision and accuracy of drift-scan astrometry. But for the less-accurate astrometry required by photometric and transient surveys, drift scanning yields a simple solution linear in time or equivalently pixel position in the drift direction, as is the case for a transit telescope.

The main distortion introduced by drift scanning, and not present in stare exposures, is a peculiar alteration to the point-spread function (PSF) dependent on declination or column position. This results from the curved path followed by sources in the image as they drift across the focal plane. Because the paths followed by stars are circles about the celestial rotation pole, their projected paths in the focal plane are also curved. The closer a star is to the pole, the greater the curvature of its projected path. Hence, no matter how well the columns of the CCD are aligned to the average drift direction of the image, the images of those stars that are closest to the pole will have a small drift component perpendicular to the column direction. This component changes sign as the image crosses the field and curves towards and then away from the field center. To make the situation worse, the drift rate in the column direction is column dependent in inverse proportion to the cosine of the declination – the closer to the pole, the slower the drift rate. With the readout rate of the CCD matched to the average drift rate of the scan, those sources closest to the outside columns of the CCD will drift at faster or slower rates. Together, these dependencies produce a C-shaped distortion to the PSF, with maximum elongations D_r and D_c along the row and column directions, respectively, given by

$$D_r = [lw/2f] \tan \delta \qquad (15.1)$$

and

$$D_c = [l^2/8f] \tan \delta \qquad (15.2)$$

Here f is the telescope focal length, d is the declination of the source, and l and w are the length and width of CCD in the row and column directions, respectively (see Baltay *et al.* 2007).

To mitigate this declination-dependent distortion, the only recourses are to keep the telescope pointed close to the equator, where the distortions are smallest, and to choose a

CCD small enough so that D_r and D_c are negligible compared to the instrumental seeing. Note, also, that irrespective of the circular motion of the stars about the pole, there is an independent path distortion due to the tangential projection of the sky on to the focal plane. This distortion scales with the array size and focal-length similarly to D_r and D_c. For a more thorough description of these and other scanning distortions see Gibson and Hickson (1992), Stone *et al.* (1996), Bastian and Biermann (2005), and Baltay *et al.* (2007).

15.3 Object detection and astrometry

The detection of point and extended sources in CCD scans is discussed in Chapter 14. Here is it sufficient to mention that the detection techniques used for drift-scanned images are similar to the techniques used for stare images. For drift scans, the bias and flat fields are one-dimensional images, since the drift scanning removes any row-dependence to the corrections and they can be obtained in the same way as for stare images. After these corrections are made as outlined in Chapter 14, the flattened drift scans are searched for sources with peak intensities above a specific threshold, normally set to be the sky background intensity plus a fixed multiple off the root-mean-square (rms) noise in the sky signal. Depending on the science goal, the flux and spatial extent of each source is determined by fitting a Gaussian profile to the source intensity, or else by measuring the average shape of the PSF for the stars near the source and fitting the amplitude and position of this average PSF to the source intensity profile. For a more detailed description of this procedure, see for example Andrews *et al.* (2007).

The first step to obtain an astrometric solution for a drift scan is to create a source catalog, listing pixel coordinates (x, y in the row and column directions, respectively) and fluxes. The astrometric transformations, $F_a(x, y)$ and $F_d(x, y)$, taking x, y to right ascension and declination (α, δ) in a given reference frame, can then be computed using a reference catalog. Initially the reference positions are precessed to the of-date, equatorial reference frame (α^*, δ^*) of the source catalog (see Chapters 7 and 8). Given the best estimate for the telescope focal length, CCD pixel scale, the telescope pointing, and the start time of the drift scan, predictions are made for the pixel locations of the reference stars in the drift-scan images. Typically, these estimates will differ from the actual coordinates of the reference stars owing to the inaccuracy of the telescope pointing and uncertainties in the other parameters. A match must be made to determine which of the detected sources are the reference stars, and thus determine the observed locations (x_i, y_i) of the reference stars. This can be a non-trivial problem if the telescope pointing error is large or if the plate scale is not well determined. A robust triangle-matching algorithm (Valdes *et al.* 1995) can be used that allows for displacement, rotation, and scale changes between the predicted and observed positions.

With observed pixel locations x_i and y_i determined for catalog reference stars, simple linear solutions can then found to transform their pixel coordinates to of-date positions, α_i^*

and δ_i^*:

$$\alpha_i^*(x_i, y_i) = A_1 x_i + B_1 y_i + C_1 \tag{15.3}$$

$$\delta_i^*(x_i, y_i) = A_2 x_i + B_2 y_i + C_2 \tag{15.4}$$

Here A_1, A_2, B_1, B_2, C_1, and C_2 are constants and i is a catalog index. As discussed above, $\alpha_i^*(x_i, y_i)$ is necessarily linear in y_i because the drift scan averages over any field distortions in the column direction. There may also be a small, linear dependence on x_i owing to a row-dependent shear. This would result from a rotation of the focal plane or any other asymmetry in the sky projection. Random telescope movements and other time-dependent factors discussed earlier can introduce significant residuals to the linear solution. However, if these residuals are slowly varying on a timescale similar to or longer than the drift time across the array, and if the number of observed reference stars is high, then the dependence of the residuals on y_i (which is equivalent to time in drift-scan image) can be fit with a higher-order smoothing function and used as a correction to the linear fit. The transformation $\delta_i^*(x_i, y_i)$ will be linear in x_i, except at high declinations or for very large fields where the image distortion caused by the curvature of the star trails is large. The dependence on y_i is, again, a shear term to account for field rotation. As with $\alpha_i^*(x_i, y_i)$, any slowly varying, time-dependent residuals can be fit and corrected. With $\alpha_i^*(x_i, y_i)$ and $\delta_i^*(x_i, y_i)$ thus determined, these same transformations can then be used to determine of-date positions $\alpha*$ and $\delta*$ for all the other non-reference sources in the scanned image. Precessing these positions to a standard reference frame thus yields the full transformations $F_a(x, y)$ and $F_d(x, y)$.

15.4 The Palomar Quest camera

The Palomar Quest Survey, in operation at the 1.2-m Samuel Oschin Schmidt Telescope at Palomar from 2003 to 2008, provides a good demonstration of drift scanning. As mentioned above, the camera consists of a 112-CCD mosaic arranged in four columns (or "fingers") in the focal plane (see Fig. 15.1). The field drifts across each of the four fingers, which are normally covered by four different filters, to allow for near-simultaneous observation in four different passbands. Because the field is large, curvature of the drift path across the array is large. To minimize the resulting distortion to the PSF, the fingers are rotated so that the path at a given declination remains parallel to the CCD columns. Also, each CCD in a given finger is read out at a different rate to account for the differential drift rates across the finger. There is still some remaining distortion owing to the curvature and differential rates across each individual CCD. The size of these distortions is limited to no more than the typical seeing by choosing CCDs with appropriate dimensions. Equating D_r to D_c in Eqs. (15.1) and (15.2) requires CCDs with aspect ratios $l/w = 4$. This also requires CCDs with $l < 3.1$ cm, given the telescope focal length (3.06 m), as well as the choice to observe at declinations as high as 25 degrees, and typical seeing of 1.25 arcseconds.

Using the USNO A-2.0 catalog as a reference, the typical astrometric precision obtained in Palomar Quest drift scans with respect to the catalog positions is 0.4 arcseconds. This

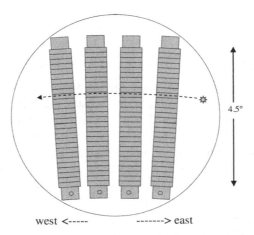

west <----- ------> east

Fig. 15.1 The CCD layout of the Palomar Quest camera. The CCDs are arranged in four columns or "fingers" of 28 CCDs each. In drift-scan mode, stars follow curved paths across the array. The fingers are rotated to keep the arcs parallel to the pixel columns on the CCDs. The CCD dimensions are also chosen to keep the size of remaining PSF distortions to less than the typical seeing.

result is limited by the precision of the USNO catalog itself. With a more precise catalog (see Chapters 20 and 3), more precise astrometry could be obtained. The limiting precision has been determined by matching pixel coordinate positions of stars observed in two different drift scans. The typical rms deviation between positions is then found to be 0.2 arcseconds, including faint stars near the limit of detection ($R \sim 21$ mag). Much better performance has been obtained with drift-scanning telescopes dedicated to precision astrometry. For example, FASTT has obtained positions for asteroids accurate to 0.060 arcseconds (Stone 2000) using more accurate astrometric catalogs as a reference.

References

Adelman-McCarthy, J. K., and 152 co-authors. (2007). The fifth data release of the Sloan Digital Sky Survey. *AJ,* **172** supp, 634.

Andrews, P., Baltay, C., Bauer, A., *et al.* (2007). The QUEST Data Processing Software Pipeline. *PASP*, **120**, 703.

Baltay, C., Rabinowitz, D., Andrews, P., *et al.* (2007). The QUEST Large Area CCD camera. *PASP*, **119**, 1278.

Bastian, U. and Biermann, M. (2005). Astrometric meaning and interpretation of high-precision time delay integration CCD data. *A&A*, **438**, 745.

Evans, D. W., Irwin, M. J., and Helmer, L. (2002). The Carlsberg Meridian Telescope CCD Drift Scan Survey. *A&A*, **395**, 347.

Gehrels, T., Marsden, B. G., McMillan, R. S., and Scotti, J. V. (1986). Astrometry with a scanning CCD. *AJ*, **91**, 1242.

Gibson, B. K. and Hickson, P. (1992). Time-delay integration CCD read-out technique: image deformation. *MNRAS*, **258**, 543.

Gunn, J. E., Carr, M., Rockosi, C., *et al.* (1998). The Sloan Digital Sky Survey Photometric Camera. *AJ*, **116**, 3040.

Howell, S. B. (2006). *Handbook of CCD Astronomy*. Cambridge: Cambridge University Press, WA.

Janesick, J. R. (2001). *Scientific Charge-Coupled Devices*. Bellingham, WA: SPIE, SPIE Press Monograph, PM83.

Lindegren, L., Babusiaux, C., Bailer-Jones, C., *et al.* (2008). The Gaia mission: science, organization and present status. *Proc. IAU Symp.*, **248**, 217.

McGraw, J. T., Angel, J. R. P., and Sargent, T. A. (1980). In *Applications of Digital Image Processing in Astronomy*, ed. D. A. Eliot. Bellingham, WA: SPIE, p. 20.

Perryman, M. A. C., de Boer, K. S., Gilmore, G., *et al.* (2001). GAIA: composition, formation and evolution of the Galaxy. *A&A*, **369**, 339.

Rabinowitz, D. (1991). Detection of Earth-approaching asteroids in near real time. *AJ*, **101**, 1518.

Stone, R. (2000). Accurate FASTT positions and magnitudes of asteroids: 1997–1999 observations. *AJ*, **120**, 2708.

Stone, R. C., Monet, D. G., Monet, A. K., *et al.* (1996). The Flagstaff Astrometric Scanning Transit Telescope (FASTT) and star positions determined in the extragalactic reference frame. *AJ*, **111**, 1721.

Valdes, F. G., Campusano, L. E., Velasquez, J. D., and Stetson, P. B. (1995). FOCAS Automatic Catalog Matching Algorithms. *PASP*, **107**, 1119.

Statistical astrometry

ANTHONY G. A. BROWN

Introduction

The term "statistical astrometry" refers to the inference of astrophysical quantities from samples (i.e. collections) of stars or other sources. Often the mean value of the astrophysical quantity or a description of its distribution is sought. The applications are numerous and include luminosity calibration, star-cluster membership determination, separating the Galaxy's structural components, detecting low-contrast substructure in the Galactic halo, etc. These studies become especially interesting now that astrometric catalogs are available with large numbers of parallaxes. However, there are a number of effects that can lead to errors in the inference of astrophysical quantities and thus complicate the interpretation of astrometric data. This chapter is focused on discussing these complicating effects with the aim of providing guidance on the optimal use of astrometric data in statistical studies.

16.1 Effects complicating the interpretation of astrometric data

The main effects that can complicate the interpretation of astrometric data for samples of objects are summarized below.

Completeness and selection effects

Any sample chosen from an astrometric catalog will suffer from incompleteness in the data and selection effects. Both issues are related to the properties of the astrometric survey and how the sample is chosen. In addition, they may depend on the location on the sky or the source properties. Hence, the chosen sample almost never represents the true distribution of sources in the astrophysical parameter space. Ignoring this fact leads to biased results in the scientific interpretation of the data.

Astrometry for Astrophysics: Methods, Models, and Applications, ed. William F. van Altena.
Published by Cambridge University Press. © Cambridge University Press 2013.

Correlated errors

In general, the errors of the astrometric parameters for a given source will not be statistically independent. Moreover, the errors for *different* sources may also be correlated. The covariance of the errors for a given source are provided, e.g. in the Hipparcos Catalogue. Ignoring these correlations may lead to spurious features in the distribution of derived astrophysical quantities. For examples of such features in the Hipparcos data see Brown *et al.* (1997).

Systematics as a function of sky position

The details of the way in which the astrometric measurements are collected will be reflected in systematic variations of the errors and the correlations over the sky. Examples are the systematic "zonal errors" from photographic-plate surveys and the systematics induced by the scanning law for the Hipparcos and Gaia missions. Taking these systematics into account is especially important for studies that make use of sources spread over large areas on the sky.

Estimating astrophysical quantities

When estimating astrophysical quantities from the analysis of samples of objects it is natural first to calculate these quantities for each individual object from the astrometric (and complementary) data and then analyze their distribution in the space of the astrophysical parameters. This allows us to work with familiar quantities such as distance, velocity, luminosity, angular momentum, etc. However, it is important to keep in mind that the actual data do not represent the astrophysical parameters in their natural coordinates. In particular, it is *not* the distance to sources which is measured directly but their parallactic displacements on the sky caused by the motion of the Earth around the Sun (listed as the parallax ϖ in the catalog). Hence, many astrophysical quantities are non-linear functions of the astrometric parameters. Examples are the distance, absolute magnitude, and angular momentum of stars, which are functions of $1/\varpi$, $\log \varpi$, and $1/\varpi^2$, respectively. Simplistic estimates of astrophysical parameters from the astrometric data can then lead to erroneous results. The only robust way around this is forward modelling of the data, as discussed in Section 16.5.

These aspects of statistical astrometry will now be discussed in more detail. Although no general solution exists for dealing with the many potential complications in the interpretation of the astrometric data, there are a number of "best practices," which are listed in the last section.

16.2 Know your catalog

In order to fully account for observational errors, completeness, and selection effects it is important to understand the contents of the astrometric catalog you are working with, including at least a basic understanding of how it was constructed. As an example, the Hipparcos Catalogue will be reviewed but similar considerations hold for any other astrometric catalog.

The Hipparcos Catalogue (ESA 1997) contains 118 218 entries, consisting primarily of single stars and about 25 000 binaries or multiple systems. In addition, the catalog contains ~ 60 Solar System objects and one quasar. For each entry, the five astrometric parameters, α, δ, ϖ, $\mu_{\alpha*}$, μ_δ, and their errors are listed, together with a rich set of complementary and auxiliary data. The latter consist of the correlation coefficients for the astrometric measurement errors (see Section 16.3), astrometric-solution quality indicators, photometry (V, $(B - V)$, $(V - I)$, and the Hipparcos H_p magnitude), variability indicators, extensive information on the astrometric solutions for multiple systems, and spectral types.

The Hipparcos mission and catalog were extensively documented both before and after the launch (see ESA 1989, 1997) and it is very instructive to go through the catalog statistics provided in Section 3 of Volume 1. From Fig. 3.2.1 therein it is clear that the distribution of catalog stars over the sky is very different from the actual distribution of stars at optical wavelengths. The Hipparcos measurements were collected on the basis of an input catalog which was constructed with input from the astronomical community. The sky distribution of Hipparcos stars thus shows a number of features that reflect this fact (for details see Turon *et al.* 1992), the most notable of which are the nearby OB associations in Gould's Belt. One has to be very careful of dealing with these selection effects when deriving average quantities (such as luminosity functions) from groups of stars taken from the Hipparcos Catalogue. Note that a specific effort was made to ensure the completeness of the Hipparcos survey to $V \sim 7$–8 mag. Astrometric data sets such as Gaia and the HST Guide Star Catalogue do not contain these explicit selection effects; nevertheless subtle selection effects may occur due to the varying completeness which will depend on the density of sources on the sky, the imaging resolution, and the behaviour of detection algorithms.

Another noteworthy feature of the Hipparcos Catalogue is the strong dependence of the errors in α, $\mu_{\alpha*}$ and ϖ on ecliptic latitude β. This can be understood from the way Hipparcos scans the sky. The scanning circles always cross the ecliptic at a large angle, which means that for sources near the ecliptic any displacement in ecliptic longitude λ projects poorly onto the measurements. The result is larger uncertainties in λ and $\mu_{\lambda*}$ which after transformation to the equatorial system mostly translate to larger uncertainties in α and $\mu_{\alpha*}$. The dependence of σ_ϖ on β follows from the fact that the parallactic displacement for sources near the ecliptic is mostly along λ. For details, see Section 2.5.1 in van Leeuwen (2007).

Finally, the catalog statistics also show that the correlations of the errors between the astrometric parameters vary systematically over the sky. Some of the systematic structure is caused by transforming from the ecliptic coordinates in which the measurements are carried out to the equatorial coordinates used in the catalog. Most of the systematics can be connected to the details of the scanning law, the distribution of observations in time, and the distribution of observations with respect to the position of the Sun.

Similar systematic variations over the sky will also occur for Gaia and other large surveys such as the Large Synoptic Survey Telescope (LSST), Pan-Starrs, and the Guide

Star Catalogue. It is important to be aware of and understand the origin of these effects, especially when looking for subtle large-scale features in the data.

16.3 Dealing with correlated errors

In general, the astrometric parameters for a source are estimated by the numerical fitting of a model for the source motion to the astrometric observations. This model will usually be the standard model for stellar motion but may be more complex in the case of, e.g., binary stars. The result of such a fit will be an estimate of the astrometric parameter vector $\mathbf{a}$ and its covariance matrix $\mathbf{C_a}$. The covariance matrix $\mathbf{C_a}$ is given by

$$\mathbf{C_a} = \mathrm{E}\left[\left(\mathbf{a}^{\mathrm{est}} - \mathbf{a}^{\mathrm{true}}\right)\left(\mathbf{a}^{\mathrm{est}} - \mathbf{a}^{\mathrm{true}}\right)'\right] = \mathrm{E}\left[\Delta\mathbf{a}\Delta\mathbf{a}'\right] \tag{16.1}$$

where $\mathrm{E}[\mathbf{x}]$ is the expectation value of $\mathbf{x}$, and $\mathbf{a}^{\mathrm{est}}$ and $\mathbf{a}^{\mathrm{true}}$ are the estimated and true astrometric parameters. The elements c_{ij} of $\mathbf{C}$ are

$$c_{ii} = \sigma_i^2, \quad c_{ij} = \rho_{ij}\sigma_i\sigma_j \tag{16.2}$$

where σ_i is the standard error on component a_i, and ρ_{ij} is the correlation coefficient for the components a_i and a_j.

The correlation coefficients will generally be non-zero because the components of $\mathbf{a}$ are obtained from the same set of measurements. Hence, it will not be sufficient just to rely on the standard errors to deal with the uncertainties in the astrometric parameters. The covariance matrix can be taken into account by using a generalized version of the "error-bar." The confidence region around $\mathbf{a}^{\mathrm{true}}$ is described by

$$z = \Delta\mathbf{a}'\mathbf{C_a}^{-1}\Delta\mathbf{a} \tag{16.3}$$

The distribution of z is described by the χ_ν^2 probability distribution, where ν, the number of degrees of freedom, is equal to the dimension of $\Delta\mathbf{a}$ (for more details see Section 15.6.5 in Press *et al.* 2007). If the vector $\mathbf{b}$ is derived from $\mathbf{a}$ via some transformation $\mathbf{f}(\mathbf{a})$, the covariance matrix of $\mathbf{b}$ is $\mathbf{C_b} = \mathbf{J}\mathbf{C_a}\mathbf{J}'$. Here $\mathbf{J}$ is the Jacobian matrix of the transformation $[\mathbf{J}]_{ij} = \partial f_i/\partial a_j$. Thus, we can calculate the covariance matrix of any set of variables derived from the observed astrometric parameters. An important example of such a transformation is the propagation of astrometric parameter errors in time. This is fully detailed in Section 1.5 in Volume 1 of ESA (1997).

A good example of the use of the covariance matrix in the astrophysical interpretation of astrometric data can be found in the study of the Hyades cluster with Hipparcos data (Perryman *et al.* 1998). The member stars of the cluster were all selected on the assumption that they move through space with a common motion vector (except for a small velocity dispersion). The covariance matrix for the space velocities for individual stars, calculated from the astrometric and radial velocity data, was used to decide whether or not a particular star has a space motion consistent with the mean cluster motion. A subsequent analysis of the velocity residuals with respect to the mean motion suggests the presence of a velocity ellipsoid with a specific orientation of its axes. In addition, the residual velocity vectors seem to indicate the presence of shear or rotation in the velocity field. These features would have interesting implications for the cluster dynamics. However, as shown by Perryman *et al.* (1998), they can all be explained as a consequence of the correlated errors in the astrometric parameters combined with the location and space velocity of the Hyades with respect to the Sun.

Correlated errors in the astrometric parameters (and correlations between astrometric parameters of different sources) should thus never be ignored but should be accounted for in the scientific analysis of the data.

16.4 Astrophysical parameter estimation through inversion

A prime motivation for the undertaking of astrometric surveys is the collection of parallaxes, which are the only distance measure to sources outside the Solar System free from assumptions about the intrinsic properties of the source. Parallaxes therefore play a fundamental role in the estimation of astrophysical parameters of interest such as the luminosities of stars, or the distribution function of stars in phase space. As noted in Section 16.1, the natural approach is to invert the astrometric observables in order to obtain, for example, distances or absolute magnitudes. However, this approach can lead to numerous complications and erroneous results. We will illustrate this by focusing on the calibration of stellar luminosities (absolute magnitudes) using trigonometric parallax data.

16.4.1 Statistical biases in simplistic luminosity calibration

We will assume here that we have conducted a parallax survey of all stars of a particular luminosity class and that these stars are distributed uniformly throughout the volume around the Sun. The question we ask is: what is the mean absolute magnitude μ_M of this class of stars and what is the variance σ_M^2 around the mean? The survey data provided are the measured parallaxes ϖ_0, apparent magnitudes m and the corresponding observational errors σ_ϖ and σ_m. The latter vary with apparent magnitude, the brighter stars having smaller measurement errors. The straightforward approach to this problem would be to estimate for each star i its absolute magnitude as $\widetilde{M}_i = m_i + 5\log\varpi_{o,i} + 5$ and then determine the values of μ_M and σ_M^2 from the resulting distribution of $\widetilde{M}_i$. In most realistic cases this will lead to erroneous results due to the following three problems.

Transformation bias

The observed parallax ϖ_0, which is estimated from the astrometric observations, is never error-free and will have some distribution, usually assumed to be Gaussian, around the true parallax ϖ:

$$P(\varpi_0|\varpi) = \frac{1}{\sigma_\varpi\sqrt{2\pi}}e^{-(\varpi_0-\varpi)^2/2\sigma_\varpi^2} \tag{16.4}$$

where $P(\varpi_o|\varpi)$ is the probability of observing a parallax ϖ_o given the true parallax ϖ. In the absence of systematic errors the measured parallax is itself an unbiased estimate of the true parallax: $E[\varpi_o] = \varpi$. This is not true of the distance, i.e. $E[1/\varpi_o] \neq 1/\varpi$. The same holds for the absolute magnitude; $E[\log \varpi_o] \neq \log \varpi$. The equations for the resulting bias in the estimated distance or absolute magnitude can be formulated (see e.g. Smith and Eichhorn 1996) but contain integrals that depend on the *true* value of the parallax and which are undefined for parallaxes near or below zero. There have been various attempts in the literature to calculate corrections for this bias (see e.g. Smith and Eichhorn 1996 and Brown *et al.* 1997) but these rely on tricks that remove zero and/or negative parallaxes in order to allow for the calculation of the integrals. These are highly dubious practices which do not get around the problem that the bias correction depends on the (unknown) value of the true parallax and for realistic cases will have a very large variance (rendering the "correction" meaningless).

The Lutz–Kelker effect

The above transformation bias occurs because we attempted to infer the distance or absolute magnitude from the observed parallax by using for any given star no other knowledge than the value of the observed parallax, its error, and the apparent magnitude. We can do better by introducing additional information that we have about the sample. Examples of such information are: the magnitude limit of the survey, which restricts the range of plausible values of M and sets a lower limit on ϖ for a given M; the distribution of stars in space, which is known to be finite and can be approximated as uniform near the Sun; and the class of stars under investigation, which provides prior information on the value of M.

This was recognized by Lutz and Kelker (1973) who wrote an influential paper in which they presented corrections for the statistical bias in luminosity calibrations with trigonometric parallaxes. The corrections were derived by taking a (partially) Bayesian approach to the luminosity-calibration problem and introducing prior information on the distribution of stars in space, namely the assumption that the space density of stars is constant. This leads to the following expression for the probability $P(\varpi|\varpi_0)$ that the true parallax of given star is ϖ given that its observed parallax is ϖ_0:

$$P(\varpi|\varpi_0) \propto P(\varpi_0|\varpi)\varpi^{-4} \qquad (16.5)$$

where $P(\varpi_0|\varpi)$ is the expression from Eq. (16.4). The above equation expresses the fact that the number of stars is a strongly increasing function of distance. This leads to a high probability that the true parallax is smaller than the observed one (more stars with small parallaxes will have observed parallaxes that are too high than vice versa). Lutz and Kelker then used this expression for the probability distribution of the true parallax to derive the expectation value of $\Delta M = M - M_0$, the difference between the true and observed absolute magnitude. These values are then tabulated as corrections to be applied to the mean absolute magnitude derived from a sample of stars with observed parallaxes.

The specific bias just described is referred to in the astronomical literature as the "Lutz–Kelker effect" and the corrections tabulated in Lutz and Kelker (1973) are often applied when absolute magnitudes for stars are calculated from parallaxes. However, these corrections are only applicable subject to the following conditions: the distribution of stars in space is uniform, the sample is volume-complete, and the corrections are applied to stars for which the ratio σ_ϖ/ϖ_0 is constant. Moreover, the luminosity function of the stars is assumed to be flat in absolute magnitude, and the corrections can only be calculated by assuming a lower limit on the ratio ϖ/ϖ_0 and restricting the sample to stars with $\sigma_\varpi/\varpi_0 < 0.175$. This combination of requirements renders the Lutz–Kelker corrections largely meaningless.

This issue was recognized after the publication of the Lutz and Kelker paper and attempts were made at deriving more sophisticated bias corrections by introducing additional assumptions about the luminosity function (e.g. Turon and Crézé 1977), or by relaxing the assumption about the spatial distribution of stars and taking the information from proper motions into account (e.g. Hanson 1979). A more extensive discussion of the literature on the Lutz–Kelker effect can be found in Smith (2003).

Malmquist bias

This bias is caused by the fact that any magnitude-limited sample will include systematically brighter objects as the sample volume increases due to the combination of selection on apparent brightness and the intrinsic spread in the absolute magnitudes of the sources. In the luminosity calibration problem stated above an apparent magnitude limit on the sample will cause the nearer and/or intrinsically brighter stars to be preferentially selected. This will thus lead to an unrepresentative distribution of absolute magnitudes (and parallaxes) in the sample leading to large systematic errors in the estimate of μ_M, the "Malmquist bias."

Just as with the Lutz–Kelker effect we can derive "corrections" for the Malmquist bias when certain assumptions are made about the sample. The most often quoted example is the correction derived by Malmquist (1922) for a stellar sample with a uniform spatial distribution and a Gaussian luminosity function. A derivation of the Malmquist bias can be found in Section 3.6.1 of Binney and Merrifield (1998) and a more general treatment of the Malmquist effect is given by Butkevich et al. (2005). The objections to using the Lutz–Kelker corrections also hold for Malmquist corrections.

16.4.2 Minimizing bias with alternative approaches

Because of the problems with direct methods discussed above much effort has been spent on developing alternative methodologies. One approach is to reformulate the problem under investigation such that the non-linear transformation of parallaxes is avoided.

Examples of the reformulation of the astrophysical question can be found in Feast and Catchpole (1997) and Arenou and Luri (1999). In Feast and Catchpole (1997) the zero-point of the Cepheid period–luminosity relation was sought and the authors re-wrote the period–luminosity relation so that an expression was obtained for the zero-point, which is linear in parallax. The zero-point can then be derived for a set of Cepheids without worrying about the transformation bias associated with first calculating values for the absolute luminosities. Likewise, Arenou and Luri (1999) suggest the use of the "astrometry based luminosity"

$a_V = \varpi_0 10^{0.2V-1}$. This measure of stellar luminosity can be used to construct Hertzsprung–Russell (HR) diagrams free of transformation or Lutz–Kelker biases and in addition allows the use of negative parallaxes or parallaxes with large σ_ϖ/ϖ_0. However, keep in mind that quantities such as a_V, although linear in parallax, are non-linear in V. This means that biases can enter via the new non-linear transformation even if the relative errors in V are typically much smaller than the relative parallax errors. In addition, formulating problems such that they are linear in parallax does not get rid of other sources of bias due to the choice of sample.

16.5 Forward modeling of the data

The approach to the interpretation of astrometric data discussed in the previous section is very much focused on the idea of inverting the observed quantities and then somehow "fixing" the resulting astrophysical parameters. The "corrections" can only be calculated by making often unrealistic assumptions and they can only be applied once stars with negative or small parallaxes have been removed from the sample. Such "censoring" of the observed data is very wasteful – negative parallaxes are perfectly legitimate measurements and contain information – and will make the bias in the results worse. This approach also implies that there is something wrong with the data. In fact, Lutz and Kelker (1973) stated in their paper: "In conclusion, we have shown that a systematic error exists for all trigonometric parallaxes". This suggests a systematic error intrinsic to the *measurement process* rather than the *interpretation* of the measurements. It should be stressed here that the biases discussed above have nothing to do with the data as such but arise entirely from the way we choose to interpret the data. They will also occur for data free from systematic measurement errors.

The "bias" problem can be avoided entirely by leaving the data as they are and instead estimating the desired astrophysical parameters through data modeling. That is, predict the astrometric (and other) observables from the model and then compare them with the actual observations in order to decide on the value of the model parameters. One can object that the estimated parameters then become model-dependent but as discussed above the direct methods are also model-dependent because numerous assumptions about the sample properties have to be introduced in order to enable the derivation of bias corrections. Sample properties and selection effects should of course also be accounted for in data modeling. To illustrate the data modeling approach we now work out a simple Bayesian luminosity-calibration example.

16.5.1 Bayesian luminosity calibration

Bayesian data analysis offers the most general approach to data modeling that takes all available information into account. We again use the simple parallax survey described at the beginning of Section 16.4 and ask the following question: what are the most likely values of μ_M and σ_M given the observations $\mathbf{o} = \{\varpi_{0,i}, m_i\}$ ($i = 0, \ldots, N-1$)? In order to answer this question we will formulate it probabilistically using Bayes' theorem:

$$P(\mu_M, \sigma_M, \mathbf{t}|\mathbf{o}) = \frac{P(\mathbf{o}|\mu_M, \sigma_M, \mathbf{t})P(\mu_M, \sigma_M, \mathbf{t})}{P(\mathbf{o})} \qquad (16.6)$$

The left-hand side of the equation is the probability of μ_M, σ_M, and the true values of the parallax and absolute magnitude of each star ($\mathbf{t} = \{\varpi_i, M_i\}$), given the observations $\mathbf{o}$. $P(\mathbf{o}|\mu_M, \sigma_M, \mathbf{t})$ is the probability of the data given the true values of the observables, i.e. the likelihood, and $P(\mu_M, \sigma_M, \mathbf{t})$ is the joint probability distribution of the luminosity distribution model parameters and the true values of the observables. This term represents our prior information on the distribution of stars in space and luminosity. The term $P(\mathbf{o})$ is the probability of obtaining the data and can be considered a normalizing constant for the optimization problem considered here.

In order to obtain estimates for μ_M and σ_M we now seek to maximize the *posterior probability* $P(\mu_M, \sigma_M, \mathbf{t}|\mathbf{o})$. We can write this as

$$P(\mu_M, \sigma_M, \mathbf{t}|\mathbf{o}) \propto$$
$$\prod_i \exp\left[-\frac{1}{2}\left(\frac{\varpi_{o,i}-\varpi_i}{\sigma_{\varpi,i}}\right)^2\right] \times \exp\left[-\frac{1}{2}\left(\frac{m_i-M_i+5\log\varpi_i+5}{\sigma_{m,i}}\right)^2\right] \qquad (16.7)$$
$$\times \varpi_i^{-4} \times \frac{1}{\sigma_M\sqrt{2\pi}}\exp\left[-\frac{1}{2}\left(\frac{M_i-\mu_M}{\sigma_M}\right)^2\right]P(\mu_M)P(\sigma_M)$$

This represents the joint probability of $(\mu_M, \sigma_M, \mathbf{t})$ given $\mathbf{o}$. The first two terms express the assumption that the errors in the measured parallaxes and apparent magnitudes are Gaussian. The ϖ_i^{-4} term expresses our assumption that the stars are distributed uniformly in space around the Sun and the third Gaussian is the assumption on the luminosity distribution of the stars in our sample. Finally, $P(\mu_M)$ and $P(\sigma_M)$ represent our prior information on plausible values of μ_M and σ_M.

Our job now is to work out the posterior likelihood from the above equation. We can then determine the probability distribution of (μ_M, σ_M) by marginalizing over the "nuisance parameters" $\mathbf{t}$ (we are not interested in the true values of the parallaxes and absolute magnitudes of the individual stars). From this distribution we can then make estimates $\widetilde{\mu_M}$ and $\widetilde{\sigma_M}$, for example by taking the mean. The problem is of course that $P(\mu_M, \sigma_M, \mathbf{t}|\mathbf{o})$ is a function over a very high-dimensional space and it will not be possible to calculate it analytically. This problem in fact is part of the reason why recourse was often sought to the "corrections" approach described in Section 16.4. There are, however, numerical methods that allow a sampling of the posterior distribution to be constructed. A popular example is the so-called Markov Chain Monte Carlo (MCMC) method, which will be used in the example discussed here. A thorough discussion of Bayesian data analysis and MCMC methods is beyond the scope of this chapter. A brief introduction can be found in Chapter 15.8 of Press *et al.* (2007). Basically, the MCMC method produces an intelligent sampling of $P(\mu_M, \sigma_M, \mathbf{t}|\mathbf{o})$ by making a controlled random walk through the parameter space $(\mu_M, \sigma_M, \mathbf{t})$. The sampling will result in a distribution of points in the parameter space for which the density is proportional to the posterior likelihood. From this distribution of points histograms of μ_M and σ_M values can be constructed. These histograms then represent a sampling of the posterior probability distributions for μ_M and σ_M *marginalized over all other model parameters*. This means that these distributions represent our posterior

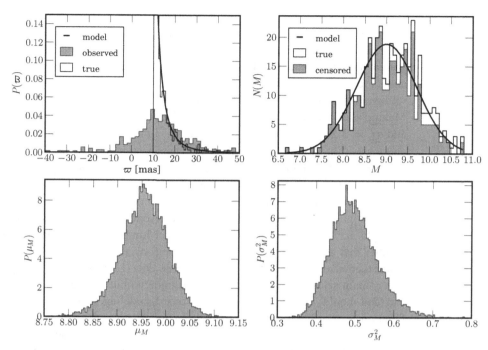

Fig. 16.1   Results of the Bayesian luminosity calibration using the Markov Chain Monte Carlo method. The top two panels show the simulated survey, with on the left the distribution of true stellar parallaxes and on the right the distribution of true absolute magnitudes (open histograms). The thick lines are the model distributions from Eq. (16.7). The shaded histogram in the top-left panel shows the observed parallaxes. Note the presence of observed parallaxes below 10 mas and below 0 mas. The effect of discarding negative parallaxes on the absolute magnitude distribution of the sample is shown by the shaded histogram in the top-right panel. The bottom panels show the sampled posterior probability distributions for μ_M and σ_M^2 obtained after running the MCMC method.

knowledge of the luminosity function parameters taking into account *all* the uncertainties due to observational errors and the vagueness of our prior information.

We will apply this now to a simulated survey containing stars of a single type. The survey contains $N = 400$ stars for which the absolute magnitudes are drawn from a Gaussian luminosity function with $\mu_M = 9$ and $\sigma_M^2 = 0.49$. The stars are uniformly distributed between distances of 1 and 100 pc (i.e. 10 milliarcseconds (mas) $\leq \varpi_i \leq 1000$ mas) and the survey is assumed to be volume-complete. The known observational errors $\sigma_{\varpi,i}$ and $\sigma_{m,i}$ vary as a function of the apparent magnitude (according to photon noise) and contain a "calibration floor" for bright stars. We furthermore assume that we know the distribution of stars to be uniform and that we know the limits on the true parallax distribution. Hence, the only unknowns to estimate from the data are μ_M and σ_M. The prior on μ_M was assumed to be flat between $\mu_M = -5$ and $+16$, while for σ_M^2 the prior distribution was proportional to $1/\sigma_M^2$ for variances between 0.01 and 1.0. The MCMC method was run on the expression for the posterior likelihood given in Eq. (16.7), using 10^6 iterations with 250 000 "burn-in" steps and storing every 150th sample. The result is summarized in Fig. 16.1.

The figure shows both the simulated survey and the sampled posterior probability distributions for μ_M and σ_M^2 obtained after running the MCMC method. Note the presence of observed parallaxes with small (< 10 mas) and negative values. These are not discarded but are all used in the MCMC method. The resulting *distributions* of μ_M and σ_M^2 are the answer to our luminosity-calibration problem as far as the Bayesian is concerned. However, we can summarize the results in this case as $\widetilde{\mu_M} = 8.96 \pm 0.05$ and $\widetilde{\sigma_M^2} = 0.50 \pm 0.06$. Making a naive estimate using only stars with positive parallaxes would lead to $\widetilde{\mu_M} = 8.8$ and $\widetilde{\sigma_M^2} = 3.2$, while using only stars with $0 < \sigma_\varpi / \varpi_o \leq 0.175$ would lead to $\widetilde{\mu_M} = 8.7$ and $\widetilde{\sigma_M^2} = 0.5$. Note that for both the erroneous estimates Lutz–Kelker corrections would not be applicable and would in any case correct $\widetilde{\mu_M}$ downward, away from the true value.

There are many aspects of this example that can be criticized. For one, the true distribution of parallaxes will not be known accurately in practice. However, there is nothing to prevent us from introducing additional parameters describing possible distributions $P(\varpi)$ (for example ϖ^{a+4} with unknown upper and lower limits on the parallax, where a is a small number that allows for deviations from a uniform distribution of stars in space). These could then also be determined from the same data. More seriously, any practical sample will contain selection effects, such as a magnitude limit. These selection effects, however, can (and should) also be included in the forward modeling, either through a known selection function or one with parameters to be estimated.

The main message is that forward modeling using probabilistic reasoning can naturally take into account many of the complications associated with astrometric survey data. In the example above negative parallaxes are used, as well as parallaxes with relatively large errors. In addition, the model has no problem handling the variation of the (relative) observational errors and correlations in the errors can easily be included in the model.

16.5.2 Other examples of astrometric data modeling

The luminosity-calibration method proposed by Luri *et al.* (1996) is an example of a maximum-likelihood data-modeling approach (see also Ratnatunga and Casertano 1991). This method takes into account not only the apparent-magnitude and parallax data but also the proper motions and radial velocities. The model takes into account the luminosity function, the kinematics, and the spatial distribution of the stars, and allows for a separation of the sample in terms of different populations. The observational errors and sample-selection effects are taken into account and all parallaxes are used, including negative and low-accuracy parallaxes. The method was applied to the luminosity calibration of the HR diagram based on Hipparcos data (Gómez *et al.* 1997). The important point here is the use of all the available data. This makes the data analysis more complex but enhances the information content of the data enormously. For example, the proper motions can be used as a proxy for distance and thus provide additional constraints on the stellar luminosities. Kinematic information from the combination of proper motion, radial velocities, and parallaxes can be used to separate populations of stars as shown very elegantly in the HR diagram constructed by Gould (2004) in which the stars are color-coded according to their tangential motion.

A different example of data modeling is the kinematic modeling of star-cluster astrometric data. In this case the cluster stars are assumed all to move with the same space motion and the data are then described in terms of this model. The internal cluster velocity field has to be taken into account. Recent examples of this type of statistical astrometry can be found in de Bruijne (1999), de Bruijne *et al.* (2001), and Lindegren *et al.* (2000). In all cases the cluster model includes the parallaxes of the stars themselves as parameters to estimate. This combination of the measured trigonometric parallaxes and the kinematic parallaxes (following from the stars' proper motions and tangential velocities) leads to improved estimates of the parallaxes (with smaller errors). The basic reason for this is the use of the distance information in the proper motions for a set of co-moving stars. The appearance of the HR diagrams for the Hyades (de Bruijne *et al.* 2001, Madsen *et al.* 2002) and the Scorpius OB2 association (de Bruijne 1999) is clearly much cleaner when using the kinematically improved parallaxes. Kinematic modeling in addition offers the very interesting possibility to measure radial velocities using non-spectroscopic methods as discussed by Dravins *et al.* (1999) and demonstrated in Lindegren *et al.* (2000) and Madsen *et al.* (2002).

Modeling as a tool for statistical astrometry is very powerful especially when making use of all available data and also complementary data from surveys other than the one being examined. Once the huge amount of astrometric data from surveys like Gaia, LSST, and Pan-Starrs become available modeling will be an indispensable tool. Efforts are already underway to develop data modeling tools for the study of Galactic structure and dynamics with future astrometric survey data (see for example Binney 2011).

16.6 Recommendations

The availability of large samples of stars and other objects with high-quality astrometric data allows us to address fundamental astronomical questions through statistical astrometry. However, the naive use of the astrometric data can lead to severe problems in the scientific interpretation. We list below a number of guidelines that can be followed when an astronomical question is to be addressed with statistical astrometry.

- Be aware that for almost any astrophysical quantity there is a non-linear transformation involved when deriving it from astrometric parameters. This is because it is the distance and not the parallax that is needed to calculate e.g. tangential velocities from proper motions, angular momenta of stellar orbits, energies of stellar orbits, action angle variables, etc.
- Consider very carefully the astronomical question to be investigated with the astrometric survey data. What exactly are the astrophysical parameters to be determined? Are direct methods appropriate or should the question be reformulated such that non-linear transformations are avoided?
- The catalog from which you are working should be well understood and samples should be selected according to well-defined criteria. Only a detailed understanding of the

properties of the sample you are working with will ultimately enable the elimination of potential biases in the results.

- Never apply "corrections" for the "Malmquist" or "Lutz–Kelker" effects to the results derived when using direct methods. Most often the two effects occur simultaneously, rendering the standard corrections invalid. In addition, the implicit assumptions are unlikely to be valid.

- Make use of data modeling by predicting the observables and judging the "goodness of fit" of the model in the data space. This entirely avoids the biases associated with non-linear transformations of the astrometric parameters and offers a natural way of taking selection effects into account. Bayesian data analysis is a very powerful tool in this respect.

- For specific samples or individual stars you can opt to retain only the stars with the best relative parallax errors, the upper limit being $\sim 10\%$. However, there will be an implicit truncation of the true parallax (stars with larger true parallaxes will tend to have smaller relative errors) as well as a bias toward brighter objects (which will have smaller absolute errors). Thus, severe biases may still occur in the results.

- Use all available information. This includes not only proper motions and radial velocities but also photometric data and data on the astrophysical properties of the stars in the sample (spectral type, luminosity class, metallicity, etc.). The complementary data will provide extra constraints on the distances and luminosities and can be used to separate populations. In addition, photometric or spectroscopic distance indicators can be used instead of relatively inaccurate parallaxes. These alternative distance indicators will be very well calibrated after the Gaia mission.

- When including additional information, as advocated in the previous item, keep in mind that then there is a danger of some circularity in the results. The calibration of distance indicators and astrophysical properties of stars is ultimately tied to parallax measurements of a small set of standard stars. Thus one should make sure that these calibrations are consistent with (i.e. derived from) the best-available parallax data for the standard stars.

- Do not ignore the correlations in the astrometric errors and how they vary as a function of location in the sky.

- Do not discard negative or inaccurate parallaxes without good reason. These data still contain information.

The reader is strongly encouraged to examine each case with these guidelines in mind in order to decide on the best way to interpret the astrometric data and thus ensure the maximum scientific return from them.

References

Arenou, F. and Luri, X. (1999). Distances and absolute magnitudes from trigonometric parallaxes. ASP Conf. Ser., **167**, p. 13.

Binney, J. (2011). Extracting science from surveys of our Galaxy. *Pramana*, **77**, 39–52.

Binney, J. and Merrifield, M (1998). *Galactic Astronomy*. Princeton, NJ: Princeton University Press.

Brown, A. G. A., Arenou, F., van Leeuwen, F., Lindegren, L., and Luri, X. (1997). Some considerations in making full use of the Hipparcos Catalogue. In *Hipparcos Venice '97*. ESA SP-402, p. 63.

de Bruijne, J. H. J. (1999). Structure and colour–magnitude diagrams of Scorpius OB2 based on kinematic modelling of Hipparcos data. *MNRAS*, **310**, 585.

de Bruijne, J. H. J., Hoogerwerf, R., and de Zeeuw, P. T. (2001). A Hipparcos study of the Hyades open cluster. Improved colour–absolute magnitude and Hertzsprung–Russell diagrams. *A&A*, **367**, 111.

Butkevich, A. G., Berdyugin, A. V., and Teerikorpi, P. (2005). Statistical biases in stellar astronomy: the Malmquist bias revisited. *MNRAS*, **362**, 321.

Dravins, D., Lindegren, L., and Madsen, S. (1999). Astrometric radial velocities. I. Non-spectroscopic methods for measuring stellar radial velocity. *A&A*, **348**, 1040.

ESA (1989). *The Hipparcos Mission: Pre-launch Status*. ESA SP-1111.

ESA (1997). *The Hipparcos and Tycho Catalogues*. ESA SP-1200.

Feast, M. W. and Catchpole, R. M. (1997). The Cepheid period-luminosity zero-point from Hipparcos trigonometrical parallaxes. *MNRAS*, **286**, L1–L5.

Gómez, A. E., Luri, X., Mennessier, M. O., Torra, J., and Figueras, F. (1997). The luminosity calibration of the HR Diagram revisited by HIPPARCOS. In *Hipparcos Venice '97*. ESA SP-402, p. 207.

Gould, A. (2004). $v_\perp$ CMD. *astro-ph/0403506*.

Hanson, R. B. (1979). A practical method to improve luminosity calibrations from trigonometric parallaxes. *MNRAS*, **186**, 875.

Lindegren, L., Madsen, S., and Dravins, D. (2000). Astrometric radial velocities. II. Maximum-likelihood estimation of radial velocities in moving clusters. *A&A*, **356**, 1119.

Luri, X., Mennessier, M. O., Torra, J., and Figueras, F. (1996). A new maximum likelihood method for luminosity calibrations. *A&AS*, **117**, 405.

Lutz, T. E. and Kelker, D. H. (1973). On the use of trigonometric parallaxes for the calibration of luminosity systems: theory. *PASP*, **85**, 573.

Madsen, S., Dravins, D., and Lindegren, L. (2002). Astrometric radial velocities. III. Hipparcos measurements of nearby star clusters and associations. *A&A*, **381**, 446.

Malmquist, K. G. (1922). *Lund Medd. Ser. I*, **100**, 1.

Perryman, M. A. C., Brown, A. G. A., Lebreton, Y., *et al.* (1998). The Hyades: distance, structure, dynamics, and age. *A&A*, **331**, 81.

Press, W. H., Teukolsky, S. A., Vetterling, W. T., Flannery, B. P. (2007). *Numerical Recipes: The Art of Scientific Computing*, 3rd edn. Cambridge: Cambridge University Press.

Ratnatunga, K. U. and Casertano, S. (1991). Absolute magnitude calibration using trigonometric parallax: incomplete, spectroscopic samples. *AJ*, **101**, 1075.

Smith, H., Jr. (2003). Is there really a Lutz–Kelker bias? Reconsidering calibration with trigonometric parallaxes. *MNRAS*, **338**, 891.

Smith, H., Jr. and Eichhorn, H. (1996). On the estimation of distances from trigonometric parallaxes. *MNRAS*, **281**, 211.

Turon, C. and Crézé, M. (1977). On the statistical use of trigonometric parallaxes. *A&A*, **56**, 273.

Turon, C. *et al.* (1992). The HIPPARCOS Input Catalogue. I – Star Selection. *A&A*, **258**, 74.

van Leeuwen, F. (2007). *Hipparcos, the New Reduction of the Raw Data*. Dordrecht: Springer.

17 Analyzing poorly sampled images: HST imaging astrometry

JAY ANDERSON

Introduction

Space observatories, such as Hubble Space Telescope (HST), Spitzer, and Kepler, offer unique opportunities and challenges for astrometry. The observing platform above the atmosphere allows us to reach the diffraction limit with a stable point-spread function (PSF). The downside is that there are limits to how much data can be downloaded per day, which in turn limits the number of pixels each detector can have. This imposes a natural compromise between the sampling of the detector and its field of view, and in order to have a reasonable field of view the camera designers often tolerate moderate undersampling in the detector pixels.

A detector is considered to be well sampled when the full width at half maximum (FWHM) of a point source is at least 2 pixels. If an image is well sampled, then the pixels in the image constitute a complete representation of the scene. We can interpolate between the image pixels using sinc-type functions and recover the same result that we would get if the pixels were smaller. This is because nothing in the scene that reaches the detector can have finer detail than the PSF. On the other hand, if the image is undersampled, then there can be structure in the astronomical scene that changes at too high a spatial frequency for the array of pixels to capture completely.

Some of the information that is lost to undersampling can be recovered. One common strategy to do this involves dithering. By taking observations that are offset from each other by sub-pixel shifts, we can achieve different realizations of the scene. These realizations can be combined to produce a single, higher-resolution representation of the scene that contains more information than any single observation. "Drizzle" is a popular tool that was produced by the Space Telescope Science Institute (STScI; Fruchter and Hook 1997, 2002) to combine multiple dithered pointings into a single composite image. It allows the user to specify a "drop size" that is used to sift the pixels of a set of exposures into a more finely sampled output frame. Drizzle rigorously preserves flux, but its algorithm can cause the light center of the output pixels to be offset from the output-pixel centers, which creates complications for pixel-fitting-type analyses. Lauer (1999) has devised a Fourier-based approach to combine exposures without compromising the sampling of the output composite image, but the approach has not yet been put together as a publicly accessible software package.

Astrometry for Astrophysics: Methods, Models, and Applications, ed. William F. van Altena.
Published by Cambridge University Press. © Cambridge University Press 2013.

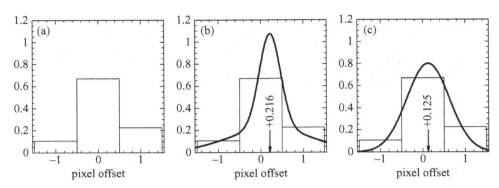

Fig. 17.1 (a) Hypothetical one-dimensional star profile; (b) the profile can be fit by the sum of a sharp Gaussian PSF with a FWHM of 0.6 pixel and a broad Gaussian halo with a FWHM of 2.4 pixels; (c) the profile can also be fit by a single Gaussian with a FWHM of 1.2 pixel. The center derived by each PSF is indicated by the arrow.

The above image-reconstruction techniques are necessary when we do not know anything *a priori* about the scene and must therefore permit any possible structure in the scene that is allowed for by the resolution of the telescope. However, if we happen to know that the scene is made up of point sources, then that changes everything. This extra information means that instead of each image being an under-constrained representation of the scene, each image is now an *over*-constrained representation of the scene. This is because point sources all have the same profile: the PSF. A point source can be fully described by three parameters: its (x, y) position in the charge-coupled device (CCD) frame and its total flux. In a sense, it can be said that a point source, like a black hole, "has no hair." Modulo these three parameters, all point sources are the same.

Typically, in an undersampled image, a well-exposed star will leave significant flux in the image's central 5×5 pixels. This means we have 25 constraints for the three unknowns, formally a very well-constrained problem. Unfortunately, simply because the problem is well constrained, that does not mean that it is easy to solve. It turns out that the (x, y) position we extract for the star will depend sensitively on the model we use for the PSF.

17.1   The need for a good PSF

In the same way that an undersampled detector gives us an incomplete picture of the astrometric scene, it also gives us an incomplete picture of the PSF. The left panel of Fig. 17.1 gives an example of an undersampled star image. On the left we see a distribution of pixel values in the x-profile of a star. About 70% of the flux shown is in the central pixel, 20% in the pixel on the right, and 10% in the pixel on the left.

It is clear that the center of the star is somewhere within the bright pixel, closer to the pixel on the right than to the one on the left, but if we want to determine a precise center for the star, we must fit this distribution with a PSF. In panels (b) and (c) we see that this distribution of pixel values can be fit exactly by two very different PSFs. The first PSF is composed of a narrow Gaussian with a broader Gaussian halo, and the second PSF is simply a single medium-width Gaussian. Although when we integrate these PSFs over the pixels, they both fit the pixels exactly, the first PSF yields a position that is almost a tenth of a pixel closer to the pixel on the right than the second PSF. Where is the star's true center? It depends on which PSF is the true PSF.

It is clear that if we fit a star with the incorrect PSF, we will end up with a bias in its measured position. This bias will be correlated with the pixel phase of the star. If the star is centered on a pixel then, by symmetry, the center of a symmetric PSF fit to the star will be at the center of the central pixel. Similarly, if the star lands on the boundary between two pixels, then there will be two equal pixels and its center can be inferred to be at the boundary. The ambiguity comes when the star's center is somewhere between the pixel center and the edge. If the PSF is very sharp, then as the star moves to the right from $\Delta x = 0$, at first there will not be much flux transfer into the right-most pixel; however, as the star's center gets closer to $\Delta x = 0.5$, there will be a rapid transfer of flux. With a broader PSF the flux transfer from the central pixel to the one on the right will be more gradual.

From this thought experiment, we see that if we fit for star positions with an incorrect PSF model, it will introduce what we will call "pixel-phase bias." This systematic error tends to resemble a sine curve, where it is zero at pixel phases $\Delta x = 0$ and $\Delta x = \pm 0.5$, and of opposite sign when the star is on the right and left sides of the pixel. In order to examine this bias, we must compare the measured position for a star against its "true" position as a function of where the star lands within a pixel. Since we do not normally have access to a catalog of accurate positions, our "true" positions come from measuring each star in different exposures at different dither positions, and transforming these measurements into the frame of a single observation, so that we can compare the measured position in the frame against an unbiased (or less-biased) reference position. Figure 2 of Anderson and King (2000) shows an example of this bias for WFPC2.

The pixel-phase-bias error can often be as large as 0.07 pixel when the PSF model is different from the actual PSF. This bias is typically largest when we use simple centroids to measure star positions, since the centroid algorithm implicitly assumes a very slowly varying PSF. The presence of this bias in measured positions tells us that there is a mismatch between the PSF model and the true PSF, but it does not tell us directly how to construct a more accurate model. It is clear that measuring systematically accurate positions from undersampled images will require careful attention to the PSF and how the pixels deliver the photons to us.

17.2   The PSF

The point-spread function is, as its name implies, the response of the detector to a point source. We can think of this mathematically as something similar to a wave function $\psi(\Delta x, \Delta y)$, which tells us what fraction of light lands at location $(\Delta x, \Delta y)$ relative to a star's center. Some detectors (such as X-ray detectors or old-fashioned photomultiplier tubes) record the arrival time and location of every detected photon. Such detectors are able to approximate the true instrumental PSF as the probability that a photon from a point source will land at a particular location relative to that point source.

Most optical imaging nowadays uses CCDs as discussed in Chapter 14 or similar detectors that collect the flux from the astronomical scene in pixels. To first order, these pixels can be thought of as light buckets that record the photons that fall within them. If we want to determine how much flux a given pixel at location (i, j) will register from a point source at (x_{star}, y_{star}), we must integrate the PSF over the pixel-response function $\Pi(\Delta x, \Delta y)$:

$$\boldsymbol{P}_{ij} = \boldsymbol{F}_{star} \int\int_{x,y} \Psi(x - x_{star}, y - y_{star}) \Pi(i - x, j - y) dx \, dy + S_{star} \tag{17.1}$$

Here S is the sky value, i.e. the background that would be recorded if the star were not present.

This integral is done over the whole image, since the pixel-response function can have a non-zero contribution from parts of the detector that are not formally contained within the pixel. There is some chance that a photon will hit near a pixel and generate an electron that eventually ends up within that pixel. (This is, in fact, the case for the Advanced Camera Survey's Wide-Field Channel (ACS/WFC) on HST. Due to thinning of the CCD, there is a significant amount of charge diffusion present in the detector.) On the other extreme, some older infrared detectors have pixels that have lower quantum efficiency at the edges of pixels, resulting in a pixel-response function that is sharper than a pixel. For these reasons, it is not always possible to treat $\Pi(\Delta x, \Delta y)$ as a perfect top-hat function.

The function $\Pi(\Delta x, \Delta y)$ has been measured empirically for a few detectors by moving focused laser spots across the detector pixels in a laboratory, but such experimental data are not in general available for most detectors. Thankfully, it is still possible to do high-precision astrometry and photometry without having detailed knowledge of the pixel-response function. This is because when we analyze actual images of the sky, it turns out that we never deal with the pixel-response function without also having to deal with the instrumental PSF: stars do not come to us as laser spots, but rather as PSFs. Conversely, we never encounter the instrumental PSF without it first being integrated over detector pixels.

There is, therefore, no need to know either ψ or Π separately. What we do encounter when we analyze the pixels in an image is actually the convolution of the two, which we call the "effective" PSF: $\psi_E \equiv \psi \otimes \Pi$. Since the effective PSF is the convolution of two continuous, two-dimensional functions it is also a smooth, continuous, two-dimensional function: $\psi_E = \psi_E(\Delta x, \Delta y)$.

17.3 The effective PSF

Whereas the instrumental PSF tells us the spatial density of photons that rain down at a particular location relative to a point source, the effective PSF tells us something much more practical. It tells us what fraction of a star's light lands in a pixel that is centered at $(\Delta x, \Delta y)$. This is a purely empirical quantity. It is easy to imagine that as a star is moved around on the pixel grid, the fraction of light in a given pixel will naturally vary smoothly in response to the location of the star's center relative to that pixel. The effective PSF is simply a direct way to describe this variation.

For the instrumental PSF, it is common to adopt a Gaussian- or a Moffat-type function, and for the pixel-response function we often assume a top hat, or some laboratory-determined kernel. But there is no obvious simple analytical formulation for the effective PSF. Its empirical nature suggests that perhaps a simple tabular representation would suffice: we could simply determine the value of ψ_E at an array of finely spaced points in its $(\Delta x, \Delta y)$ domain. This table should be sampled finely enough to capture all the intrinsic structure within the two-dimensional ψ_E function. Sampling it once every quarter pixel in each dimension is usually sufficient, since it can only be as sharp as the broadest of the functions that comprise it (ψ_{INST} and Π_{DET}).

Typically, for HST detectors, we want to represent the PSF over a 21×21-pixel domain, so that we can model the PSF out to a radius of 10 pixels. By sampling every quarter pixel, this would require an array of 81×81 elements. The central element $(41, 41)$ would correspond to $\psi_E(0.00, 0.00)$ and would tell us what fraction of a star's light will fall within a pixel if the star is centered on that pixel. The next grid-point to the right represents $\psi_E(0.25, 0.00)$, the fraction of light that falls in a pixel if the pixel center is a quarter pixel left of the star's center. Stars will of course not always fall at these convenient fiducial points within pixels. They could lie anywhere within a pixel. To evaluate what fraction of light will fall in a pixel that is located at an arbitrary location, such as $(0.35, -1.25)$, relative to a star's center, we use bi-cubic interpolation within the ψ_E grid.

Dealing with the PSF in this empirical way has two advantages. First, the PSF model is easier to solve for. Instead of trying to determine which instrumental PSF, when integrated over the pixels, best matches the observations, here we simply document what fraction of light lands in pixels as a function of where the pixels are relative to the star's center. For this, we simply need to know the position and total flux of the stars that will be used to construct the PSF. The process of constructing a PSF model involves collating many observations of many different stars in different dithered images, each at a different pixel phase, into a single coherent two-dimensional function $\psi_E(\Delta x, \Delta y)$.

The second advantage of the empirical effective-PSF approach is in fitting stars. A stellar image consists of an array of pixel values. If we want to simulate what the image of a star would look like if its center were at a particular location in a pixel, we simply interpolate the ψ_E table at the appropriate array of locations. This tells us what fraction of light would fall in each of the observed pixels, given the assumed center. We find the best-fitting center by identifying which trial location for the star best matches the observed distribution of flux among the pixels. This is much easier than the instrumental-PSF approach, where we have to evaluate the PSF function over a pixel face each time we want to model observations. The details of fitting stars with effective PSFs can be found in Anderson and King (2000).

17.4 An accurate PSF model

The effective-PSF approach clearly makes it easier to solve for the PSF and easier to use the PSF, but it does not necessarily guarantee that our PSF will be accurate. The reason we want an improved representation of the PSF is to be able to model it more accurately in order to remove the bias that can arise when there are many PSFs that could fit the undersampled observations.

The key to arriving at an accurate PSF model is the same as that for arriving at a higher-resolution model of the scene: dithering. By dithering, we can observe the same star at a variety of pixel phases. Since we know the offsets between dithers quite accurately (from the many stars in each image), we can use this knowledge to correct the observed star positions for any pixel-phase bias in their initial measurement. These improved positions will then allow us to construct a valid, unbiased PSF model. This is clearly an iterative procedure, since we cannot measure positions without a PSF, and cannot construct a PSF without good positions. As we improve the PSF, we improve the positions of stars and the transformations between frames, which in turn enables us to improve the PSF even more. Again, Anderson and King (2000) discuss this procedure in detail.

It is worth noting that the PSF often varies with position across the detector, so rather than using the above techniques to solve for a single PSF, we often have to solve for an array of PSFs across the detector. Each Wide-Field Planetary Camera 2 (WFPC2) chip required an array of 3×3 PSFs, and each ACS/WFC detector required an array of 9×5 PSFs, across its 4096×2048 pixels.

17.5 The ultimate limitations caused by undersampling

We have shown above that undersampling requires careful attention to the details of how a star's flux gets distributed among its pixels, but with such careful attention systematically accurate astrometry can be achieved. We typically find that for stars with signal-to-noise ratios greater than 200 in HST detectors we can measure positions precise to 0.01 pixel in a random sense, and calibrate the systematic error due to pixel-phase bias to better than 0.003 pixel.

Of course not all degrees of undersampling can be remedied in this way. If the FWHM of the instrumental PSF is much smaller than the pixel size, then a star can be moved around within its central pixel without transferring much flux into its neighboring pixels. This flux transfer is the handle we have on the star's position, and without it there is no way to measure accurate positions. Therefore, the above approach can measure good positions for mild to moderate undersampling, but no method can measure good positions when the detector is pathologically undersampled.

17.6 Distortion

Measuring systematically accurate positions on detectors is only half the challenge of space-based astrometry. The other part of the challenge concerns relating the positions measured in one exposure to those in another. The same stable observing platform that makes it possible to construct exquisitely accurate PSFs and measure precise positions also allows for a high-precision calibration of distortion and other image artifacts.

Distortion results when there is a non-linear mapping from the sky to the detector pixels. There are two components of distortion: optical field-angle distortion, or OFAD, from the telescope and camera, and distortions introduced within the detector itself. The cameras on-board HST were designed to maximize throughput, with the compromise that they suffer from significant OFAD. The square ACS detectors actually map to rhombuses on the sky, with interior angles of $81°$ and $99°$ and with non-linear components of OFAD that can move a star by more than 50 pixels from where it would otherwise be. In addition to these optics-based distortions, the filters also introduce local distortions of up to 0.15 pixel, which are coherent over 100 pixels or so. Finally, the detectors themselves can have irregularities in the pixel grid due to imperfections in the manufacturing process (see Anderson and King 1999). The stability of the observing platform and the precision with which we can measure positions allows all these distortions to be accurately modeled and removed from our measurements, so that it is possible to relate positions measured in one pixel of one exposure with positions measured in another pixel of another exposure.

17.7 Applications

The imaging astrometry we have discussed here for space-based missions is all *differential* astrometry: we are measuring positions of one star relative to those of its nearby neighbors. While positions can be measured to better than a milliarcsecond in single exposures, the absolute pointing accuracy relies on the guide-star positions, which can have errors of more than an arcsecond. In order to do absolute astrometry we must know the absolute coordinates of some of the objects in our field. It is rare to have astrometric standard stars in an HST field of view, partly because of the small field of view and partly because stars bright enough to be in a catalog are often saturated in HST images. Therefore, most astrometry with HST is necessarily differential, unless the extra effort is taken to establish an intermediate magnitude reference system to form a "bridge" between the bright standards and the faint objects in the target exposure. The above issues are discussed in detail in Chapter 19.

Nonetheless, even with its "differential" focus there are still many astrometric projects that HST and other similar space-based telescopes can pursue. One of the challenges in stellar-population studies of stars in clusters is how to differentiate member stars from field stars (see Chapter 25), so that we work with a pure sample of cluster stars. Proper motions of HST enable membership to be determined definitively for clusters well out to 10 kpc with only a few years' baseline. Exquisite cluster-field separation allows us to study less-populated regions of the color-magnitude diagram, such as the Horizontal Branch, the Sub-Giant Branch, the photometric binaries, cataclysmic variables, and the lower main sequence near the hydrogen-burning limit. The resolution of HST also allows us to measure precise positions for stars well into the centers of most globular clusters, so that we can examine how the individual stars are moving relative to each other. Such motions can reveal a central massive black hole, the velocity-distribution function, and allow a direct distance determination (via comparison of the proper-motion dispersions with radial-velocity dispersions, see van der Marel and Anderson 2010 for an example).

Important differential astrometry can also be done for field stars. Parallax measurements (Chapter 21) of the Pleiades, field white dwarfs, and neutron stars have been measured by comparing the displacement of the target stars against field stars whose distances we can estimate. Astrometry can be used to help constrain the properties of stars undergoing gravitational lensing (Bennett *et al.* 2007).

Finally, although it is difficult for HST to do absolute astrometry, it *can* measure absolute proper motions, by making differential measurements where one of the objects is known to be extragalactic. In fact, HST has an advantage in this respect. Whereas ground-based observations rely on bright galaxies to tie their proper-motion frames to an inertial one, HST can use the much more plentiful faint galaxies. These galaxies are only a bit broader than stars, and there are hundreds of them in almost every deep exposure. The challenge is that since galaxies are resolved and are not simple point sources, a separate GSF (galaxy-spread function) must be constructed for each one, so that a consistent position can be measured for each one in each exposure. Thus far, this technique has been used to measure absolute motions for a few globular clusters, a few dwarf spheroidals, and the Large and Small Magellanic Clouds, but there is potential for much more absolute proper-motion work.

References

Anderson, J. and King, I. R. (1999). Astrometric and photometric corrections for the 34th-row error in HST's WFPC2 Camera. *PASP*, **111**, 1095.

Anderson, J. and King, I. R. (2000). Towards high-precision astrometry with WFPC2. I. Deriving an accurate point-spread function. *PASP*, **112**, 1360.

Bennett, D. P., Anderson, J., and Gaudi, B. S. (2007). Characterization of gravitational microlensing planetary host stars. *ApJ*, **660**, 781.

Fruchter, A. S. and Hook, R. N. (1997). Novel image reconstruction method applied to deep Hubble Space Telescope images. *SPIE*, **3164**, 120.

Fruchter, A. S. and Hook, R. N. (2002). Drizzle: a method for the linear reconstruction of undersampled images. *PASP*, **114**, 144.

Lauer, T. (1999). Combining undersampled dithered images. *PASP*, **111**, 227L.

van der Marel, R. P. and Anderson, J. (2010). New limits on an intermediate-mass black hole in Omega Centauri. II. Dynamical models. *ApJ*, **710**, 1063.

Image deconvolution

JORGE NÚÑEZ

Introduction

The techniques of image deconvolution can increase your effective telescope aperture by 40% without decreasing the astrometric precision or introducing artificial bias. Some studies also show that appreciable gain in astrometric accuracy can be obtained.

18.1 Theory of deconvolution

18.1.1 The imaging equation

In several parts of this book it has been pointed out that astrometry, as part of astronomy, is an observational science in which the unknown physical basis is, in our observations, convolved with the structure of the source, the emission process, the atmosphere, the telescope detector interaction, etc. In a typical exposure of the sky taken from the ground this convolution makes our point-like stars appear as pixelized extended spots of light of about one arcsecond (or more) in size. The light in the star images shows, in general, a Gaussian-like pattern but it can vary across the frame. In all types of observations (optical imaging from the ground or space, optical and radio interferometry, etc.), the process can be mathematically described as an imaging equation which is a relationship (with an integral operator) between the distribution of the source and the distribution of the observational data. We can write the imaging equation as a convolution

$$\mathbf{F}(x, \xi)^* a(\xi) = p(x) \tag{18.1}$$

where $a(\xi)$ and $p(x)$ are the unknown source and observed data functions and the kernel $F(x, \xi)$ is known as the point-spread function (PSF). Taking into account the noise caused by both the detection system and the signal itself (shot noise) we can write the equation

$$\mathbf{F}a(\xi) + n(x) = p(x) \tag{18.2}$$

Astrometry for Astrophysics: Methods, Models, and Applications, ed. William F. van Altena.
Published by Cambridge University Press. © Cambridge University Press 2013.

In practice, we separate the problem in data space and in the unknown function space. In the linear case we obtain the typical imaging equation

$$\mathbf{F}^*\mathbf{a} + \mathbf{b} + \mathbf{n} = \mathbf{p} \tag{18.3}$$

where $\mathbf{a}$ and $\mathbf{p}$ are the unknowns and data vectors respectively, $\mathbf{n}$ and $\mathbf{b}$ are vectors representing the noise and the background in both object and data spaces, and $\mathbf{F}$ is a sparse matrix representing the PSF. Note that $\mathbf{a}$, $\mathbf{p}$, $\mathbf{n}$, and $\mathbf{b}$ represent, in general, two-dimensional images but since they are separated, they can be represented as vectors. Thus, image deconvolution consists in obtaining the best approximation of the true object $\mathbf{a}(x, y)$ from the inversion of the imaging equation. However, the problem represented by the imaging equation is an ill-posed noisy inverse problem. This is because the PSF is band-limited and the convolution with it makes the spatial higher frequencies in the data zero, or very small. Since noise is present at all frequencies, the higher spatial frequencies of the data are lost in the noise. Thus, the imaging equation cannot be correctly solved by linear methods such as matrix inversion or direct Fourier inversion since these methods magnify the noise giving unacceptable results. The goal of image deconvolution is to extract from the data $\mathbf{p}(x, y)$ an approximation to the object $\mathbf{a}(x, y)$ with reduced ripple, with improved resolution (if possible) and making due allowance for noise. This involves some form of interpolation and extrapolation in the Fourier domain. A good algorithm should also give less weight to high-spatial-frequency data that are corrupted by the noise and instead give more importance to extrapolated values obtained from better data at lower frequencies. In image deconvolution, as in other sciences, the nature of the problem (including the nature of the noise) and the data itself (including the method of observation) will lead to the use of different methods. Thus, there is no unique preferred algorithm to solve the image-deconvolution problem and different algorithms for deconvolution are based on different instrumental noise models, different methods of computation (maximum-likelihood, Bayesian, maximum-entropy, etc.), the nature of the object (pointlike or extended), and even on the available computational resources. In this chapter we will suggest useful references commenting and giving examples of some of the most widely used algorithms. From all the above, it is clear that the success of the deconvolution process depends both on a good knowledge of the PSF and of the nature of the noise present in the data.

18.1.2 Noise sources

From previous chapters we know that the knowledge and minimization of noise is fundamental in astrometry. This is especially true in the case of deconvolution since deconvolution is an ill-posed problem in which we want to correct the degradation caused by the PSF. Indeed the PSF is the result of the superposition of several effects such as diffraction, optical aberrations, and, most important for ground observations, refraction noise produced by atmospheric turbulence. In addition to this, we must consider noise intrinsic to the signal (shot noise), which is due to the arrival of the individual photons at the detector and the noise introduced by the detector electronics (readout noise). In astronomical imaging, the shot noise is in general governed by a Poisson process while the readout noise obeys a Gaussian process. In image pulse-counting cameras (IPCSs) the Poisson noise dominates since the system records directly the arrival of the individual photons and there is no readout noise. However, in CCD cameras Gaussian readout noise must be added to the intrinsic Poisson noise giving mixed statistics (see Núñez and Llacer 1993 for a detailed description). As stated above, different noise models lead to different deconvolution algorithms.

18.1.3 Dealing with aberrated PSFs

The PSF can be approximated by the image of a point source (a star) observed thorough the entire system (atmosphere, telescope, filter, sampling, and detector). In the case of space-based telescopes such as the Hubble Space Telescope (HST) this is approximately the diffraction figure of the optical system (telescope and filters) sampled and observed by the detector and can be computed using ray-tracing programs such as the HST TinyTim software (www.stsci.edu/software/tinytim/tinytim.html). In the case of space-based observations, this is the best way to compute the PSF. However, even in this case, the PSF can be noticeably different for different parts of the detector. In the case of ground-based images the situation is worse because the PSF is dominated by the time-variant atmospheric seeing. To model the PSF in the case of ground-based astrometry most authors describe the turbulence-generated core of the PSF using an elliptical two-dimensional (2D) Gaussian model or a Moffat profile (see Fors 2006 for details). There is, also, an external aureole about 10 magnitude fainter than the core that can be ignored in PSF fitting and deconvolution. Thus, for ground-based observations, the best way to obtain the PSF is to use several real stars to compute the PSF based in an elliptical two-dimensional Gaussian or a Moffat profile. Astronomical software such as the Image Reduction and Analysis Facility (IRAF) offers tools for this purpose. Of course, it is always possible to use one real star (or the mean of several stars) as an approximation for the PSF. Some authors have developed algorithms known as "blind deconvolution" (White 1994) in which the PSF is estimated along with the object $a(x, y)$. However, for astrometry these algorithms can introduce more problems than benefits and they are not recommended.

Besides the space and time variability of the PSF, another common problem with the PSF is the asymmetry of the figure. This asymmetry is present in both space and ground-based observations and is especially important in time-delayed integration (TDI) astrometry. As described in Chapter 15, in TDI observations the PSF has a form of a small "banana" with no symmetry and of different shape depending on the row of the charge-coupled device (CCD) being scanned. For this type of astrometric observation, the best way to compute the PSF is to use the mean of several real stars.

As will be discussed later, almost all deconvolution algorithms are iterative and in each iteration several convolutions are computed. Given the size of astronomical image frames, the convolutions need to be computed using the fast Fourier transform (FFT). However, to use the FFT a space-invariant PSF needs to be used, i.e. a PSF valid for the entire frame. In the case of strongly space-variant PSFs the best way to deal with the problem is to cut the frame into smaller sub-frames in which the PSF can be considered as constant.

18.1.4 Algorithms

For many years, image restoration was considered a luxury in optical astronomy. However, since the discovery in 1990 of a severe problem of spherical aberration in the mirror of the HST, a substantial amount of work has been done in image deconvolution directed towards optical astronomy, covering the different types of data noise and proposing dozens of algorithms. Since in this chapter it is not possible to discuss all of them, we direct the reader to excellent published proceedings (White and Allen 1990, Hanisch and White 1994), special issues (Núñez 1995), and reviews (Molina *et al.* 2001, Starck *et al.* 2002). In these works you can find, among many others, diverse algorithms based in maximum-likelihood, maximum-entropy and the Bayesian paradigm. Astrometric application of those algorithms were studied by Girard *et al.* (1995), Prades and Núñez (1997), and Fors (2006), and the examples presented below are taken from those papers. At the present time, the most widely used algorithms for astrometry are the Richardson–Lucy (R–L) algorithm with versions for Poisson data (Lucy 1974) and CCD cameras (Nuñez and Llacer 1993, 1998) and the wavelet-decomposition-based maximum-likelihood (AWMLE) method (Otazu 2001, Starck *et al.* 2002, Fors 2006). We will describe them briefly.

18.1.5 The Richardson–Lucy algorithm

The R–L algorithm is obtained by maximizing the likelihood in the case of Poisson noise. The formula can also be obtained using the expectation-maximization algorithm. It is an iterative algorithm with the form

$$a_i^{(k+1)} = a_i^{(k)} \sum_j F_{ji} \frac{p_j}{(\mathbf{a}^{(k)} * \mathbf{F})_j} = a_i^{(k)} \mathbf{F} * \frac{\mathbf{p}}{\mathbf{a}^{(k)} * \mathbf{F}} \tag{18.4}$$

In this formula $\mathbf{F}$, $\mathbf{p}$, and $\mathbf{a}$ are defined as above. In each iteration, the R–L algorithm computes the convolution of the present approximation $a^{(k)}$ with the PSF (denominator of the equation), then the image of the deviations between the data $\mathbf{p}$ and the convolution (quotient of the equation) is again convolved with the transposition of the PSF and the result is used to update the present approximation. Since the original R–L algorithm is for Poisson data, several modifications have been developed. For CCD data it has been shown (Núñez and Llacer 1993) that if σ is the standard deviation of the readout noise, the data $\mathbf{p}$ should be changed by a filtered version $\mathbf{p}'$ defined as

$$p_j' = \sum_{k=0}^{\infty} \left(k e^{-\frac{(k-p_j)^2}{2\sigma^2}} \frac{(a*F)_j^k}{k!} \right) \bigg/ \sum_{k=0}^{\infty} \left(e^{-\frac{(k-p_j)^2}{2\sigma^2}} \frac{(a*F)_j^k}{k!} \right) \tag{18.5}$$

The last algorithm is implemented in the IRAF package and in many other software packages (see Molina *et al.* 2001 for a more comprehensive list of software packages). In IRAF, the software for image deconvolution can be found in the STSDAS package stsdas/analysis/restore. There, the reader will find implementations for the R–L method (lucy) besides other algorithms such as: an implementation of the maximum-entropy method (MEM), the primitive Wiener filtering Fourier linear method (wiener), and the CLEAN method of PSF fitting and removal (sclean). The R–L algorithm has a number of desirable characteristics: it solves the cases of both pure Poisson data ($\sigma = 0$) and Poisson data with

Gaussian readout noise (CCDs); it maintains the positivity of the solution; and it is easy to implement. Some implementations (Prades and Núñez 1997) include flat-field corrections, removal of background, and that it can be accelerated. Since the algorithm can be applied to a large number of imaging situations, including CCD and pulse-counting cameras, we recommend the interested reader to start any deconvolution study using the R–L algorithm or any of its extensions.

It is important to note that the R–L solution is an unconstrained "classical" solution of the ill-posed Fredholm integral imaging equation. The result is that the iterative R–L algorithm produces solutions that are highly unstable, with high peaks and deep valleys. Thus, it is necessary to stop the process before reaching convergence. To stop the algorithm several approaches have been proposed. In the above cited reviews and proceedings, the reader can find most of them (from pure visual inspection to highly mathematical methods). Probably, the most robust method to stop the algorithm is the cross-validation tests, which compute the likelihood of the solution with respect to an alternative set of data (cross-likelihood). The stopping point is given by the maximum of the cross-likelihood against the iteration number (Núñez and Llacer 1993). However, even when stopping the R–L algorithm at the optimum point it is easy to find that some zones (particularly the stars) are underdeveloped while other areas (particularly the background) are overdeveloped. To avoid this, several extensions of the R–L algorithm (known as PLUCY, CPLUCY, and GIRA) can also be used. Currently, some of the most promising extensions are based on wavelet decomposition. Otazu (2001) developed the algorithm AWMLE. In this algorithm the stopping point of the R–L algorithm varies from point to point depending on the spatial frequency content of the area computed by the discrete wavelet transform. Note, however, that in this case, the algorithm should be driven to full convergence, thus increasing the central processing unit (CPU) time cost. Fors (2006) used the AWMLE algorithm to restore successfully astrometric frames (see Figs. 18.3 and 18.4 below). A similar wavelet-based algorithm can be found in Starck *et al.* (2002).

As discussed above, other approaches to control the deconvolution are based in the maximum-entropy and the Bayesian paradigm. There is no room in this chapter to describe the dozens of approaches and algorithms that have been developed and we direct the reader to the reviews, proceedings, and special issues cited above and references therein. It is interesting to note that some of the algorithms, although developed in a very different way, are closely related in computation to the R–L algorithm. For example, introducing the entropy as prior information in the Bayesian framework, and maximizing the Bayesian posterior probability, Núñez and Llacer (1993, 1998) obtained algorithms called FMAPE and FMAPEVAR (for space-variant prior information balance) with a main loop which is identical to the loop of the R–L. A similar situation can be found in most maximum-entropy algorithms.

18.2 Correcting for atmospheric degradation

Image reconstruction of astrometric frames makes substantial improvements in the qualitative appearance and recovery of faint star images in terms of the number of stars that

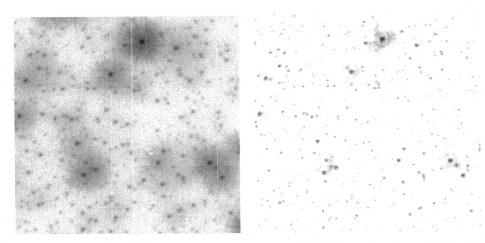

Fig. 18.1 Raw image of an HST PC astrometric frame (left) and its restoration (right) after 30 iterations using the R–L algorithm. From Prades and Núñez (1997).

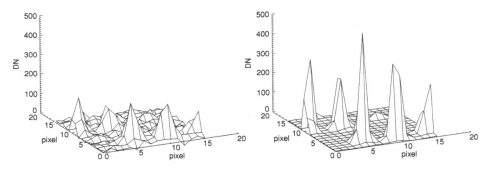

Fig. 18.2 Plot showing part of a raw frame (left) and one with 20 R–L iterations deconvolved (right) astrometric frame from a CCD Meridian Circle (Prades and Núñez, 1997).

can be detected. As an example, we present the reconstruction of three astrometric frames: one obtained from space and the others from the ground. To show also the versatility of the restorations, we used two different algorithms (R–L and AWMLE both for data with Poisson noise and then with Gaussian readout noise) and synthetic, real, and semi-synthetic (fitted) PSFs.

In the first example (Prades and Núñez 1997), the reconstruction is shown of an astrometric image obtained with the Planetary Camera (PC) of the aberrated HST by van Altena in 1990. Figure 18.1 shows the raw data (left) and its restoration (right) after 30 iterations using the R–L algorithm and a synthetic PSF generated by the TinyTim software. Even with the limited quality of this figure the improvement in both appearance and number of stars seen is evident.

Deconvolution is also efficient in ground-based images. In the second example (Prades and Núñez 1997), a 512×512 astrometric frame obtained from the ground is shown. The image was obtained using a CCD camera in TDI mode (see Chapter 15) attached to a meridian circle. To show the improvement attainable by deconvolution, Fig. 18.2 shows

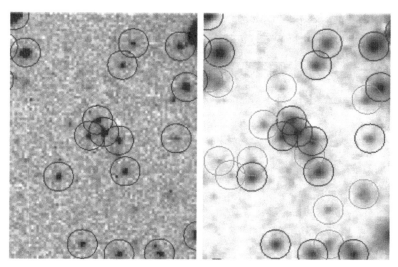

Fig. 18.3   Raw frame (left) and restoration (right) of a CCD exposure obtained with a Baker–Nunn camera. In the restored frame the newly detected stars by the Sextractor algorithm (not detected in the raw frame) are circled in grey. The AWMLE algorithm at convergence was used with a semi-synthetic Moffat PSF (Fors, 2006).

a three-dimensional plot of a part of the raw frame (left) and its restoration (right) after 20 R–L iterations using a real image of a star for the PSF. Again, the improvement of the reconstructed frame with respect to the raw data is evident.

In the third example (Fors 2006), a part of the restoration of a large CCD frame (4096 × 4096 pixels) obtained with the retrofitted Baker–Nunn camera of the Rothney Astrophysical Observatory in Canada (NESS-T BNC program) is shown. The Baker–Nunn cameras are modified Super-Schmidt telescopes of 50 cm aperture working at focal ratio $f/1$. Figure 18.3 shows the raw data (left) and its restoration (right) using the AWMLE algorithm at convergence (140 iterations) and a semi-synthetic (Moffat) PSF. The Sextractor algorithm, with the same parameters, was used to detect the stars present in both frames. In the raw frame of Fig. 18.3 (left) the detected stars are circled in black. In the restored image (right) the newly detected stars that were not detected in the raw image are circled in grey. In the detail shown, a very important increase in detected stars was achieved since 12 newly detected stars were added to the 18 stars detected in the raw image. In the whole image (not shown) a total of 2644 stars were detected and matched to a catalog in the restored image versus only 1724 stars detected in the original.

18.3 Potential gains in precision and limiting magnitude

18.3.1 Gain in limiting magnitude

The gain in limiting magnitude can be estimated using the magnitude histograms of matched detections for the original frame and for the deconvolution. In Fig. 18.4 (Fors 2006) the

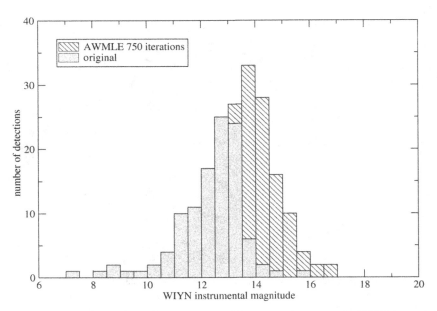

Magnitude histograms of matched detections from Fors (2006) for an original image and the AWMLE deconvolution at convergence of a QUEST image.

magnitude histogram of the original frame is over-plotted with the one from an AWMLE deconvolution at convergence. The data belong to a frame obtained during the QUasar Equatorial Survey Team (QUEST) program using the 1 m Venezuelan Schmidt Telescope working in scanning (TDI) mode (see Chapter 15). The dashed area in Fig. 18.4 represents the stars detected only in the restored frame. In this example a total of 197 stars were detected and matched to a substantially deeper exposure taken with the WIYN 3.5-m telescope at Kitt Peak versus only 109 stars detected and matched in the raw frame. If the area of each histogram (which coincides with the number of matched detections) is considered, a limiting magnitude gain of $\Delta m = 0.64$ magnitudes is derived using the simple equation

$$\Delta m = 2.5 \log(N_{\text{restored}}/N_{\text{raw}}) \tag{18.6}$$

It is important to point out that a gain of $\Delta m \simeq 0.6$ magnitudes means an increment of 80% in available detections or, equivalently, an increase of 32% in telescope aperture or, in telescope cost, a decrease of 2.3 times without losing astrometric performance as is demonstrated in Section 18.3.3.

18.3.2 Gain in limiting resolution and blended images

We define the limiting resolution $\Delta \varphi_{\text{lim}}$ of a given exposed CCD frame as the smallest separation detected in it. Image deconvolution can also increase the limiting resolution of the deconvolved images with respect to the raw ones. In this sense, image deconvolution can help to effectively separate blended images of stars. Using QUEST and NESS-T data,

Fors (2006) performed an in-depth study of the possible gain in limiting resolution using the AWMLE image-deconvolution algorithm. He obtained identical values of $\Delta\varphi_{lim} \simeq 1$ pixel for QUEST and NESS-T data, corresponding to $\Delta\varphi_{lim} \simeq 1''$ and $\Delta\varphi_{lim} \simeq 3.9''$ respectively. In other words, the resolution limits following deconvolution were found to be improved in each case by about 1 pixel and only slightly modulated by other factors such as drift-scanning systematics or limited knowledge of PSF modeling. In conclusion, image deconvolution (AWMLE in this case) has shown its powerful deblending capabilities, which could be of interest for many projects such as, for example, quasi-stellar object (QSO) lensing searches (QUEST) or new near-Earth object (NEO) discoveries (NESS-T). For example, as Fors (2006) pointed out, the limiting resolution after deconvolution of QUEST images is $3.9''$, which is within the cutoff value of the separation distribution of the 82 gravitational lenses currently known. This means that QUEST data can potentially be used for resolving lensed QSOs directly from deconvolved data. Of course, high-resolution imaging continues to be necessary for confirming lens geometry.

Despite the good results cited above for deblending stellar images, an important warning should be issued here. To study the effect of deconvolution on the astrometric accuracy of blended images, Prades and Núñez (1997) performed a series of tests consisting of generating artificial pairs of stars with decreasing separation. The centering algorithm was modified to fit both stars simultaneously (instead of centering each one separately) using two two-dimensional elliptic Gaussian functions. The pairs of blended star images were deconvolved using the R–L method and the centering algorithm was applied to both the raw and deconvolved images. The positions obtained using the raw data show no bias with respect to the true values. However, the positions obtained using the deconvolved images seem to show a bias consisting of a shift towards the centroid of the blended image. This result is not definitive and more studies should be done to confirm or reject this effect and, if confirmed, to assess whether it is it is due to the nature of the problem or to the reconstruction algorithm itself. Thus, as a conclusion, it is possible to say that image deconvolution can increase the limiting resolution and help to deblend images of stars but, also, we would warn against use of deconvolution techniques for blended star images if the astrometric positions of the components are of importance.

18.3.3   Gain in astrometric precision

In an earlier work, Girard *et al.* (1995) performed a series of tests to assess whether image deconvolution can retain or improve the astrometric information to be extracted from an image. The tests involved two 800×800 pixel HST-WFPC1 frames of NGC 6752 (40-s and 500-s exposure respectively). Both frames were deconvolved using the R–L algorithm with 100 iterations. Given the space-variant character of the PSF of the HST, Girard *et al.* used three different PSFs centered at pixels $(200, 600)$, $(400, 400)$, and $(600, 200)$ respectively. The reconstructed intensity profiles were centered, and the long-exposure positions were transformed into those of the short exposure to determine the unit weight measuring error. The results are given in Table 18.1 and suggest that 1 milliarcsecond (mas) positional precision can be obtained using reconstructed images versus 2 mas using the raw images. In a later study, Prades and Núñez (1997) carried out another experiment

Table 18.1 Positional accuracy (long- to short-exposure) from Girard *et al.* (1995)

	σ_x(mas)	σ_y(mas)
Raw	1.4	2.1
PSF(200, 600)	1.4	1.2
PSF(400, 400)	1.3	1.0
PSF(600, 200)	0.8	0.8

Table 18.2 Standard deviations of the Gaussian-fitted profiles from Prades and Núñez (1997)

Image	$\sigma_x('')$	$\sigma_y('')$
Raw	0.84	1.11
Deconvolved (R–L 20 its.)	0.40	0.56
Deconvolved (R–L 100 its.)	0.26	0.41

using a two-dimensional Gaussian fitting to obtain the positions of the stars in both the raw image and the reconstruction of the meridian circle CCD image of Fig. 18.2. Table 18.2 lists the means of the standard deviation in x and y coordinates of the fitted Gaussian profiles. The results show that after reconstruction, the standard deviation of the Gaussian-fitted stellar intensity profiles are about half those obtained using the raw data. Given that the σ of the fitted stellar profile is directly related to the seeing present during the exposure, the results show that the reconstructed image can be considered equivalent to an image obtained with a seeing that is one-half of the observed FWHM. See Chapter 9 for a detailed discussion of how astrometric precision varies with the observed FWHM.

18.3.4 Absence of bias

Perhaps the most important problem in the use of image deconvolution for astrometry is the presence of any systematic bias in the positions of the deconvolved stars. In a first study, Girard *et al.* (1995) noticed indications that, in the R–L deconvolved images of the HST PC, the derived image centers of the stars were biased toward the center of the brightest pixel. This effect could be due to several factors such as the use of a constant PSF to deconvolve images that have a known space-variant PSF or to the deconvolution algorithm itself.

However, in a later study, Prades and Núñez (1997) investigated this possible bias further, carrying out several tests using computer-generated data. They found no bias at all in the simulations. In addition, Fors (2006), in a recent study using data obtained from the Flagstaff Astrometric Transit Telescope (FASTT) also found no systematic bias in the position of stars after the deconvolution. It appears that the Girard *et al.* (1995) results were spurious and may have been the result of small-number statistics.

18.4 CPU cost

Since almost all of the deconvolution algorithms are iterative and based on the FFT, it is better to estimate the CPU cost in the number of FFTs or even in the number of floating-point operations. For example, the most widely used algorithm, the R–L, needs four FFTs per iteration (all the other computations are vector operations that have little impact on CPU time). For a frame of N total pixels the number of operations to perform an FFT is $O(N \log_2 N)$. Thus, the total number of operations to perform an n-iteration R–L deconvolution is proportional to about $4nN \log_2 N$. Other algorithms such as AWMLE, FMAPE, or the maximum-entropy method take more time since they also use four FFTs per iteration but need to reach convergence and that can be at 500 or more iterations. Finally, at each iteration the AWMLE performs a wavelet decomposition that adds appreciable computer time, but that limitation is largely compensated by the increased scientific return that deconvolution can offer.

18.5 Summary

Given any combination of telescope and detector, the techniques of image deconvolution can increase the scientific return of the observations. In this sense, the effect of the deconvolution can be equivalent to using a telescope with up to 40% larger aperture without affecting the astrometric precision or introducing artificial bias. Some studies indicate that the astrometric precision is also increased. Deconvolution can also help to deblend images of close pairs but, in this case, the astrometric properties of the stars can be compromised. The CPU cost can be high but with continuously faster and cheaper computers, image deconvolution is worthy of consideration in most cases. It is important to point out that the success of the deconvolution process is highly dependent on the correct choice of the PSF and the noise model.

References

Fors, O. (2006). New observational techniques and analysis tools for wide field CCD surveys and high resolution astrometry. PhD Thesis. University of Barcelona. See: www.tdx.cat

Girard, T. M., Li, Y., van Altena, W. F., Núñez, J., and Prades, A. (1995). Astrometry with reconstructed HST Planetary Camera (WF/PC 1) images. *Int. Jo. Imaging Syst. Technol.* (Special Issue on Image Reconstruction and Restoration in Astronomy), **6**, 395.

Hanisch, R. J. and White, R. L., eds. (1994). *The Restoration of HST Images and Spectra II*. Baltimore, MD: STScI.

Lucy, L. B. (1974). An iterative technique for the rectification of observed distributions. *AJ*, **79**, 745.

Molina, R., Núñez, J., Cortijo, F., and Mateos, J. (2001). Image restoration in astronomy. A Bayesian perspective. *IEEE Signal Process. Mag.*, **18**, 11.

Núñez, J., ed. (1995). Image reconstruction and restoration in astronomy. *Int. J. Imaging Syst. Technol.* (Special Issue on Image Reconstruction in Astronomy). Special Issue **6**.

Núñez, J. and Llacer, J. (1993). A general Bayesian image reconstruction algorithm with entropy prior. Preliminary application to HST data. *PASP*, **105**, 1192.

Núñez, J. and Llacer, J. (1998). Bayesian image reconstruction with space-variant noise suppression. *A&AS*, **131**, 167.

Otazu, X. (2001). Algunes aplicacions de les Wavelets al procés de dades en astronomía i teledetecció. PhD Thesis. University of Barcelona. See: www.tdx.cat.

Prades, A. and Núñez, J. (1997). Improving astrometric measurements using image reconstruction. In *Visual Double Stars. Formation, Dynamics and Evolutionary Tracks*, ed. J. A. Dacobo, A. Elipe, and H. McAlister. Dordrecht: Kluwer Academic, p. 15.

Starck, J. L., Pantin, E., and Murtagh, F. (2002). Deconvolution in astronomy: a review. *PASP*, **114**, 1051.

White R. L., (1994). Better HST point-spread functions: phase retrieval and blind deconvolution. In *The Restoration of HST Images and Spectra II*, ed. R. J. Hanisch and R. L. White. Baltimore: STScI., p. 198.

White, R. L. and Allen, R. J., eds. (1990). *The Restoration of HST Images and Spectra*. Baltimore, MD: STScI.

19 From measures to celestial coordinates

ZHENG HONG TANG AND WILLIAM F. VAN ALTENA

Introduction

The goal of this book is to present an introduction to the techniques of astrometry and to highlight several applications of those techniques to the solution of current problems of astrophysical interest. In some cases we require the absolute positions of objects to establish reference frames and systems, while in others we need the change in position with time to obtain the distances and tangential velocities of the objects. The astrometric procedures necessary to solve those problems have been laid out in considerable detail by Konig (1933, 1962), Smart (1931), and others, so we will summarize the methods used to transform raw coordinate measurements of celestial objects on photographic plates and CCDs to their corresponding coordinates on the celestial sphere. We recommend that the reader consult the above references where clarification is needed. Once we have the desired celestial coordinates and their changes with time that yield parallaxes and proper motions, we can then create catalogs of those quantities and apply them to the solution of problems in galactic structure, the masses of stars, membership in star clusters, dynamical studies of objects in the Solar System, and extrasolar planets, and to help to set limits to some cosmological models.

19.1 Telescope and detector alignment

We normally assume that our telescope and detector have been carefully aligned so that the image quality will be optimum over the field of view (FOV). However, even if the apparent image quality is good over the FOV, it may be that residual misalignments remain that complicate the transformation from detector to sky. Quite often, high-order polynomials are used to absorb the effects of those misalignments. Unfortunately, that procedure obscures the interpretation of the transformation terms and can lead to spurious and/or lower-accuracy results.

Most modern large telescopes and detector systems make it rather difficult for the user to easily verify their alignment; hence the only alternative is to consult the engineers in charge

Astrometry for Astrophysics: Methods, Models, and Applications, ed. William F. van Altena.
Published by Cambridge University Press. © Cambridge University Press 2013.

of telescope operations to verify its current state of alignment and the tolerances to which it was aligned. Section 13.2.5 discusses the misalignments in telescopes and the types of images produced by those misalignments (see Fig. 13.6 in particular). For small telescopes, it is often possible to place a piece of glass over the entrance aperture, replace the detector with a piece of flat reflecting glass and point a small laser from the entrance aperture of the telescope towards the center point of the glass in the detector plane. Symmetry of the laser-beam entrance point with respect to the exit beam about the geometric center of the telescope aperture will verify the perpendicularity of the detector plane to the central axis of the telescope. Having checked the perpendicularity of the detector plane with respect to the telescope tube axis and corrected any residual tilt, it is now possible to check for possible tilt of the objective or primary mirror's optical axis with respect to the telescope tube axis using the relatively simple test device described by Schlesinger and Barney (1925). Residual tilts, centering errors, and improper spacing of the optics can at times be diagnosed by characteristic aberrations described in Chapter 13 and by Schroeder (2000).

The other major alignment issue is the pointing of the polar axis in equatorial-mounted telescopes. Polar-axis alignment and flexure are critical to avoid image degradation in wide-field telescopes due to field rotation, but less important for most of the small-field telescopes that use single charge-coupled devices (CCDs). Telescopes with mosaics of CCDs covering more than one degree begin to have more critical polar-axis-alignment tolerances. We note here that due to atmospheric refraction, the optimum pointing of the instrumental polar axis depends on the declination of the field being observed. Most of the modern alt-azimuth telescopes require image rotators on the detector-mounting adaptor, and accurate programming of that device is critical for the maintenance of high image quality. Details on alignment procedures can be found in the monographs by Arend (1951), Wallace (1979), Wallace and Tritton (1979), and Meeks *et al.* (1968).

19.2 Transforming from the detector plane to the celestial sphere

In this section we image a portion of the celestial sphere onto a perfectly flat detector plane that is perpendicular to the line $T'OT$ that connects our target to the detector plane while passing through a pinhole camera O. In Fig. 19.1, T' is the target, O the pinhole camera and T the projected image of T'.

This imaging process is called a gnomonic projection. In Fig. 19.1, vector TO represents the optical axis of a telescope that is perpendicular to the ideal focal plane and has focal length F. The equatorial coordinates of point T' are (α_0, δ_0). The "tangential coordinate system" on the tangential plane has T as its origin with the declination circle projecting onto the tangential plane as the axis η, which is positive in the direction of increasing declination, while axis ξ is perpendicular to η and is positive in the direction of increasing right ascension.

In Fig. 19.1, suppose that S' is a star on the celestial sphere that is imaged onto the tangential plane at S, while P is the north celestial pole. In the following we take the focal

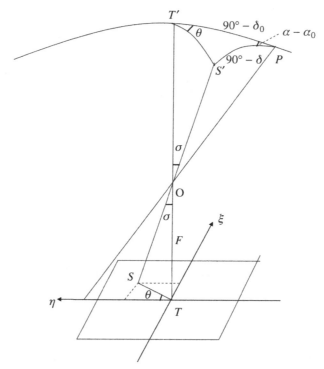

Fig. 19.1 Schematic representation of the gnomonic projection.

length F of the telescope as unity and in the triangle $\triangle OST$, the length is

$$TS = \tan \sigma \tag{19.1}$$

with tangential coordinates

$$\begin{cases} \xi = \tan \sigma \cdot \sin \theta \\ \eta = \tan \sigma \cdot \cos \theta \end{cases} \tag{19.2}$$

In spherical triangle $PT'S'$ we have

$$\cos \sigma = \sin \delta \sin \delta_0 + \cos \delta \cos \delta_0 \cos(\alpha - \alpha_0) \tag{19.3}$$
$$\sin \sigma \sin \theta = \cos \delta \sin(\alpha - \alpha_0) \tag{19.4}$$

and

$$\sin \sigma \cos \theta = \sin \delta \cos \delta_0 - \cos \delta \sin \delta_0 \cos(\alpha - \alpha_0) \tag{19.5}$$

Substituting Eqs. (19.3)–(19.5) into Eq. (19.2) we have

$$\begin{cases} \xi = \dfrac{\cos \delta \sin(\alpha - \alpha_0)}{\sin \delta_0 \sin \delta + \cos \delta_0 \cos \delta \cos(\alpha - \alpha_0)} \\[3mm] \eta = \dfrac{\cos \delta_0 \sin \delta - \sin \delta_0 \cos \delta \cos(\alpha - \alpha_0)}{\sin \delta_0 \sin \delta + \cos \delta_0 \cos \delta \cos(\alpha - \alpha_0)} \end{cases} \tag{19.6}$$

which represent the tangential coordinates of star image S on the tangential plane. If the tangential point (α_0, δ_0) and celestial coordinates (α, δ) of an object are known, then its tangential coordinates (ξ, η) can be calculated. Likewise, Eq. (19.6) can be inverted to yield Eq. (19.7) and then solve for the celestial coordinates of an object given its tangential coordinates. The advantage of the second formulation for obtaining the declination is that the right-ascension difference does not appear.

$$
\begin{cases}
\alpha = \alpha_0 + \arctan\left(\dfrac{\xi \sec \delta_0}{1 - \eta \tan \delta_0}\right) \\[2ex]
\delta = \arctan\left(\dfrac{\eta + \tan \delta_0}{1 - \eta \tan \delta_0} \cos(\alpha - \alpha_0)\right) \text{ or} \\[2ex]
\delta = \arcsin\left(\dfrac{\sin \delta_0 + \eta \cos \delta_0}{\sqrt{1 + \xi^2 + \eta^2}}\right)
\end{cases}
\tag{19.7}
$$

19.3 Transforming between two tangential coordinate systems

Our tangential coordinate system is centered at (α_0, δ_0); however, the center of our detector field of view, which should be on the optical axis of the telescope, is unlikely to be pointed exactly at the above tangential point. Therefore, we need to transform our detector coordinate system, which is centered at a slightly different tangential point (α'_0, δ'_0), to that of the tangential coordinate system. These two tangential coordinate systems have slightly different orientations, σ and σ', with respect to the celestial pole P, and their respective tangential points are separated by an arc of length s. If s were small, then the transformation between the two coordinate systems would be a simple translation and rotation; however, the translation is over the surface of a sphere, which complicates the transformation a bit. Assuming that $\Delta\sigma = \sigma' - \sigma$ and s are small and following the development of Konig (1962), we can write the relationships between the two tangential coordinate systems as

$$
\begin{aligned}
\xi &= +\xi' + \Delta\sigma \cdot \eta' - s \sin\sigma - (s \sin\sigma \cdot \xi' + s \cos\sigma \cdot \eta')\xi' \\
\eta &= -\Delta\sigma \cdot \xi' + \eta' + s \cos\sigma - (s \sin\sigma \cdot \xi' + s \cos\sigma \cdot \eta')\eta'
\end{aligned}
\tag{19.8}
$$

The first three terms in Eq. (19.8) represent the usual transformation between two coordinate systems on the same plane, where $\Delta\sigma$ is the rotation between the two frames, and $s \sin\sigma$ and $s \cos\sigma$ are the offsets in (α, δ), respectively, while the quadratic terms correct for translation over the surface of a sphere, which introduces a "tilt" between the two tangential planes. The latter quadratic terms are often referred to as "plate tilt." If we let $p = s \sin\sigma$ and $q = s \cos\sigma$, then Eq. (19.8) can be rewritten as

$$
\begin{aligned}
\xi &= +\xi' + \Delta\sigma \cdot \eta' - p - (p \cdot \xi' + q \cdot \eta')\xi' \\
\eta &= -\Delta\sigma \cdot \xi' + \eta' + q - (p \cdot \xi' + q \cdot \eta')\eta'
\end{aligned}
\tag{19.9}
$$

where all coordinates are expressed in radians.

19.4 Correcting the (x, y) measures for various errors and systematic effects

We now need to transform the (x, y) measures to the (ξ', η') tangential coordinate system. The (x, y) measures contain errors and/or systematic effects introduced by (1) the detector and/or measuring machine, (2) misalignment of the detector and optics, and (3) systematic effects such as refraction, aberration, precession, and nutation. We will deal with each of these in the following subsections.

19.4.1 Detector and measuring machine errors

Most astrometry today utilizes CCDs; however, for many proper-motion investigations, the first-epoch observations still come from measures of old photographic plates, which emphasizes the importance of preserving those plates in proper archives. In some cases the plates have been measured and those measures are available, but that is not the case for the large majority of old, historically important plates. Very few institutions maintain precision measuring machines in operational condition, so it can be a task to gain access to one for the measurement of your old plates. At measurement time it is important to calibrate the errors of the machine to avoid the unnecessary introduction of additional complications into the analysis process. Operation manuals are often useful for obtaining the calibrated errors due to coordinate scale errors and non-orthogonal measuring axes. It is often the case that the measuring machine will introduce systematic errors that are a function of the brightness of the object being measured or density of the photographic emulsion. Those errors can normally be eliminated by measuring the plate in the normal position and then rotating it by 180 degrees the second time. On the other hand, those magnitude equation systematic errors that are intrinsic to the emulsion, such as errors in guiding the telescope, cannot be eliminated in this way.

CCDs are not without errors, especially when you are attempting to achieve the highest-accuracy positions. Many of the potential errors are described in Chapter 14, but to highlight the most common we note that the pixel pattern laid down on a CCD during its fabrication process depends on the geometry of the mask and the accuracy of the measuring machine, and camera optics used to repetitively cover the whole CCD chip (see for example Anderson and King 2000). The second major error encountered in CCDs is due to charge-transfer inefficiency (see Section 14.2.6), which introduces systematic errors as a function of the number of counts in a pixel and the location of the image with respect to the readout register of the CCD. Zacharias *et al.* (2000) describe in detail how you can correct for this type of problem. The third complication with CCDs comes from the desire to cover the largest possible field of view with what are normally small CCDs. As a consequence, the pixels often fail to properly sample the observed image, especially those on-board satellites such as the Hubble Space Telescope (HST). Such "undersampled" images can lead to substantial systematic errors depending on where the point-spread function of a particular image is centered in relation to a pixel center. The analysis process for dealing with undersampled images is quite complicated, but the systematic errors can be eliminated by using the techniques described in Chapter 17.

Filters and CCD windows are also a part of the detector system and flatness errors and variations in the index of refraction will result in positional errors that depend on the separation of the components from the final focal plane as discussed in Section 13.2.4 and in greater detail in van Altena and Monnier (1968). Correction for these small-scale systematic errors is usually made through the use of a vector plot of x, y residuals on the focal plane. The first use of this correction scheme (see Taff *et al.* 1990) came with the creation of the HST Guide Star Catalog and attempts to model the large systematic errors found on Schmidt plates. Additional references for preparing such a mask are by Zacharias *et al.* (1997), Girard *et al.* (2004) and Hanson *et al.* (2004).

19.4.2 Instrumental errors

Instrumental errors include those introduced through optical aberrations (see Section 13.2.3), as well as misalignment of the optics (see Section 19.1) and detector. The detector alignment errors generally include corrections to the zero point of the coordinate system, plate scale, rotation, and tilt of the detector with respect to the optical axis of the telescope. The classical optical aberrations in the limit of small angles θ are summarized in Eq. (13.14) (repeated below as Eq. (19.10)) and include spherical aberration (term 1), coma (term 2), astigmatism (term 3), and OFAD (term 4).

$$AA3 = a_3 \frac{y^3}{R^3} + a_2 \frac{y^2\theta}{R^2} + a_1 \frac{y\theta^2}{R} + a_0\theta^3 \qquad (19.10)$$

In Eq. (19.10), AA3 represents the third-order angular aberrations, R the radius of curvature of the mirror, y the height above the optical axis of the incoming ray at the optical surface (telescope aperture) while θ is the angle of the ray with respect to the optical axis. Quoting from Chapter 13, we see the following.

Spherical aberration does not depend on θ, hence is constant over the object field. A change in the sign of y changes the sign of this aberration and thus rays from opposite sides of the aperture are on opposite sides of the chief ray at the image. When rays over the entire aperture are included, the image is seen to be circularly symmetric about the chief ray, as shown in Fig. 4.6 in Schroeder (2000). We note that in this chapter since spherical aberration is not dependent on field location, the primary effect is only a blurring of the image with no dependence on the coordinate in the detector plane.

Coma is proportional to $y^2\theta$ and hence is changed in sign when θ changes sign. Coma is invariant to the sign of y and therefore rays from opposite sides of the surface are on the same side of the chief ray at the conjugate image. The result is an asymmetric image that looks like a comet, hence the name coma. A set of through-focus spot diagrams is shown in Fig. 13.6(a) (Fig. 5.9 in AO2). We note in this chapter that since coma is a linear function of the field angle, its size changes with the coordinate in the detector plane and it can mimic a change in the plate scale. For a detector with a linear response to light, the centroid of a comatic image should not change with intensity; however, photographic emulsions are non-linear and CCDs can be non-linear at some intensities, so that the centroid of a comatic image can be seen to change with both light intensity and coordinate in the detector plane. The commonly used correction term for coma is the product of the image brightness, or magnitude, times the coordinate, x or y.

Astigmatism is proportional to $y\theta^2$ and hence is unchanged by a sign change in θ. Comments about the sign of y for spherical aberration are true for astigmatism as well, but the character of the image is quite different. This is easily seen in Fig. 13.6(b) (Fig. 5.3 in AO2). In going through focus of an astigmatic image we find two orthogonal line images with a circular blur located midway between the line images. Each of the line images lies on its own curved focal surface, as does the circular blur. Each of these surfaces has its own radius of curvature and is related to an aberration that does not appear in Eq. (13.14): *curvature of field*. In this chapter we note that given the even power of the field angle, astigmatism results only in an elongation of the images that grows with coordinate on the detector plane. However, the complication is that each of the two line images referred to above lies on its own focal surface, each of which is curved. Normally the telescope is focused at the mean of the two surfaces and this aberration does not affect astrometric observations.

Optical field-angle distortion (OFAD) is proportional to θ^3 (and θ^5 in many commonly used correctors) but does not depend on y. Thus, this aberration, if it is the only one present, does not affect the image quality, only its position. For a set of point objects equally spaced perpendicular to the optical axis, the set of images would not be equally spaced if OFAD is present or, looked at in a different way, the plate scale varies with position in the field of view. Figure 13.6(c) shows the patterns on a flat detector of a square object field by systems with OFAD; one is called barrel, the other pincushion. In most modern telescopes of the Ritchey–Chrétien design a field corrector is designed so that the dominant remaining distortion is OFAD, and thus the best possible images are obtained over the whole field of view. As a result, the OFAD can be very strong and it must be calibrated and corrected for all astrometric investigations. Guo *et al.* (1993) developed procedures to calibrate the OFAD in prime-focus cameras for the 4-m telescopes and Jefferys *et al.* (1994) for the HST Fine Guidance Sensor (FGS) instrument; the details of those methods can be found in the cited publications. We note here that it is very important to locate the OFAD center accurately for each exposure, otherwise systematic errors in the coordinates will be introduced into the image positions. The unknown difference between the true and assumed center of optical distortion and possible variations of this distance as a function of telescope pointing and other factors render telescopes with large optical distortions very difficult to use for high-accuracy, wide-field astrometry. For classical astrograph-type refracting telescopes and modern designs of dedicated, astrometric telescopes (de Vegt *et al.* 2003) the optical distortion is very small.

19.4.3 Spherical corrections

Under the terminology of spherical corrections, we include: atmospheric refraction, velocity aberration, precession, and nutation. The reason for classifying them as "corrections" is that their values can in general be calculated to relatively high accuracy and it makes sense to eliminate their effects from the measured positions before proceeding with the

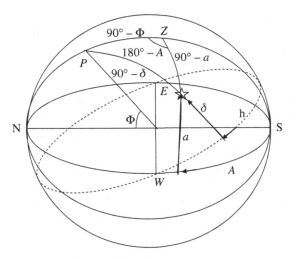

Fig. 19.2 Relation between the equatorial and horizon–zenith coordinate systems.

transformation to the celestial sphere. It is common practice for some researchers to simply include higher-order terms in the transformations and not pre-correct for the spherical terms; however, it then becomes more difficult to interpret the significance of the transformation coefficients and separate problems due to detector and optics alignment from truly astronomical effects. We will deal with each of the spherical corrections in this section.

Atmospheric refraction of light

In Fig. 19.2 we illustrate the relationship between the equatorial and horizon–zenith coordinate systems. A more detailed discussion of coordinate systems may be found in Chapter 7. For the purposes of this section, we note that refraction of light amounts to a contraction of the zenith angle $z = 90 - a$, which is the great circle connecting the target of interest and the zenith, Z, where a is the altitude of the target above the horizon.

The contraction of the zenith angle due to refraction is given by Eq. (9.10) and its origin is discussed in detail in Chapter 9; for convenience we repeat it here as Eq. (19.11), where z_0 is the true zenith distance.

$$R \approx \beta_0 \tan z_0 + \beta' \tan^3 z_0 + \beta_c (B - V) \tan z_0 + \dots \qquad (19.11)$$

We note here that the refraction R is a function of the zenith angle as well as the wavelength, atmospheric temperature, pressure, and humidity, and the spectral-energy distribution of the target, or targets. Since the detector FOV may span one or more degrees on a side, the direction towards the zenith as seen by the various objects will vary across the FOV and therefore results in a slight contraction of the FOV. At the zenith the contraction is equal in both coordinates, while on the meridian the contraction in azimuth is proportional to the field size and independent of the zenith angle, and the contraction in the zenith coordinate is proportional to $(1 + z^2)$. The contraction across a 1-degree FOV amounts to $2''$ (sea level) at a zenith angle of 45°, so it is large and needs to be corrected. To first order and for small FOV these effects are similar to that produced by thermal variations in the telescope.

The most straightforward method to correct for these effects is to rotate the detector coordinates so that the positive y-axis points towards the zenith, transform the new (x, y) to approximate tangential coordinates with the zenith Z as the pole, correct the zenith–horizon tangential coordinates for refraction, transform the new tangential coordinates back to a set of refraction-corrected detector coordinates, and finally rotate the tangential coordinates back to the original detector angle. Assuming that the detector y-axis is pointed towards the north celestial pole, NCP, and that the angle opposite the great circle are NZ % as seen from the field center (a "star" in Fig. 19.2) is χ, we have

$$\begin{cases} x'_m = x_m \cos \chi - y_m \sin \chi \\ y'_m = x_m \sin \chi + y_m \cos \chi \end{cases} \tag{19.12}$$

We now transform to the zenith tangent plane using

$$\begin{cases} \cot \Delta a = \dfrac{-x'_m}{\sin z_0 - y'_m \cos z_0} \\[3mm] \cos z_m = \dfrac{\cos z_0 + y'_m \sin z_0}{\sqrt{\left(1 + (x'_m)^2 + (y'_m)^2\right)}} \end{cases} \tag{19.13}$$

where (a_0, z_0) is the true azimuth and zenith distance of the tangent point, and $\Delta a = a - a_0$ is the difference between the azimuth of the measured object and that of the tangent point. We then correct for refraction relative to the plate center using

$$z' = z_m + \Delta R, \quad \text{where } \Delta R = R(z_m) - R(z_0) \tag{19.14}$$

where z_m is the measured zenith distance and refraction R is calculated from Eq. (19.11). We next compute the new zenithal tangential coordinates z', z_0, and Δa and transform our refraction corrected coordinates back to our detector-based system, (x'_m, y'_m)

$$\begin{cases} x'_m = \dfrac{-\sin z' \sin \Delta a}{\cos z' \cos z_0 + \sin z' \sin z_0 \cos \Delta a} \\[3mm] y'_m = \dfrac{\cos z' \sin z_0 - \sin z' \cos z_0 \cos \Delta a}{\cos z' \cos z_0 + \sin z' \sin z_0 \cos \Delta a} \end{cases} \tag{19.15}$$

and then rotate back to the NCP:

$$\begin{cases} x = x'_m \cos \chi + y'_m \sin \chi \\ y = -x'_m \sin \chi + y'_m \cos \chi \end{cases} \tag{19.16}$$

Equations (19.12) through (19.16) will correct for the majority of the systematic effects of atmospheric refraction; however, there will always remain small errors due to the approximations used in Eq. (19.14) to express the shift. The dominant errors include our lack of knowledge of the actual spectral-energy distribution of each star as well as the temperature, pressure, and humidity distribution of the troposphere along the path that the light travels on its way from the star to the detector. More details can be found in Chapter 9. Because of these limitations it is always good practice to carefully analyze your results to check for systematic trends as a function of zenith distance and color of the star as is discussed later in this chapter.

Velocity aberration

Velocity aberration is a consequence of the finite velocity of light and involves a small contraction along the great circle towards the apex of the observer's motion. The correction for aberration is discussed in detail in Chapters 7 and 8, for observations with the FGSs of the HST in Jefferys *et al.* (1994) and for "time-delayed integration" in Chapter 15, so here we will only summarize a practical procedure for implementing the corrections for ground-based observations. The correction is less important than for refraction, since the coefficient $(20.5'')$ is about one-third that for refraction and the correction is proportional to the sine of the angle instead of the tangent. Therefore, instead of $\Delta z = R = \beta \tan z$ for refraction, we have $\Delta\theta = k \sin\theta$, where θ is the great-circle angle from the apex of the observer's motion to the target. The maximum contraction across a 1-degree FOV will be at the apex and only amounts to one-third of an arcsecond. Because of the smaller maximum shift and its sinusoidal dependence, aberration differential corrections are usually ignored for most ground-based astrometry; however, they become important for space applications such as the HST. If needed, the corrections can be applied using the formalism developed in the preceding subsection for refraction if we locate the tangential plane at the apex of the observer's motion instead of the zenith. We then substitute θ for ζ, let Δa be the angle between the target and the NCP as seen from the apex in Eqs. (19.13) through (19.15), and substitute the aberration correction, $\Delta\theta = k \sin\theta$, for Eq. (19.11).

Precession and nutation of the NCP

Since both precession and nutation, which are discussed in detail in Chapter 7, are rotations of the coordinate systems, there are no scale changes and hence no effect on the differential astrometry other than to necessitate a rotation to some selected reference frame. That rotation is normally done automatically with the selection of a reference catalog (see Chapter 20), which to first order gives positions at the equator and equinox of J2000. For higher-level accuracy see Chapters 7 and 8 for details.

19.5 Transforming the (*x*, *y*) measures to the tangential coordinate system (ξ, η)

19.5.1 Image centering

The (x, y) measures will have come from any of a variety of sources depending on the detector used for the observations. The fundamental limitations on the image center precision (see Section 9.4 and Lindegren 2010) depend on the signal-to-noise ratio in the image, i.e.

the number of collected photons in the image, and for ground-based observations the turbulence characteristics of the atmosphere. In addition to the above fundamental limitations, centers derived from photographic observations will have a precision that depends on the characteristics of the photographic emulsion (see Lee and van Altena 1981), the measuring machine and the algorithm used to derive the center. The highest precision is obtained using profile-fitting algorithms, such as those developed by Lee and van Altena (1981), DAOPHOT by Stetson (1987) and the (XWIN_IMAGE, YWIN_IMAGE) Gaussian centers in SExtractor (v2.13) from Bertin and Anouts (1996). For CCD observations, the optimum centering of undersampled images is discussed in Chapter 17 while for well-sampled images some procedures are discussed in Section 14.5 and Lindegren (2010).

19.5.2 Transformation models

We now correct the (x, y) measures for the various errors and systematic effects discussed in Sections 19.3 and 19.4, noting that to first order the measured coordinates (x, y) differ from the tangential coordinates (ξ, η) by a scale value. We can therefore replace the unknown tangential coordinates, say in Eq. (19.9), without any loss in accuracy by the measured (x, y) coordinates. In the following subsections we discuss the various terms used to transform the measured to tangential coordinates.

Zero point

The coordinates measured on the measuring machine or detector have their own zero points and need to be corrected to the tangential point (optical axis of the telescope), which is normally not known with sufficient accuracy and can vary with time and position in the sky due to telescope flexure. This term is a simple constant, which will differ in x and y. This constant amounts to a collection of leftover zero-point differences from the various model parameters.

Scale

Our detector coordinates (mm or pixels) need to be converted to radians. The nominally adopted scale in arcsec/mm or arcsec/pixel will differ from the actual scale on the exposure due to thermal variations in the telescope, incomplete correction for atmospheric refraction, velocity aberration, and inaccurate modeling of the higher-order model terms discussed below; therefore the scale must always be a part of the transformation model. While the scale values in x and y would normally be identical in most telescopes, incomplete correction for atmospheric refraction and inaccurate modeling of the higher-order terms generally cause the actual x and y scale values to differ significantly.

Orientation

Our measured coordinate system is never aligned perfectly with the tangential coordinate system, so the orientation terms are always required. A pure rotation to the tangential coordinate system is rarely encountered, due to incomplete calibration of the plate-measuring machine and inaccurate modeling of the higher-order terms that generally cause the actual x and y orientation values to differ significantly. The model that includes *zero point, scale,* and *orientation* terms for both x and y coordinates is generally referred to as the "full linear model."

Plate tilt

As we saw in Eqs. (19.8) and (19.9), the translation from one tangential coordinate system to an adjacent one over the surface of the celestial sphere introduces quadratic terms into the transformation equations. Similar quadratic terms are introduced when we transform from a coordinate system that is tilted with respect to the optical axis to the correct tangential coordinate system. In addition, inaccuracies in locating the OFAD center will introduce quadratic and higher-order terms since the differential of the cubic OFAD is a quadratic. These three effects generally lead to the use of a three-term quadratic transformation instead of the simpler two-term quadratic that arises from plate tilt.

Optical field-angle distortion (OFAD)

As noted in the previous discussion, inaccuracies in the OFAD coefficients will introduce quadratic and higher-order terms in the transformation to the tangential coordinate system. A separate complication with OFAD is that we need to know the location of the telescope's optical axis and the origin of the OFAD. In practice, most modern telescopes have very little OFAD, but reflecting telescopes with prime-focus correctors typically have large OFAD and hence the location of the corrector with respect to the telescope's optical axis becomes a significant factor. A related systematic effect on the detected target positions may be due to the presence of other optical components such as a "corrector plate" in Schmidt telescopes. The formalism for the correction is discussed by Stock (1981) and Taff (1989).

Magnitude equation

The "magnitude equation" is a systematic shift in the image position that is a function of the image's brightness. The three principal sources for the introduction of a magnitude equation include: (1) telescope guiding errors combined with a non-linear detector such as a photographic emulsion; (2) charge-transfer inefficiency (see Chapter 14 for a discussion of CTI) in a CCD; and (3) measuring machines and their image-centering devices both automatic and visual. The third source can be eliminated by measuring a plate in both direct (x_d, y_d) and reverse (x_r, y_r) orientations (i.e. rotated by 180 degrees) and then writing $x = (x_d - x_r)/2$ and $y = (y_d - y_r)/2$. If the characteristics of the measuring machine and/or measurer remain constant then this procedure will eliminate any magnitude equation introduced from that source. The astrometric effects of CTI can often be calibrated and removed, but they can change with time as the detector ages, especially with space-borne CCDs. The magnitude equation is very dangerous in astrometry, since spurious proper motions introduced by a magnitude equation can mimic those due to cosmic effects such as secular proper motions (see Chapter 22) and introduce offsets in the zero points of secondary reference systems. Finally, it is important to understand that while the magnitude equation is normally approximated by a linear term in magnitude (see Table 19.1 and Eq. (19.17)), the systematic effects are often non-linear and their removal can only be accomplished through a laborious graphical precorrection of the trends, if a sufficient number of highly accurate reference stars is available.

Coma

In Section 19.4.2 the effect of coma on the measured position of a star was discussed and found to be the product of the image brightness, or magnitude, times the coordinate, x or y.

Color

The color of the target object can have an effect on its measured position. In general, the cause of such a systematic shift is due to incomplete correction for atmospheric refraction since refraction is a function of the wavelength of light and objects with different spectral-energy distributions will be refracted by differing amounts. The problems associated with correcting completely for refraction are discussed in Section 9.1.

Color magnification

The term "color magnification" refers to plate scale being a function of the color of the target. This effect is usually encountered in older wide-field refracting telescopes and cluster proper-motion investigations (see Chapter 25) in the comparison of old photographs taken with different telescopes.

Residual coordinate-dependent shifts

In this category we include essentially everything that cannot be dealt with analytically by the above categories. For example, OFAD is normally modeled as a cubic dependence on the coordinate, but small fifth-order terms are present in some corrector systems and often ignored. In addition, incomplete correction for a Schmidt corrector and small localized shifts due to optical problems and telescope flexure are often lumped together in this category. The elimination of this category of errors is usually done through the application of a correction "mask," which is a large table of corrections to be applied that are a function of the coordinates (x, y). Examples of masks are given by Zacharias *et al.* (1997), Girard *et al.* (2004), and Hanson *et al.* (2004).

Transforming between two tangential systems

In Table 19.1 we summarize the model parameters from the above discussion that are commonly used in the transformation of the measured coordinates to the tangential coordinate system. If we rewrite Eq. (19.9) in terms of the differences between two tangential coordinate systems that are close together, then we can replace the tangential coordinates on

Table 19.1 Model parameters for transforming from measured to tangential coordinates

Measured to tangential	Term in x	Term in y
Zero point	c_x	c_y
Scale	$a_x \cdot x$	$a_y \cdot y$
Orientation	$b_x \cdot y$	$b_y \cdot x$
Plate tilt	$p \cdot x(x+y)$	$q \cdot y(x+y)$
OFAD	$x \cdot (d_1 x^2 + d_2 xy + d_3 y^2)$	$y \cdot (d_1 x^2 + d_2 xy + d_3 y^2)$
Magnitude equation	$m_x \cdot \mathrm{mag}$	$m_y \cdot \mathrm{mag}$
Coma	$e_x \cdot \mathrm{mag} \cdot x$	$e_y \cdot \mathrm{mag} \cdot y$
Color index (ci)	$f_x \cdot ci$	$f_y \cdot ci$
Color magnification	$g_x \cdot x \cdot ci$	$g_y \cdot y \cdot ci$
Mask	$\mathrm{mask}_x(x, y)$	$\mathrm{mask}_y(x, y)$

the right-hand side of Eq. (19.9) by their corresponding measured rectangular coordinates since they differ largely by only a scale factor. This has the advantage that the measured coordinates are known for all objects, while the tangential coordinates are known only for those objects with previously known celestial coordinates. Rewriting Eq. (19.9) in this form and including the terms from Table 19.1, we have Eq. (19.17), where the coefficients in x and y are indicated by their corresponding subscripts. Coefficients common to x and y are the plate tilt p and q and those for the OFAD, d_1, d_2 and d_3.

$$\Delta x = \xi - x = a_x \cdot x + b_x \cdot y + c_x + p \cdot x(x+y) + x \cdot (d_1 x^2 + d_2 xy + d_3 y^2)$$
$$+ m_x \cdot mag + e_x \cdot mag \cdot x + f_x \cdot ci + g_x \cdot x \cdot ci + \mathrm{mask}_x(x, y)$$
$$\Delta y = \eta - y = b_y \cdot x + a_y \cdot y + c_y + q \cdot y(x+y) + y \cdot (d_1 x^2 + d_2 xy + d_3 y^2)$$
$$+ m_y \cdot mag + e_y \cdot mag \cdot y + f_y \cdot ci + g_y \cdot y \cdot ci + \mathrm{mask}_y(x, y)$$

(19.17)

Estimating the coefficients

We estimate the coefficients of these model parameters (plate constants) by selecting "reference stars" (standard stars) on the detector field with good images that already have celestial coordinates in a high-quality astrometric catalog (see Chapter 20). The tangential coordinates (ζ, η) of the reference stars are then calculated using an adopted tangential point that is close to the actual, but unknown, tangential point and then converted to millimeters or pixels in our detector system using an adopted plate scale. A least-squares solution is then made on the differences between the converted tangential coordinates and their measured (x, y) coordinates to estimate the coefficients, or unknowns. Finally, the estimated coefficients are used to calculate the desired tangential coordinates of the remaining stars, which are called "field stars." We now discuss the individual steps in more detail, since the care we take in estimating the most appropriate model and estimating the coefficients can have an important impact of the final accuracy of our astronomical results.

19.5.3   Reference stars

Reference stars and source catalogs

At this point we need to consider carefully how many of the unknowns in Table 19.1 we will attempt to estimate. There are 11 terms in Table 19.1 in x and another 11 in y, in addition to the mask tables that must be determined from a different set of observations. Depending on the size of the detector we may not have a sufficient number of reference stars to estimate accurately the desired number of unknowns. A useful rule of thumb is that you need at least three times the number of reference stars as coefficients that you wish to determine. Given the brightness limits (detector saturation *vs.* exposure signal-to-noise ratio for the faintest objects) and detector sizes available on the telescope, it may be difficult to find an adequate number of reference stars in a particular catalog. For example, the approximate limiting magnitude and number of accurate reference stars per square degree available in several accurate astrometric catalogs listed in Chapter 20 are respectively: Hipparcos (8.5, 2.5), Tycho-2 (12.5, 25), UCAC3 (16, 2500) and south of −20 declination only, SPM4 (17.5, 7000). For the large telescopes, it is often necessary to utilize deep-survey catalogs such as the USNO-B (21, 25 000); however, the accuracy of the coordinates in those catalogs will limit the derived astrometric results. An alternative is to develop a set of secondary reference stars, usually through observations made on one or more separate telescopes where a series of successively deeper exposures enables us to extend the reference system of an accurate astrometric catalog to fainter and fainter levels so as to reach beyond the detector saturation limit of the large telescope's observations. Normally this procedure works well but considerable care needs to be taken so that systematic errors due to the propagation of magnitude equation errors do not dominate the zero-point of the derived secondary reference-star system, otherwise significant offsets of the secondary from the catalog reference system can occur. An example of the procedures used to develop a secondary reference system can be found in de Vegt *et al.* (2001).

The epoch and equinox of nearly all of the catalogs that might be used for the selection of reference stars are on the J2000 system. However, our observations will have been obtained at some other epoch than 2000.0; therefore we need to use the proper motions and parallaxes (in rare cases where the stars may be very nearby) of the reference stars to transform them to the date of the observation. In this process we introduce errors into the positions of the reference stars since the proper motions have their own measurement errors. This systematic degradation of the reference-star system with time is one of the principal reasons for continuing the observation of reference stars so that our reference system improves instead of degrading with time.

Distribution of standard stars

In order to accurately estimate the various coefficients we need uniform distributions of the reference stars over the parameters being determined. Otherwise very dangerous extrapolations beyond the reference-star parameter range to that of the target stars will be required. For example, the spatial distribution over the detector should be uniform to determine the corrections to the adopted zero-points (c, c'), plate scales (a, a'), orientations (b, b'), and plate tilt (p, q). The reference-star magnitude (m) and color (ci) ranges should encompass those of the target stars as should the compound quantities coma $(mag \cdot x, mag \cdot y)$, color magnification $(x \cdot ci, y \cdot ci)$, etc. The situation is more difficult for the higher-order terms, such as distortion, where accurate estimation requires large numbers of reference stars at the extreme edges of the detector. It often happens that we end up with an inadequate number of reference stars and compromises need to be made in the number of coefficients to be determined and/or the final positional accuracy desired.

Coefficient estimation

Estimation of the coefficients can be made using standard least squares or more robust procedures such as GaussFit (Jefferys *et al.* 1988 and Chapter 21), which adopts Bayesian statistics and the use of priors. The latter approach is preferable, since it enables you to include previously determined parameters such as the OFAD and its errors as a part of the model. In either case, the proper approach is to begin a solution with the minimum number of parameters and search for outliers and correlations with higher-order parameters. One useful procedure is to plot the residuals from the lower-order solution (say for the scale, orientation, and zero-points) versus the coordinates, magnitude, compound, and higher-order parameters. Normally the trends of the residuals against the various additional parameters make it easy to isolate the additional parameters needed. Outliers in the residuals due to poor measures and images disturbed by a nearby image, cosmic ray, or plate defect can also be identified and deleted if necessary. At times the outliers may be due to a poorly determined proper motion used to transfer the reference-star catalog epoch to the epoch of the exposure being analyzed. We then make successive solutions including the next most important coefficients until no further trends are apparent in the residual plots. This approach is time-consuming, but it is necessary if accurate astrometry is desired and a high level of quality control is to be maintained on the final product.

Some of the coefficients are essentially constant in time, since they characterize the optical telescope (coma and OFAD) and the detector (plate tilt), while others can vary slowly with temperature and telescope flexure (plate scale and orientation). If some parameters stay constant over a series of observations these parameters should be constrained by additional observation equations in the least-squares solution. An alternative approach is to pre-correct the data for the effect of those parameters and then use a simpler model without those parameters in the least-squares solution of individual observations. An example of this approach is to use a "mask" to correct x, y data for "higher-order terms" and not use optical-distortion terms in the solution. If a series of observations is spatially overlapping, a "block adjustment" type of solution can be performed, which solves for all parameters of all observations in a single least-squares solution, utilizing the fact that there is only one position for any given star to be estimated. This approach constrains the individual observation model parameters to their optimal solution, if the data are "clean" and the correct physical model is used. For details see Eichhorn (1960), Googe *et al.* (1970), Taff (1988), and Zacharias (1992).

Tangential coordinates for the unknown objects

Once we are satisfied with the transformation coefficients estimated above, we use those coefficients to calculate the tangential coordinates for the objects with unknown positions using Eq. (19.17), solving for (ζ, η) followed by the equatorial coordinates with Eq. (19.7).

19.6 Mosaics of CCDs

One serious limitation of CCDs is their very small physical size (currently up to 95 by 95 mm), which limits the area of the sky that can be covered in one exposure and therefore the number of reference stars available for the astrometric reduction of measured to tangential coordinates. One solution to this limitation has been to make a mosaic of CCDs in the focal plane of the telescope, such as the mosaic camera of the 4-m Mayall reflector at the Kitt Peak National Observatory and for time-delayed-integration (Chapter 15) observations, the Sloan Digital Sky Survey (Gunn *et al.* 1998), and the Gaia satellite (Lindegren 2010). These mosaics of CCDs have improved the sky coverage of single exposures by factors of 100 and more, but they create enormous data-handling problems and are an astrometric nightmare to deal with properly. The principal problems with mosaics are that each CCD has its own photometric and astrometric properties and move around in the focal plane as if mounted on a flexible substrate. As a result, the user needs to determine the photometric and astrometric characteristics of each CCD and monitor it with time. In addition, the relative positions and orientations of the CCDs must be monitored with time so that the calibration of the mosaic becomes a major part of any observational program. However, the observational gain of being able to increase the throughput of your time on the telescope by a factor of 100 or so makes these new instruments extremely valuable. Procedures for the calibration of mosaic cameras are discussed in papers by Platais *et al.* (2002, 2006).

19.7 Differential astrometry

Up to this point we have considered only the transformation of observed focal-plane image positions to their equatorial coordinates on the celestial sphere. This process requires the careful calibration and monitoring of the detailed characteristics of the optics, detectors, and the telescope. A much simpler problem is presented to us if we look for small *changes* in the apparent positions of celestial objects instead of their positions. For example, if we are interested in the relative parallaxes or proper motions of objects, then we need only determine changes in the position of the object over a period of months or years and the position of the object is of secondary importance. As a result, we need to consider changes in the plate tilt, OFAD, refraction, etc., and the higher-order terms and "ripples" that are often seen in the OFAD are minimized since we are only looking for changes in those effects. Even though we are interested only in changes in the optics, detectors, and telescope, it is still good practice to remove the known effects of aberration, refraction, plate tilt, and OFAD before proceeding with the differential-astrometry solutions so that the positions used in the solutions are at least corrected for the major instrumental and spherical effects. The gain in relative positional precision over that of absolute positional accuracy can be substantial and is discussed in detail in Section 9.4. Applications of relative astrometry to the determination of stellar parallaxes can be found in Chapter 21, while the isolation of star-cluster members based on their relative proper motions is discussed in Chapter 25, and the determination of the relative orbits of binary stars can be found in Chapters 23 and 24, and the discovery and characterization of exoplanets in Chapter 27.

Differential-astrometry solutions can be made using the same Eq. (19.17) as before, but the "coefficients" determined will then be changes in those coefficients between the two exposures being examined.

19.8 Summary

In this chapter we have studied the effects that alter the positions of celestial objects as they pass from the celestial sphere to the detector, and developed procedures to make those transformations. In succeeding chapters we will examine the application of these astrometric techniques to the solution of a variety of current astrophysical research problems.

References

Anderson, J. and King, I. R. (1999). Astrometric and photometric corrections for the 34th-row error in HST's WFPC2 Camera. *PASP*, **111**, 1095.

Anderson, J. and King, I. R. (2000). Toward high-precision astrometry with WFPC2. I. Deriving an accurate point-spread function. *PASP*, **112**, 1360.

Arend, S. (1951). *Theorie de l'equatorial visuel et de l'equatorial photographique. Reglage pratique de l'equatorial visuel et de l'astrographe*. Brussels: Observatoire Royal de Belgique.

Bertin, E. and Arnouts, S. (1996). SExtractor: software for source extraction. *A&AS*, **117**, 393.

de Vegt, C., Hindsley, R., Zacharias, N., and Winter, L. (2001). A catalog of faint reference stars in 398 fields of extragalactic radio reference frame sources (ERLcat). *AJ*, **121**, 2815.

de Vegt, C., Laux, U., and Zacharias, N. (2003). A dedicated 1-meter telescope for high precision astrometric sky mapping of faint stars. In *Small Telescopes in the New Millennium II. The Telescopes We Use*, ed. T. Oswalt, Dordrecht: Kluwer, p. 255.

Eichhorn, H. K. (1960). Über die Reduktion von photographischen Sternpositionen und Eigenbewegungen. *AN*, **285**, 233.

Girard, T. M., Dinescu, D. I., van Altena, W. F., Platais, I., Monet, D. G., and López, C. E. (2004). The Southern Proper Motion Program. III. A near-complete catalog to $V = 17.5$. *AJ*, **127**, 3060.

Girard, T. M., van Altena, W. F., Zacharias, N., *et al.* (2011). The Southern Proper Motion Program. IV. The SPM4 Catalog. *AJ*, **142**, 15.

Googe, W. D., Eichhorn, H., and Lukac, C. F. (1970). The overlap algorithm for the reduction of photographic star catalogues. *MNRAS*, **150**, 35.

Guo, X., Girard, T., van Altena, W. F., and López, C. E. (1993). Space velocity of the globular cluster NGC 288 and astrometry with the CTIO 4 meter telescope. *AJ*, **105**, 2182.

Gunn, *et al.* (1998). The Sloan Digital Sky Survey Photometric Camera. *AJ*, **116**, 3040.

Hanson, R. B., Klemola, A. R., Jones, B. F., and Monet, D. G. (2004). Lick Northern Proper Motion Program. III. Lick NPM2 Catalog. *AJ*, **128**, 1430.

Jefferys, W. H., Fitzpatrick, M. J., and McArthur, B. E. (1988). GaussFit – a system for least squares and robust estimation. *Celestial Mechan.*, **41**, 39.

Jefferys, W., Whipple, A., Wang, Q., *et al.* (1994). Optical field-angle distortion calibration of FGS3. In *Calibrating Hubble Space Telescope*, ed. J. C. Blades and S. J. Osmer., Baltimore, MD: Space Telescope Science Institute, p. 353.

Konig, A. (1933). Reduktion photographischer Himmelsaufnahmen. In *Handbuch der Astrophysik*. Berlin: Springer Verlag, vol. 1, ch. 6.

Konig, A. (1962). Astrometry with astrographs. In *Astronomical Techniques*, ed. W. A. Hiltner. Chicago, IL: University of Chicago Press, ch. 20, p. 461.

Lee, J.-F. and van Altena, W. F. (1983). Theoretical studies of the effects of grain noise on photographic stellar astrometry and photometry. *AJ*, **88**, 1683.

Lindegren, L. (2010). High-accuracy positioning: astrometry. In *Observing Photons in Space*. ed. M. C. E. Huber, A. Pauluhn, J. L. Culhane, *et al.* Bern: International Space Science Institute, ISSI Scientific Reports Series, p. 279.

Meeks, M. L., Ball, J. A., and Hull, A. B. (1968). The pointing calibration of the Haystack antenna. *IEEE Trans. Antennas Propag.*, **AP-16**, 746.

Platais, I., Kozhurina-Platais, V., Girard, T. M., *et al.* (2002). WIYN Open Cluster Study. VIII. The geometry and stability of the NOAO CCD Mosaic Imager. *AJ*, **124**, 601.

Platais, I., Wyse, R. F. G., and Zacharias, N. (2006). Deep astrometric standards and Galactic structure. *PASP*, **118**, 107.

Schlesinger, F. and Barney, I. (1925). *Trans. Astron. Obs. Yale Univ.*, **4**, 5.

Schroeder, D. (2000). *Astronomical Optics*, 2nd edn. San Diego, CA: Academic Press.

Smart, W. M. (1931). *Spherical Astronomy*. Cambridge: Cambridge University Press.

Stetson, P. (1987). DAOPHOT – a computer program for crowded-field stellar photometry. *PASP*, **99**, 191.

Stock, J. (1981). Block adjustment in photographic astrometry. *Revista Mexicana de Astronomia y Astrofisica*, **6**, 115.

Taff, L. G. (1988). The plate-overlap technique – a reformulation. *AJ*, **96**, 409.

Taff, L. G. (1989). Schmidt plate astrometry – subplate overlap. *AJ*, **98**, 1912.

Taff, L. G., Lattanzi, M. G., and Bucciarelli, B. (1990). Two successful techniques for Schmidt plate astrometry. *ApJ*, **358**, 359.

van Altena, W. F. and Monnier, R. C. (1968). Astrometric accuracy and the flatness of glass filters. *AJ*, **73**, 649.

Wallace, P. T. (1979). *Telescope Pointing Investigations at the Anglo-Australian Observatory*. Epping: Anglo-Austrian Society.

Wallace, P. T. and Tritton, K. P. (1979). Alignment, pointing accuracy and field rotation of the UK 1.2 m Schmidt telescope. *MNRAS*, **189**, 115.

Zacharias, N. (1992). Global block adjustment simulations using the CPC 2 data structure. *A&A*, **264**, 296.

Zacharias, N., de Vegt, C., and Murray, C. A. (1997). CPC2 plate reductions with HIPPAR-COS stars: first results. In *Proceedings of the ESA Symposium "Hipparcos – Venice '97."* ESA SP-402, 85.

Zacharias, N., Urban, S. E., Zacharias, M. I., *et al.* (2000). The first US Naval Observatory CCD astrograph catalog. *AJ*, **120**, 2131.

20

Astrometric catalogs: concept, history, and necessity

CARLOS E. LÓPEZ

Introduction

A catalog is defined as a list of items, of some class, usually in systematic order, and with a description of each. In almost every catalog, it is possible to distinguish three types of data: primary, pseudoprimary, and secondary. For example, in a catalog devoted to large proper-motion stars, the primary data are the proper-motion data while spectral types or magnitudes or any additional information would be regarded as secondary data. In other words, secondary data complement the primary data. Pseudoprimary data are less important than the primary data, but for example may enable the user to uniquely identify the primary data.

The compilation of catalogs including positions, proper motions, and parallaxes is among the best-known activities of astrometry. In the early years of astronomy, providing positions of celestial bodies was a necessity for many of the ancient cultures. With roots dating back to the pre-Greek civilizations, astrometry provided the information for computing solar and lunar eclipses and the determination of time.

Catalogs and databases follow a recycling process in the sense that as new data or information are published, or better detectors become available, new comparisons, improved coordinates, proper motions, or parallaxes can be obtained. The catalogs that nowadays are considered obsolete were in earlier times widely used by the entire community and are now the basis for improved versions. In fact, probably more than any other science, astronomy is based on the analysis of data obtained through the years by different and generally unrelated research groups. Probably one of the best examples in this sense is the Astrographic Catalogue (see Section 20.3 below).

20.1 History of astrometric catalogs

Since the times of Hipparchus, most of the new accomplishments in almost every area of our science have been made through the re-analysis of previously existing data plus the incorporation of "modern" values obtained with "new" and more sophisticated instruments.

Astrometry for Astrophysics: Methods, Models, and Applications, ed. William F. van Altena.
Published by Cambridge University Press. © Cambridge University Press 2013.

Hipparchus himself benefited greatly from the use of observations acquired more than a century before his time. The comparison of his own (modern) observations with those obtained by Aristillus and Timocharis allowed him to discover precession, one of the most important parameters in positional astronomy. In a similar way, but almost two thousand years later, in 1718, Halley noticed the existence of proper motions when comparing his own position determinations with previous catalogs. From these two examples, it can easily be seen that some of the most important discoveries in astronomy were made when new observations were compared with old determinations. History shows us that for astronomy to develop new concepts and to expand into new domains, the preservation of old data, in any form and at any cost, is paramount. Old data does not compete with new measurements; on the contrary, old and historic data must be regarded as an invaluable complement to new determinations. Both old and new data are vital to a better understanding of the astronomical evolution of celestial objects.

The history of astrometric catalogs, taking the detector used in the observational process as a parameter, can be divided in three different periods: visual (with or without instrument), photographic, and charge-coupled device (CCD) (Earth- or space-based instruments). The unaided human eye was used as the detector from the very beginning of astronomy until the year 1609 when Galileo Galilei started to use the telescope during his observations; since then, most of the astrometric surveys were made using meridian circles or some other type of optical instrument. This approach lasted for over 300 years.

Hipparchus completed one of the first catalogs around 129 BC. There is debate concerning the number of objects listed by Hipparchus: some historians have argued that there were 1080 stars while others think that there were only 850 stars. Hipparchus' catalog was included in Ptolemy's *Almagest*, the biggest compendium of astronomy that was in use for about 14 centuries. Previous compilations by Greek and Chinese astronomers date back to around 300 BC or 360 BC (Eichhorn 1992).

During the Middle Ages the most important compilations were the *Alfonsine Tables*, compiled under the reign of Alfonso the Wise and probably first published around 1252. (Actually, the real purpose of these tables was to provide a means to calculate the positions of the planets based on Ptolemy's geocentric Solar System model, so they may not qualify as a catalog in the current context.) One of the main users of the Alfonsine Tables was Copernicus (Gingerich 1992).

Ulugh-Beg (1393–1449) may be considered as the last astronomer of the Middle Ages. He was born and spent much of his life in Persia (Iran). At the age of sixteen he became the governor of Samarkan (now Uzbekistan) where he built his own observatory. In order to improve the accuracy he decided to build a very large instrument, like a sextant with a radius of 36 meters. Using this sextant he compiled, in 1437, a catalog listing 994 stars, which is considered the greatest star catalogue between Ptolemy and Tycho Brahe. Another very important accomplishment was the determination of the length of the sidereal year with an error of 58 seconds (for a detailed analysis of Ulugh-Beg's observations see Krisciunas 1993).

Tycho Brahe is among the first astronomers who started to make accurate systematic observations using his mural quadrant (an instrument of his own design) during the pre-telescope years. Among his works are the observations of the supernova of 1572, the comet of 1577, and a large collection of positions of Mars. After Brahe's death in 1601 these positions were used by his assistant Johannes Kepler and were critical to the derivations of his three laws.

20.2 Visual catalogs

With the introduction of the telescope as an astronomical instrument, the compilation of astrometric catalogs started to grow in both quantity and quality. Galileo's observation of Jupiter's four largest moons should be considered among the first astrometric observations ever made with a telescope. Even though these observations were not made in the modern sense, it was the variation of their positions relative to Jupiter that allowed Galileo to conclude that these four bodies orbited Jupiter.

The pursuit of better accuracy led Ole Rømer to develop the meridian circle in 1690 which, with modifications, is still in use today. The meridian circle or transit circle (conceived at the end of the seventeenth century) was a combination of a quadrant (a wall in the plane of the meridian) and a telescope. Although this new instrument allowed the observer to determine both right ascension and declination, the meridian circle was not widely accepted until the nineteenth century.

With meridian circles as the most important instrument for astrometry the observation of many "zone catalogs" was initiated. Most of these surveys (early to mid 1800s) were limited to specific areas and there was no agreement on the "system" that should be used (the convenience of having a system, in the modern sense, to which the star positions were referred was proposed in the late 1800s). One of the first attempts to make a comprehensive survey was initiated by Argelander with the observation of a zone from +80 to –2 degrees. The zones from –2 degrees to the South Pole were added later. This first survey, known as the Durchmusterungs, was made in three parts, from three different locations and, in one case (Cape) photographic plates were used. They include the Bonner Durchmusterung, the Córdoba Durchmusterung, and the Cape Photographic Durchmusterung.

Another very important achievement around the middle of the nineteenth century was the determination of the first trigonometric parallax by Bessel. Bessel's results on 61 Cygni were published at the end of 1838, a few months before Struve's result on Vega.

The most important work of the early twentieth century was the General Catalogue (GC) published by Boss in 1937. The work on this project was started by Lewis Boss in 1910 with the publication of a preliminary version called Preliminary General Catalogue (PGC) and completed in 1937 with the publication of the GC including over 33 000 stars. It is important to note that Boss's ultimate aim in the PGC/GC project was the construction of a fundamental system (which was finally accomplished by his son, Benjamin Boss). A second attempt to create a fundamental system based on very different principle was the N30 catalog compiled by Morgan (1952).

For a detailed historical account of astrometric catalogs see Eichhorn (1974, pp. 101–278) as well as Jaschek (1989) for a discussion of different catalog types.

20.3 Photographic catalogs

Photographic astrometric surveys began after the photographic plate was accepted by the astronomical community as a detector (around 1880). The first observing program with the newly introduced astrographic telescope, which was built to take advantage of the photographic plate, was the largest observational international project at the end of the nineteenth century: the Astrographic Catalogue (AC). The decision to embark on the AC was criticized as many considered it too soon in the development of astrographic telescopes and photographic plates to undertake such a large project using a rather small (2×2 degrees) field of view. However, from today's perspective, the launch of the AC can be regarded as a positive initiative. Thanks to that decision astrometry has a marvelous set of over 4 million first-epoch positions (about 100 years old by now) for high-precision proper-motion determinations.

The AC was never used as such until very recently. Two major catalogs have now been produced using the published coordinates of the AC: the 4 Million (Gulyaev and Nesterov 1992) and the AC2000 (Urban *et al.* 1998). The former combined the AC with the GSC (Guide Star Catalogue) to compute proper motions; both sets of positions (AC and GSC) were reduced into the PPM (Positions and Proper Motions) system. In addition to positions and proper motions, the 4 Million also provides cross-identifications with other surveys like the Durchmusterungs, PPM, SAOC (Smithsonian Astrophysical Observatory Catalogue), and variable stars. The AC2000 contains over 4.5 million positions (and magnitudes) with an average epoch of 1907. In order to reduce the published x, y coordinates of the AC, Corbin and Urban (1990) first compiled the ACRS (Astrographic Catalogue Reference Stars) to be used as the reference frame. The positions of the AC2000 are on the Hipparcos/International Celestial Reference System (ICRS). A more recent re-analysis of the AC2000 has yielded the AC2000.2 (Urban *et al.* 2001). For a marvelous and detailed history of the AC see Eichhorn (1974, pp. 279–323).

While the observations for the AC were underway, other photographic surveys were also in progress, such as the Yale Zones and the AGK2 (it must be remembered that the AGK1 was observed with meridian circles; AGK stands for Astronomische Gesellschaft Katalog). In total, Yale published 30 catalogs with epochs ranging from 1928 to 1942. The first volume of the Yale zones to appear (Schlesinger *et al.* 1926) included over 5000 stars in the equatorial zone. The last zone (−60 to −70 degrees) was delayed for various reasons and was not published until 1983 (Fallon and Hoffleit 1983).

Due to the development of artificial satellites and the urgent need for tracking them in the mid 1960s, it was felt that having a general comprehensive compilation embracing all current catalogs in a single one was critical. This project was undertaken by the Smithsonian Astrophysical Observatory and, in 1966, the "Star Catalog" (subtitled "Positions and Proper Motions of 257 997 Stars for the Epoch and Equinox of 1950") became available to the astronomical community. The catalog, known as SAOC, covered from +90 to −90 degrees and for many years it was the standard reference frame for the reduction of photographic plates. Thousands of positions of asteroids, comets, major planets, and natural satellites were obtained using it as a primary reference catalog for stellar positions and proper motions. But time is the enemy of all catalogs through the errors in the proper motions, which gradually degrade the usefulness of any astrometric catalog. The SAOC was no exception and after more than 25 years in service it was replaced by the PPM (Röser and Bastian 1993). The PPM also covered the entire sky and lists over 375 000 positions and proper motions in the J2000/FK5 system. Typical values of the root-mean-square (rms) errors in the northern hemisphere are 0.27 arcsec in the positions at epoch 1990, and 0.42 arsec/century in the proper motions, while the corresponding figures in the southern sky are much better (Röser *et al.* 1994). The main improvement of the PPM over the SAOC was made by the inclusion of the century-old AC data. After its publication, the PPM replaced the SAOC as the reference frame for most photographic astrometry programs. Due to the discontinuation of photographic plates and the advent of CCD detectors, the PPM has now been replaced by much higher-density and fainter catalogs.

20.4 Hipparcos and the modern catalogs

In the last twenty years, the history of astrometry was highlighted by two main events: the first was the launch of Hipparcos in 1989 and the second was the establishment in 1998 of the ICRS thus bringing to an end the reign of the FK series. With this major change, and for the first time, astrometry's reference frame, the International Celestial Reference Frame (ICRF), to which positions have to be referred, is based on 212 extragalactic radio sources observed with very long baseline interferometry (VLBI) techniques (see Ma *et al.* 1998). In the optical domain, the ICRS is realized by the Hipparcos stars. An important review concerning the adoption and implications of this change has been given by Feissel and Mignard (1998).

Today, thanks to the use of a new generation of extremely high-quality Earth- and space-based instruments, astrometry is undergoing new and profound changes. From the multiple fiber-fed instruments on Earth to the most advanced and sophisticated satellites there is a common need for better positions of celestial objects. Astrometry, once again, has to provide accurate positions, proper motions, and stellar parallaxes. The importance of the new astrometry is such that some of the coming space missions need to rely on ground-based astrometric observations for success.

In order to meet the requirements imposed by today's astronomy new astrometric surveys oriented to providing the classical parameters are underway. Owing to the increasing power of both computers and detector development, the current projects have grown in size (more than 500 million entries in general) and accuracy (a few milliarcseconds). Some of these new surveys are based on new observations while many others are making use of the extensive material provided by the databases, thus bringing up a new way of doing astronomy: the Virtual Observatory. An excellent account of some specific items related to the history of catalogs and astrometry in general has been presented by Hoeg (2008).

Table 20.1 lists the major compilations that for different reasons became landmarks in the development of astrometric catalogs. Catalogs covering a large portion of the sky are

Table 20.1	The major astrometric catalogs	
Year	Catalog	Number of objects
360 BC	Chinese	?
260 BC	Aristillus & Timocharis	850?
129 BC	Hipparchus	850 (1080?)
150	Ptolomey's Almagest	1080
1252	Alfonsine Tables	?
1437	Ulugh-Beg	1018
1594	Tothman & Whilhem	1004
1601	Tycho Brahe	1005
1690	Hevelius	1564
1725	Flamsteed	3310
1751	Lacaile	9766
1760	Lalande	50 000
1792	Piazzi	7646
~1846	Bessel	75 000
1847	British Associacion	47 390
1850	Durchmusterung	300 000
1887	Astrographic Catalogue	~4 000 000
1910	Preliminary General Catalogue	30 000
1926	Yale Zones	~150 000
1937	General Catalogue	33 000
1950	N30	5000
1960	SAOC	257 997
1975	AGK3	183 145
1984	FK5	1535
1990	GSC 1.0	~20 000 000
1992	4 Million	4 000 000
1993	PPM	350 000
1994	NPM1	149 000
1995	Yale Trigonometric Parallaxes	8112
1996	USNO A1.0	488 000 000
1997	Hipparcos	118 218
1997	Tycho	1 058 332
1998	USNO A2.0	500 000 00
1998	AC2000	4 500 000
2000	GCS II	1 000 000 000
2000	Tycho-2	2 500 000
2001	SPM	30 000 000
2002	UCAC 1	40 000 000
2002	SuperCosmos Sky Survey	~1 000 000 000
2003	USNO B1.0	~1 000 000 000
2003	NPM2	400 000
2004	Pulkovo Catalog	58 483
2004	UCAC2	48 330 571
2004	UCAC2 Bright Star Supplement	430 000

(*cont.*)

Table 20.1 (cont.)		
Year	Catalog	Number of objects
2005	NOMAD	~1 000 000 000
2006	GSC 2.3	945 592 683
2006	CMC14	95 858 475
2007	Hipparcos New Reduction	117 955
2008	PPMX	~18 000 000
2009	UCAC3	~100 000 000
2009	SPM4	~100 000 000
2009	XPM	~280 000 000
2010	PPMXL	~900 000 000

Table 20.2 Photographic wide-field surveys and catalogs

Project name	Epoch year	Positional error. (mas)	Magnitude range	Sky area	Literature reference
AC	1890–1930	200	4–13	All	[1]
NPM	1947–1987	100	5–18	North	[2], [3]
SPM	1965–2008	40–150	5–18	South	[4], [5]
USNO-B	1949–2002	250	12–21	All	[6]
GSC 2.3	1949–2002	250	12–21	All	[7]
NOMAD	....	....	0–21	All	[8]

[1] Urban *et al.* 1998; [2] Klemola *et al.* 1987; [3] Hanson *et al.* 2004; [4] Girard *et al.* 1998a; [5] Girard *et al.* 1998b, Girard *et al.* 2011; [6] Monet *et al.* 2003; [7] Bucciarelli *et al.* 2008; [8] Zacharias *et al.* 2004.

included. Updated or new versions of a given catalog are listed only if there is an important increment in the number of objects between the old and the new edition. Most of the catalogs listed in Table 20.1 are available at the Centre de Données Astronomiques de Strasbourg.

Tables 20.2–20.4 summarize some of the most important photographic surveys already completed and those under development (in different bandpasses) or to be started in the near future. Links to the major programs are also included (see also McLean 2001).

20.5 Space-based astrometric surveys

A number of space-based astrometric surveys are being planned for the near future: Gaia, Nano-JASMINE, JASMINE, SIM, and J-MAPS. Due to the objectives and the impact on almost every branch of astronomy, ESA's Gaia mission is the expected to be the Hipparcos successor.

The goal of Gaia is to perform a detailed census of our Galaxy through the measurements of high-accuracy astrometry, radial velocity, and multi-color photometry of around 1 billion stars down to the 21st magnitude. Gaia's survey is expected to be complete to magnitude

Table 20.3 Wide-field optical and near-infrared surveys

Survey name	Operation year	Sky area	Bandpass	Aperture (m)	R magnitude range	Scale ("/px)	Fi (d
			Current optical surveys				
SDSS	2005 ...	1/4	u, g, r, i, z	2.5	15–22	0.40	1..
CMC	~2000	North	r	0.2	9–17	0.70	Sc
CMASF	~1995	South	r	0.2	9–17	0.7	Sc
PM2000	~2000		r	0.2	9–17	0.70	Sc
UCAC	1998–2004	North	579–643 nm	0.2	8–16	0.90	1.
Pan-STARRS	2008 ...	70%	Optical	1.8	15–24	0.30	7.
			Future optical surveys				
URAT	2010 ...	All	670–750 nm	0.2	7–18	0.90	28
LSST	2014 ...	South	320–1080 nm	6.7	16–25	0.20	9.
			Near-infrared surveys				
2MASS	1997–2001	All	J, H, Ks	1.3	0–19	2.00	S

Table 20.4 Links to survey projects

Survey	Address
SDSS	www.sdss.org
SPM	www.astro.yale.edu/astrom/spm4cat/spm4.html
CMC	www.ast.cam.ac.uk/~dwe/SRF/camc.html
CMASF	www.roa.es
PM2000	vizier.hia.nrc.ca/ftp/cats/I/300/ReadMe
UCAC	www.usno.navy.mil/USNO/astrometry/optical-IR-prod/ucac
URAT	www.usno.navy.mil/USNO/astrometry/optical-IR-prod/urat
Pan-STARRS	pan-starrs.ifa.hawaii.edu/public/home.html
LSST	www.lsst.org/lsst/science
2MASS	www.ipac.caltech.edu/2mass

19–20. Almost every branch of modern astronomy will be benefited by Gaia, including stellar populations in the Local Group, distance scale and age of the Universe, dark-matter distribution, reference frame (quasars, astrometry), Solar System research (Mignard *et al.* 2007), and different aspects of planetary systems (Sozzetti 2010). But above all, Gaia will provide a new view of our own Galaxy (Brown 2008). Gaia is scheduled for launch in 2013.

20.6 The Virtual Observatory

Even though observational astronomy is undergoing major changes thanks to a combination of technological advances in both detectors and telescope design, the ability and capability of researchers in processing and analyzing the overwhelming amount of data from ground and space observations is lagging. For that reason, a large amount of information remains hidden in a growing number of collections of CCD images and archives spread throughout almost all astronomical institutions. The first attempt to accomplish worldwide dissemination and easy access to large databases (mainly catalogs of different types) was the creation of the data centers, with the CDS at Strasbourg being one of the first.

To overcome the problem of the growing accumulation of unused data, the Virtual Observatory (VO) was proposed about a decade ago as a way of handling all this information and allowing the astronomical community easy access to multiple archives in multiple wavelengths. With this in mind, it is possible to say that the VO is meant to transform the many terabytes of data buried in astronomical archives, image banks, and sky surveys into useful and valuable scientific information. Thus, the VO can be defined as a set of tools that will enable the user to extract information from different astronomical catalogs and image banks located at physically different data centers in a manner that is as seamless and transparent as browsing the World-Wide Web. Needless to say, if the entire community wants to have a successful VO, then any restriction to open access to databases should be avoided, otherwise the VO will be a failure. Since almost every archive has its own architecture and organization, the most important challenge for the VO is to extract information from as many archive sources and surveys containing data of various types and differing quality as possible. The very large size of most databases and the variable quality (uncertainty) of the data over the entire electromagnetic spectrum pose unique and very specific problems to the many VO tools which are being developed in the form of software packages.

A concept closely related to the VO is that of *data mining*, which is the process through which data is extracted from the source catalog or catalogs. Probably in the same sense we could define *image mining* as the process through which the user is able to extract information from images (and the subsequent processing) available at different image banks. One of the best examples of the image-mining process is the location of satellite number V of M31 described by Armandroff *et al.* (1998). Good examples of data mining are the recent investigations of López Martí *et al.* (2011) and Jiménez-Esteban *et al.* (2011). The new terminology related to the VO also includes *federation of data*, which is in fact the combination of multiple databases, generally giving information in different wavelengths, in order to collect all the available information on a given target (see Williams 2001 for a detailed discussion). Some VO initiatives include: National Virtual Observatory (NVO; www.us-vo.org/), Russian Virtual Observatory (RVO; www.inasan.rssi.ru/eng/rvo/), European Virtual Observatory (EURO-VO; www.euro-vo.org/pub/), etc. These initiatives are now being collected into the International Virtual Observatory Alliance (IVOA; www.ivoa.net). The VO has become so important for the international community that the IAU established a Working Group (WG) on Virtual Observatory, Data Center, and Networks under Commission 5 at the Prague General Assembly in 2006. This WG coordinates activities closely with other WGs of Commission 5, namely: Astronomical Data, Designations, Libraries, and (mainly) the flexible image transport system (FITS) working group.

20.7 Radio astrometry

The diffraction limit of a telescope is inversely proportional to the wavelength used for the observations. Since radio wavelengths are about 10^3 to 10^5 times longer than those of visible light, no high-resolution imaging or high-precision angular measures (astrometry) can be performed with a single radio telescope operated like an optical telescope. Interferometry (Hecht and Zajac 1982) overcomes this problem, in particular the VLBI technique (Thompson *et al.* 1991 and Chapter 12), which connects radio telescopes across the globe, results in angular measures far more precise (order 0.1 mas) than obtained by traditional ground-based optical telescopes. A detailed discussion of radio interferometry is given in Chapter 12. Observations by VLBI of about 700 compact, extragalactic sources define the current ICRF (Fey *et al.* 2004), and the ICRF2 in the near future, after centuries of optically defined fundamental systems (Walter and Sovers 2000). VLBI reductions have to solve for Earth orientation, plate tectonics, and geophysical and other parameters simultaneously with the astrometric parameters (source positions) and assume a mean zero motion and rotation of the ensemble of extragalactic sources. Observations now push for higher frequencies to mitigate this problem (Fey *et al.* 2009).

The next step in radio reference-frame densification is the VLBA calibrator catalog (www.vlba.nrao.edu/astro/calib, http://gemini.gsfc.nasa.gov/vcs/) of about 4000 sources with median positional error below 1 mas. The Large Quasar Astrometric Catalog (ftp://syrte.obspm.fr/pub/LQAC) contains over 100 000 quasars with best available astrometric data, and cross-correlating radio and optical information. Extremely accurate observations (approaching 10 microarcseconds) can be performed with phase-referencing techniques. Popular targets are radio stars (Boboltz *et al.* 2007) and the galactic center and distance scale (Reid 2008). Structure analysis, proper motions, and trigonometric distance programs for specific targets are great opportunities for radio astrometry. Proposals can be directed through the National Radio Astronomy Observatories NRAO (www.nrao.edu), the European VLBI Network (www.evlbi.org), or the Japanese VLBI Exploration of Radio Astrometry, VERA, project (www.miz.nao.ac.jp/en/content/project/vera-project).

References

Armandroff, T. E., Davies, J. E., and Jacoby, G. H. (1998). A survey for low surface brightness galaxies around M31. I. The newly discovered Dwarf Andromeda V. *AJ*, **116**, 2287.

Boboltz, D. A., Fey, A. L., Puatua, W. K., *et al.* (2007). Very Large Array plus Pie Town astrometry of 46 Radio Stars. *AJ*, **133**, 906.

Boss, B. (1937). *General Catalogue of 33342 Stars for the Epoch 1950*. Washington, DC: Carnegie Institution of Washington, Publication 486.

Brown, A. G. A. (2008). Learning about Galactic structure with Gaia astrometry. *AIP Conf. Proc.*, **1082**, 209.

Bucciarelli, B., Lattanzi, M. G., McLean, B., *et al.* (2008). The GSC-II catalog release GSC 2.3: description and properties. *Proc. IAU Symp.*, **248**, 316.

Corbin, T. and Urban, S. (1990). Faint reference stars. *Proc. IAU Symp.*, **141**, 433.

Eichhorn, H. (1974). *Astronomy of Star Positions*. New York, NY: Ungar Publishing.

Eichhorn, H. (1992). Star catalogue, historic. In *The Astronomy and Astrophysics Encyclopedia*, ed. S. P. Maran, p. 661.

Fallon, F. and Hoffleit, D. (1983). Catalogue of the positions and proper motions of stars between declinations −60° and −70°. *Trans. Yale Univ. Obs.*, **32**, No. 2.

Feissel, M. and Mignard, F. (1998). The adoption of ICRS on 1 January 1998: meaning and consequences. *A&A*, **331**, L33.

Fey, A., Ma, C., Arias, E. F., *et al.* (2004). The second extension of the International Celestial Reference Frame. *AJ*, **127**, 3587.

Fey, A., Boboltz, B., Charlot, P., *et al.* (2009). Extending the ICRF to higher radio frequencies. *AAS Bull.*, **41**, 676.

Gingerich, O. (1992). *The Great Copernicus Chase and Other Adventures in Astronomical History*. Cambridge: Cambridge University Press.

Girard, T. M., Platais, I., Kozhurina-Platais, V., and van Altena, W. F. (1998a). The Southern Proper Motion Program I. *AJ*, **115**, 855.

Girard, T. M., Dinescu, D. I., van Altena, W. F., *et al.* (1998b). The Yale/San Juan Southern Proper Motion Catalog 3. *ASP Conf.*, **317**, 206.

Girard, T. M., van Altena, W. F., Zacharias, N., *et al.* (2011). The Southern Proper Motion Program. IV. The SPM4 Catalog. *AJ*, **142**, 15.

Gulyaev, A. and Nesterov, V. (1992). On the 4 Million Star Catalogue (in Russian). Moscow: Moscow University Press, p. 1.

Hanson, R. B., Klemola, A. R., Jones, B. F., and Monet, D. G. (2004). Lick Northern Proper Motion Program III. *AJ*, **128**, 1430.

Hecht, E. and Zajac, A. (1982). *Optics*. Reading, MA: Addison-Wesley.

Hoeg, E. (2008). Astrometry and optics during the past 2000 years. See: www.astro.ku.dk/∼erik/HistoryAll.pdf.

Jaschek, C. (1989). *Data in Astronomy*. Cambridge: Cambridge University Press, p. 52–54.

Jiménez-Esteban, F., Caballero, J., and Solano, E. (2011). Identification of blue high proper motion objects in the Tycho-2 and 2MASS catalogues using Virtual Observatory tools. *A&A*, **525**, A29. (arXiv:1009.3466).

Klemola, A. R., Jones, B. F., and Hanson, R. B. (1987). Lick Northern Proper Motion Program. *AJ*, **94**, 501.

Krisciunas, K. (1993). A more complete analysis of the errors in Ulugh Beg's star catologue. *J. History Astron.*, **24**, 269.

López Martí, B., Jiménez-Esteban, F., and Solano, E. (2011). A proper motion study of the Lupus clouds using Virtual Observatory tools. *A&A*, **529**, A108.

Ma, C., Arias, E. F., Eubanks, T. M., *et al.* (1998). The International Celestial Reference Frame as realized by very long baseline interferometry. *AJ*, **116**, 516.

McLean, B. (2001). An overview of existing ground-based wide-area surveys. *ASP Conf.*, **225**, 103.

Mignard, F., Cellino, A., Muinonen, K., *et al.* (2007). The Gaia mission: expected applications to asteroid science. *Earth, Moon, and Planets*, **101**, 97.

Monet, D. G., Levine, S. E., Canazian, B., *et al.* (2003). The USNO-B Catalog. *AJ*, **125**, 984.

Morgan, H. R. (1952). Catalog of 5268 standard stars, based on the normal system N30. *Astron. Papers Am. Eph. And Naut. Alm.*, **13**.

Reid, M. (2008). Micro-arcsecond astrometry with the VLBA. *Proc. IAU Symp.*, **248**, 474.

Röser, S. and Bastian, U. (1993). The final PPM Star Catalogue for both hemispheres. *Bull. Inform. CDS*, **42**, 11.

Röser, S., Bastian, U., and Kuzmin, A. (1994). PPM Star Catalogue: the 9000 stars supplement. *A&AS*, **105**, 301.

Schlesinger, F. F., Hudson, C. J., Jenkins, L., and Barney, J. (1926). Catalogue of 5833 stars −2° to +1°. *Trans. Yale Univ. Obs.*, **5**.

Sozzetti, A. (2010). The Gaia Astrometric Survey. *Highlights of Astronomy*, **15**, 716.

Thompson, A. R., Moran, J. M., and Swenson, G. W. (1991). *Interferometry and Synthesis in Radio Astronomy*. Malabar, FL: Krieger Publications.

Urban, S., Corbin, T., and Wycoff, G. (1998). The AC 2000: the Astrographic Catalogue on the system defined by the Hipparcos Catalogue. *AJ*, **115**, 2161.

Urban, S., Corbin, T. E., Wycoff, G. L., *et al.* (2001). The AC 2000 Catalogue. *Bull. Am. Astron. Soc.*, **33**, 1494.

Walter, H. G. and Sovers, O. J. (2000). *Astrometry of Fundamentals Catalogues*. New York, NY: Springer.

Williams, R. (2001). Approaches to federation of astronomical data. *ASP Conf. Ser.*, **225**, 302.

Zacharias, N., Urban, S. E., Zacharius, M. I., *et al.* (2004). The Second US Naval Observatory CCD Astrograph Catalog. *AJ*, **127**, 3043.

21

Trigonometric parallaxes

G. FRITZ BENEDICT AND BARBARA E. McARTHUR

Introduction

One consequence of observing from a moving platform is that all objects exhibit parallax. The measurement of parallax yields distance, a quantity useful in astrophysics. In particular, with distance we can determine the absolute magnitude of any object, a primary parameter in two of the most useful "maps" in astronomy: the Hertzsprung–Russell diagram (e.g. Perryman *et al.* 1997, Fig. 3), showing the relation between absolute magnitude (luminosity) and color (temperature); and the mass–luminosity relation (e.g. Henry 2004, Fig. 3), a tool for turning luminosity into mass, a stellar attribute which determines the past and future aging process for any star. Another example of the utility of absolute magnitudes is the Cepheid period–luminosity relation (PLR). The example used here to illustrate parallax determination had improving that relationship as its ultimate goal.

The technology used to generate parallaxes has proceeded from naked-eye measurements with mechanical micrometers (Bessel 1838), through hand measurements of photographic plates (Booth and Schlesinger 1922), through computer-controlled plate scanners (Auer and van Altena 1978), through computer-controlled CCD cameras (Henry *et al.* 2006, Harris *et al.* 2007), through the triumph of the Hipparcos astrometric satellite (Perryman *et al.* 1997), to space-borne optical interferometers (Benedict *et al.* 2007, 2009) and extremely long baseline radio interferometers (Reid *et al.* 2009). Each stage of this historical sequence is characterized by improvements in both the centering of the images of the target and reference stars and the mathematical challenge in distilling the final parallax from those centers.

As an example of a parallax determination we outline an approach to obtain as accurate as possible parallaxes from astrometric measurements secured with the Fine Guidance sensors (FGSs) on the Hubble Space Telescope (HST) (the aforementioned space interferometers). This recipe was first used in Harrison *et al.* (1999) and refined in subsequent papers (Benedict *et al.* 1999, 2002a, 2002b, 2003, 2007, 2009; McArthur *et al.* 1999, 2001, and Soderblom *et al.* 2005). Many of the aspects of this approach have been independently derived and used by other practitioners of the art, in particular the Georgia State group (Jao *et al.* 2005, Henry *et al.* 2006) and the US Naval Observatory (e.g. Harris *et al.* 2007). Where we produce our centers by measuring a fringe zero-crossing position with a shearing interferometer (basically a wavefront-tilt sensor), others use charge-coupled device (CCD)

Astrometry for Astrophysics: Methods, Models, and Applications, ed. William F. van Altena.
Published by Cambridge University Press. © Cambridge University Press 2013.

cameras and centering codes (e.g. SExtractor, Bertin and Tissier 2007; DAOPHOT, Stetson 1987). At each step in the sample analysis presented here we point out similarities with ground- or space-based CCD parallax techniques.

21.1 The mathematical foundation of parallax

Following Murray (1983), we can represent the topocentric coordinate direction $\tilde{r}$ at epoch $\tilde{T}$ by

$$\mathbf{r} = \tilde{\mathbf{r}}_{I} + \mu(\tilde{T} - \tilde{T}_{I}) - a_{u}^{-1}(\sin\tilde{\omega})\tilde{\mathbf{r}}_{I} \times (\mathbf{b}_{0} \times \tilde{\mathbf{r}}_{I}) \tag{21.1}$$

where μ is the proper-motion vector, $\mathbf{b}_0$ is the barycentric coordinate vector to the observer, and $\tilde{\omega}$ is the parallax. In terms of standard coordinates $\vec{\xi}$, $\vec{\eta}$ we have

$$\begin{bmatrix} \vec{\xi} \\ \vec{\eta} \end{bmatrix} = \begin{bmatrix} \bar{\mathbf{u}}_{s}'\vec{t} \\ \bar{\mathbf{v}}_{s}'\vec{t} \end{bmatrix} \tag{21.2}$$

with $\bar{\mathbf{u}}_s'$ and $\bar{\mathbf{v}}_s'$ the mutually orthogonal unit vectors in the plane perpendicular to some tangential direction unit vector $\mathbf{w}$. Setting the standard coordinates corresponding to the barycentric direction at epoch $\tilde{T}$ to $(\tilde{\xi}_I + \Delta\tilde{\xi}_I, \tilde{\eta}_I + \Delta\tilde{\eta}_I)$, where $(\tilde{\xi}_I, \tilde{\eta}_I)$ are initial approximations and $(\Delta\tilde{\xi}_I, \Delta\tilde{\eta}_I)$ are small, we then have the components of Eq. (21.1) along the standard coordinates axes

$$\begin{bmatrix} \vec{\xi} - \tilde{\xi}_I \\ \vec{\eta} - \tilde{\eta}_I \end{bmatrix} = \begin{bmatrix} \Delta\tilde{\xi}_I \\ \Delta\tilde{\eta}_I \end{bmatrix} + (\tilde{T} - \tilde{T}_I)\begin{bmatrix} \mu_{\xi} \\ \mu_{\eta} \end{bmatrix} + \begin{bmatrix} P_{\xi} \\ P_{\eta} \end{bmatrix}\tilde{\omega} \tag{21.3}$$

where

$$\begin{bmatrix} \mu_{\xi} \\ \mu_{\eta} \end{bmatrix} = (1 + \tilde{\xi}^2 + \tilde{\eta}^2)^{1/2}\mu'\begin{bmatrix} \bar{\mathbf{u}}_s - \tilde{\xi}_I\mathbf{w} \\ \bar{\mathbf{v}}_s - \tilde{\eta}_I\mathbf{w} \end{bmatrix} \tag{21.4}$$

and parallax factors are

$$\begin{bmatrix} P_{\xi} \\ P_{\eta} \end{bmatrix} = (1 + \tilde{\xi}^2 + \tilde{\eta}^2)^{1/2}a_{u}^{-1}\mathbf{b}_0'\begin{bmatrix} \bar{\mathbf{u}}_s - \tilde{\xi}_I\mathbf{w} \\ \bar{\mathbf{v}}_s - \tilde{\eta}_I\mathbf{w} \end{bmatrix} \tag{21.5}$$

Let $\mathbf{R} \equiv [\mathbf{p}\,\mathbf{q}\,\mathbf{r}]$ denote the normal triad corresponding to direction $\mathbf{r}$. Thus the components of displacement due to parallax along the $\mathbf{p}$ and $\mathbf{q}$ axes are $(-\tilde{\omega}a_u^{-1}\mathbf{b}_0'\mathbf{p}, -\tilde{\omega}a_u^{-1}\mathbf{b}_0'\mathbf{q})$, where $-\tilde{\omega}$ is the parallax, a_u^{-1} is a scaling factor (the mean Sun–Earth distance in AU), and $\mathbf{b}_0'$ the barycentric coordinate vector. The challenge is to measure these displacements, either relative to nearby (narrow-angle, small-field astrometry) or angularly distant (global astrometry) reference stars.

21.2 Differential *versus* global astrometry

When obtaining a parallax through differential means, great care must be taken to remove the motions of the reference stars, including parallax and proper motion. The corrected reference frame then serves as a stationary "background" so that the parallax (and proper motion) of the scientifically interesting target can be measured. The correction from relative to absolute parallax would be unnecessary if we could measure against reference stars that had angular separations of 90°. Such reference stars would have no parallax. This idealized technique is called global astrometry. The best recent results from this technique are the catalogs generated by the Hipparcos satellite (Perryman *et al.* 1997, van Leeuwen 2007). Hipparcos measured the angular separations between a target star and reference stars that were at distances of 58°. The recent Hipparcos re-reduction by van Leeuwen (2007) has yielded parallaxes that typically have formal $1 - \sigma$ errors of 0.8 milliarcseconds (mas).

Given that the Hipparcos errors increase for fainter targets, and that there are scientifically interesting targets for which even higher precision would have great value, ground-based CCD cameras and the FGSs (white-light shearing interferometers) on HST have been used to mine previously unexploited regions of parallax phase-space.

21.3   Differential parallaxes with HST

In this section we use the Galactic Cepheid ℓ Carinae (ℓ Car) as an example of the techniques developed to obtain an absolute parallax from differential, narrow-angle astrometry. Additional details can be found in Benedict *et al.* (2007), while the general problems encountered in astrometric analyses are discussed in Chapter 19.

Nelan (2011) provides an overview of the FGS instrument and Benedict *et al.* (2002b) describe the fringe tracking (POS) mode astrometric capabilities of an FGS, along with typical data-acquisition and -reduction strategies.

Analogous to a CCD exposure, each individual FGS data set (a collection of wavefront-tilt measurements for the science target and astrometric reference stars) required approximately 33 minutes of spacecraft time. Eleven sets of astrometric data were acquired with HST FGS 1r for ℓ Car. We obtained most of these 11 sets in pairs near maximum parallax factor typically separated by a week, a strategy designed to protect against unanticipated HST equipment problems. A few single data sets were acquired at various minimum parallax factors to aid in separating parallax and proper motion. The ℓ Carinae complete data aggregate spans 1.95 years. A typical ground-based CCD campaign might have a similar duration (e.g. Gizis *et al.* 2007).

The data were reduced and calibrated as detailed in McArthur *et al.* (2001), Benedict *et al.* (2007), and Soderblom *et al.* (2005). The FGS, unlike a CCD camera, does not observe target and reference stars simultaneously. Consequently, each data set contains multiple measurements of reference stars, this to correct for intra-orbit drift of the type seen in the cross-filter calibration data shown in Fig. 1 of Benedict *et al.* (2002a).

Data are downloaded from the Space Telescope Scientific Institute (STScI) archive and passed through a pipeline processing system. This pipeline extracts the astrometry measurements (typically 1 to 2 minutes of fringe-position information acquired at a 40 Hz rate, which yields several thousand discrete measurements), extracts the median (which we have found to be the optimum estimator), and corrects for the optical field angle distortion

Table 21.1 ℓ Car reference stars: visible and near-infrared photometry

ID	FGS ID	V	$B - V$	$U - B$	$V - I$	K	$J - K$	$V - K$
4273957	2	14.32	0.71	0.30	0.89	12.52	0.30	1.80
4273905	4	13.53	0.95	0.63	1.13	11.20	0.62	2.33
2M	5	13.23	1.18	1.08	1.33	10.29	0.78	2.95
4066585	8	10.77	1.58	1.95	1.87	6.56	1.10	4.22
4066439	9	13.49	0.57	0.08	0.72			
4066556	10	13.01	0.60	−0.04	0.80	11.48	0.34	1.53

(OFAD). This correction for OFAD (McArthur *et al.* 2002) reduces as-built HST telescope and FGS 1r distortions with amplitude $\sim 1''$ to below 2 mas over much of the FGS 1r field of regard. This correction is analogous to the distortion mapping and corrections carried out for large-format CCD cameras (e.g. Anderson *et al.* 2006). Lastly, the pipeline corrects for velocity aberration and spacecraft jitter and attaches all required time tags and parallax factors.

21.3.1 Reference stars

Any parallax determined from small-field astrometry will be measured with respect to reference-frame stars, each with their own parallax. We must either apply a statistically derived correction from relative to absolute parallax (van Altena *et al.* 1995, hereafter YPC95) or estimate the absolute parallaxes of the reference-frame stars. In principle, the colors, spectral type, and luminosity class of a star can be used to estimate the absolute magnitude M_V and V-band absorption A_V. The absolute parallax for any reference star is then simply

$$\pi_{\mathrm{abs}} = 10^{-(V - M_V + 5 - A_V)/5} \tag{21.6}$$

The luminosity class is generally more difficult to estimate than the spectral type (temperature class). However, the derived absolute magnitudes M_V are critically dependent on the luminosity class. To confirm the luminosity classes we obtain 2MASS[1] photometry and UCAC2 (Zacharias *et al.* 2004) or PPMXL (Roeser *et al.* 2010) proper motions for a field centered on any parallax target, and iteratively employ the technique of reduced proper motion (Yong and Lambert 2003, Gould and Morgan 2003) to confirm our giant/dwarf classifications.

Reference-star photometry

Our bandpasses for reference-star photometry typically include BVI and JHK (from 2MASS). In addition, Washington–DDO photometry (Majewski *et al.* 2000) is used to confirm the luminosity classifications for the later-spectral-type reference stars. Table 21.1

---

[1] The Two Micron All Sky Survey is a joint project of the University of Massachusetts and the Infrared Processing and Analysis Center/California Institute of Technology.

lists the photometry used in our estimates of reference-star spectrophotmetric parallax. The ID comes from the UCAC2 catalog, except for the star labeled 2M, which is 09454541-6230004 from the 2MASS catalog.

Reference-star spectroscopy

The spectra from which we estimated ℓ Car reference-star spectral type and luminosity class for this particular field come from a spectrograph with resolution 3.5 Å/ full width at half maximum (FWHM) with wavelength coverage from 3750 Å $\leq \lambda \leq$ 5500 Å. Classifications used a combination of template matching and line ratios. Spectral types for the stars are generally better than ± 2 subclasses. Our luminosity-class uncertainty is reflected in the final-input spectrophotometric parallax errors (Table 21.2).

Interstellar extinction

Assuming an $R = 3.1$ Galactic reddening law (Savage and Mathis 1979), we derived A_V values for each reference star by comparing the measured colors (Table 21.1) with intrinsic $(V-K)_0, (B-V)_0, (U-B)_0, (J-K)_0$, and $(V-I)_0$ and colors from Bessell and Brett (1988) and Cox (2000). We estimated A_V from $A_V = 1.1 E(V-K) = 5.8 E(J-K) = 2.77 E(U-B) = 3.1 E(B-V) = 2.26 E(V-I)$, where the ratios of total to selective extinction were derived from the Savage and Mathis (1979) reddening law and a reddening estimate in the direction of ℓ Car from Schlegel *et al.* (1998), via NED.[2] We then calculated a field-wide average A_V to be used in Eq. (21.6). For the ℓ Car field $\langle A_V \rangle = 0.52 \pm 0.06$ magnitude. In this case our independent determination is in good agreement with the David Dunlap Observatory online Galactic Cepheid database,[3] which averages seven measurements of color excess to obtain $\langle E(B-V) \rangle = 0.163 \pm 0.017$, or $\langle A_V \rangle = 0.51 \pm 0.05$.

Estimated reference-frame absolute parallaxes

We derived absolute parallaxes for each reference star using M_V values from Cox (2000) and the $\langle A_V \rangle$ derived from the photometry. Our adopted errors for $(m - M)_0$ are 0.5 magnitude for all reference stars. This error includes uncertainties in $\langle A_V \rangle$ and in the spectral types used to estimate M_V. Our reference-star parallax estimations from Eq. (21.6) are listed in Table 21.2. For the ℓ Car field individually, no reference-star absolute parallax is better determined than $\sigma_\pi / \pi = 23\%$. The average absolute parallax for the reference frame is $\langle \pi_{abs} \rangle = 0.85$ mas. We compare this to the correction to absolute parallax discussed and presented in YPC95 (Section 3.2, Fig. 2). Entering YPC95, Fig. 2, with the ℓ Car Galactic latitude, $l = -7°$, and average magnitude for the reference frame, $\langle V_{ref} \rangle = 13.0$, we obtain a correction to absolute of 1 mas. This gives us confidence in our spectrophotometric determination of the correction to absolute parallax. We introduce into our reduction model our spectrophotometrically estimated reference-star parallaxes as observations with

---

[2] NASA/IPAC Extragalactic Database.
[3] See www.astro.utoronto.ca/DDO/research/cepheids/cep! heids.html.

Table 21.2 ℓ Car astrometric reference-star spectrophotometric parallaxes

ID	V	Spectral type	M_V	A_V	$m - M$	$\tilde{\omega}_{abs}$(mas)
2	14.32	G0V	4.4	0.33	9.89±0.5	1.2±0.3
4	13.53	G8III	0.9	0.14	12.63±0.7	0.3±0.1
5	13.23	K0III	0.7	0.72	12.48±0.5	0.4±0.1
8	10.77	K4III	−0.1	1.02	10.72±0.5	1.1±0.3
9	13.49	F3V	3.2	0.45	10.24±0.5	1.1±0.3
10	13.01	F4V	3.3	0.46	9.67±0.5	1.4±0.3

error, The use of spectrophotometric parallaxes offers a more direct (less Galaxy model-dependent) way of determining the reference-star absolute parallaxes. See Jao *et al.* (2005) for an alternative approach, establishing reference-star parallaxes using multi-band photometry.

21.3.2 Modeling

Ancillary calibrations

Ground-based astrometry must deal with differential color refraction (DCR), position shifts caused by the fact that the amount of refraction due to the atmosphere depends on the color of a star. If astrometric reference stars and scientifically interesting target stars were all the same color, this would not be a problem. Consider planetary nebulae central stars, with surface temperatures in excess of 50 000 K. Compare these to the typical G or K reference star. In Johnson UBV bandpasses, the $B - V$ color indices can differ by over a magnitude. Jao *et al.* (2005) contains a useful discussion of DCR and calibration approaches for ground-based parallax work.

Obtaining the parallax of ℓ Car is, unfortunately, similarly afflicted; not by the atmosphere, but by the fact that our astrometer, the FGS, contains refractive optical components. The FGS also contains a 1% transmission neutral density filter, used in this case to permit observations of ℓ Car, with an apparent magnitude $V = 3.72$. No two faces of any filter are perfectly parallel, hence they introduce (again, color-dependent) position shifts. The first of these effects we call "lateral color," the second "cross-filter." Calibrations of these specific to the FGS are detailed in Benedict *et al.* (2002b), while for the general case, see van Altena and Monnier (1968).

The astrometric model

With the positions measured by FGS 1r we determine the scale, rotation, and offset "plate constants" relative to an arbitrarily adopted constraint epoch (the so-called "master plate") for each observation set (the data acquired at each epoch). Our ℓ Car reference frame contains six stars. The modeling approach outlined in Benedict *et al.* (2002b) is used. It employs corrections for both cross-filter and lateral-color positional shifts, using values specific to FGS 1r determined from previous calibration observations with that FGS.

We employ GaussFit (Jefferys *et al.* 1988) to minimize χ^2. The solved equations of condition for the ℓ Car field are

$$x' = x + lc_x(B - V) - \Delta XFx \tag{21.7}$$

$$y' = y + lc_y(B - V) - \Delta XFy \tag{21.8}$$

$$\xi = Ax' + By' + C - \mu_x \Delta t - P_\alpha \tilde{\omega}_x \tag{21.9}$$

$$\eta = Dx' + Ey' + F - \mu_y \Delta t - P_\delta \tilde{\omega}_y \tag{21.10}$$

where x and y are the measured coordinates from HST; lc_x and lc_y are the lateral color corrections; ΔXFx and ΔXFy are the cross-filter corrections in x and y, applied only to the observations of ℓ Car; and $B - V$ are the $B - V$ colors of each star. Note that the color of ℓ Car changes with pulsation phase. This, too, was taken into account. The parameters A, B, D, and E are scale and rotation plate constants; C and F are offsets; μ_x and μ_y are proper motions; Δt is the time difference from the mean epoch; P_α and P_δ are parallax factors; and $\tilde{\omega}_x$ and $\tilde{\omega}_y$ are the parallaxes in x and y. We obtain the parallax factors from a Jet Propulsion Laboratory (JPL) Earth orbit predictor (Standish 1990), upgraded to version DE405.

Prior knowledge and modeling constraints

In a quasi-Bayesian approach the reference-star spectrophotometric absolute parallaxes (Table 21.2) and UCAC2 proper motions (we now use the PPMXL) were input as observations with associated errors, not as hardwired quantities known to infinite precision. Input proper-motion values have typical errors of 4–6 mas yr^{-1} for each coordinate. The lateral-color and cross-filter calibrations and the $B - V$ color indices are also treated as observations with error. We employ the technique of reduced proper motions H_k to provide a confirmation of the reference-star estimated luminosity class listed in Table 21.2. We obtain proper motion and J, K photometry from UCAC2 and 2MASS for a $1/3° \times 1/3°$ field centered on ℓ Car. Plotting $H_K = K + 5\log(\mu)$ against $J - K$ color index for those 436 stars, reference stars ref-4, -5, and -8 remain clearly separated from the others, supporting their classification as giants.

We stress that no previously measured ℓ Car parallax was used as prior knowledge in the modeling. Only reference-star prior knowledge was so employed. Our ℓ Car parallax result is blind to previous parallax measures from Hipparcos and/or parallaxes from surface-brightness estimates.

Assessing reference-frame residuals

From the Gaussian nature of the histograms of the ℓ Car field astrometric residuals we conclude that we have obtained satisfactory OFAD correction, and that the lateral-color and cross-filter calibrations have been applied successfully. The resulting reference-star "catalog" in ξ and η standard coordinates was determined with average position errors $\langle \sigma_\xi \rangle = 0.50$ and $\langle \sigma_\eta \rangle = 0.62$ mas.

An absolute parallax for ℓ Car

For the ℓ Car astrometric analysis we reduced the number of modeling coefficients in Eqs. (21.9) and (21.10) to four, as done for our previous work on the Pleiades (Soderblom *et al.* 2005). We constrained the fit to have a single scale term by imposing $D = -B$ and $E = A$. Final model selection was based on reference-star placement relative to the target, the total number of reference stars, reduced χ^2 (χ^2/DOF, where DOF = degrees of freedom), and parallax error. For ℓ Car we find $\tilde{\omega} = 2.01 \pm 0.20$ mas.

HST parallax accuracy

Our parallax precision, an indication of our internal, random error, is ~ 0.2 mas. To assess our accuracy, or external error, we have in the past (Benedict *et al.* 2002b, Soderblom *et al.* 2005) compared our parallaxes with results from independent measurements from Hipparcos (Perryman *et al.* 1997). We now compare with the new reduction (van Leeuwen 2007). In Fig. 21.1 we plot Hipparcos and HST parallaxes for 29 objects in common. These range from nearby M-dwarf stars and binaries to distant Cepheids.

We have no large systematic differences with Hipparcos for any objects with $\sigma_{\tilde{\omega}}/\tilde{\omega} < 10\%$. However, small changes in parallax can have significant impact. For example, the 2007 Hipparcos Pleiades parallax is $\tilde{\omega}_{\text{abs}} = 8.18 \pm 0.13$ mas, compared to the FGS value, $\tilde{\omega}_{abs} = 7.43 \pm 0.17$ mas (Soderblom *et al.* 2005). The latter agrees with a number of independent distance determinations as summarized by Soderblom *et al.*

21.4 The utility of parallaxes

21.4.1 From parallaxes to absolute magnitudes

Once we have a parallax often the ultimate goal is the intrinsic brightness of the star of interest, its absolute magnitude in some bandpass, M_V for V. However, when using a trigonometric parallax to estimate the absolute magnitude of a star, a correction should be made for the Lutz–Kelker (L–K) bias (Lutz and Kelker 1973) as modified by Hanson (1979). See also Chapter 16 for a detailed discussion of statistics used in astrometry.

We appeal to Bayes theorem to justify the application of a Lutz–Kelker–Hanson (LKH) bias correction to a single star, rather than a class of stars. See Barnes *et al.* (2003), Section 4, for an accessible introduction to Bayes theorem as applied to astronomy as well as Chapter 16. Invoking Bayes theorem to assist with generating an absolute magnitude from our ℓ Car parallax, we would say, "what is the probability that a star from this population with this position would have parallax $\tilde{\omega}$ (as a function of $\tilde{\omega}$), given that we haven't yet measured $\tilde{\omega}$?" In practice we would use the space distribution of the population to which ℓ Car presumably belongs. This space distribution is built into the prior $p(\tilde{\omega})$ for $\tilde{\omega}$, and used to determine

$$\text{p}(\tilde{\omega}|\tilde{\omega}_{\text{observed}}\&K) \sim p(\tilde{\omega}_{\text{observed}}|\tilde{\omega}\&K)p(\tilde{\omega}|K) \tag{21.11}$$

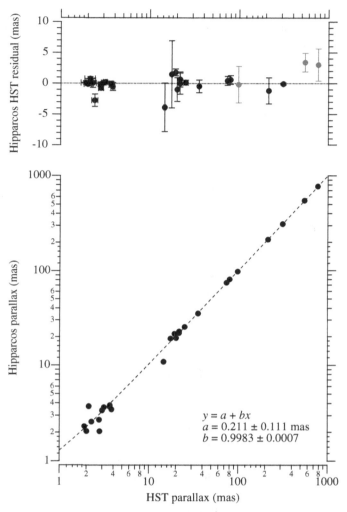

Fig. 21.1　Hipparcos parallaxes from the re-reduction of van Leeuwen (2007) plotted against published HST parallaxes in common (through May 2011, from McArthur *et al.* 2011). The linear fit is an impartial regression that properly treats errors in both HST and Hipparcos parallaxes (GaussFit; Jefferys *et al.* 1988). Each residual is tagged with the published *x* (HST) and *y* (Hipparcos) error bars. The average and median parallax errors are for HST 0.24, 0.20 mas and for Hipparcos 1.22, 0.71 mas. The largest difference is for Feige 24, a faint (for Hipparcos) white-dwarf–M-dwarf short-period binary star (Benedict *et al.* 2000).

where K is prior knowledge about the space distribution of the class of stars in question and '&' is an 'and' operator. The function $p(\tilde{\omega}_{\text{observed}}|\tilde{\omega}\ \&\ K)$ is the standard likelihood function, usually a Gaussian normal with variance $\sigma_{\tilde{\omega}}$. The "standard" L–K correction has $p(\tilde{\omega}|K) \sim \tilde{\omega}^{-4}$. Looking at a star in a disk population close to the Galactic plane requires $\tilde{\omega}^{-3}$ (ignoring spiral structure), which is the prior we use for ℓ Car. The LKH bias is proportional to $(\sigma_{\tilde{\omega}}/\tilde{\omega})^2$. Presuming that ℓ Car belongs to the same class of object (evolved

Main Sequence stars), we scale the LKH correction determined in Benedict *et al.* (2002b) for δ Cephei (Cep) and obtain for ℓ Car an LKH bias correction, LKH $= -0.08$ magnitude. We identify the choice of prior for this bias correction as a possible contributor to systematic errors in the zero-points of our PLR, at the 0.01 magnitude level.

With an average apparent V magnitude, $\langle V \rangle = 3.724$, and the absolute parallax, 2.01 ± 0.20, we determine a distance modulus for ℓ Car. We derived a field-average absorption, $\langle A_V \rangle = 0.52$. With this $\langle A_V \rangle$, the measured distance to ℓ Car, and the LKH correction we obtain $M_V = -5.35 \pm 0.22$ and a corrected true distance modulus, $(m - M)_0 = 8.56$. The M_V error has increased slightly by combining the $\langle A_V \rangle$ error and the raw distance modulus error in quadrature.

21.4.2 The Leavitt law for Galactic Cepheids

As first discovered by Henrietta Leavitt (Leavitt and Pickering 1912), Cepheid period and luminosity are related. In honor of that discovery we suggest that the period–luminosiy relation be renamed the Leavitt law. With 10 relatively precise parallaxes for Galactic Cepheids from our FGS astrometry we generated 10 absolute magnitudes. Our Leavitt law zero-point in the V band now has a $1 - \sigma$ error of 0.02 magnitude with a slope error of 0.12 magnitude. A recent use of these parallaxes was to densify the Leavitt law by calibrating other less-direct distance estimators (Fouqué *et al.* 2007), thereby reducing the slope error.

21.5 The future of parallax determinations

Again, as an example, additonal geometrical parallaxes for Cepheids will come from the space-based, all-sky astrometry mission Gaia (Mignard 2005) with $\sim$20 microsecond of arc precision parallaxes. Particularly important would be additional parallaxes of long-period Cepheids, those most often used in establishing Leavitt laws (hence, distances) for external galaxies. Final results are expected by the end of this decade.

Ground-based parallax work will likely increase by orders of magnitude as wide-field, high-resolution camera/telescope combinations (e.g. PAN-STARRS, Magnier *et al.* 2008) come on line, flooding our community with hundreds of observations per year of most of the sky. Parallax is a natural byproduct of any such series of observations. Lastly, any object sufficiently puzzling will require a parallax. In the coming age of giant telescopes, the faint limit at which parallax work can be done will open up the sky to many magnitudes below our present knowledge.

References

Anderson, J., Bedin, L. R., Piotto, G., Yadav, R. S., and Bellini, A. (2006). Ground-based CCD astrometry with wide field imagers. I. Observations just a few years apart allow decontamination of field objects from members in two globular clusters. *A&A*, **454**, 1029.

Auer, L. H. and van Altena, W. F. (1978). Digital image centering. II. *AJ*, **83**, 531.

Barnes, T. G., Jefferys, W. H., Berger, J. O., *et al.* (2003). A Bayesian analysis of the Cepheid distance scale. *ApJ*, **592**, 539.

Benedict, G. F., McArthur, B. E., Chappell, D. W., *et al.* (1999). Interferometric astrometry of Proxima Centauri and Barnard's Star using Hubble Space Telescope Fine Guidance Sensor 3: Detection limits for substellar companions. *AJ*, **118**, 1086.

Benedict, G. F., McArthur, B. E., Franz, O. G., *et al.* (2000). Interferometric astrometry of the detached white dwarf-M dwarf binary Feige 24 using HST Fine Guidance Sensor 3: white dwarf radius and component mass estimates. *AJ*, **119**, 2382.

Benedict, G. F., McArthur, B. E., Fredrick, L. W., *et al.* (2002a). Astrometry with the Hubble Space Telescope: a parallax of the fundamental distance calibrator RR Lyrae. *AJ*, **123**, 473.

Benedict, G. F., McArthur, B. E., Fredrick, L. W., *et al.* (2002b). Astrometry with the Hubble Space Telescope: a parallax of the fundamental distance calibrator δ Cephei. *AJ*, **124**, 1695.

Benedict, G. F., McArthur, B. E., Fredrick, L. W., *et al.* (2003). Astrometry with the Hubble Space Telescope: a parallax of the central star of the planetary nebula NGC 6853. *AJ*, **126**, 2549.

Benedict, G. F., McArthur, B. E., Feast, M. W., *et al.* (2007). Hubble Space Telescope Fine Guidance Sensor parallaxes of Galactic Cepheid variable stars: period–luminosity relations. *AJ*, **133**, 1810.

Benedict, G. F., McArthur, B. E., Nopiwotzki, R., *et al.* (2009). Astrometry with the Hubble Space Telescope: trigonometric parallaxes of planetary nebula nuclei NGC 6853, NGC 7293, Abell 31, and DeHt 5. *AJ*, **138**, 1969.

Bertin, E. and Tissier, G. (2007). VOTables in TERAPIX Software. *Astronomical Data Analysis Software and Systems XVI*, **376**, 507.

Bessel, F. W. (1838). On the parallax of 61 Cygni. *MNRAS*, **4**, 152.

Bessell, M. S. and Brett, J. M. (1988). JHKLM photometry – standard systems, passbands, and intrinsic colors. *PASP*, **100**, 1134.

Booth, M. and Schlesinger, F. (1922). The parallaxes of fifty-seven stars. *AJ*, **34**, 31.

Cox, A. N., ed. (2000). *Allen's Astrophysical Quantities*, 4th edn. New York, NY: AIP Press, Springer.

Fouqué, P., Arriagada, P., Storm, J., *et al.* (2007). A new calibration of Galactic Cepheid period–luminosity relations from B to K bands, and a comparison to LMC relations. *A&A*, **476**, 73.

Gizis, J. E., Jao, W.-C., Subasavage, J. P., and Henry, T. J. (2007). The trigonometric parallax of the brown dwarf planetary system 2MASSW J1207334-393254. *ApJ*, **669**, L45.

Gould, A. and Morgan, C. W. (2003). Transit target selection using reduced proper motions. *ApJ*, **585**, 1056.

Hanson, R. B. (1979). A practical method to improve luminosity calibrations from trigonometric parallaxes. *MNRAS*, **186**, 875.

Harris, H. C., Dahn, C. C., Canzian, B., *et al.* (2007). Trigonometric parallaxes of central stars of planetary nebulae. *AJ*, **133**, 631.

Harrison, T. E., McNamara, B. J., Szkody, P., *et al.* (1999). Hubble Space Telescope Fine Guidance Sensor Astrometric parallaxes for three dwarf novae: SS Aurigae, SS Cygni, and U Geminorum. *ApJ*, **515**, L93.

Henry, T. J. (2004). The mass–luminosity relation from end to end. *ASP Conf. Ser.*, **318**, 159.

Henry, T. J., Jao, W.-C., Subasavage, J. P., *et al.* (2006). The solar neighborhood. XVII. Parallax results from the CTIOPI 0.9 m Program: 20 new members of the RECONS 10 parsec sample. *AJ*, **132**, 2360.

Jao, W.-C., Henry, T. J., Subasavage, J. P., *et al.* (2005). The solar neighborhood. XIII. Parallax results from the CTIOPI 0.9 Meter Program: stars with $\mu \geq 1.0''\mathrm{yr}^{-1}$. *AJ*, **129**, 1954.

Jefferys, W., Fitzpatrick, J., and McArthur, B. (1988). GaussFit – a system for least squares and robust estimation. *Celest. Mech.*, **41**, 39.

Leavitt, H. S. and Pickering, E. C. (1912). Periods of 25 variable stars in the Small Magellanic Cloud. *Harvard College Observatory Circular*, **173**, 1.

Lutz, T. E. and Kelker, D. H. (1973). On the use of trigonometric parallaxes for the calibration of luminosity systems: theory. *PASP*, **85**, 573.

Magnier, E. A., Liu, M., Monet, D. G., and Chambers, K. C. (2008). The extended solar neighborhood: precision astrometry from the Pan-STARRS 1 3π survey. *Proc. IAU Symp.*, **248**, 553.

Majewski, S. R., Ostheimer, J. C., Kunkel, W. E., and Patterson, R. J. (2000). Exploring halo substructure with giant stars. I. Survey description and calibration of the photometric search technique. *AJ*, **120**, 2550.

McArthur, B. E., Benedict, G. F., Lee, J., *et al.* (1999). Astrometry with Hubble Space Telescope Fine Guidance Sensor 3: the parallax of the cataclysmic variable RW Triangulum. *ApJ*, **520**, L59.

McArthur, B. E., Benedict, G. F., Lee, J., *et al.* (2001). Interferometric astrometry with Hubble Space Telescope Fine Guidance Sensor 3: the parallax of the cataclysmic variable TV Columbae. *ApJ*, **560**, 907.

McArthur, B., Benedict, G. F., Jefferys, W. H., and Nelan, E. (2002). The optical field angle distortion calibration of HST Fine Guidance Sensors 1R and 3. In *The 2002 HST Calibration Workshop*, Proceedings of a Workshop held at the Space Telescope Science Institute, Baltimore, Maryland, October 17 and 18, 2002, ed. S. Arribas, A. Koekemoer, and B. Whitmore. Baltimore, MD: Space Telescope Science Institute, p. 373.

McArthur, B., Benedict, G. F., Harrison, T. E., and van Altena, W. F. (2011). Astrometry with the Hubble Space Telescope: trigonometric parallaxes of selected hyads. *AJ*, **141**, 172.

Mignard, F. (2005). The Gaia mission: science highlights. *ASP Conf. Ser.*, **338**, 15.

Murray, C. A. (1983). *Vectorial Astrometry*. Bristol: Adam Hilgar.

Nelan, E. P. (2011). *Fine Guidance Sensor Instrument Handbook*, version 19.0. Baltimore, MD: Space Telescope Science Institute.

Perryman, M. A. C., Lindegren, L., Kovalevsky, J., *et al.* (1997). The HIPPARCOS Catalogue. *A&A*, **323**, L49.

Reid, M. J., Menten, K. M., Brunthaler, A., *et al.* (2009). Trigonometric parallaxes of massive star-forming regions. I. S 252 & G232.6+1.0. *ApJ*, **693**, 397.

Roeser, S., Demleitner, M., and Schibach, E. (2010). The PPMXL Catalog of Positions and Proper Motions on the ICRS. Combining USNO-B1.0 and the Two Micron All Sky Survey (2MASS). *AJ*, **139**, 2440.

Savage, B. D. and Mathis, J. S. (1979). Observed properties of interstellar dust. *ARA&A*, **17**, 73.

Schlegel, D. J., Finkbeiner, D. P., and Davis, M. (1998). Maps of dust infrared emission for use in estimation of reddening and cosmic microwave background radiation foregrounds. *ApJ*, **500**, 525.

Soderblom, D. R., Nelan, E., Benedict, G. F., *et al.* (2005). Confirmation of errors in Hipparcos parallaxes from Hubble Space Telescope Fine Guidance Sensor astrometry of the Pleiades. *AJ*, **129**, 1616.

Standish, E. M., Jr. (1990). The observational basis for JPL's DE 200, the planetary ephemerides of the Astronomical Almanac. *A&A*, **233**, 252.

Stetson, P. B. (1987). DAOPHOT – a computer program for crowded-field stellar photometry. *PASP*, **99**, 191.

van Altena, W. F. and Monnier, R. C. (1968). Astrometric accuracy and the flatness of glass filters. *AJ*, **73**, 649.

van Altena, W. F., Lee, J. T., and Hoffleit, E. D. (1995). *Yale Parallax Catalog*, 4th edn. New Haven, CT: Yale University Observatory (YPC95).

van Leeuwen, F. (2007). Validation of the new Hipparcos reduction. *A&A*, **474**, 653.

Yong, D. and Lambert, D. L. (2003). Finding cool subdwarfs using a $V - J$ reduced proper-motion diagram: stellar parameters for 91 candidates. *PASP*, **115**, 796.

Zacharias, N., Urban, S. E., Zacharias, M. I., *et al.* (2004). The Second US Naval Observatory CCD Astrograph Catalog (UCAC2). *AJ*, **127**, 3043.

PART V

APPLICATIONS OF ASTROMETRY TO TOPICS IN ASTROPHYSICS

22

Galactic structure astrometry

RENÉ A. MÉNDEZ

Introduction

As discussed in detail in Chapter 2, two emblematic (space-based) astrometric missions to be launched in the near future, namely, Gaia[1] and SIM Lite,[2] will have a profound impact in many branches of astrophysics, including Galactic structure and stellar populations. It is therefore interesting to mention, first, some of the expected capabilities of these astrometric missions. In a nutshell, SIM Lite (recently deferred) will be an optical interferometer operating in an Earth-trailing solar orbit. The goal is to measure stellar positions, trigonometric parallaxes, and proper motions down to magnitude 20 with an accuracy of 4 microarcseconds (μas). This breakthrough in capabilities will be possible because SIM will use optical interferometry. On the other hand, the primary goal of Gaia is to investigate the origin and subsequent evolution of our Galaxy through a census of more than 10^9 stars (both in our Galaxy and in nearby members of the Local Group), measuring their positions with a best accuracy of 7 μas, 24 μas at magnitude 15 with decreasing accuracy to magnitude 20. Together with stellar distances and motions, this mission will allow astronomers to build the most accurate three-dimensional map to date of the stellar constituents of our Galaxy. Gaia will also perform low-spectral-resolution and photometric measurements for all objects. With an approximate launch date on late 2013, the final catalog is expected to be released by 2022.

Both, the Gaia[3] and SIM[4] official web pages provide excellent up-to date information about what is to be expected from these missions. The reader is referred to the review papers (Gaia: Perryman *et al.* 2001, Lindegren 2010; SIM: Unwin *et al.* 2008) published by the leading science teams involved in these projects, where many of the capabilities, expected results, research opportunities, as well as the preliminary observations required in support of both missions are described in detail. On the other hand, a detailed description of the Gaia

---

[1] Originally meaning "Global Astrometric Interferometer for Astrophysics" but, as the project evolved, the interferometer concept was replaced by a new design, preserving the name of the mission.
[2] Originally meaning "Space Interferometric Mission", now known as "SIM Lite Astrometric Observatory," the "Lite" reflects revisions of the original instrumentation requirements, without seriously compromising the expected science outcome (NASA has now deferred this mission).
[3] http://www.esa.int/science-e/www/area/index.cfm?fareaid=26.
[4] http://sim.jpl.nasa.gov/.

Astrometry for Astrophysics: Methods, Models, and Applications, ed. William F. van Altena.
Published by Cambridge University Press. © Cambridge University Press 2013.

project and science case is contained in the Gaia "red-book,"[5] and the equivalent "SIM Lite Astrometric Observatory."[6] You must bear in mind that both projects are technologically very challenging, and are thus in constant evolution and re-evaluation. As a consequence, their final performance may actually be different both prior to launch, and once in orbit.

In the specific area of Galactic structure and resolved stellar populations, there is a recent in-depth review by Turon *et al.* (2008),[7] describing the broad impact that the Gaia mission will have on this branch of astronomical research, and the synergies required to make optimum usage of the Gaia data using existing or planned ground- and space-based facilities. Similarly, a specific science driver for SIM is discussed by Majewski *et al.* (2006, but see, in more general terms, the SIM Galactic science through a SIM key project led by Dr. Steven R. Majewski[8]).

In the following sections we will describe *some* of the currently open questions that these (and other) future missions will address, and the methodology of data analysis which will be used to, hopefully, solve them using astrometric observations in conjunction with ground- and space-based complimentary data. I emphasize that I will *not* do a thorough review of all possible interesting aspects of relevant Galactic research (due to space limitations), but I will rather focus on a few selected pedagogical examples that may illustrate what challenges and opportunities lie ahead, with the hope to motivate future researchers into this branch of astronomy.

22.1 Going Galactic: equatorial vs. Galactic coordinates

Proper motions are usually measured in the equatorial system. However, for comparison purposes with Galactic models, or to derive relevant Galactic parameters (see the following sections), it is necessary to compute proper motions in a coordinate system that is more appropriate for such purposes. For this we introduce a (right-handed) spherical coordinate system, usually called the "Galactic system," given by the triad of unitary vectors $(\hat{e}_r, \hat{e}_l, \hat{e}_b)$ along the (heliocentric) radial and Galactic longitude and latitude directions respectively (see Fig. 22.1, where we also introduce a Cartesian system for the velocities (U, V, W), which will be further explored in the next section).

In order to convert equatorial proper motions into "Galactic" proper motions, we note that the north celestial pole (NCP) and the north Galactic pole (NGP) do not point in the same direction in the sky (Fig. 22.2). If we observe an object S with equatorial coordinates (α, δ), and the right ascension and declination of the north Galactic pole are (α_p, δ_p), then, application of the cosine relationship to the spherical triangle S–NGP–NCP easily leads to the angular distance NGP–S through

$$\cos \chi = \sin \delta_p \sin \delta + \cos \delta_p \cos \delta \cos(\alpha - \alpha_p) \tag{22.1}$$

---

[5] Available at http://www.esa.int/esapub/br/br163/br163.pdf.

[6] Available at http://sim.jpl.nasa.gov/keyPubPapers/simBook2009.

[7] Available at http://www.stecf.org/coordination/eso-esa/galpops.php.

[8] For details, see http://www.astro.virginia.edu/rjp0i/takingmeasure/index.shtml.

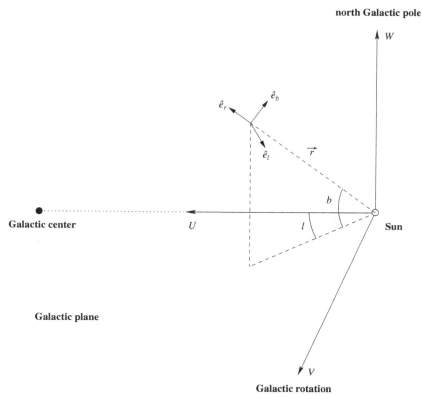

north Galactic pole

W

$\hat{e}_b$

$\hat{e}_r$

$\hat{e}_l$

$\vec{r}$

b

Galactic center U l Sun

Galactic plane

V

Galactic rotation

Fig. 22.1 Definition of the Galactic system for an object at distance r from the Sun located at Galactic longitude and latitude (l, b), given by the unitary vectors pointing in the radial (line-of-sight) direction $(\hat{e}_r)$, in the direction of increasing Galactic longitude $(\hat{e}_l)$, and increasing Galactic latitude $(\hat{e}_b)$, respectively. The heliocentric (U, V, W) system further discussed in Section 22.2 is also indicated.

This expression allows us to compute the distance χ without any quadrant ambiguity since $0 \le \chi \le 180°$ (incidentally we note that $\chi = 90° - b$, where b is the Galactic latitude of the target S).

If we now introduce the "angle of the star" η (as defined originally by Smart 1977, page 276, exercise 8[9]), Fig. 22.2 (inset) shows that

$$\begin{pmatrix} \mu_l \\ \mu_b \end{pmatrix} = \begin{pmatrix} \cos \eta & \sin \eta \\ -\sin \eta & \cos \eta \end{pmatrix} \cdot \begin{pmatrix} \mu_\alpha \\ \mu_\delta \end{pmatrix} \qquad (22.2)$$

where the sine and cosine formulae allow us to compute η without any quadrant ambiguity

$$\sin \eta = \frac{\sin \alpha - \alpha_p \cos \delta_p}{\sin \chi}$$

$$\cos \eta = \frac{\sin \delta_p - \sin \delta \, \cos \chi}{\sin \chi \, \cos \delta} \qquad (22.3)$$

[9] The first edition of this classical astrometry text book appeared in 1931.

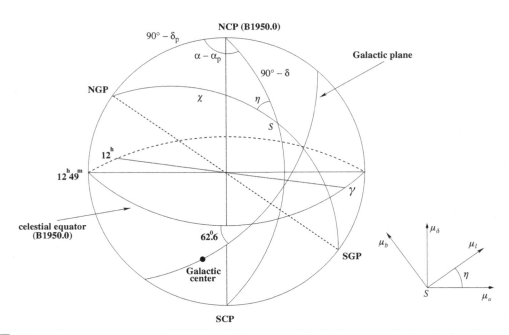

Fig. 22.2 Relationship between the equatorial (B1950.0) and Galactic systems, including the equator of both systems, and the origin of the RA system (vernal equinox, at γ). The IAU adopted values for the NGP are $(\alpha_p, \delta_p) = (12^h49^m, +27°24')$ (note that the figure has an *approximate* orientation). In the inset we show the relationship between the equatorial and Galactic proper motions (note that the proper motions in right ascension μ_α, as well as in Galactic longitude, μ_l, are in units of angular displacement (great circle) on the sky per unit time).

If we have proper motions in the B1950.0 system, then we can directly use the International Astronomical Union (IAU) adopted values for the position of the NGP (α_p, δ_p) $(12^h49^m, +27°24')$ (Blaauw *et al.* 1959, 1960; see also IAU 1959) to compute the angle of the star (Eqs. (22.1) and (22.3)), and then Eq. (22.2) to get the Galactic components (μ_l, μ_b).

The situation is however different if the proper motions are in the J2000.0 as is the case if the motions are referred to the International Celestial Reference Frame (ICRS) system. In principle, in this case, a precise calculation requires that we must first transform the coordinates and the motions to the B1950.0 (see e.g. Murray 1989 or Soma and Aoki 1990), and then use the procedure outlined above to get the Galactic proper motions. The reason for this is that the Galactic system was defined in the B1950.0, and the transformation to/from the J2000.0 system is a point-to-point transformation; we cannot go from one system to the other through e.g. a simple rigid rotation of the fundamental (equatorial) plane and the pole (as in the case of the equatorial B1950 to/from the Galactic system, see e.g. Lane 1979). When using transformations such as those presented by Soma and Aoki (1990), we need to be careful since in the old B1950.0 system the proper motions had fictitious components due to the variation of the elliptic terms of stellar aberration and the systematic proper-motion errors in the FK4 system (further discussion of this topic can be found in Perryman 1997, Section 1.5). Therefore, if we intend to analyze the intrinsic proper motions of stars, we should not include such fictitious components when converting proper motions from J2000.0 to B1950.0 (Soma, private communication). For most applications we can, however, transform proper motions directly from the J2000.0 coordinates to Galactic coordinates using the following equatorial coordinates of the Galactic pole (α_p, δ_p) $(12^h51^m26.2754^s, +27°07'41.705'')$ referred to the J2000.0 coordinate system. These values were derived from the IAU defined values for (α_p, δ_p) referred to the B1950.0 coordinate system (for details see Miyamoto and Soma 1993, especially their equation 29). For further details regarding the conversion to/from Galactic coordinates and equatorial coordinates, the reader is referred to Chapters 4, 7, and 8 (see also Johnson and Soderblom 1987).

22.2 Galactic rest frame vs. Galactocentric velocities

Different authors, and depending on the specific application, use various reference systems to study the Galactic motions. It is therefore desirable to define them and to uncover their relationships. We start from the simple equation

$$\vec{R} = \vec{R}_\Theta + \vec{r} \tag{22.4}$$

where $\vec{R}$ is a vector pointing to an object whose heliocentric position (i.e. as seen from the actual position of the Sun) is $\vec{r}$. We note here that, depending on the application, the origin chosen for $\vec{R}$ is either the instantaneous (i.e. fixed or "at rest") position of the Sun, or the Galactic center.

The vector $\vec{R}_\Theta$ represents the position of the Sun as seen from the origin chosen for $\vec{R}$. Since in Eq. (22.4) all terms could obviously be functions of time, we can write

$$\frac{d\vec{R}}{dt} = \frac{d\vec{R}_\Theta}{dt} + \frac{d\vec{r}}{dt} \tag{22.5}$$

From this basic equation we can identify various terms, usually adopted in the literature. The first term on the right-hand side is formally split into two terms, having quite different physical interpretation, namely

$$\frac{d\vec{R}_\Theta}{dt} = \vec{V}_{LSR}(\vec{R}_\Theta) + \vec{V}_\Theta \tag{22.6}$$

The first (and dominant) term on the right-hand side of Eq. (22.6) corresponds to the motion of the so-called "local standard of rest" (LSR) at the Sun's position, and represents the (pure circular) motion of an object in exact centrifugal equilibrium with the Galactic potential at the (instantaneous, i.e. present) position of the Sun. This quantity is formally equivalent to the rotation curve of the Galaxy, evaluated at the Sun's position (see Eqs. (22.14) and (22.16)), adopted as 220 km/s by the IAU (Kerr and Lynden-Bell 1986[10]). The second term

---

[10] Note that recent results, using μas-accuracy very long baseline interferometry (VLBA) astrometry, suggest, however, a larger value (Reid 2008). See also Mendez *et al.* (1999).

on the right-hand side of Eq. (22.6) is the "solar peculiar motion" and corresponds to the velocity of the Sun with respect to the LSR due to the fact that the Galactic stars have a finite velocity dispersion, and thus the Sun is not exactly in circular motion around the Galactic center.

On the other hand, in Eq. (22.5) we identify $\frac{d\vec{r}}{dt} \equiv \vec{V}_{\text{Hel}}$ as the velocity of the object as seen, or measured, from the Sun, i.e., it is its heliocentric velocity. Combining these definitions into Eq. (22.5) we thus have

$$\frac{d\vec{R}}{dt} = \vec{V}_{\text{Hel}} + \vec{V}_{\text{LSR}}(\vec{R}_\odot) + \vec{V}_\odot \tag{22.7}$$

The left-hand side of Eq. (22.7) has *two* different meanings, depending on the origin adopted for the vector $\vec{R}$. If its origin is the instantaneous position of the Sun, then the term $\frac{d\vec{R}}{dt} \equiv \vec{V}_{\text{GRF}}$ is sometimes referred to as "Sun-centered" or "Galactic rest frame" (GRF) velocity, and it represents the velocity of an object as seen from a point located at the (instantaneous) position of the Sun (*not* the Galactic center!), but at rest with respect to the Galactic center (unlike the actual Sun, which revolves around it). On the other hand, if the origin adopted for $\vec{R}$ is the Galactic center, then the term $\frac{d\vec{R}}{dt} \equiv \vec{V}_{\text{GC}}$ is the velocity as seen from an observer in the Galactic center, and it is usually termed "Galactocentric" (GC) velocity. Given the ambiguity existing in the choice for the origin of $\vec{R}$ it is important to specify what quantity we are calculating, or deriving, in a specific application. Fortunately, the transformation from one origin to another is very simple. If we use, as is customary in Galactic studies, a right-handed Cartesian coordinate system, oriented so that one of the axes points towards the Galactic center (U-axis), another axis points towards Galactic rotation (V-axis), and the last axis points towards the north Galactic pole (W-axis), then we can express $\vec{V}_{\text{GRF}} = (U, V, W)^{11}$ (see Fig. 22.1). We introduce another Cartesian system $\vec{V}_{\text{GC}} = (\Pi, \Theta, Z)$ oriented such that the Π component is parallel to the vector from the Galactic center to the Sun, and points in the direction opposite to the Galactic center, the Θ component is parallel to the Galactic plane and points in the direction of rotation of the Galactic disk, and the Z component points in the direction of the north Galactic pole, then the geometry of the situation (see Fig. 22.3) implies that $(\Pi, \Theta, Z) = (-U, V, W)$.

It is also customary to compute radial ($V_{\text{GC},r}$) and tangential ($V_{\text{GC},l}$, $V_{\text{GC},b}$) velocities as seen from the Galactic center through the (Π, Θ, Z) velocities, by introducing a (left-handed) spherical coordinate system given by the unitary vectors $(\hat{e}'_r, \hat{e}'_l, \hat{e}'_b)$ (in analogy with the corresponding vectors as seen from the Sun (see Fig. 22.1); note, however, that in that case we have a right-handed coordinate system, see Section 22.1). In this case one can easily show that

$$\begin{pmatrix} V_{\text{GC},r} \\ V_{\text{GC},l} \\ V_{\text{GC},b} \end{pmatrix} = \begin{pmatrix} \cos b_{\text{GC}} \cos l_{\text{GC}} & \cos b_{\text{GC}} \sin l_{\text{GC}} & \sin b_{\text{GC}} \\ -\sin l_{\text{GC}} & \cos l_{\text{GC}} & 0 \\ -\sin b_{\text{GC}} \cos l_{\text{GC}} & -\sin b_{\text{GC}} \sin l_{\text{GC}} & \cos b_{\text{GC}} \end{pmatrix} \cdot \begin{pmatrix} \Pi \\ \Theta \\ Z \end{pmatrix} \tag{22.8}$$

where the angles (l_{GC}, b_{GC}), equivalent to Galactic longitude and latitude, but now as seen from the Galactic center (Fig. 22.3), are related to the heliocentric distance r, and the

---

[11] For example, in this system, Dehnen and Binney (1998) have found from Hipparcos data that the solar peculiar velocity in Eq. (22.6) is $\vec{V}_\odot = (10.00 \pm 0.36, 5.25 \pm 0.62, 7.17 \pm 0.38)$ km/s.

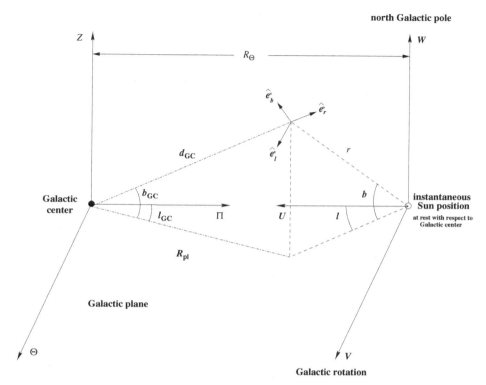

Fig. 22.3 Relationship between the Galactic rest frame velocities (U, V, W), and the Galactocentric velocities (Π, Θ, Z). We also introduce the Galactocentric radial and tangential velocities, represented by their unit vector vectors $(\hat{e}'_r, \hat{e}'_l, \hat{e}'_b)$. Note that, unlike in Fig. 22.1 (which is for a heliocentric observer), the origin of the (U, V, W) Galactic rest frame system is stationary with respect to the Galactic center, and therefore it *does not* move with the Sun *nor* with the LSR.

Galactic latitude b and longitude l, by

$$\cos l_{GC} = \frac{R_\Theta - r \cos b \cos l}{R_{pl}}, \ \sin l_{GC} = \frac{r \cos b \sin l}{R_{pl}} \tag{22.9}$$

$$\cos b_{GC} = \frac{R_{pl}}{d_{GC}}, \ \sin b_{GC} = \frac{r \sin b}{d_{GC}} \tag{22.10}$$

where R_Θ is the solar Galactocentric distance (whose value, adopted by the IAU, is 8.5 kpc; see Kerr and Lynden-Bell 1986). The parameter R_{pl} is the distance to the target projected onto the Galactic plane, whereas d_{GC} is the line-of-sight distance, both measured from the Galactic center, and given by

$$R_{pl} = \sqrt{R_\Theta^2 + r^2 \cos^2 b - 2Rr \cos b \cos l}, \ \text{and} \ d_{GC} = \sqrt{R_{pl}^2 + r^2 \sin^2 b} \tag{22.11}$$

For a specific application of these transformations, the reader is referred to Costa *et al.* (2009). In this paper, starting from (heliocentric) measurements of proper motions, Galactocentric radial and tangential velocities for the Large and Small Magellanic Cloud (LMC and SMC) satellite galaxies are computed using the above equations to compare them with theoretical predictions from dynamical models of the interaction between the Milky Way and the LMC–SMC system.

22.3 Grand-scale Galactic motions

A particular case of Eq. (22.7) refers to the case when the object in question shares some form of Galactic motion (i.e. it belongs to one of its *stellar populations*) for, in this case, in analogy (but not exactly similar!) with (22.6), it is customary to write

$$\frac{d\vec{R}}{dt} = \langle \vec{V}(\vec{R}) \rangle + \vec{V}_{\text{pec}} \tag{22.12}$$

where $\langle \vec{V}(\vec{R}) \rangle$ is the mean (pure) *rotational* velocity for a star of that Galactic population at the Galactic location $\vec{R}$, and $\vec{V}_{\text{pec}}$ is the peculiar velocity of the object considered (including e.g. radial and off-the-plane motions) with respect to its own mean Galactic rotational speed (somewhat equivalent to the Solar peculiar motion, but now for the object(s) under study). We must note that $\langle \vec{V}(\vec{R}) \rangle$ is, in general, *different* from the Galactic rotation curve, especially for large velocity dispersion (or large asymmetric drift) systems (e.g. for a halo star it would be nearly zero if the halo has no net rotation about the Galactic center). For details, connecting $\langle \vec{V}(\vec{R}) \rangle$ to Galactic dynamics and structure, the reader is referred to Mendez and van Altena (1996, especially their equation 4).

From Eqs. (22.7) and (22.12) we then conclude that the heliocentric velocities (directly related to proper motions, as we shall see in a moment) are given in the (U, V, W) reference system described in the previous section by

$$\vec{V}_{\text{Hel}} = (U_{\text{Hel}}, V_{\text{Hel}}, W_{\text{Hel}}) = \langle \vec{V}(\vec{R}) \rangle + \vec{V}_{\text{pec}} - \vec{V}_{\text{LSR}}(\vec{R}_\odot) - \vec{V}_\odot \tag{22.13}$$

Each of the terms in the right-hand side of Eq. (22.13) are given by

$$\langle \vec{V}(\vec{R}) \rangle = \begin{pmatrix} \langle V(\vec{R}) \rangle \cdot \sin l_{\text{GC}} \\ \langle V(\vec{R}) \rangle \cdot \cos l_{\text{GC}} \\ 0 \end{pmatrix} = \begin{pmatrix} \langle V(\vec{R}) \rangle \cdot \dfrac{r}{R_{\text{pl}}} \cos b \sin l \\ \langle V(\vec{R}) \rangle \cdot \dfrac{(R_\odot - r \cos b \cos l)}{R_{\text{pl}}} \\ 0 \end{pmatrix} \tag{22.14}$$

$$\vec{V}_{\text{pec}} = \begin{pmatrix} U' \cos \beta \cos \alpha + V' \sin \alpha + W' \sin \beta \cos \alpha \\ -U' \cos \beta \sin \alpha + V' \cos \alpha - W' \sin \beta \sin \alpha \\ -U' \sin \beta + W' \cos \beta \end{pmatrix} \tag{22.15}$$

$$\vec{V}_{\text{LSR}}(\vec{R}_\odot) = \begin{pmatrix} 0 \\ V_{\text{LSR}}(R_\odot) \\ 0 \end{pmatrix} \tag{22.16}$$

$$\vec{V}_\odot = \begin{pmatrix} U_\odot \\ V_\odot \\ W_\odot \end{pmatrix} \tag{22.17}$$

The velocities (U', V', W') in Eq. (22.15) are the peculiar velocities with respect to the principal axes of the so-called "velocity ellipsoid," which describes the velocity distribution of the particular Galactic component (or stellar population) under study. The angles (α, β) (as measured from the Galactic center, see Fig. 22.3) correct for the tilt of the velocity ellipsoid with respect to the local (U, V, W) system and are given by $\alpha = R_1 \cdot l_{GC}$, $\beta = R_2 \cdot b_{GC}$, where the angles l_{GC} and b_{GC} are given by Eq. (22.9) through (22.11), and the numerical factors R_1 and R_2 (in the range $0 \leq R_i \leq 1$) indicate whether there is any "vertex deviation" (in which case $R_1 \neq 1$), and whether the velocity ellipsoid has cylindrical ($R_2 = 0$) or spherical ($R_2 = 1$) symmetry.

Note that, in general, we can compute transverse motions in *each* of the components (U, V, W). These heliocentric proper motions would be given by

$$\begin{pmatrix} \mu_U \\ \mu_V \\ \mu_W \end{pmatrix} = K \cdot r^{-1} \cdot \begin{pmatrix} U_{\text{Hel}} \\ V_{\text{Hel}} \\ W_{\text{Hel}} \end{pmatrix} \tag{22.18}$$

where K is a conversion factor between the proper units for the velocities and the proper motions. For example, if the velocities are in km/s, the heliocentric distance r is in pc, and the proper motions are in arcsec/yr, then K is approximately equal to 4.74. A specific application in this case could be to compare the predictions of measured proper motions in the direction of the Galactic poles, in which case (see below) the (U, V) velocities are almost completely resolved by the measured proper motions, without the need to use radial velocities.

The classical radial (V_r) and tangential (V_l, V_b) heliocentric velocities in the (right-handed) spherical coordinate system given by the unitary vectors $(\hat{e}_r, \hat{e}_l, \hat{e}_b)$ defined in Section 22.1 (Fig. 22.1) are related to the heliocentric velocities in the (U, V, W) Cartesian system through a transformation matrix given by

$$\begin{pmatrix} V_r \\ V_l \\ V_b \end{pmatrix} = \begin{pmatrix} \cos b \cos l & \cos b \sin l & \sin b \\ -\sin l & \cos l & 0 \\ -\sin b \cos l & -\sin b \sin l & \cos b \end{pmatrix} \cdot \begin{pmatrix} U_{\text{Hel}} \\ V_{\text{Hel}} \\ W_{\text{Hel}} \end{pmatrix} \tag{22.19}$$

while the proper motions in longitude (μ_l) and latitude (μ_b) would be given by[12]

$$\begin{pmatrix} \mu_l \\ \mu_b \end{pmatrix} = K \cdot r^{-1} \cdot \begin{pmatrix} V_l \\ V_b \end{pmatrix} \tag{22.20}$$

The inverse of the transformation matrix (which corresponds to its transpose) allows us to write

$$\begin{pmatrix} U_{\text{Hel}} \\ V_{\text{Hel}} \\ W_{\text{Hel}} \end{pmatrix} = \begin{pmatrix} \cos b \cos l & -\sin l & -\sin b \cos l \\ \cos b \sin l & \cos l & -\sin b \sin l \\ \sin b & 0 & \cos b \end{pmatrix} \cdot \begin{pmatrix} V_r \\ V_l \\ V_b \end{pmatrix} \tag{22.21}$$

From this expression we note that, when $\cos b \approx 0$ (i.e. near the Galactic poles), the (U, V) proper motions (see Eq. (22.18)) are nearly independent of the radial velocity, and depend, instead, almost entirely on the proper motions in Galactic longitude and latitude (which

---

[12] Both in this equation and in Eq. (22.22) the motions are in great circle units, i.e. the μ_l already incorporates the $\cos b$ factor.

can, as shown in Section 22.1, be computed from measured proper motions in equatorial coordinates (see also Chapters 4, 7, and 8)). This fact allows the statistical analysis of large samples of stars located near the Galactic poles using proper motions alone to tackle various Galactic-structure problems (see e.g. Mendez *et al.* 2000, Girard *et al.* 2006).

22.4 Kinematics in the solar neighborhood

If the peculiar velocities of the stars and the solar peculiar motion are neglected, it can be shown (see e.g. Mihalas and Binney 1981, pages 472–476) that Eqs. (22.13) through (22.17) yield, for $r/R \ll 1$, the classical "double sine/cosine" Oort result

$$\begin{pmatrix} \mu_l \\ \mu_b \end{pmatrix} = \begin{pmatrix} \cos b \cdot (A \cdot \cos 2l + B) \\ -\dfrac{1}{2} A \cdot \sin 2l \cdot \sin 2b \end{pmatrix} \tag{22.22}$$

where the Oort constants are defined by

$$A = \frac{1}{2} \left(\frac{V(\vec{R}_\odot)}{R_\odot} - \frac{dV}{dR}\bigg|_{\vec{R}_\odot} \right)$$

$$B = -\frac{1}{2} \left(\frac{V(\vec{R}_\odot)}{R_\odot} + \frac{dV}{dR}\bigg|_{\vec{R}_\odot} \right) \tag{22.23}$$

Kerr and Lynden-Bell (1986) provide a nice summary of (obviously pre-Hipparcos) values determined for A and B in comparison with the IAU adopted values in 1964. From their Table 10 they obtain unweighted means of $A = (14.4 \pm 1.2)$km/s/kpc, and $B = (-12.0 \pm 2.8)$km/s/kpc. The Hipparcos value, obtained from a sample of 1352 early-type stars with spectroscopic (not trigonometric, since the trigonometric errors were too large) distances and proper motions from the Hipparcos Catalogue itself lead to a rather large value of $A - B = 31.6 \pm 1.4$ (Miyamoto and Zhu 1998), in comparison with the IAU adopted values ($A - B = 25$). As pointed out by Binney *et al.* (1997), the overall Hipparcos Catalogue itself is kinematically biased, so that special care has to be taken when inferring kinematic properties from it. Indeed, a kinematically unbiased subsample extracted from the Hipparcos Catalogue by Dehnen and Binney (1998), which contains about 14 000 main-sequence and giant stars with a maximum and average distance of about 100 and 80 pc, has been used to study the (very complex) phase space of positions and velocities of these nearby stars, revealing a lot of substructure (see also Famaey *et al.* 2005a,b). However, Olling and Dehnen (2003) have estimated that, for the Dehnen and Binney sample, the uncertainty in the derived Oort constants would be unacceptably large: 4 km/s/kpc. For this reason, Olling and Dehnen have instead used proper motions for about 10^6 stars from a combination of the Tycho and ACT catalog (see Chapter 21), obtaining (their Eqs. (31) and (32)) $A = (9.6 \pm 0.5)$km/s/kpc, and $B = (-11.6 \pm 0.5)$km/s/kpc from young main-sequence stars, and $A = (15.9 \pm 1.2)$km/s/kpc, and $B = (-16.9 \pm 1.2)$km/s/kpc from older red giants. The large differences found for the Oort constants using different methods and stellar samples are disappointing. It is clear that well-controlled samples over large volumes of the Galactic disk, such as those that will eventually be provided by Gaia, should help resolve this issue (see also next section).

22.5 The solar Galactocentric distance and the LSR speed

As can be seen from Eqs. (22.13) through (22.17), the key parameters in the interpretation of Galactic kinematic data are the solar Galactocentric distance and the speed of the LSR, both connected to the Oort constants A and B as determined from local kinematics (see Eq. (22.23)). There are, however, other methods to measure these Galactic constants using astrometric data. If we measure the "reflex" transverse motion of a source at (or near) the Galactic center, and assume that the source is itself stationary (small $\vec{V}_{pec}$, null $\langle \vec{V}(\vec{R}) \rangle$) with respect to the reference frame used to determine such motion (e.g. a background quasi-stellar object (QSO)), then the V-component of Eq. (22.13) leads to a (heliocentric) proper motion given by

$$\mu_{V\,Hel} = -(V_{LSR}(\vec{R}_\Theta) + V_\Theta)/R_\Theta.$$

By measuring the left-hand side of the above equation, and adopting a value for the V-component of the solar peculiar motion V_Θ (see Section 22.2), we can in turn directly measure the ratio $[V_{LSR}(\vec{R}_\Theta)/R_\Theta] = A - B$ (see Eq. (22.23)). This has most recently been done by Bedin *et al.* (2003) using the Hubble Space Telescope (HST) ($[V_{LSR}(\vec{R}_\Theta)/R_\Theta] = 27.6 \pm 1.7$km/s/kpc), and by Reid and Brunthaler (2004; see also Reid 2008) using VLBA observations ($[V_{LSR}(\vec{R}_\Theta)/R_\Theta] = 29.45 \pm 0.15$km/sec/kpc). Note the very much smaller error using VLBA on Sgr A*, and the fact that the HST result (using field stars near the globular cluster M4, and a QSO as a zero-point) is within 2σ of the radio result. The problem remains to be the decoupling of $V_{LSR}(\vec{R}_\Theta)$ from R_Θ to have an independent estimate of both quantities. This is, actually, one of the main goals of the Gaia and SIM projects, by directly measuring trigonometric parallaxes of stars towards the Galactic center, although the near-infrared astrometric survey JASMINE[13] will be more suitable for this purpose, reaching 10 µas astrometric accuracy for stars brighter than magnitude 14 in the z-band (versus Gaia's 275 µas at magnitude 20 in the visible; Lindegren *et al.* 2008, Table 1), thus providing astrometry of the Galactic center and bulge that cannot be carried out with Gaia.

22.6 From radial velocities to proper motions, and back – Part I: systemic motions

From Eq. (22.19), we see that

$$V_r = \cos b \cos l \cdot U_{Hel} + \cos b \sin l \cdot V_{Hel} + \sin b \cdot W_{Hel} \tag{22.24}$$

[13] Japan Astrometry Satellite Mission for Infrared Exploration, Yano *et al.* 2008, to be launched by 2017. For more information about JASMINE, visit www.jasmine-galaxy.org/index.html.

Therefore, if we have an extended object (e.g. a nearby galaxy) with a certain bulk heliocentric velocity (usually referred to as its "systemic motion"), $(U_{Hel}, V_{Hel}, W_{Hel})$, and assume that there are no intrinsic velocity gradients (e.g. rotation, internal streaming motion, or kinematic subgroups), then by measuring the gradient of V_r as a function of (l, b) (e.g. from the center of the Galaxy), we could in turn infer the values $(U_{Hel}, V_{Hel}, W_{Hel})$ from Eq. (22.24). In this context it is common to use, instead of the (U, V, W) system introduced before, the equatorial components of the motion. In this case, we introduce a Cartesian set of orthogonal axes (X, Y, Z) oriented such that the X-axis points towards the vernal equinox (point γ), the Y-axis towards $(\alpha, \delta) = (6^h, 0°)$ (due east), and the Z-axis towards the north Celestial pole ($\delta = +90°$; see e.g. Fig. 3.5, page 34, in Mueller 1977). In this case, the (V_X, V_Y, V_Z) (heliocentric) velocities projected onto this system are related to the radial velocities and equatorial proper motions through (compare to Eq. (22.21))

$$\begin{pmatrix} V_X \\ V_Y \\ V_Z \end{pmatrix} = \begin{pmatrix} \cos \delta \cos \alpha & -\sin \alpha & -\sin \delta \cos \alpha \\ \cos \delta \sin \alpha & \cos \alpha & -\sin \delta \sin \alpha \\ \sin \delta & 0 & \cos \delta \end{pmatrix} \cdot \begin{pmatrix} V_r \\ V_\alpha \\ V_\delta \end{pmatrix} \qquad (22.25)$$

where (compare to (22.20)):

$$\begin{pmatrix} \mu_\alpha \\ \mu_\delta \end{pmatrix} = K \cdot r^{-1} \cdot \begin{pmatrix} V_\alpha \\ V_\alpha \end{pmatrix} \qquad (22.26)$$

(Notice that in this equation[14] μ_α is measured as an angular displacement on the sky per unit time.) The direction $\hat{r}$ (unitary vector) of an object at heliocentric distance r and equatorial coordinates (α_*, δ_*) in this system is given by

$$\frac{\vec{r}}{r} = \hat{r} = (\hat{r}_X, \hat{r}_Y, \hat{r}_Z) = (\cos \alpha_* \cos \delta_*, \sin \alpha_*, \cos \delta_*, \sin \delta_*) \qquad (22.27)$$

Therefore, for a galaxy with *center of mass* velocity $(V_r, V_\alpha, V_\delta)$ located at (α, δ), the observed radial velocity of an individual star (member of the galaxy) located at (α_*, δ_*) due to the bulk spatial motion of the galaxy itself would be given by

$$V_r(\alpha_*, \delta_*) = \vec{V} \cdot \hat{r} = V_X \hat{r}_X + V_Y \hat{r}_Y + V_Z \hat{r}_Z \qquad (22.28)$$

As can be seen from Eqs. (22.25) and (22.26) this expression has contributions not only from the center of mass radial velocity of the galaxy V_r, but it also involves the proper motions in right ascension and declination (compare to Eq. (22.24)), from which these quantities can be determined.

The methodology described above requires large samples of stars (to properly deal with outliers, foreground objects, and the galaxies' own internal velocity dispersion) with measured radial velocities, which until recently were lacking even for the nearest dwarf spheroidal satellite galaxies of the Milky Way. However, with the availability of multi-object fiber systems and dedicated surveys which collect large numbers of high-precision radial velocities ($\sigma(V_r)$ few km/s), this technique has recently become feasible, and it has been used for the first time by Walker *et al.* (2008), yielding significant proper motions (with uncertainties that are only slightly larger, but of the same order, as those of dedicated

---

[14] As before, the motions are in great circle units, i.e. the μ_α already incorporates the $\cos \delta$ factor.

astrometric measurements using ground- and HST-based imaging data) for the Carina, Fornax, Sculptor, and Sextans dwarf spheroidal (dSph) galaxies. A sophisticated analysis technique has been developed by Walker *et al.* (2008) to detect and deal with possible streaming motions, non-galaxy members, etc. (see also Walker *et al.* 2009). This example illustrates how astrometry and other complementary techniques become very tightly linked in the study of the astrophysics of these nearby objects. However, we will only be able to fully study the kinematics of these galaxies, free from the uncertainties and assumptions underlying these methods, by combining the spectroscopic data with the high-accuracy proper motion expected from Gaia and SIM.

22.7   From radial velocities to proper motions, and back – Part II: perspective effects

The discussion outlined in the previous section opens up an interesting aspect of the increased accuracy that the Gaia and SIM proper motions will provide, which is the following. Since the proper motion of an object depends on its distance, and this distance changes as a function of time depending on the object's radial velocity (and parallax), we should actually expect a relationship between an object's radial velocity and its proper motion, to the point that we could actually turn the problem around and obtain radial velocities *from* proper motions (and parallaxes) alone: These are the so-called "astrometric radial velocities" (including the "perspective acceleration", described by van de Kamp 1967, ch. 9, and van de Kamp 1981, ch. 8). In recent times, this has been more extensively and systematically discussed by Dravins *et al.* (1999); see specially their Equations (1) and (4) (for an application to moving clusters see Lindegren *et al.* 2000, and also Madsen *et al.* 2002). As emphasized by these authors, reaching $\sigma(V_r) < 1$ km/s with these methods requires astrometric observations on the μas accuracy level, currently not available (except for VLBA sources), but very likely to be achieved by Gaia and SIM. A didactical derivation of the equations involved is given by Green (1985, ch. 11), where it is assumed that there is no acceleration on the target. In this formulation we start by developing the variation in the right ascension and declination coordinates of a star as a function of time from an initial instant t_0, in a series expansion of the form

$$
\begin{aligned}
\alpha(t) &= \alpha(t_0) + \mu_\alpha(t_0) \cdot (t - t_0) + \dot{\mu}_\alpha(t_0) \cdot \frac{(t - t_0)^2}{2} \\
\delta(t) &= \delta(t_0) + \mu_\delta(t_0) \cdot (t - t_0) + \dot{\mu}_\delta(t_0) \cdot \frac{(t - t_0)^2}{2}
\end{aligned}
\tag{22.29}
$$

where the first-order terms correspond to the classical proper motions in right ascension and declination, respectively. The second-order terms can be evaluated, obtaining (see e.g. Green 1985, Equation 11.17)

$$
\begin{aligned}
\dot{\mu}_\alpha(t_0) &= 2\mu_\alpha(t_0) \cdot (-\Pi V_r + \mu_\delta(t_0)\tan\delta) \\
\dot{\mu}_\delta(t_0) &= -2\mu_\delta(t_0)\Pi V_r - \mu^2{}_\alpha(t_0)\sin\delta\cos\delta
\end{aligned}
\tag{22.30}
$$

where the tangential component of the velocity V_t (km/s) is given by $V_t = K \cdot \mu/\Pi$ if $\mu = \sqrt{\mu_\alpha^2 \cos^2 \delta + \mu_\delta^2}$ is in arcsec/yr and the parallax Π is in arcsec (such that if the heliocentric distance r is in pc then $\Pi = 1/r$). It can be seen from these equations that, by measuring positions, proper motions, and their first derivative, we could in principle derive both parallaxes and radial velocities or, if parallaxes are also available, we could derive radial velocities in two independent ways. By providing the required accuracy to evaluate $\dot{\mu}_\alpha(t_0)$ and $\dot{\mu}_\delta(t_0)$, Gaia and SIM will open a fully new way to obtain kinematical information for big samples of stars, even for those for which accurate radial velocities are very difficult to obtain because of their spectral features (e.g. late-type stars with broad absorption bands, or early-type stars with very few and broad lines).

22.8   Nearby galaxies: Local Group motions

Determining the kinematics of satellite galaxies of the Milky Way is crucial to estimate the mass of our Galaxy, to understand its formation process and that of its satellites, to explain the origin of stellar streams in the Milky Way's halo that seem to be related to these satellites, and to understand the role of tidal interactions in the evolution and star-formation history of low-mass galaxies and of the halo of our Galaxy. Due to their proximity to the Sun, the Magellanic Clouds have been intensively studied in this respect. For a recent astrometric study, comparing also all high-accuracy published ground- and space-based (HST) determinations of the proper motions for both the Large and Small Magellanic Clouds, including a discussion of the prevailing uncertainties, the reader is referred to Costa *et al.* (2009, 2011). It is interesting to note that one of the few constraints on the gravitational potential of our Galaxy at large galactocentric distances comes from the velocities of its satellite galaxies and halo globular clusters (Lynden-Bell *et al.* 1983, Zaritsky *et al.* 1989; see, however, Bellazini 2004 for a different approach), via the virial theorem. Radial velocities alone determine only one component of the space velocity vector, requiring the *assumption* of a velocity ellipsoid (i.e. the shape of the orbits). This introduces an uncertainty of at least a factor of two to three into the estimated mass of the Milky Way, M (e.g. Binney and Tremaine 1987, pages 595–597). Hence, the importance of having reliable proper motions for satellite galaxies, when combined with existing radial velocities, is that this eliminates the need to make assumptions regarding the orbits. Once all components of the velocity are known, we can directly estimate the enclosed mass M(see Costa *et al.* 2009, 2011). Current accuracies imply uncertainties of about 15–30% for M.

Apart from the two Magellanic Clouds, the Galaxy possesses several other low-mass satellite galaxies. These objects have been discovered in the last few decades mostly as slight over-densities of stars on wide-angle Schmidt plates. Despite their feebleness they are of vital importance for our understanding of the structure and evolution of the Galactic halo, and presumably also of the Galaxy as a whole. These dwarf galaxies are concentrated near the major galaxies of the Local Group (see e.g. Fig. 3 of Grebel 1999), which means that they have probably had a large influence from interactions with these galactic halos.

Until relatively recently, these extragalactic objects have been beyond reach as far as the measurement of proper motions is concerned. In recent times, however, the high-angular resolution and fine pixel-scale images acquired with the HST, and the introduction of modern instruments on sites with excellent conditions (seeing typically sub-arcsec), have made it feasible to measure proper motions for at least the closest objects outside the Milky Way. In a pioneering work, Schweitzer (1996, Schweitzer *et al.* 1995, 1997) measured proper motions for the Sculptor (87 kpc, van den Bergh 1999, Table 2) and Ursa Minor (63 kpc) dSph galaxies using photographic plates spanning more than 50 years of baseline. Her measurements indicate an absolute motion of $(\mu_\alpha, \mu_\delta) = (0.71 \pm 0.22, -0.06 \pm 0.25)$ mas/yr for Sculptor, implying a tangential velocity of 220 ± 125 km/s. Because of the difficulties inherent in doing astrometry with photographic plates, and the fact that extended objects (galaxies) rather than QSOs were used in her study to define the zero-point for the absolute motions, the true systematic errors of her determination could be much larger, especially considering the fact that most of the motion is in right ascension, where guiding errors, coupled with the non-linearity of photographic plates, could introduce a magnitude-dependent spurious proper motion, which is difficult to account and correct for. Recently, Piatek *et al.* (2006) have re-measured the proper motion for Sculptor using Space Telescope Imaging Spectrograph (STIS)–HST data obtaining $(\mu_\alpha, \mu_\delta) = (0.90 \pm 0.13, 0.02 \pm 0.13)$ mas/yr on a time base of only 2 years. Even though the errors went down almost a factor of two between the studies by Schweitzer and Piatek, some of the caveats described below regarding the proper-motion value for Sculptor are also applicable to this case, and advocates for future independent measurements.

Piatek *et al.* (2002) have published a value for the proper motion of Fornax $((\mu_\alpha, \mu_\delta) = (0.49 \pm 0.13, -0.59 \pm 0.13)$ mas/yr, distance 138 kpc) based on planetary Camera (PC)–+STIS–HST data with a baseline of only 2 years from four independent measurements on three distinct quasar fields. This unprecedented astrometric precision on such a short timescale is the result of small pixel size and nearly diffraction-limited images, in combination with a sophisticated effective-point-spread function technique developed to deal with the undersampled HST data by Anderson and King (2000) (see also Chapters 10 and 17). While these results are no doubt of high internal precision, it must also be noted that doing astrometry with the severely undersampled HST images is a difficult task. A perhaps even more severe problem has been the rapid degradation of the charge transfer efficiency (CTE) on the HST cameras (now referred to as "charge transfer inefficiency", CTI). This effect has a negative impact on the astrometric accuracy with which relative astrometry can be performed between different epochs, by introducing a coordinate-dependent shift in the measured coordinates, which is also a function of the background counts, the brightness of the object being measured, and the time of the exposure in a way which is not yet well understood (see below). Furthermore, shortly after the appearance of these results, a study with new distortion coefficients for the HST was published (Anderson and King 2003) indicating that those on which Piatek *et al.*'s result relies are, perhaps, not optimal. Indeed, Piatek *et al.* (2007) revised their methodology, obtaining $(\mu_\alpha, \mu_\delta) = (0.476 \pm 0.046, -0.360 \pm 0.041)$ mas/yr using the very same observational data. Dinescu *et al.* (2004) determined the proper motion of Fornax using a combination of photographic plates and WFPC2 HST images, obtaining $(\mu_\alpha, \mu_\delta) = (0.59 \pm 0.13, -0.15 \pm 0.13)$ mas/yr. The discrepancy in μ_δ is rather large, and emphasizes the difficulties inherent in making these subtle measurements. More recently, Mendez *et al.* (2010, 2011) have re-determined the proper motion of Fornax using ground-based charge-coupled device (CCD) data exclusively, obtaining a mean value, based on five quasar fields, of $(\mu_\alpha, \mu_\delta) = (0.62 \pm 0.16, -0.53 \pm 0.15)$ mas/yr, again especially discrepant with the delta proper motion from Dinescu, but quite close to the HST value. These discrepancies show how important it is to have completely independent proper-motion determinations based on different data sets, which might help to better understand the exact cause of these differences. We also note that the adopted value for the proper motion has a large impact on the derived orbits and orbital parameters (see Table 4 of Dinescu *et al.* 2004, or Table 9, Figs. 8 and 9 in Mendez *et al.* 2011).

As for Carina (100 kpc), the only available proper motion is that by Piatek *et al.* (2003) who have used PC and STIS imaging, on a baseline of 3 years, and three epochs, to derive $(\mu_\alpha, \mu_\delta) = (0.22 \pm 0.09, 0.15 \pm 0.09)$ mas/yr, using two QSOs. We note, however, the rather large discrepancy in the proper motion for the two Carina QSOs as reported in Table 1 of Piatek *et al.* (2003), especially in μ_δ, namely 0.06 ± 0.12 and 0.25 ± 0.13 mas/yr respectively.

While HST imaging instruments undoubtedly provide a unique opportunity to measure small proper motions on very short timescales, we must be careful with issues such as those reported by Bristow *et al.* (2005, Fig. 4), which show how important (and difficult) it is to deal with CTI effects. They cite as an example the change in the derived motion for Ursa Minor originally reported by Piatek *et al.* (2005) of $(\mu_\alpha, \mu_\delta) = (-0.50 \pm 0.17, 0.22 \pm 0.16)$ mas/yr as a weighted average of two QSO fields from a set of WFPC2 and STIS images, respectively (the latter without correction for CTI effects), and the value after correcting for CTI effects on STIS. Clearly, given the precision of these measurements, the CTI has a large impact on the derived motions, but unfortunately there are a number of complications associated with a good modeling of this effect (Bristow 2004, Piatek *et al.* 2005, Bristow *et al.* 2005), which make this correction rather uncertain. Indeed, as pointed out by Bristow (2004), the spurious proper motion introduced by CTI effects may be comparable to the actual proper motion of a dSph, given the small baselines (displacements) involved. This discussion shows how Gaia and SIM, with proper-motion accuracies of a few μas per year, will provide a quantum leap in our ability to study the kinematics and dynamics of the Local Group, with impact also on a full description of the kinematics of streams associated with disrupted/accreted satellite galaxies that populate the halo of our Galaxy (Belokurov *et al.* 2007, Fig. 7) and which can actually help trace and define the shape of the dark-matter halo of the Milky Way. A glimpse of the kind of questions that could be addressed is provided by the recent high-astrometric-accuracy VLBA measurements of some satellite galaxies (Brunthaler *et al.* 2007), measurements which are, however, limited to galaxies with detected maser sources and a nearby QSO reference point.

22.9 Concluding remarks

In this chapter we have touched upon some aspects that show how Gaia, SIM, JASMINE, and other similar projects, will help measure the structure and dynamics of our Milky Way, and open a new era in this research area. Our review is by no means complete, and many interesting aspects of current and future research with these (and other) facilities have not been covered. Just to indicate a few areas of current astrometric research with direct impact on studies of Galactic structure and kinematics, and which will receive a strong boost from data provided by these satellite missions, we can mention:[15]

(1) studies of the rotation rate of the Galactic bulge and kinematics of the bar (Zoccali *et al.* 2001, Vieria *et al.* 2007, Minniti and Zoccali 2008);

(2) rotation of the Galactic halo and kinematic characterization of its substructures (Majewski 2008), and distances and ages of globular clusters (Chaboyer 2008);

(3) the properties of the Galactic disk (Sumi *et al.* 2009) and the thick disk (Bochanski *et al.* 2007);

(4) tracing of the inner and outer spiral structure (Carraro *et al.* 2007) and the properties of the system of open clusters (Chen *et al.* 2008);

(5) accurate orbits and astrometric membership for globular clusters; systematic efforts in this direction have been conducted in this area only rather recently (see e.g. papers I to V in the series "Space Velocities of Globular Clusters," the last installment of which is in Casetti-Dinescu *et al.* 2007). The search for tidal structures related to globular clusters has also become an important aspect of research in this area (Casetti-Dinescu *et al.* 2009).

Based on this listing as well as on the discussions presented in this chapter, and considering the expected performance for Gaia and SIM and other future astrometric missions, it might be fair to say that astrometry lives now in a golden age, and that, as a technique, it will be extremely useful to address some of the most challenging questions of modern astrophysics.

The late Professor Claudio Anguita was an early source of inspiration. I am especially grateful to my colleagues Professors William F. van Altena and Edgardo Costa, and Doctors Mario Pedreros, Giovanni Carraro, Martin Altmann, and Maximiliano Moyano. The original idea for Fig. 22.2 is courtesy of Professor E. Costa.

References

Anderson, J. and King, I. R. (2000). *PASP*, **112**, 1360.

Anderson, J. and King, I. R. (2003). *PASP*, **115**, 113.

Bedin, L. R., Piotto, G., King, I.R., and Anderson, J. (2003). *AJ*, **126**, 247.

Bellazini, M. (2004). *MNRAS*, **347**, 119.

Belokurov, V., Zucker, D., Evans, N. W., *et al.* (2007). *ApJ*, **654**, 897.

Binney, J. and Tremaine, S. (1987). *Galactic Dynamics*. Princeton, NJ: Princeton University Press.

---

[15] For a detailed review of many of these aspects, see Jin *et al.* (2008).

Binney, J. J., Dehnen, W., Houk, N., Murray, C. A., and Penston, M. J. (1997). In *Hipparcos Venice '97*, ed. B. Battrick (ESA Special Publication SP-402), p. 473.

Blaauw, A., Gum, C. S., Pawsey, J. L., and Westerhout, G. (1959). *MNRAS*, **119**, 422.

Blaauw, A., Gum, C. S., Pawsey, J. L., and Westerhout, G. (1960). *MNRAS*, **121**, 123.

Bochanski, J. J., Munn, J. A., Hawley, S. L., *et al.* (2007). *AJ*, **134**, 2418.

Bristow, P. (2004). *ST-ECF Instrument Science Report*, CE-STIS-ISR 2004-003.

Bristow, P., Piatek, S., and Pryor, C. (2005). *Space Telescope European Coordinating Facility Newsletter*, **38**, 12.

Brunthaler, A., Reid, M. J., Falcke, H., Henkel, C., and Menten, K. M. (2007). *A&A*, **462**, 101.

Carraro, G., Geisler, D., Villanova, S., Frinchaboy, P. M., & Majewski, S. R. (2007). *A&A*, **476**, 217.

Casetti-Dinescu, D. I., Girard, T. M., Herrera, D., *et al.* (2007). *AJ*, **134**, 195.

Casetti-Dinescu, D. I., Girard, T. M., Majewski, S. R., *et al.* (2009). *ApJL*, **701**, L29.

Chaboyer, B. (2008). *Proc. IAU Symp.*, **248**, 440.

Chen, L., Hou, J. L., Zhao, J. L., and de Grijs, R. (2008). *Proc. IAU Symp.*, **248**, 433.

Costa, E., Mendez, R. A., Pedreros, M. H., *et al.* (2009). *AJ*, **137**, 4339.

Costa, E., Mendez, R. A., Pedreros, M. H., *et al.* (2011). *AJ*, **141**, 136.

Dehnen, W. and Binney, J. J. (1998). *MNRAS*, **298**, 387.

Dinescu, D. I., Keeney, B. A., Majewski, S. R., and Girard, T. M. (2004). *AJ*, **128**, 687.

Dravins, D., Lindegren, L., and Madsen, S. (1999). *A&A*, **348**, 1040.

Famaey, B., Jorissen, A., Luri, X., *et al.* (2005a). In *The Three-Dimensional Universe with Gaia*. ESA Special Publication SP-576, p. 129.

Famaey, B., Jorissen, A., Luri, X., *et al.* (2005b). *A&A*, **430**, 165.

Girard, T. M., Korchagin, V. I., Casetti-Dinescu, D. I., *et al.* (2006). *AJ*, **132**, 1768.

Grebel, E. K. (1999). *Proc. IAU Symposium*, **192**, 17.

Green, R. M. (1985). *Spherical Astronomy*. Cambridge: Cambridge University Press.

IAU (1959). IAU Information Bulletin, No. 1, 1959, 4.

Jin, W. J., Platais, I., and Perryman, M. A. C., eds. (2008). *Proc. IAU Symp.* **248**.

Johnson, D. R. H. and Soderblom, D. R. (1987). *AJ*, **93**, 864.

Kerr, F. J. and Lynden-Bell, D. (1986). *MNRAS*, **221**, 1023.

Lane, A. P. (1979). *PASP*, **91**, 405.

Lindegren, L. (2010). *Proc. IAU Symp.*, **261**, 296.

Lindegren, L., Madsen, S., and Dravins, D. (2000). *A&A*, **356**, 1119.

Lindegren, L., Babusiaux, C., Bailer-Jones, C., *et al.* (2008). *Proc. IAU Symp.* **248**, 217.

Lynden-Bell, D., Cannon, R. D., Godwin, P. J., *et al.* (1983). *MNRAS*, **204**, 87.

Madsen, S., Dravins, D., and Lindegren, L. (2002). *A&A*, **381**, 446.

Majewski, S. R., Law, D. R., Polak, A. A., and Patterson, R. J. (2006). *ApJ*, **637**, L25.

Majewski, S. R. (2008). *Proc. IAU Symp.*, **248**, 450.

Mendez, R.A. and van Altena, W.F. (1996). *AJ*, **112**, 655.

Mendez, R. A., Platais, I., Girard, T. M., Kozhurina-Platais, V., and van Altena, W. F. (1999). *ApJ*, **524**, L39.

Mendez, R. A., Platais, I., Girard, T. M., Kozhurina-Platais, V., and van Altena, W. F. (2000). *AJ*, **119**, 813.

Mendez, R. A., Costa, E., Pedreros, M. H., *et al.* (2010). *PASP*, **122**, 853.

Mendez, R.A., Costa, E., Gallart, C., *et al.* (2011). *AJ*, **142**, 93.

Mihalas, D. and Binney, J. (1981). *Galactic Astronomy*, 2nd edition. New York, NY: W. H. Freeman and Company.

Minniti, D. and Zoccali, M. (2008). *Proc. IAU Symp.*, **245**, 323.

Miyamoto, M. and Soma, M. (1993). *AJ*, **105**, 691.

Miyamoto, M. and Zhu, Z. (1998). *AJ*, **115**, 1483.

Mueller, I. I. (1977). *Spherical and Practical Astronomy*. New York, NY: Frederick Ungar Publishing Co. (second printing).

Murray, C. A. (1989). *A&A*, **218**, 325.

Olling, R. P. and Dehnen, W. (2003). *ApJ*, **599**, 275.

Perryman, M. A. C. (1997). *The Hipparcos and Tycho Catalogues*, ESA Special Publication SP-1200, vol. 1.

Perryman, M. A. C., de Boer, K. S., Gilmore, G., *et al.* (2001). *A&A*, **369**, 339.

Piatek, S., Pryor, G., Olszewski, E. W., *et al.* (2002). *AJ*, **124**, 3198.

Piatek, S., Pryor, C., Olszewski, E. W., *et al.* (2003). *AJ*, **126**, 2346.

Piatek, S., Pryor, C., Bristow, P., *et al.* (2005). *AJ*, **130**, 95.

Piatek, S., Pryor, C., Bristow, P., *et al.* (2006). *AJ*, **131**, 1445.

Piatek, S., Pryor, C., Bristow, P., *et al.* (2007). *AJ*, **133**, 818.

Reid, M. J. (2008). *RevMexAA Conf. Ser.*, **34**, 53.

Reid, M. J. and Brunthaler, A. (2004). *ApJ*, **616**, 872.

Schweitzer, A. E. (1996). Proper motion studies of the dwarf spheroidal galaxies Sculptor and Ursa Minor. Ph.D. Thesis. Madison: University of Wisconsin.

Schweitzer, A. E., Cudworth, K. M., Majewski, S. R., and Suntzeff, N. B. (1995). *AJ*, **110**, 2747.

Schweitzer, A. E., Cudworth, K. M., and Majewski, S. R. (1997). *ASP Conf. Ser.*, **127**, 103.

Smart, W. M. (1977). *Textbook on Spherical Astronomy*. Cambridge: Cambridge University Press.

Soma, M. and Aoki, S. (1990). *A&A*, **240**, 150.

Sumi, T., Johnston, K. V., Tremaine, S., Spergel, D. N., and Majewski, S. R. (2009). *ApJ*, **699**, 215.

Turon, C. Primas, F., Binney, J., *et al.* (2008). *The Messenger*, **134**, 46.

Unwin, S. C., Shao, M., Tanner, A. M., *et al.* (2008). *PASP*, **120**, 38.

van de Kamp, P. (1967). *Principles of Astrometry*. San Francisco: W. H. Freeman & Co.

van de Kamp, P. (1981). *Stellar Paths*. Dordrecht: Reidel.

van den Bergh, S. (1999). *Proc. IAU Symp.*, **192**, 3.

Vieira, K., Casetti-Dinescu, D. I., Mendez, R. A., *et al.* (2007). *AJ*, **134**, 1432.

Walker, M. G., Mateo, M., and Olszewski, E. W. (2008). *ApJ*, **688**, L75.

Walker, M. G., Mateo, M., Olszewski, E. W., Sen, B., and Woodroofe, M. (2009). *AJ*, **137**, 3109.

Yano, T., Gouda, N., Kobayashi, Y., *et al.* (2008). *Proc. IAU Symp.*, **248**, 296.

Zaritsky, D., Olszewski, E. W., Schommer, R. A., Peterson, R. C., and Aaronson, M. (1989). *ApJ*, **345**, 759.

Zoccali, M., Renzini, A., Ortolani, S., Bica, E., and Barbuy, B. (2001). *AJ*, **121**, 2638.

23

Binary and multiple stars

ELLIOTT HORCH

Introduction

Two or more stars that are located close together in space interact gravitationally, causing deviations from linear motion as each star is accelerated. If we consider the case of two stars with a physical separation of many times the radius of either star (but still close enough to generate significant accelerations), it is sufficient to consider the stars as point masses. The equations of motion for such a system can be solved by assuming the inverse-square law of gravity and applying Newton's laws of motion. Newton's solution elegantly explained Kepler's laws of planetary motion, since one of the general solutions of motion is an ellipse with the more massive body (the Sun, in the case of the Solar System) at one focus.

Kepler's third law of planetary motion (i.e. the harmonic law) as applied to the binary-star situation can be written

$$m_1 + m_2 = \frac{a^3}{P^2} \tag{23.1}$$

where m_1 and m_2 are the masses of the two stars in solar units, a is the semi-major axis of the relative orbital ellipse in astronomical units, and P is the orbital period of the system in years. If you can only apply this formula, then it is not possible to obtain individual masses from the observables on the right-hand side, nor is the mass sum possible without an estimate of the parallax of the system (which allows for the conversion of a from an angular measure to astronomical units). Furthermore, while it is usually possible to measure the orbital period to high precision, the application of the formula is complicated by the fact that the semi-major axis, and implicitly the parallax, is raised to the third power. You must have very high-precision quantities in these cases in order to obtain a mass sum which is sufficiently precise for meaningful comparisons with theory. Fortunately, modern observational methods make such comparisons possible in increasingly many cases.

Astrometry for Astrophysics: Methods, Models, and Applications, ed. William F. van Altena.
Published by Cambridge University Press. © Cambridge University Press 2013.

23.1 A brief review of the dynamical problem

The solution to the two-body problem in classical mechanics begins with Newton's central force law of gravity:

$$\mathbf{F} = -G\frac{m_1 m_2}{|\mathbf{r}|^2}\hat{\mathbf{r}} \tag{23.2}$$

where $\mathbf{r}$ is the relative coordinate vector $\mathbf{r}_1 - \mathbf{r}_2$ and $\hat{\mathbf{r}}$ is the unit vector in the same direction. The form of this equation shows that the force is directed along a line joining the two masses. Since the forces acting on these masses are radial in the relative coordinate, there is no net torque in the absence of external forces on the system. Therefore, the angular momentum vector

$$\mathbf{L} = \mathbf{r} \times m_{\mathrm{TOT}}\mathbf{v} \tag{23.3}$$

is a conserved quantity. If K_{TOT} is defined as the total kinetic energy of the two bodies and U is the potential energy of the system, then the total energy, which may be written

$$E_{\mathrm{TOT}} = K_{\mathrm{TOT}} + U = \left(\frac{1}{2}m_1 v_1^2 + \frac{1}{2}m_2 v_2^2\right) - G\frac{m_1 m_2}{|\mathbf{r}|} \tag{23.4}$$

is also conserved. Defining the center of mass of the system as

$$\mathbf{r}_{\mathrm{CM}} = \frac{m_1}{m_1 + m_2}\mathbf{r}_1 + \frac{m_2}{m_1 + m_2}\mathbf{r}_2 \tag{23.5}$$

it is relatively easy to show that the velocity of the center of mass remains constant (and can be assumed to be zero without loss of generality) in the absence of external forces. Because the only two forces, $\mathbf{F}_{12}$ and $\mathbf{F}_{21}$, are equal and opposite according to Newton's third law of motion, then the above equation can be used to show that the velocities and accelerations of the two bodies are related as follows:

$$\mathbf{v}_2 = -\frac{m_1}{m_2}\mathbf{v}_1 \text{ and } \mathbf{a}_2 = -\frac{m_1}{m_2}\mathbf{a}_1 \tag{23.6}$$

At this point it becomes possible after a few lines of algebra to rewrite the expression for the total kinetic energy in Eq. (23.4) above so that

$$K_{\mathrm{TOT}} = \frac{1}{2}\frac{m_1 m_2}{m_1 + m_2}v_{\mathrm{TOT}}^2 \tag{23.7}$$

where

$$\mathbf{v}_{\mathrm{TOT}} = \frac{d\mathbf{r}_1}{dt} - \frac{d\mathbf{r}_2}{dt} = \frac{d\mathbf{r}}{dt} \tag{23.8}$$

and $\mathbf{r}_1$ and $\mathbf{r}_2$ are the distance vectors from the center of mass to m_1 and m_2 respectively and $\mathbf{r}$ is the relative coordinate vector. Finally, rewriting Eq. (23.4) in total:

$$E_{\mathrm{TOT}} = \frac{1}{2}\frac{m_1 m_2}{m_1 + m_2}\left(\frac{d\mathbf{r}}{dt}\right)^2 - G\frac{\frac{m_1 m_2}{m_1 + m_2}(m_1 + m_2)}{r} \tag{23.9}$$

Defining the quantity $\mu = m_1 m_2/(m_1 + m_2)$ as the reduced mass, and recognizing $m_1 + m_2$ as the total mass of the system, m_{TOT}, Eq. (23.9) can be written in final form as

$$E_{TOT} = \frac{1}{2}\mu \left(\frac{dr}{dt}\right)^2 - G\frac{\mu m_{TOT}}{r} \qquad (23.10)$$

The motion, originally written in terms of the two individual particles, is rewritten here in terms of the relative motion. It can be identified as exactly the same motion as a particle of mass μ moving in a potential well, the shape of which is determined by m_{TOT}. Since the angular momentum is conserved, the motion is confined to the plane perpendicular to $\mathbf{L}$, and polar coordinates in this plane are the most convenient coordinate system for obtaining the solutions. There are three types of solutions, depending on the total energy. If E_{TOT} is positive, then the path is unbounded and hyperbolic; if E_{TOT} is exactly zero, then the motion follows a parabolic path; and if E_{TOT} is less than zero, then the motion follows an elliptical path. The elliptical solutions predict Kepler's laws, including Eq. (23.1). For a more detailed analysis, many texts exist, e.g. Thornton and Marion (2004). Calculations involving three or more stars show that in general multiples are not stable systems, the exception being so-called "hierarchical" multiples, where each successive companion has a separation from the primary star of many times the separation of the adjacent interior companion.

23.2 Computing the apparent orbit

To approach the study of binary stars through astrometry, we can either measure the positions of the two stars relative to one another (if both stars are detected in the observations), or measure the position of the brighter of the two stars relative to a background field (if the companion star is undetected). In the former case, the system is referred to as a "visual" binary, and in the latter case an "astrometric" binary. In either case, many observations are generally made of the system over a period of time comparable to or exceeding its orbital period before a high-quality orbit can be determined.

In the case of visual binaries, where relative positions are sufficient, the standard practice is to measure the separation of the two stars in arcseconds, the orientation of the system relative to celestial coordinates, and the time of the observation. The separation is usually written as ρ, and the orientation is usually defined as follows. If we draw a line from the brighter of the two stars through the fainter (known as the primary and secondary stars), then this line will make some angle with the line drawn from the brighter star, through the north celestial pole. This angle is known as the position angle, usually written θ. The sense of the angle is defined as positive in the direction starting from north through east, so that, if the secondary star lies due east of the brighter star, for example, the position angle is said to be +90 degrees. Over time, the path traced out by the (ρ, θ) coordinate pair is that of the relative apparent orbit between the two stars. In the case of astrometric binaries, the observed quantities are defined in much the same way as described above. However, assuming that the background field is chosen and measured appropriately, the path traced out by the star is the orbit of the detected star about the center of mass combined with any proper motion of the system relative to the field. If the components are both present and highly blended, then the path traced out is the called the photocentric orbit, which is in general different from the orbit of either component. To convert from the photocentric orbit to the true orbit of one or both companions, an assumption must be made regarding the luminosity of each component.

Seven orbital parameters are needed to characterize the orbital motion of the secondary star relative to the primary on the plane of the sky. These are P, the period of the system (in days or years); a, the semi-major axis of the system, in arcseconds; i, the inclination angle relative to the plane of the sky; Ω, the position angle of the nodal point lying between 0 and 180 degrees (a node being one of two points on the orbit where the plane of the sky intersects the plane of the orbit); T, the time of periastron passage (in units matching the period, either Julian date or Besselian year); e, the eccentricity of the orbital ellipse; and ω, the angle in the plane of the true orbit between the line of nodes and the semi-major axis. The three angular quantities, i, Ω, and ω, are generally reported in degrees. Three of these elements are essentially physical parameters of the system (P, T, and e), three are geometrical parameters related to the line of sight to the system (i, ω, and Ω), and in the case of the semi-major axis a the distance is needed to convert the units from an angular measure to a physical distance.

The seven orbital parameters can be shown to be related to the observable quantities (t, ρ, and θ) according to the following equations (see e.g. Aitken 1964):

$$\mu = \frac{360°}{P} \tag{23.11}$$

$$M = \mu(t - T) = E - e\sin E \tag{23.12}$$

$$r = a(1 - e\cos E) \tag{23.13}$$

$$\tan\left(\frac{1}{2}v\right) = \frac{\sqrt{1+e}}{\sqrt{1-e}}\tan\left(\frac{1}{2}E\right) \tag{23.14}$$

$$\tan(\theta - \Omega) = \pm\tan(v + \omega)\cos i \tag{23.15}$$

$$\rho = r\frac{\cos(v + \omega)}{\cos(\theta - \Omega)} \tag{23.16}$$

In the above, five other intermediate quantities appear: μ, M, E, v, and r. The first of these is simply the average number of degrees through which the companion travels per year, and the other four are defined as the mean anomaly, the eccentric anomaly, the true anomaly, and the radius vector, respectively.

A typical approach to computing the orbit is to guess initial orbital elements and then to compute the reduced chi-squared value when comparing ephemeris predictions of the elements (computed from the above formulas) against a given sequence of observational data. The orbital parameter values are then changed and the reduced chi-squared value recomputed. Based on such guesses, a downhill simplex or other minimization routine can compute orbital parameters in an iterative way to minimize the reduced chi-squared value. This represents the best-fit orbit. One difficulty is that the chi-squared parameter space for the above system of equations is usually quite non-linear and, if the initial parameters are not close enough to the final orbit, the procedure may not converge to the global minimum of the reduced chi-squared function. For this reason, techniques have been devised to

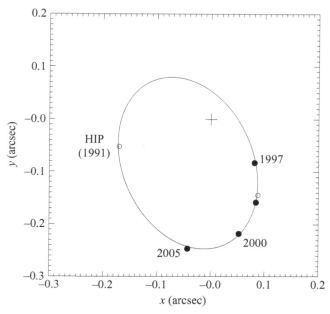

Fig. 23.1 The orbit of the binary star HDS 521 computed using a method described above. The first observation was in 1991 by Hipparcos, and subsequent observations have been obtained with ground-based techniques.

search larger sections of the parameter space prior to minimization, in order to insure proper convergence. Examples include simulated annealing (Pourbaix 1998) and various grid-searching techniques combined with an appropriate minimization method (Hartkopf *et al.* 1989). Fig. 23.1 shows an example of a binary orbit obtained using a method similar to the latter reference.

If the companion star of a binary system is not detected, it is still possible to make an estimate of the semi-major axis of the relative orbit based on a guess of the magnitude difference of the system and the mass–luminosity relation. This would most commonly be used in studying single-lined spectroscopic binary systems to determine their feasibility for studies with high-resolution imaging techniques, as described in McAlister (1976).

23.3 Observational techniques

While an abundance of data on binary stars has been collected with the important technique of visual micrometry in the last century, modern ground-based observational methods have dramatically improved the resolution and astrometric precision obtainable in binary- and multiple-star observations. They share one important aspect: they seek to overcome the limitations placed on direct imaging by the turbulence in the atmosphere above the telescope, which limits the resolution obtained to a size of the order of 1 arcsecond. This is done by compensating for the fluctuations in the atmosphere, which occur on time-scales of roughly 10 to 50 milliseconds. Above the atmosphere wavefronts coming from a stellar source are nearly planar, but after passing through the turbulent layers, they become corrugated. Bringing the light to a focus with a large telescope, we find that the effect of the turbulence is to spread out the light into a patch of typical size 1 arcsecond, while within the patch, bright and dark regions exist. The bright regions (which have a typical size set by the diffraction limit of the telescope) are called speckles.

23.3.1 Speckle interferometry

Speckle interferometry has been a very successful technique that has produced the lion's share of precise binary-star astrometry over the last 40 years. In this situation, the light from both stars comes down through nearly the same column of air above the telescope aperture. Therefore, the speckle patterns of the primary and secondary stars are nearly identical, although offset on the focal plane by an amount proportional to their separation on the sky. Because the speckles in each pattern are small compared to the overall size of the envelope of light, double speckles can be identified even when the two stars are very close together, as close as the diffraction limit.

A speckle observation consists of a series of many short exposure images of the target, where the exposure time of each frame is comparable to the timescale of atmospheric fluctuations. The standard analysis method is to examine the average autocorrelation function of the sequence of frames, where the average autocorrelation is defined by

$$\langle A_n(x, y) \rangle = \left\langle \iint dx' dy' I_n(x', y') I_n(x' + x, y' + y) \right\rangle \qquad (23.17)$$

where $I_n(x, y)$ represents the nth frame in the speckle sequence, and $\langle \ \rangle$ indicates the average over all frames in the observation. It can be shown that this function is equivalent to the histogram of the distance vectors between photon pairs so that if double speckles are present in the images, these will show up as peaks in the histogram (an excess of photons at a given separation). The averaging over many frames leads to higher signal-to-noise ratio in determining the location and height of such peaks. On the other hand, if the autocorrelation function is not computed prior to averaging and we merely average the speckle images themselves, these peaks do not build up in the same way, since the speckles occur at different places on the image plane and the average leads to a seeing-limited image of low resolution. However, the use of the autocorrelation function comes with a limitation: the symmetric nature of the integral places peaks at the positive and negative vector separation of the secondary star on the image plane. This leaves an ambiguity in the position angle of the secondary, since no distinction is made between the positions (ρ, θ) and $(\rho, \theta + 180°)$. This can be resolved with the use of more specialized correlation functions or other image products, such as the directed-vector autocorrelation (Bagnuolo *et al.* 1992), shift-and-add algorithms (Lynds *et al.* 1976), or the triple-correlation approach, also known as bispectral analysis (Lohmann *et al.* 1983).

Speckle interferometry is usually carried out with one of three different detectors: charge-coupled devices (CCDs), intensified CCDs, or electron-multiplying CCDs. In the case of

CCDs, you either have to use a device which reads out very quickly (and such devices are increasingly available), or overcome the requirement for recording many frames a second in some other way, such as with galvanometer scanning mirrors. The advantage of using a CCD is that the speckle patterns are detected on a linear device, and therefore accurate estimates of the relative photometry of the object being observed can be made in addition to high-precision astrometry. Intensified CCDs, or ICCDs, have been used over the last 20 years by the most productive speckle programs, namely the Georgia State University CHARA program and the US Naval Observatory program. These devices utilize an image intensifier to amplify the speckle pattern prior to detection by a fast-readout CCD. The advantage of this approach is that the speckles, once amplified, can easily be seen above the read noise of the CCD imager. Generally, ICCDs have a fainter limiting magnitude than bare CCDs, but, due to non-linearities in the amplification process, it is extremely difficult to obtain accurate photometry from such a device. Electron-multiplying CCDs, or EMCCDs, are a new type of detector that appears to be extremely promising for speckle imaging. The speckle pattern is detected on a CCD which has a special series of gain registers where charge is accelerated through a large voltage as it passes from one register to the next. The effect of this is to produce an amplification of the signal prior to readout as secondary electrons are dislodged. Although the properties of EMCCD speckle data are probably not known as well as for the other detectors mentioned here, it seems clear that these devices can deliver near photon-counting performance at extremely high quantum efficiency and can retrieve good relative astrometry and photometry of binary stars (Tokovinin and Cantarutti 2008; Maksimov *et al.* 2009; Docobo *et al.* 2010; Tokovinin *et al.* 2010; Horch *et al.* 2011). This may then lead to fainter objects being successfully observed in the coming years with the speckle technique. Regardless of the choice of detector, single-observation uncertainties in position with speckle-based astrometry are on the order of 1–3 milliarcseconds (mas) for a 4-m class telescope.

23.3.2 Adaptive optics

Adaptive optics systems have come into common use at large telescopes over the last 15 years. These systems seek to produce a high-resolution image at the telescope by compensating for atmospheric effects in real time. This is usually done with the use of a deformable mirror, which is sent control signals to change its surface so that it compensates for the corrugation of the incoming wavefront. After reflection, the resulting wavefront is much flatter than before and, as a result, when the light is subsequently brought to a focus, the point-spread function has much smaller width than the seeing-limited case. A key difference between adaptive optics and speckle interferometry is that long exposures can be taken before the imaging camera is read out, leading to the possibility of very faint targets being successfully observed.

The resolution and quality of the resulting images is related to the efficiency of the control loop. The control signals are generated by obtaining fast images of a bright point source (a guide star or laser point source) and then calculating from them the surface pattern needed to correct the wavefront with the deformable mirror. The result is generally diffraction-limited with good infrared systems, but obtaining truly diffraction-limited results is much harder in the visible, due to the loss of correlation (both spatially and temporally) in the speckle patterns. ten Brummelaar *et al.* (2000) have produced some photometric and astrometric results of binary stars using this technique.

23.3.3 Long baseline optical interferometry

The highest astrometric precision and resolution of binary-star observations has been achieved with interferometrically linked telescopes such as the CHARA Array, Navy Prototype Optical Interferometer, Very Large Telescope Interferometer, or the Keck Interferometer. These instruments consist of two to several telescopes, where the light from each telescope is directed towards a beam-combining facility. The phase delays between telescopes due to their relative positions is removed, and then the light is made to interfere. If the object being viewed is a binary, this will be detected as a fringe pattern that develops in the overall visibility function as different baselines are evaluated. The resolution obtainable is set by the largest baseline of the instrument, making it possible to resolve many spectroscopic binaries. This is important because the addition of the spectroscopic orbital information gives a way to determine individual masses of binary systems, not merely the mass sum, as in Eq. (23.1). Examples of this type can be found in e.g. Hummel *et al.* (2003) and Bagnuolo *et al.* (2006).

23.3.4 Space-based astrometry

Significant high-precision astrometry of binary stars has been accomplished from space in recent years. The two most important examples of this in recent years are observations with the Fine Guidance Sensor system on the Hubble Space Telescope and observations made by the Hipparcos satellite. The latter will be discussed in the next chapter. The FGS system consists of two orthogonal Koesters prisms, which act as amplitude-splitting interferometers in each of two orthogonal spatial directions, read out in each case by two phototubes. In the so-called TRANS mode of the instrument, it scans across a target in a direction x by turning a mirror within the optics system placed before the entrance face of the Koesters prism. Using the signals $A(x)$ and $B(x)$ in each arm of the interferometer, the FGS transfer function is defined as

$$T(x) = \frac{A(x) - B(x)}{A(x) + B(x)} \tag{23.18}$$

The transfer function takes value zero when the two arms of the interferometer have equal signal, and can range between -1 and 1 depending on the values of A and B. The system is designed in such a way that a small wavefront tilt in a given direction (i.e. a positional offset of the target from the center of the field of view) will result in constructive interference in one arm of the interferometer at the same time as destructive interference occurs in the other arm. For monochromatic light, we would therefore expect a series of maxima and minima in the transfer function extending away from the center of the field. In practice, the FGS system is nearly unfiltered, and the fringes die away quickly due to color effects. The detected pattern usually contains one maximum near $+1$ and one minimum near -1. For this reason, the transfer function is sometimes referred to as an "S-curve."

The S-curve has two important features relevant for high-precision astrometry. First, it is effectively a point-spread function, meaning that if the target is a double star, two copies of the S-curve will be superimposed on the image plane, with the relative heights determined by the relative brightness of the two stars and the separation related to the angular separation on the sky. For a binary star, it is sufficient to obtain two orthogonal scans of the target to determine the position angle and separation of the target, and that the relative brightness measured on both axes should in theory be the same. For more complex objects, these data would not be sufficient for uniquely reconstructing the object intensity distribution. Second, even when the primary and secondary S-curves are highly blended, the well-known nature of the S-curve permits separations and position angles of binary stars to be measured with high precision. The smallest separations that have been successfully measured by FGS are on the order of 10 mas, despite the fact that the light is collected by a 2.4-m telescope, where the diffraction limit is roughly six times larger. For a description of the FGS system, the instrument manual is most instructive (www.stsci.edu/hst/fgs/documents/instrumenthandbook/), and an important example of work done with the instrument is shown in Henry *et al.* (1999). The astrometric precision of the FGS system is approximately 1 mas under normal operating conditions.

23.4   Relating the orbit to astrophysics

The use of modern measurement techniques gives very high astrometric precision to the relative positions of the companions of binary-star systems. However, returning to Eq. (23.1), the level of precision needed to obtain mass information of sufficient precision for astrophysical studies is also high. Usually, the limitation is the uncertainty in the parallax, even for nearby systems. However, space-based missions such as Gaia give hope that within the next decade this will no longer be the case. Many more high-quality masses can therefore be expected in the coming years. Component masses, however precise, do not by themselves permit comparison with astrophysical models of stars. They must be combined with other observed parameters, such as component luminosities, effective temperatures, and metallicities. Photometry and spectroscopy are also required to take full use of the astrometric work. Because of the traditional limitations in obtaining good component magnitudes, there are many fewer systems which have been compared with stellar models than have high-quality orbit determinations. Data from Hipparcos has given good photometry of many systems in the so-called H_p filter; however, no color information was available from this mission. Some progress has been made in approaching the photometric work using adaptive optics, though this has not yielded results on a large number of systems and has been harder to calibrate than originally thought. Speckle interferometry using CCDs has yielded the largest number of relative photometry measures and has been shown to be consistent with Hipparcos magnitude differences. Several investigators have used these to obtain Hertzsprung–Russell diagram positions for the components of a few systems using spectral fitting.

One case where astrometry alone has yielded a result of great importance is that of the orbits of stars near the Galactic center. High-resolution imaging has shown this region to be populated by stars that are orbiting around a non-luminous object (Ghez *et al.* 1998; Schödel *et al.* 2002), with mass above 4 million solar masses – proof of a supermassive black hole at the center of the Galaxy.

References

Aitken, R. G. (1964). *The Binary Stars*. New York, NY: Dover Publications.

Bagnuolo, W. G., Jnr., Mason, B. D., Barry, D. J., *et al.* (1992). *AJ*, **103**, 1399.

Bagnuolo, W. G., Jnr., Taylor, S. F., McAlister, H. A., *et al.* (2006) *AJ*, **131**, 2695.

Docobo, J. A., Tamazian, V. S., Balega, Y. Y., and Melikian, N. D. (2010). *AJ*, **140**, 1078.

Ghez, A. M., Klein, B. L., Morris, M., and Becklin, E. E. (1998). *ApJ*, **509**, 678.

Hartkopf, W. I., McAlister, H. A., and Franz, O. G. (1989). *AJ*, **98**, 1014.

Henry, T. J., Franz, D. G., Wasserman, L. H., *et al.* (1999). *ApJ*, **512**, 864.

Horch, E. P., Gomez, S. C., Sherry, W. H., *et al.* (2011). *AJ*, **141**, 45.

Hummel C. A., Benson, J. A., Hutter, D. J., *et al.* (2003). *AJ*, **125**, 2630.

Lohmann, A. W., Weigelt, G. P., and Wirnitzer, B. (1983). *Appl. Opt.*, **22**, 4028.

Lynds, C. R., Worden, S. P., and Harvey, J. W. (1976). *ApJ*, **207**, 174.

Maksimov, A. F., Balega, Y. Y., Dyachenko, V. V., *et al.* (2009). *Astrophys. Bull.*, **64**, 296.

McAlister, H. A. (1976). *PASP*, **88**, 317.

Pourbaix, D. (1998). *A&AS*, **131**, 377.

Schödel, R., Ott, T., Genzel, R., *et al.* (2002). *Nature*, **419**, 694.

ten Brummelaar, T. A., Mason, B. D., McAlister, H. A., *et al.* (2000). *AJ*, **119**, 2403.

Thornton, S. T. and Marion, J. B. (2004). *Classical Dynamics of Particles and Systems*. Monterey, CA: Brooks-Cole.

Tokovinin A. and Cantarutti, R. (2008). *PASP*, **120**, 170.

Tokovinin, A., Mason, B. D., and Hartkopf, W. I. (2010). *AJ*, **139**, 743.

24

Binaries: HST, Hipparcos, and Gaia

DIMITRI POURBAIX

Introduction

Hipparcos (see ESA 1997), launched in 1989, repeatedly observed the whole sky for about 3.5 years but only measured a pre-selected sample of stars ($\sim 120\,000$). The typical positional precision at the end of mission is of the order of one milliarcsecond (mas). Although it was primarily an astrometric mission, its on-board photometer also yielded some sparse light curves of eclipsing binaries. All astrometric observations are one-dimensional measurements, whether the observations are those of a resolved or unresolved binary, or a single star. That measurement is the separation between some reference point and the target projected along the scanning direction (which changes continuously). With a primary mirror of 30 cm operating in the visible (up to $V \sim 12$ mag), its resolving power was not that impressive (about $0.5''$).

The Hubble Space Telescope (HST) is a pointing instrument launched by NASA in 1990. The main instrument for binary observation (and astrometry in general) is the Fine Guidance Sensor (FGS, see Benedict *et al.* 2008). Two out of the three FGSs lock onto guide stars while the science FGS observes the target with a single-measurement precision of 1 mas.

Gaia, yet another European Space Agency (ESA) mission, will be launched in 2013. Although it will include a spectrograph, primarily for radial-velocity determination, as well as a photometer operating in two distinct bands, Gaia will be based on the same principles as Hipparcos, namely a spinning and precessing instrument with two pointing directions with a large angular separation. Like Hipparcos, Gaia will be a whole-sky mission but, unlike its predecessor, it will not rely upon any input catalog. Instead, it will be magnitude-limited: any object brighter than 20th magnitude will be observed and analyzed.

24.1 The HST, Hipparcos, and Gaia

24.1.1 Hipparcos binaries

As already stated, Hipparcos used an input catalog (HIC) of preselected stars to be observed. Among the 120 thousand entries, about 10% were resolved binaries (or higher

Astrometry for Astrophysics: Methods, Models, and Applications, ed. William F. van Altena.
Published by Cambridge University Press. © Cambridge University Press 2013.

multiplicity) taken from a preliminary version of the CCDM by Dommanget and Nys (1994). Owing to the characteristics of the instruments, no system closer than 0.1″ nor with a difference of magnitude larger than 4.0 magnitudes was selected. For those objects, listed in the C part of the Double and Multiple Star Annex (DMSA/C) of the Hipparcos and Tycho Catalogues, the separation and angular position are given for the mean epoch only, i.e. 1991.25.

Besides the resolved systems, about 3000 putative binaries emerged from the default single-star processing, i.e. they had a poor single star fit due to the noticeable motion of their photocentre with respect to their centre of mass over the mission duration. For 90% of those systems, only some curvature of the relative motion was modeled (DMSA/G). That is typically the case of binaries with long orbital periods. For the remaining 10%, an orbital solution was derived (DMSA/O), usually with the help of an already known spectroscopic orbit.

A much less conventional class of binaries was also discovered by Hipparcos, namely the variability-induced movers (VIMs) described in Wielen (1996). They are systems with a variable component where the motion of the photocentre (along the line between the primary and secondary) is totally correlated with the variability of the system. Hipparcos discovered 288 of those (DMSA/V) but a large fraction of them were discarded later on after a flaw in the chromaticity correction of very red stars was noticed by Pourbaix *et al.* (2003). Despite the variability in the V band, and therefore in $V - I$ too, the reduction adopted a constant $V - I$ for the chromaticity correction.

Finally, there is the garbage collector (DMSA/X), i.e. about 1600 stars for which no satisfactory model could fit the motion of their photocentre. That is typically a reservoir of objects deserving further investigation, e.g. spectroscopic monitoring looking for a trend (or better) of the radial velocity. The only thing the reduction pipeline provided was an estimate of the noise level consistent with the residuals of the observations.

That was essentially the situation when the output catalog was released in 1997. It is presently rather different. Besides the Hipparcos results, ESA also released the partly pre-processed and calibrated Intermediate Astrometric Data, which are one-dimensional observations that make it possible to update the adopted model to account for new ground-based results, typically new or revised spectroscopic orbits. The benefit of such a combination is obvious: astrometry supplies the inclination that spectroscopists are desperately looking for.

Detailing all the investigations which have made use of these data would go far beyond the page limit of this introduction; however, a good description of research with Hipparcos is given by Perryman (2008). To summarize, there are those where the lack of noticeable astrometric wobble caused by the companion gives an upper limit on the mass of that companion (Perryman *et al.* 1996) and those where the wobble is indeed present, especially with the help of some external orbital information (Fekel *et al.* 2005). Other investigations rely upon the comparison of the short-term proper motion based on just 3 years of observations by Hipparcos and a long-term one, especially from Tycho-2 (Høg *et al.* 2000). Long-period binaries are indeed likely to show some discrepancy between the two proper motions (Makarov and Kaplan 2005).

24.2 HST missions

Unlike Hipparcos, HST works like a genuine ground-based instrument, with scientists competing for some observing time. The number of binary studies with HST therefore reflects the low popularity of binary-star astronomy among the committees rather than the actual capabilities of the instruments in that area. A few programs nevertheless passed through the time-allocation-committee net.

With the FGS, for separations down to 10 mas (respectively 20 mas), the two components can be resolved if the magnitude difference does not exceed 1.5 mag (respectively 4.0 mag). Here, by resolved, we mean the individual point-spread functions can be recovered despite their high overlap below the theoretical diffraction limit of 60 mas (see also Chapter 23). Besides resolving the components (Henry *et al.* 1999, Benedict *et al.* 2001), HST was also used to derive the parallax of specific binaries (Gies *et al.* 1997) like any other star.

In addition to observing binaries (Bean *et al.* 2007), HST can also target stars known to host extrasolar planets (Benedict *et al.* 2006 and Chapter 27).

24.3 Gaia missions

The content of this section is still quite speculative as Gaia has not yet been launched. From what is presently foreseen, binary investigations will proceed as with Hipparcos but will go far beyond resolved and astrometric binaries and will indeed supply its own batch of binaries. On average, 80 data points will be available, with a temporal distribution imposed by the scanning law.

Eclipsing binaries will pass through two filters before their observations are eventually fit to derive some orbital and physical parameters. The first filter will be located in the default photometric pipeline where some variability has to be noticed first. Based on the scanning law and the duration and depth of the eclipses, some eclipsing binaries might be lost in the process. The second filter classifies the variability, which must be of eclipsing type for the object to be processed by the eclipsing binary pipeline. Once again, depending on the distribution of the observations in and out of the eclipses, it is possible to assign the wrong type and prevent the object from being processed as an eclipsing binary.

From the previous paragraph, you could get the impression that most eclipsing binaries will likely be missed but that will not be the case. In addition to its photometers, Gaia will also be equipped with a radial-velocity spectrometer which will find many spectroscopic binaries (in a much narrower range of magnitude than eclipsing binaries). Despite the poor precision of the velocities with respect to current ground-based data, Gaia will detect many more spectroscopic binaries than ever before (Pourbaix *et al.* 2005). Unlike eclipsing binaries, there will be only one screening before entering the spectroscopic-binary pipeline: the variability of the radial velocity. As soon as a change in the radial velocity is noticed, the star will be processed as a spectroscopic binary.

Astrometric binaries will naturally be discovered as well. As soon as the default single-star astrometric processing yields a poor fit, the star will enter a special reduction pipeline similar to that of Hipparcos. All types of binaries will be considered: acceleration and orbital with or without the VIM effect, multiple companions with or without non-Keplerian effects, etc. In terms of numbers, most of the Gaia binaries will be resolved but without noticeable displacement over the mission lifetime. For those, as was the case for Hipparcos, only relative polar coordinates of the secondary with respect to the primary at the mean epoch will be listed. For the unresolved and a few lucky resolved ones, the modeling will go as close as possible to the full orbital solution.

Stopping here would, however, underuse the capabilities of the instrument. Instead, once a solution has been obtained, the reduction pipeline will try to combine it with other data sets. For instance, once a spectroscopic orbit has been obtained, even a poor one, the photometric data will be screened for a signature of the same companion. If successful, the two data sets will be combined and a better simultaneous solution derived, thus reducing the number of parameters to be fit. So, even if an eclipsing solution was not initially sought due to poor sampling, it might be obtained in this second stage of processing. The same will be attempted for the combined spectroscopic–astrometric solution.

24.4 Conclusion

Over the past two decades, two space missions have shown the respective advantages of pointing (HST) and survey (Hipparcos) instruments. For a scientist interested in binaries, the former is better when a specific object is subject to investigation (e.g. stars with extrasolar planets).

During the preparation of this chapter, a new potential player entered the field. If it ever flies, JMAPS (US Department of Navy) will extend the Hipparcos faint end to 14th magnitude while matching its precision (1 mas) during a 3-year mission. At the time of writing, the status of the accessibility of the data, and therefore the potential scientific outcome of the mission, is still unsettled.

The rewards of working with a survey instrument are much more difficult to estimate. No doubt, *your* favorite object will not be observed in the best way, as it would with a pointed instrument, but thousands of others, with unexpected features, will also be observed. Instead of studying one object in particular, the statistical properties of millions of systems can be studied. For instance, even if the precision of Gaia radial velocities is not the best ever, the sheer quantity of spectroscopic orbits produced by Gaia will overwhelm current catalogs of orbits by at least one order of magnitude.

References

Bean, J. L., McArthur, B. F., Benedict, G. F., *et al.* (2007). The mass of the candidate exoplanet companion to HD 33636 from Hubble Space Telescope astrometry and high-precision radial velocities. *AJ*, **134**, 749.

Benedict, G. F., McArthur, B. E., Franz, O. G., *et al.* (2001). Precise masses for Wolf 1062 AB from Hubble Space Telescope interferometric astrometry and McDonald Observatory radial velocities. *AJ*, **121**, 1607.

Benedict, G. F., McArthur, B. E., Gatewood, F., *et al.* (2006). The extrasolar planet Epsilon Eridani b: orbit and mass. *AJ*, **132**, 2206.

Benedict, G. F., McArthur, B. E., and Bean J. L. (2008). HST FGS astrometry – the value of fractional millisecond of arc precision. In *A Giant Step: from Milli- to Micro-arcsecond Astrometry*, ed. W. J. Jin, I. Platais, and M. A. C. Perryman. Cambridge: Cambridge University Press.

Dommanget, J. and Nys, O. (1994). Catalogue of the Components of Double and Multiple Stars (CCDM). First edition. *Comm. Obs. Roy. Belgique, Serie A*, **115**, 1.

ESA (1997). *The Hipparcos and Tycho Catalogues*. ESA Special Publication SP-1200.

Fekel, F., Barlow, D. J., Scarfe, C. D., *et al.* (2005). HD 166181 = V815 Herculis, a single-lined spectroscopic multiple system. *AJ*, **129**, 1001.

Gies, D. R., Mason, B. D., Bagnuolo, W. G., *et al.* (1997). The O-type binary 15 Monocerotis nears periastron. *ApJ*, **475**, L49.

Henry, T. J., *et al.* (1999). The optical mass–luminosity relation at the end of the Main Sequence (0.08-0.20 $M_\odot$). *ApJ*, **512**, 864.

Høg, E. Fabricius, C., Markarov, V. V., *et al.* (2000). The Tycho-2 catalogue of the 2.5 million brightest stars. *A&A*, **355**, L27.

Makarov, V. V. and Kaplan G. H. (2005). Statistical constraints for astrometric binaries with nonlinear motion. *AJ*, **129**, 2420.

Perryman, M. A. C. (2008). *Astronomical Applications of Astrometry: A Review Based on Ten Years of Exploitation of the Hipparcos Data*. Cambridge: Cambridge University Press.

Perryman, M. A. C., Lindegren, L., Arenou, F., *et al.* (1996). HIPPARCOS distances and mass limits for the planetary candidates: 47 Ursae Majoris, 70 Virginis, and 51 Pegasi. *A&A*, **310**, L21.

Pourbaix, D., Platais, I., Detournay, S., *et al.* (2003). How many Hipparcos variability-induced movers are genuine binaries? *A&A*, **399**, 1167.

Pourbaix, D., Knapp, G. R., Szkody, P. *et al.* (2005). Candidate spectroscopic binaries in the Sloan Digital Sky Survey. *A&A*, **444**, 643.

Wielen, R. (1996). Searching for VIMs: an astrometric method to detect the binary nature of double stars with a variable component. *A&A*, **314**, 679.

25

Star clusters

IMANTS PLATAIS

Introduction

To describe the Milky Way Galaxy, it is convenient to divide the entire collection of Galactic stars into components with broadly consistent properties (e.g. age, chemical composition, kinematics). The main idea behind this division is that those components represent stars of a loosely common origin and, thus, are an important means to understanding the formation and evolution of galaxies. Thus, the Milky Way can be "partitioned" into the four main stellar populations: the young thin and the older thick disks, the old and metal-poor halo, and the old and metal-rich bulge. Each of these populations has its own characteristics such as the size, shape, stellar density distribution, and the internal velocity distribution. The latter can be as low as $\sim$15 km/s in one coordinate (for the thin disk) and as high as $\sim$100 km/s (for the halo). It is important to realize that along any direction in our Galaxy, there is always a juxtaposition of these populations, which can be described only in a statistical sense.

Star clusters are another distinct population of objects permeating the entire Galaxy. Our Milky Way Galaxy hosts a large number of recognized open ($\sim$1800) and globular ($\sim$160) clusters that are extremely valuable tracers of the main four Galactic populations. These stellar systems are gravitationally bound and this property along with virtually no dispersion in metallicity and age within an individual system sets them apart from the other Galactic stellar populations. It is not surprising that star clusters are commonly called "laboratories" – as stellar evolution theory is firmly rooted in the star clusters. For the sake of astrometry, there is no principal difference between the open and globular clusters, although the latter are more amenable to kinematic and dynamical studies owing to their high masses. The richest globular clusters contain up 10^6 solar masses ($M_\odot$), while the poorest open clusters may contain only $\sim$100 $M_\odot$.

For a star cluster, its parallax (ϖ), position on the sky (α, δ), and proper motion ($\mu_{\alpha*}$, μ_δ) constitute the five fundamental astrometric parameters. Because these parameters can only be measured for individual stars, they must be somehow averaged to derive their mean values characterizing the entire cluster. This requires specialized group-solution techniques discussed by van Leeuwen and Evans (1998). In practice, a rigorous five-parameter solution for stars has been achieved only in the Hipparcos Catalogue for a limited but

Astrometry for Astrophysics: Methods, Models, and Applications, ed. William F. van Altena.
Published by Cambridge University Press. © Cambridge University Press 2013.

carefully selected sample of stars. The underlying principle of global space astrometry is measuring precise angular distances between groups of stars separated by many degrees (e.g. $58°$ for Hipparcos) over the span of a few years. A revolutionary expansion of five-parameter solutions in terms of the number of stars and accuracy of measurements is expected with the Gaia space astrometry mission, planned for launch in 2013. Measuring long arcs from the ground with accuracies better than ~ 100 milliarcseconds (mas) is practically impossible in optical astrometry. However, ground-based and also space-borne imaging astrometry is efficient over a restricted area of the sky (on the order of $1°$, depending on the detector size) and relative to anonymous or specially selected reference stars and extragalactic objects such as quasi-stellar objects (QSOs) and compact galaxies. In most cases, the output of such measurements are proper motions, permitted by repeated observations over at least a few years.

The knowledge of accurate positions, parallaxes, proper motions, and radial velocities provide opportunities for studies of the kinematics and dynamics of star clusters. The principal areas, where the knowledge of proper motions is key for stellar clusters, are the following: (1) cluster membership; (2) Galactic orbits of clusters; (3) internal kinematics and dynamics of clusters; (4) kinematic distances of clusters.

25.1 Cluster membership

Star clusters have been discovered as projected star-density enhancements on the sky. An imaginary circle can be drawn on the sky around such an enhancement and all stars within the circle considered as cluster members. This simplistic approach works reasonably well for rich globular clusters. However, it frequently fails in the case of open clusters, especially with poorly populated clusters at fainter magnitudes. The main contributor to this failure are the Galactic thin disk stars, which lie along the same line of sight as an open cluster being studied. Both disk stars and the members of open clusters are, in general, quite similar with the notable exception of their respective luminosity functions. On the other hand, the gravitational field of a cluster shapes its velocity field in such a way that there is always a small dispersion about the cluster's mean velocity in both radial-velocity and proper-motion spaces. The existence of this clump is the basis of cluster membership by kinematic means.

Suppose we have measured proper motions of stars with reasonable precision within some area around the star cluster (Fig. 25.1). Following Balaguer-Núñez et al. (1998), the general frequency function for an i-th cluster member can be written

$$\Phi_c^\mu = \frac{\exp\left\{-\dfrac{1}{2}\left[\dfrac{(\mu_{x_i} - \mu_{x_c})^2}{\sigma_{x_c}^2 + \epsilon_{x_i}^2} + \dfrac{(\mu_{y_i} - \mu_{y_c})^2}{\sigma_{y_c}^2 + \epsilon_{y_i}^2}\right]\right\}}{2\pi(\sigma_{x_c}^2 + \epsilon_{x_i}^2)^{1/2}(\sigma_{y_c}^2 + \epsilon_{y_i}^2)^{1/2}} \tag{25.1}$$

where (μ_{x_i}, μ_{y_i}) is the proper motion of the i-th cluster member along the x and y axes (normally, aligned to right-ascension and declination); $(\epsilon_{x_i}, \epsilon_{y_i})$ stand for the proper-motion

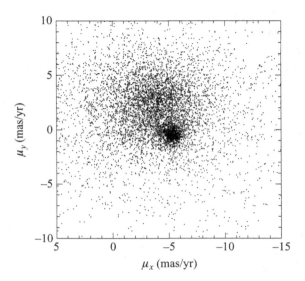

Fig. 25.1 A vector-point diagram of NGC 188 (Platais *et al.* 2003). The tight clump at $\mu_x = -5.3$ and $\mu_y = -0.4$ mas/yr is the cluster location in the vector point diagram, while the broad distribution of field stars is centered at $\mu_x = -4$ and $\mu_y = +3$ mas/yr and covers the whole figure.

error of the i-th star; (μ_{x_c}, μ_{y_c}) is the cluster's proper motion and $(\sigma_{x_c}, \sigma_{y_c})$ is the cluster's intrinsic proper-motion dispersion.

A similar field-star frequency function for the i-th star is

$$\Phi_f^\mu = \frac{\exp\left\{-\dfrac{1}{2(1-\gamma^2)} \cdot \left[\dfrac{(\mu_{x_i} - \mu_{x_f})^2}{\sigma_{x_f}^2 + \epsilon_{x_i}^2} - \dfrac{2\gamma(\mu_{x_i} - \mu_{x_f})(\mu_{y_i} - \mu_{y_f})}{(\sigma_{x_f}^2 + \epsilon_{x_i}^2)^{1/2}(\sigma_{y_f}^2 + \epsilon_{y_i}^2)^{1/2}} + \dfrac{(\mu_{y_i} - \mu_{y_f})^2}{\sigma_{y_f}^2 + \epsilon_{y_i}^2}\right]\right\}}{2\pi(1-\gamma^2)^{1/2}(\sigma_{x_f}^2 + \epsilon_{x_i}^2)^{1/2}(\sigma_{y_f}^2 + \epsilon_{y_i}^2)^{1/2}}$$

(25.2)

where (μ_{x_f}, μ_{y_f}) is the mean proper motion (center) of field stars and $(\sigma_{x_f}, \sigma_{y_f})$ is the intrinsic dispersion of field-star proper motions. The correlation coefficient γ is necessary to add because the field-star frequency function is never circular, hence

$$\gamma = \frac{(\mu_{x_i} - \mu_{x_f})(\mu_{y_i} - \mu_{y_f})}{\sigma_{x_f}\sigma_{y_f}}$$

(25.3)

To be complete, we should also consider the star spatial surface density on the sky as a function of distance from the cluster center. For cluster stars this function can be approximated with a Gaussian

$$\Phi_c^r = \frac{\exp\left\{-\dfrac{1}{2}\left[\left(\dfrac{x_i - x_c}{r_c}\right)^2 + \left(\dfrac{y_i - y_c}{r_c}\right)^2\right]\right\}}{2\pi r_c^2}$$

(25.4)

where r_c is the characteristic radius of cluster; (x_i, y_i) are tangential coordinates of the i-th cluster member, and (x_c, y_c) is the cluster center in the same coordinate system. For the field, a uniform distribution is adopted

$$\Phi_f^r = \frac{1}{\pi r^2} \tag{25.5}$$

The total distribution Φ of all stars can then be written

$$\Phi = n_c \cdot \Phi_c^\mu \cdot \Phi_c^r + n_f \cdot \Phi_f^\mu \cdot \Phi_f^r = \Phi_c + \Phi_f \tag{25.6}$$

where n_c and n_f are the normalized total numbers of field stars and cluster members ($n_c + n_f = 1$). There are several unknown parameters for the cluster ($n_c, x_c, y_c, r_c, \mu_{x_c}, \mu_{y_c}, \sigma_{x_c}, \sigma_{y_c}$) and field ($n_f, \mu_{x_f}, \mu_{y_f}, \sigma_{x_f}, \sigma_{y_f}, \gamma$) distributions, which should be solved for or found prior to solution for the remaining parameters. One way to obtain the solution is using the method of maximum likelihood

$$\sum_{i=1}^{N} \frac{\partial \ln \Phi(\mu_{x_i}, \mu_{y_i}, r_i)}{\partial u_j} = 0, \ j = 1, 2, 3, ..., 13 \tag{25.7}$$

where u_j is the unknown j-th parameter. For details of the maximum-likelihood method, refer to Sanders (1971). Once the distribution parameters are derived, the membership probability of the i-th star to be a cluster member is defined

$$P_\mu = \frac{\Phi_c(i)}{\Phi_c(i) + \Phi_f(i)} \tag{25.8}$$

While providing a general outline for calculating membership probabilities, we have ignored a possible dependence of distribution parameters on apparent magnitude.

The expressions (25.1)–(25.6) are written for the general case. In practice, they can be substantially simplified if, for instance, $\sigma_c \ll \epsilon$ or $\sigma_{x_c} \simeq \sigma_{y_c}$ or $\epsilon_x \simeq \epsilon_y$. There is no universal recipe for all possible cases. Especially vulnerable in this sense are the poorly populated open clusters, for which a 13-parameter solution is not feasible. In addition to this difficulty, the calculated probability in Eq. (25.8) is biased because the luminosity functions for star clusters and field stars differ, especially at the faint end. While at the bright end the cluster stars dominate, this pattern changes towards the fainter magnitudes where field stars heavily dominate. The net result is that membership probabilities for bright cluster members are too low, whereas at fainter magnitudes the probabilities are inflated. One way to cope with this problem is to divide the entire sample of stars into the subsamples at fixed ranges of magnitude and to derive the distribution parameters for each subsample. Due to the small number of cluster members at the faint magnitudes, this may not work very well.

An alternative to a full-fledged membership calculation is the so-called local sample method (Kozhurina-Platais *et al.* 1995). The basic idea of this method is to simplify the frequency functions and for each target star define a local sample of stars with properties as close as possible to the target. For instance, in most cases it is possible for field stars to replace the complicated form in Eq. (25.2) with a very simple flat distribution in the vicinity of the cluster centroid in the vector-point diagram. We note that frequency functions should then be replaced with density functions which, unfortunately, require the binning of proper-motion data. In addition, a number of distribution parameters can be estimated prior to solution, thus significantly lowering the number of unknowns to solve for and reducing the degrees of freedom.

A concluding remark about membership probabilities is related to their interpretation. If subsamples are used, it is mandatory to find the maximum membership probability P_μ for each subsample and then define what fraction from this maximum P_μ is predominantly populated by cluster stars. The point of this exercise is to make sure that, for example, the faint cluster members are not trimmed unnecessarily, which may happen when a single P_μ limiting cutoff is applied to all stars.

25.2 Absolute proper motions and Galactic orbits

For cluster membership determination, the type of proper motion (relative or absolute) is irrelevant. An exception is the Galactic orbit calculation where the proper motions must be absolute, i.e. very distant objects such as QSOs should have no apparent motion in the tangential plane. An example of absolute proper motions is the Hipparcos Catalogue where the listed proper motions are absolute to within ~0.25 mas/yr, if we neglect the measuring errors. These proper motions were used to derive accurate mean absolute motions for 18 open clusters at distances out to 500 pc (Robichon *et al.* 1999). None of the globular clusters were measured by Hipparcos. How can we derive absolute proper motions of globular clusters and distant open clusters? The basic technique is to measure relative proper motions of cluster stars and the extremely remote background galaxies and QSOs, and then subtract from all proper motions the mean "motion" of these background objects (e.g. Dinescu *et al.* 1997).

The calculation of a cluster's Galactic orbit requires the current position of that cluster, its distance, and the three-dimensional velocity vector. While the mean radial velocity can be measured with high accuracy (to within 0.1 km/s), even a very good knowledge of the cluster's absolute proper motion cannot ensure a comparable accuracy of the linear velocities in the tangential plane, because the distance of a cluster is usually obtained by indirect methods having poorly constrained systematic errors.

The algorithm for calculating the Galactic orbit, the models of Galactic gravitational potential, and the issues of the orbit's angular momentum and total energy are all beyond the scope of this chapter. The reader is advised to consult the paper by Dinescu *et al.* (1999) for details (see also Chapter 22).

25.3 Internal velocity dispersion

Using proper motions to measure the internal velocity dispersion in star clusters is a delicate matter. While in the center of globular clusters the velocity dispersion can reach nearly 20 km s^{-1} (van de Ven *et al.* 2006), in open clusters it is an order of magnitude smaller – typically, only ~0.5 km s^{-1}. At 1 kpc this translates to σ_μ ~0.1 mas yr^{-1} of proper-motion dispersion. Detecting such a small dispersion is often below the capacity of even the best ground-based proper-motion studies. In order to obtain a reliable estimate of σ_μ, the precision of the proper-motion measures, ϵ_i, should be on the order of or smaller than the internal velocity dispersion.

The observed proper-motion dispersion of n cluster members in one coordinate is

$$\sigma_{\mu,0} = \left(\frac{1}{n-1} \sum_{i=1}^{n} \mu_i^2 \right)^{\frac{1}{2}}, \tag{25.9}$$

where μ_i represents the i-th star's proper motion relative to the mean cluster motion. To obtain the true (intrinsic) proper-motion dispersion, $\sigma_{\mu,T}$, the observed dispersion must be corrected for the errors of the proper-motion measurements

$$\sigma_{\mu,T} = \left[\sigma_{\mu,0}^2 - \frac{1}{n} \sum_{i=1}^{n} \epsilon_i^2 \right]^{\frac{1}{2}}. \tag{25.10}$$

Taking the total differential, the uncertainty in $\sigma_{\mu,T}$ can be expressed

$$\epsilon_{\sigma_{\mu,T}} = \frac{1}{\sqrt{2n}\, \sigma_{\mu,T}} \sigma_{\mu,0}^2 + \frac{1}{n\sigma_{\mu,T}} \epsilon_i \epsilon_{\epsilon_i}, \tag{25.11}$$

where ϵ_{ϵ_i} is the uncertainty in the proper-motion uncertainty estimate of the i-th star. The error of $\sigma_{\mu,T}$ is a compound of two sources of errors. The first is the statistical error due to the finite number of stars in the sample,

$$\epsilon_{\sigma_{\mu,0}} = \frac{1}{\sqrt{2n}} \sigma_{\mu,0}. \tag{25.12}$$

The form of the second component, that due to the uncertainty of the proper motion uncertainty, ϵ_{ϵ_i}, will depend on the manner in which the proper motions and proper-motion uncertainties were determined. In the simple case of μ_i being the mean of m_i independent plate-pair determinations, the second term in Equation 25.11 becomes

$$\frac{1}{n\sigma_{\mu,T}} \epsilon_i \epsilon_{\epsilon_i} = \frac{1}{n\sigma_{\mu,T}} \sum_{i=1}^{n} \sqrt{\frac{1}{2m_i}}\, \epsilon_i^2. \tag{25.13}$$

If a cluster is relatively nearby, the motion projection effects must be carefully considered, including the relative secular parallax (a star on the far side of a cluster has a smaller proper motion than that on the near side) and perspective motion towards the cluster's convergent point. A globular cluster such as ω Centauri may show a complicated structure of its internal velocity field, which should be corrected accordingly (van de Ven et al. 2006). While the expected high-precision Gaia proper motions will enable internal velocity dispersion determination for hundreds of star clusters, caution should be exercised with respect to unresolved long-period binaries which may compromise the estimate of σ_μ.

25.4 Virial theorem and kinematic distance

For spherical quasi-equilibrium systems, such as star clusters, there is a balance between the system's kinetic and potential energies. Suppose that all N stars have equal mass m and are distributed within the sphere of a radius r and moving along random orbits with the average velocity v. The total kinetic energy K of such a system is

$$K = \frac{Nmv^2}{2}$$ (25.14)

while the total gravitational energy W of this system is

$$W = -\frac{1}{2}\frac{GNm^2}{r}$$ (25.15)

and the virial theorem states that

$$2K + W = 0$$ (25.16)

Hence, the total virial mass $M = Nm$ of the system is uniquely related to r and v

$$M = \frac{2rv^2}{G}$$ (25.17)

If we measure radial velocities of stars, V_r, in an isotropic and non-rotating cluster, then the dispersion of radial velocities σ_r is related to the average internal velocity as $v^2 = 3\sigma_r^2$. In real life, our ideal star cluster with a total mass M has a characteristic radius set by the Galactic tidal forces. Omitting details (Innanen *et al.* 1983), the tidal radius of a cluster r_c is

$$r_t = \frac{2}{3}R_p \left(\frac{M}{(3+e)M_p}\right)^{\frac{1}{3}}$$ (25.18)

where R_p is the cluster's distance from the Galactic center assuming that the orbit is nearly circular, M_p is the mass of Galaxy within the R_p and e is the orbital eccentricity ($e \approx 0$) of its orbit.

An interesting consequence of an isotropic velocity distribution in a star cluster is the relationship between the kinematic distance d to the cluster, radial-velocity dispersion, and proper-motion dispersion

$$d(\text{kpc}) = \frac{\sigma_r\ (\text{km/s})}{4.74\sigma_\mu\ (\text{mas/yr})}$$ (25.19)

The anisotropy of the velocity distribution and possible rotation of a real cluster require some adjustments to the observed dispersions as shown by McLaughlin *et al.* (2006) for the example of 47 Tucanae, also known as NGC 104.

References

Balaguer-Núñez, L., Tian, K. P., and Zhao, J. L. (1998). Determination of proper motions and membership of the open clusters NGC 1817 and NGC 1807. *A&AS*, **133**, 387.

Dinescu, D. I., Girard, T. M., van Altena, W. F., *et al.* (1997). Space velocities of southern globular clusters. I. Astrometric techniques and first results. *AJ*, **114**, 1014.

Dinescu, D. I., Girard, T. M., and van Altena, W. F. (1999). Space velocities of globular clusters. III. Cluster orbits and halo substructure. *AJ*, **117**, 1792.

Innanen, K. A., Harris, W. E., and Webbink, R. F. (1983). Globular cluster orbits and the Galactic mass distribution. *AJ*, **88**, 338.

Kozhurina-Platais, V., Girard, T. M., Platais, I., *et al.* (1995). A proper-motion study of the open cluster NGC 3680. *AJ*, **109**, 672.

McLaughlin, D. E., Anderson, J., Meylan, G., *et al.* (2006). Hubble Space Telescope proper motions and stellar dynamics in the core of the globular cluster 47 Tucanae. *ApJS*, **166**, 249.

Platais, I., Kozhurina-Platais, V., Mathieu, R. D., *et al.* (2003). WIYN Open Cluster Study. XVII. Astrometry and membership to $V = 21$ in NGC 188. *AJ*, **126**, 2922.

Robichon, N., Arenou, F., Mermilliod, J.-C., and Turon, C. (1999). Open clusters with Hipparcos. I. Mean astrometric parameters. *A&A*, **345**, 471.

Sanders, W. L. (1971). An improved method for computing membership probabilities in open clusters, *A&A*, **14**, 226.

van de Ven, G., van den Bosch, R. C. E., Verolme, E. K., and de Zeeuw, P. T. (2006). The dynamical distance and intrinsic structure of the globular cluster ω Centauri. *A&A*, **445**, 513.

van Leeuwen, F. and Evans, D. W. (1998). On the use of the Hipparcos intermediate astrometric data. *A&AS*, **130**, 157.

FRANÇOIS MIGNARD

Introduction

This chapter focuses on the peculiarities of astrometry applied to Solar System objects. This is a whole subject by itself and within the scope of this book we will restrict ourselves to an overview, emphasizing the features that make Solar System observations definitely distinct from stellar astrometry.

A beginner would naively think that Solar System astrometry deals primarily, if not only, with the major planets. Actually, the field is much broader and the planets are only a very small subset of the whole subject.

The list of potential sources is conveniently broken down into large categories as follows:

- The major planets from Mercury to Neptune. They are much too big, let alone too bright, to be observed by conventional astrometric techniques (meaning to determine their positions) in a direct way, although this remark does not fully apply to transit instruments. Their true positions are eventually obtained from the positions of their faint satellites against the background stars combined with the theory of their motion. In this chapter we omit physical observations of the planets necessary to determine their rotation and establish a local reference frame on their surfaces.
- The planetary satellites, Table 26.1, comprising three subcategories:
    - (i) The small inner satellites, orbiting very close to the planetary surface and, for the giant planets, connected to their ring system.
    - (ii) The classical satellites, usually the largest, orbiting at a few planetary radii, with low inclination to the planet's equatorial plane and nearly circular orbits. These are essentially the satellites discovered before the space era.
    - (iii) The faint outer satellites, orbiting on eccentric irregular orbits, prograde or retrograde.
- The minor planets, comprising primarily the more than 400 000 asteroids lying between Mars and Jupiter, including the ∼3000 known Jupiter trojans.
- The Kuiper-belt objects (KBOs), including the Pluto–Charon system.
- The comets.

When observed with the naked eye, we do not see any difference between the observations of stellar sources and that of Solar System objects. This similarity does not hold when we

Astrometry for Astrophysics: Methods, Models, and Applications, ed. William F. van Altena.
Published by Cambridge University Press. © Cambridge University Press 2013.

Table 26.1 Number of known natural satellites as of January 1, 2011				
Method	Inner	Classical	Outer	Total
Mercury	0	0	0	0
Venus	0	0	0	0
Earth	0	1	0	1
Mars	0	2	0	2
Jupiter	4	4	56	64
Saturn	14	8	39	61
Uranus	13	5	9	27
Neptune	6	1	6	13

come to accurate astrometric observations at high resolution. In this case, the astrometry of Solar System objects differs from its stellar counterpart in several respects following from the nature of the sources observed, primarily their apparent size and their motion. Eventually, this has led to the development of specific techniques not relevant for stellar sources.

- Solar System bodies are moving significantly against the background stars. A planet, with an exposure of moderate duration (few minutes) keeping the star images fixed on the photographic plate or on a CCD frame, shows up as a little streak and not as a fuzzy circular spot.
- Many of the Solar System bodies (planets, satellites, comets, ...) are resolved by even a small telescope of a few tens of centimeters aperture. This means that the planetary images are no longer a pure diffraction pattern or seeing disk on the focal plane.
- Faint planetary inner satellites (those orbiting at a few planetary radii) must be observed in the glare of their parent planet, increasing the noise and making their detection and astrometric measurement more difficult.
- For the stars, the difference between the direction observed from the Earth (corrected for the annual aberration) and a standard direction with its origin at the Solar System barycenter is very small, less than 1 arcsec, and equal to the parallactic effect at the observation time. For Solar System sources, the barycentric or geocentric (even topocentric) directions are completely different things.
- Star positions are referred to the barycenter of the Solar System at the time the light reaches the observer and not to the position they actually have in the Galaxy at this time. In contrast, Solar System object positions are given by allowing for the time taken for the light to propagate from their location at the emission time to the observer at the reception time. (See Chapters 5, 7, 8, and 19 for computation of the apparent position.)

26.1 The purposes of Solar System astrometry

26.1.1 Solar System mapping

A very first objective is to identify the moving objects and determine their orbits in order to predict their positions in the past and into the future. This is a very similar goal to the sky mapping carried out with stellar catalogs considered as being just a list of stars with their positions on the sky. The stellar catalogs provide a description of the sky (where the stars are and how bright they are) at a particular epoch. Given the very slow true motion of the stars on the celestial sphere, this kind of description has a rather long-term validity, several years or decades, according to the accuracy requirements. Since the Solar System objects have no fixed position, a similar position catalog would just give a snapshot of the Solar System of little interest. Therefore, what most closely resembles a static stellar catalog is a set of orbital elements for every planet, satellite, and comet allowing us to compute their geometric and apparent positions at any time through a geometrical model (before Newton) or a dynamical model (after Newton).

26.1.2 Solar System dynamics

These orbital elements, or equivalently the position and velocity vector of a moving body at a particular time, are derived from the combination of astrometric observations carried out from the Earth or from a space observatory. Beyond the first approximation of Keplerian motion, the true motions are very complex with numerous periodic inequalities over a wide range of periods. Constraining the dynamical model, and therefore the underlying physics (gravitation theory, resonance trapping, shape of the central planets, tides between planets and satellites, masses of satellites, thermal properties of asteroids, etc.) can only be done by adjusting the model parameters over a large set of observations rather well distributed over time. The prediction of the existence of Neptune by Le Verrier and Adams in 1846 or the evidence of an anomalous perihelion precession of Mercury established by Le Verrier and Newcomb are typical examples of the use of Solar System astrometry to astronomy and fundamental physics. Several orbits of well-suited minor planets, like 3200 Phaethon, 1566 Icarus or 37924 2000 BD19 (highly eccentric, small semi-major axes) are carefully monitored to measure their orbital precession and check against the predicted values resulting from Einstein's gravitational theory and solar flattening. In the case of minor planets, accurate orbital elements corrected from the planetary perturbations allow members of dynamical families to be identified, which in return supports the common origin from the disruption of a parent body of all the members. The membership assignment depends critically on the quality of the reconstructed orbit and therefore on the underlying astrometry.

26.1.3 Reference frame

The dynamical theory of the Solar System is based on equations (Newton's law or general relativity) which are simple in a preferred reference frame, non-rotating with respect to distant extragalactic sources. Confronting the observations with the theory provides a way to tell whether the frame in which the positions are determined is accelerating or not. To achieve this goal we need to be sure that the impossibility of fitting a model to the observations does not result from a deficiency in the dynamical model, which is naturally hard to prove with a single category of data. However, the lack of inertiality of the reference frame would show up in the residuals with specific features and it is easy to check whether adding an acceleration or a rotation significantly decreases the residuals to a level compatible with the random errors of the observations. Therefore, the combination of astrometric observations of moving bodies with their dynamical theory permits us to construct a dynamical reference frame largely independent of the kinematical frame defined by the quasars.

26.1.4 Asteroid masses

Mass determination of the largest asteroids is only possible with the help of precise asteroid observations. During their motions it happens from time to time that two minor planets come close enough to each other to perturb their respective motions. The disturbance is a function of the smallest distance during the approach, the relative velocity, and the sum of the masses of the two bodies. Accurate observations before and after the passage allows us to reconstruct the path and to assess the sum of the masses. In most cases of interest, one of the planets is much larger (the perturber) than the other (the target), allowing us to determine the mass of the perturber. The first successful application of the method goes back to 1966 with the estimate of the mass of 4 Vesta during a close approach with 197 Arete. Less than 30 masses are known today and very few with a relative accuracy better than 10%. Space astrometry with Gaia will soon drastically change the situation thanks essentially to the higher astrometric accuracy allowing us to detect much smaller perturbations between two planets than with ground-based astrometry. Therefore, the number of efficient (meaning detectable by astrometric measurements) close approaches is much larger and smaller orbital arcs can be used to retrieve the orbital elements and the perturber's mass.

26.1.5 Planetary physics

For Solar System objects, astrometry goes far beyond positional observations as we will see later in this chapter. The accurate timing of stellar occultations (Section 26.3.3) is the most efficient way (at least with ground-based observations) to determine the size and shape of small bodies. Combined with their masses, this leads to the bulk density of the planet, a key parameter to constrain the interior of the planets and the processes leading to the formation and evolution of the different populations of the minor bodies. Correlation with photometric and spectral properties further constrains the surface optical and mineralogical properties.

26.2 Motion of Solar System objects

26.2.1 The apparent motion

Celestial mechanics is the basic tool needed to understand the absolute and relative motion of Solar System objects. They are all orbiting the Sun or a parent body, in the case of a planetary satellite or a binary asteroid. The detection of the motion during a long exposure, or between two well-separated observations, is the best way to establish that the source is a Solar System object. The apparent diameter could be an alternative method in principle, given the apparent diameter of a planet or satellite, but in a small telescope and normal seeing conditions, the vast majority of the Solar System bodies do not show a discernible disk. A large asteroid in the main belt, ~ 100 km in diameter, has an apparent size of 100 milliarcseconds (mas) at opposition, much below the seeing disk. Even a highly skilled naked-eye observer could not distinguish its image from that of a star with a large telescope.

To estimate qualitatively the geocentric motion of a minor planet a very simple model with circular and coplanar orbits is sufficient. Let a_0 and a be respectively the orbital radii of the Earth ($a_0 = 1$ AU) and the planet. The linear velocities are $v_0 \simeq 30$ km/s and $v \simeq 30(a_0/a)^{\frac{1}{2}}$ km/s. When the planet is observed close to opposition, its distance is $\rho = a - a_0$ and its velocity relative to the Earth is $v_0\left((a_0/a)^{\frac{1}{2}} - 1\right)$. For an external planet with $a > a_0$ a negative value indicates that the motion is retrograde at opposition. Finally, the magnitude of the angular motion is given by

$$\mu = \frac{v_0\left(1 - (\frac{a_0}{a})^{\frac{1}{2}}\right)}{a - a_0} \tag{26.1}$$

With $a_0 = 1$ AU and $a = 2.7$ AU, we get $\mu \sim 10$ mas/s, which can be considered typical for main-belt asteroids. It will be several times larger for near-Earth objects (NEOs; asteroids crossing the orbit of the Earth) and thirty times smaller for a KBO with $a > 50$ AU.

During the interval of a synodic period of a planet (interval of time between two consecutive oppositions) the distance to the Earth, the radial velocity, and the apparent motion vary significantly as shown in Fig. 26.1 for an orbit similar to that of Ceres. On the ground, routine observations and surveys take place near oppositions, but occultations may happen at any time and, in space, other constraints may force us to observe at quadrature, as with Hipparcos or Gaia.

26.2.2 The planetary phase

A Solar System body seen from the Earth shows a disk usually partially illuminated. The angle seen from the planet between the direction of the Sun and that of the Earth is the phase angle ϕ. The center of illumination is the subsolar point while the center of the body for an Earth observer is the subearth point. For the same simple circular model of an external planet, the phase is readily computed with (see notations in Fig 26.2),

$$\rho \cos \phi = a - a_0 \cos \psi \tag{26.2a}$$
$$\rho \sin \phi = a_0 \sin \psi \tag{26.2b}$$

from which one can eliminate the geocentric distance and get ϕ without ambiguity. The phase is zero at opposition and maximum at quadrature with $\phi_m = \sin^{-1}(a_0/a)$. For a main-belt object it reaches 20 degrees, but can be much larger for an NEO at its closest approach to the Earth. For the satellites of Jupiter and Saturn, the maximum phases are respectively

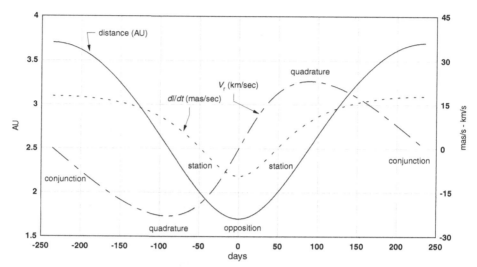

Fig. 26.1 Typical geocentric variations of the distance, radial velocity, and angular rate for a minor planet of the main belt. The plots are based on a circular orbit of 2.7 AU in radius in the plane of the ecliptic seen from the center of the Earth. The left scale gives the geocentric distance and the right scale applies to both the radial velocity (in km/s) and the angular motion (mas/s). The important phases (opposition, conjunction, ...) over a synodic cycle are also indicated.

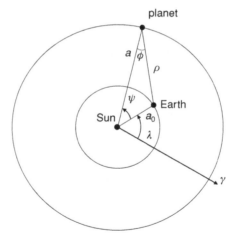

Fig. 26.2 Motion of an external planet. The phase ϕ is the angle measured from the planet between the direction of the Sun and that of the Earth.

10 and 6 degrees. For a circular disc of radius R_p the illuminated area is composed of a semicircle and the half of an ellipse of semi-minor axis $R_p \cos \phi$. Therefore, the fraction of disk illuminated is $(1 + \cos \phi)/2$. For the astrometric reduction of objects with a sizeable diameter, the planetary phase is the source of a systematic shift δ between the observed direction ($\sim$ the center of light) and the center of figure of the body which is directly related to the direction of the center of mass. The qualitative pattern is illustrated in Fig. 26.3 with

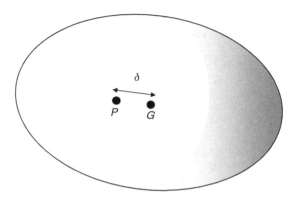

Fig. 26.3 Astrometric phase shift for an extended body. There is an offset of size δ between the center of figure G and the photocenter P.

the photocenter close to the center of the illuminated regions. The accurate computation of this shift can be very complex and depends strongly on the optical properties of the surface and on how the light is reflected and scattered as a function of the phase. Again, a simple model can give a rough approximation of the magnitude of the effect by taking the photocenter as the geometric barycenter of the illuminated area. For a body with angular diameter θ we get

$$\delta = \frac{2}{3\pi}(1 - \cos\phi)\theta \tag{26.3}$$

A more realistic approach with a Lambert light scattering leads to (Lindegren 1977)

$$\delta = \frac{3\pi}{32}\frac{(1 + \cos\phi)}{(\pi - \phi)\cot\phi + 1}\theta \tag{26.4}$$

which is approximated at small phase angle by $\delta \approx (3/16\phi)\,\theta$. For a main-belt asteroid at maximum phase, this amounts to about 6% of its apparent diameter. The detailed modeling becomes really crucial for high-precision astrometry achievable in space with Gaia or for the Galilean satellites. In the case of Gaia, the individual position measurements of rather bright asteroids ($V < 15$ mag), are expected to have an astrometric accuracy of 0.2 mas. Taking for the diameter $\sim 20\,\mathrm{km}$ (appropriate for $V \sim 15$ mag) and a distance of 2.5 AU gives $\theta \sim 10\,\mathrm{mas}$ and a phase shift $\delta \sim 0.6\,\mathrm{mas}$, since Gaia observes on the average minor planets at quadrature. Therefore, the modeling constraints are very demanding if we are to extract the direction of the center of mass without loss of accuracy (Kaasalainen and Tanga 2004). This gets worse for brighter planets, which are simultaneously better observed and bigger.

26.3 Observational techniques

Given the variety of observable objects it is not surprising to see a number of different techniques applied to determine the position, rotation, or size of Solar System

Table 26.2 Observational techniques relevant for Solar System objects and their astrometric accuracies (from Arlot 2008). On the tangent plane, 1 mas $\sim$ 1 km at 1 AU.

Method	Accuracy angular	Accuracy range	Accuracy size	Sources
Photographic plates	100–500 mas			planets, satellites, asteroids
Transit instruments	50–300 mas			planets, asteroids, satellites
Optical CCD	20–100 mas			asteroids, satellites
Stellar occultation	50–100 mas		5 km	asteroids, satellites
Mutual phenomena	10 mas			satellites, binary asteroids
Radar		100–500 m	1 km	Venus, Mercury, asteroids
Lunar laser ranging		3 cm		the Moon
VLBI	5 mas			link with an orbiting spacecraft
Space astrometry	100 μas		10 km	asteroids, satellites

objects. None of the techniques is universal. The standard methods (see Table 26.2) are all inherited directly from classical stellar astrometry using transit instruments or photographic plates, while more specialized methods have been developed just for satellites or asteroids.

26.3.1 Photographic plates

For about a century this has been the best method to discover new members of the Solar System and to obtain positional information on moving bodies. The detection is achieved by taking photographs with wide-angle cameras using long exposures, so that moving objects become visible from the trail left on the plate, or with repeated short exposures yielding several images of moving objects on the plate. Even for a single exposure of several minutes, the trail remains very small (a few millimeters with a long-focus instrument) for an average main-belt asteroid moving at opposition at about 10 mas/s. So the method was efficient only with exposures above 1 hour. Automatic image processing allows the detection of these patterns on a plate, or today on a charge-coupled device (CCD) frame. The astrometry proceeds in the same way (Chapter 19) as for the secondary star catalogs with the position measured relative to reference stars with known positions and proper motions. The use of photographic plates came to an end with the advent of electronic detectors in the early 1980s. However, a very large archive of plates is stored in many observatories in the world and could be remeasured with fast automatic machines and reduced by using modern astrometric star catalogs. This is a way to observe in the past with today's technology and an international collaboration has started between the IMCCE in Paris, the Royal Observatory of Belgium and the US Naval Observatory in the USA. This will come to full fruition with the Gaia catalog, giving many more reference stars with nearly perfect positions compared to the plate-measurement errors.

26.3.2 CCD astrometry

Imaging on CCDs is covered in Chapters 14, 15, and 19 and is not very different for Solar System bodies. The main departure happens when objects are big enough to be resolved by the telescope and produce an extended image on the CCD. The centering estimate is degraded and special care must be exercised, but the steps to obtain the celestial coordinates are similar to that used for the stars and presented in Chapter 19. The motion of the sources (planets, satellites, . . .) against background stars generates a streak on the CCD in the same way as for the photographic plate. Depending on the purpose of the observations, the telescope guiding can be set to the planet's motion to maintain its image fixed on the CCD while the stars produce elongated images. The accuracy of astrometric CCD observations depends on the field size, the number of reference stars in the frame, the image sampling by the pixels (allowing for the atmospheric seeing), and the source brightness. Accuracy lies in the range of 20 to 100 mas at the moment, with the main limitation coming from the accuracy of the reference stars visible in the frame. Again, post-Gaia CCD astrometry will improve significantly (below 10 mas) with many more faint stars having accurate astrometry available.

26.3.3 Stellar occultations

Stellar occultations by minor planets or satellites are special events occurring along narrow strips on the Earth's surface. They provide direct determinations of the shape and size of the occulting bodies.

An occultation occurs for an Earth-based observer when the path of a Solar System object passes in front of a star and blocks its light. The star is occulted by the moving body or, stated differently, the observer stands in the shadow cast by the body. The geometry is very simple and has been exploited since the early 1980s to determine accurately either the apparent direction of the planet at the time of the disappearance or the size of the chord drawn by the intercepted light ray on the shape of the planet projected on the sky. The latter technique is today the best way (aside from direct, but very rare, flybys) to determine the shape and size of small Solar System bodies. The shadow on the Earth is just a cylindrical projection of the actual shape since the stellar source has no apparent size. Its displacement on the Earth's surface is primarily due to the relative motion of the planet with respect to the Earth rather than to the Earth's rotation. This means that on the Earth the width of the shadow strip is not larger than the planet's size, that is to say a few tens of kilometers for a typical asteroid. For a minor planet of the asteroid main belt, the relative velocity is about 15 km/s near opposition and 30 km/s at quadrature. Hence, for a planet with a diameter of 50 km, the occultation will not last more than a few seconds, and will be less than 1 second for the smaller bodies. Therefore, the measurement requires high-speed photometry to record the disappearance of the starlight. As the light recorded before and after the occultation is the sum of the light produced by the star and the asteroid, while only the asteroid is seen during the occultation, the method is more efficient when the occulted star is brighter than the occulting body. For the maximum yield, occultations must be observed within coordinated campaigns to produce several chords allowing us to reconstruct the overall shape and size of the occulting bodies, and also to cope with the uncertainty of predictions.

Actually, aside from the technical aspects of the observation, one of the current limitations of this technique is the ability to predict accurately the path of the shadow on the Earth, so that potential observers can be alerted and be in place with the observing equipment during the occultation. With poor predictions, most of the observers will be outside the occultation strip and will see nothing (this is also useful to assess the prediction accuracy and also supplies an upper bound of the planet size). The accurate prediction depends on both the quality of the ephemeris and that of the star positions. To illustrate the point, it is worth mentioning that a difference in the apparent direction of just half of 1 arcsecond translates into almost 1000 km on its shadow path on the Earth. Gaia will improve the accuracy of the predictions in a dramatic way (Tanga and Delbo 2007) with a much better star catalog and orbits of more than 300 000 bodies, typically 100 times better than today. A complete set of size measurements down to ~ 10 km main-belt asteroids could be obtained in a few years, provided that a small network of ground-based 1-m telescopes is devoted to occultation studies.

26.3.4 Mutual events

When the Sun or the Earth crosses the equatorial plane of Jupiter (once every 6 years), mutual eclipses and occultations can be observed between a pair of the Galilean satellites. An *eclipse* occurs when a satellite goes through the shadow cone brought about by the Sun and another satellite. This is a phenomenon independent of the location of the observer and results from the alignment Sun–satellite A–satellite B. Technically, this is similar to a lunar eclipse. This can only happen when the jovicentric declination of the Sun is close to zero. An *occultation* is the passage of one satellite behind another as seen from the Earth and results from the alignment Earth–satellite A–satellite B. This can only happen when the jovicentric declination of the Earth is close to zero. Given the small inclinations of the orbits of the Galilean satellites to the equator of Jupiter, both phenomena occur only in the vicinity of the Jovian equinox, once every 6 years, and over a period of about 6 months.

This happened in 1997, 2003, and 2009 (Fig. 26.4). While similar phenomena have been known (and observed) with Jupiter and its satellites soon after their discovery by Galileo, the mutual phenomena have only been predicted with sufficient accuracy rather recently (mid 1970s) and are extremely valuable in astrometry because these are well defined events generating neat light curves thanks to the lack of significant atmosphere surrounding the satellites. With a relative velocity between satellites of about 10 km/s and a diameter of ~ 3000 km (Europa) to ~ 5000 km (Ganymede) typical phenomena last less than 10 minutes. The central time of the event can be recorded with a timing accuracy of ~ 3– 4 s, equivalent to 30 km in the position of the satellite. This gives about 10 mas in the geocentric angular position at the opposition of Jupiter, an order of magnitude better than the equivalent observations from the eclipses by Jupiter and about four times better than direct CCD astrometric observations.

The actual photometric modeling to achieve this accuracy is extremely complex and depends on the phase, the surface properties, and the varying albedos and there is still a

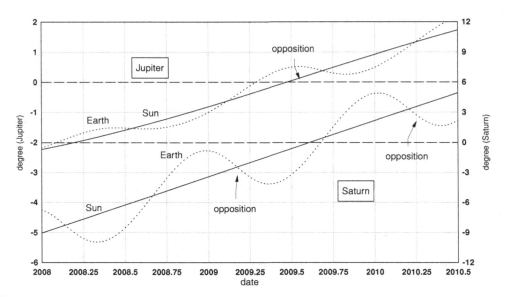

Fig. 26.4 Jovicentric (left scale) and saturnocentric (right scale) latitude of the Sun and the Earh. Mutual phenomena occur only when the Sun (respectively the Earth) crosses the equatorial plane of the planets and lasts about 6 months around opposition.

margin for improvement. Similar phenomena are now routinely predicted and observed for the Saturn system (every 15 years in 1994, 2009, ..., at the same period as the ring-system disappearance) and Uranus (every 42 years, 2007, 2049, ...) and have similar accuracy in kilometers.

This quick overview does not exhaust, by far, the methods specifically applied to Solar System astrometry, but at least gives an overview of the problems and of the main issues. A good up-to-date reading at a more technical level can be found in Arlot (2008).

References

Arlot, J.-E. (2008). Astrometry of the Solar System: the ground-based observations. *Proc. IAU Symp.*, **248**, 66.

Kaasalainen, M. and Tanga, P. (2004). Photocentre offset in ultraprecise astrometry: implications for barycentre determination and asteroid modelling. *A&A*, **416**, 367.

Lindegren, L. (1977). Meridian observations of planets with a photoelectric multislit photometer. *A&A*, **57**, 55.

Tanga, P. and Delbo, M. (2007). Asteroid occultations today and tomorrow: toward the Gaia era. *A&A*, **474**, 1015.

27

Extrasolar planets

ALESSANDRO SOZZETTI

Introduction

The Doppler detection of the Jupiter-mass planet around the nearby, solar-type star 51 Pegasi (Mayor and Queloz 1995) heralded the new era of discoveries of extrasolar planets orbiting normal stars. Four different techniques have been successfully used for the purpose of exoplanet detection. Decade-long, high-precision (1–5 m/s) radial-velocity surveys of ~3000 F-G-K-M dwarfs and subgiants (e.g. Butler *et al.* 2006, Udry and Santos 2007, Eggenberger and Udry 2010) in the solar neighborhood ($d \leq 50$ pc) have yielded so far the vast majority of the objects in the present sample (a total of over 760 planets in ~600 systems, as of June 2012). Ground-based photometric transit surveys (e.g. Charbonneau *et al.* 2007, Collier Cameron 2011) are uncovering new transiting systems at a rate of ~30 per year, while space-borne observatories such as CoRoT and particularly Kepler hold promise of increasing by an order of magnitude the present yield (over 230 transiting systems known to date). Finally, over three dozen sub-stellar companions have also been detected so far by means of gravitational microlensing (e.g. Bond *et al.* 2004, Beaulieu *et al.* 2006, Gaudi *et al.* 2008, Muraki *et al.* 2011), direct-imaging surveys (e.g. Chauvin *et al.* 2005, Kalas *et al.* 2008, Marois *et al.* 2008, 2010), and timing techniques (Silvotti *et al.* 2007, Lee *et al.* 2009, Beuermann *et al.* 2010).

The sample of planetary systems in the solar neighborhood ($d \leq 200$) is amenable to follow-up studies with a variety of indirect and direct techniques. In particular, for planets found transiting their parent stars, precision follow-up studies of such objects during primary and secondary eclipse using high-resolution imaging and stellar spectroscopy at visible and infrared wavelengths (notably with the space-based platforms of the Hubble and Spitzer Space Telescopes, respectively, but also from the ground with 8- and 10-m class telescopes) have enabled direct observations of their transmission spectra, emitted radiation, and large-scale atmospheric circulation. High-resolution stellar spectra obtained during transits have been used to determine the degree of alignment of the planet's orbital angular momentum vector with the stellar spin axis (the Rossiter–McLaughlin effect), while perturbations in the timing of planetary transits have been used to infer the presence of satellites or additional planetary companions, not necessarily transiting (for reviews of the field of transiting-planet astrophysics, see for example Baraffe *et al.* 2010 and references therein, Seager and Deming 2010 and references therein).

Astrometry for Astrophysics: Methods, Models, and Applications, ed. William F. van Altena.
Published by Cambridge University Press. © Cambridge University Press 2013.

27.1 Emerging properties of planetary systems

The observational data on extrasolar planets show such striking properties that we must infer that planet formation and evolution is a very complex process.

Doppler-detected exoplanets have minimum masses[1] from $\sim$2 Earth masses ($M_\oplus$) up to the 13 Jupiter masses (M_J) theoretical dividing line between planetary bodies and brown dwarfs (Oppenheimer et al. 2000), orbital periods P from less than one day up to $\sim$15 years, and eccentricities as large as $e \simeq 0.95$ (except for planets within 0.1 AU, which are found nearly always in circular orbits, presumably due to tidal circularization).

The present-day estimate of the frequency of giant planets around solar-type (late F, G, and early-K spectral types) stars in the solar neighborhood ($d \leq 50$ pc) is $f_p \approx 10\%$ for planets with masses in the range $0.3M_J \leq M_p \leq 10M_J$ and orbital periods $P < 2000$ days. Based on the evidence for longer-period companions in radial-velocity data sets, $f_p \approx 20\%$ is inferred for $a < 20$ AU (e.g. Cumming et al. 2008, Wittenmyer et al. 2011). Short-period ($P < 50$ days), low-mass planets (super-Earths and Neptunes in the approximate mass range $2M_\oplus \leq M_p \leq 30M_\oplus$) also appear rather abundant ($f_p \approx 15\%$) around the same stellar sample (e.g. Howard et al. 2010, 2011). The increasing number of detections allows the statistical properties of extrasolar planetary systems to be characterized, such as the mass, period, and eccentricity distributions (Tabachnik and Tremaine 2002, Butler et al. 2003, Fischer et al. 2003, Lineweaver and Grether 2003, Jones et al. 2003, Udry et al. 2003, Gaudi et al. 2005, Ford and Rasio 2006, Jones et al. 2006, Ribas and Miralda-Escudé 2007, Howard et al. 2010, Pont et al. 2011), and the incidence of giant planets as a function of host-star metallicity (Fischer and Valenti 2005, Santos et al. 2004, Sozzetti et al. 2009) and mass (Butler et al. 2004, 2006; Endl et al. 2006; Johnson et al. 2007, 2010). These observational data have important implications for the proposed models of formation and early evolution of planetary systems (Pollack et al. 1996, Boss 2001, 2002, 2006, 2011, Mayer et al. 2002, Ida and Lin 2004a, 2004b, 2005, 2008, 2010, Alibert et al. 2005, 2011, Lissauer and Stevenson 2007, Durisen et al. 2007, Mordasini et al. 2009).

About 50 multiple-planet systems are confirmed to-date (with up to seven components!), with many other single-planet systems showing a long-term velocity trend likely indicating a second planet with a long orbital period (Wright et al. 2009). It is estimated that $\sim$30% of stars with planets have at least one more planetary-mass companion, integrating over all spectral types and for nearby stars within 200 pc (Wright et al. 2009). Furthermore, several important patterns are beginning to emerge in present-day data, such as multiple-planet systems seemingly harboring preferentially low-mass objects, rather than gas giants (Latham et al. 2011), or statistically significant differences in orbital-element distributions between single- and multiple-planet systems (Wright et al. 2009, Tremaine and Dong 2011). Multiple systems exhibit great dynamical diversity, with at least three main families identified (hierarchical systems, secularly interacting systems, systems in mean motion resonances; see for example Barnes and Quinn 2004, Beaugé et al. 2008, Goździewski

---

[1] The radial-velocity wobble induced by an orbiting companion about the primary has a semi-amplitude $K = 28.4 \left(\frac{P}{1\,\mathrm{yr}}\right)^{-\frac{1}{3}} \left(\frac{M_\star}{1\,M_\odot}\right)^{-\frac{2}{3}} \frac{M_c \sin i}{1\,M_J} \frac{1}{\sqrt{(1-e^2)}}$ m/s. The intrinsically one-dimensional measurement only permits a lower limit to the companion mass to be derived through the product $M_c \sin i$.

et al. 2008 and references therein, Lissauer *et al.* 2011b). Present-day data on multiple systems, including very "flat" systems of transiting planets such as those uncovered by the Kepler mission (Holman *et al.* 2010, Lissauer *et al.* 2011a) provide important clues to the relative importance of several proposed mechanisms of dynamical interactions between forming planets, gaseous/planetesimal disks, and distant companion stars (Papaloizou *et al.* 2007, Ford and Rasio 2008 and references therein, Fabrycky 2010), and allow us to measure the likelihood of formation and survival of terrestrial planets in the habitable zone[2] of the parent star (Menou and Tabachnik 2003, Jones *et al.* 2005, Hinse *et al.* 2008 and references therein, Jones and Sleep 2010, Dvorak *et al.* 2010).

As for the class of close-in exoplanets found eclipsing their parent stars, photometric transit observations reveal a planet's radius,[3] and in combination with radial-velocity measurements, permit a determination of the planet's mass. This combination of measurements provides the only available direct constraint on the density and hence bulk composition of exoplanets.[4] The early evidence for very under-dense transiting hot Jupiters constitutes the first important challenge to the proposed evolutionary models of the internal structure of strongly irradiated exoplanets (Marley *et al.* 2007 and references therein, Burrows *et al.* 2008, Fortney and Nettelmann 2010, Batygin *et al.* 2011, Laughlin *et al.* 2011), while the over two dozen direct measurements of planetary emission and absorption to-date (e.g. Seager and Deming 2010 and references therein) have allowed initial comparisons with atmospheric models for giant planets (Baraffe *et al.* 2008, Fortney *et al.* 2008) and super-Earths (Valencia *et al.* 2007, 2010; Valencia 2011).

When the observational data on exoplanets are taken as a whole, we then realize how we're now witnessing the beginning of a new era of comparative planetology, in which our Solar System can finally be put in the broader context of the astrophysics of planetary systems.

27.2 The future of direct and indirect detection techniques

The comparison between theory and observation shows that several difficult problems are limiting our ability at present to elucidate in a unified manner the various phases of the complex processes of planet formation and evolution. Rather, we often resort to investigate

---

[2] For any given star, the region of habitability is defined as the range of orbital distances at which a potential water reservoir, the primary ingredient for the development of a complex biology, would be found in liquid form (e.g. Kasting *et al.* 1993).

[3] The periodic dimming in flux in the light curve of the primary due to a transiting planet has a characteristic fractional depth $\frac{\Delta F}{F} \approx \left(\frac{R_p}{R_\star}\right)^2$. The planet's radius is then obtained if a reasonable estimate of the stellar radius is available. Assuming random orientations of planetary orbits, transits occur with a probability $p_{tr} = 0.0045 \left(\frac{a}{1\,\mathrm{AU}}\right)^{-1} \left(\frac{R_\star + R_p}{R_\odot}\right) \left(\frac{1 + e\cos(\frac{\pi}{2} - \omega)}{1 - e^2}\right)$, and a duration $t_{tr} = 13 \left(\frac{R_\star}{R_\odot}\right) \left(\frac{P}{1\,\mathrm{yr}}\right)^{\frac{1}{3}} \left(\frac{M_\star}{M_\odot}\right)^{\frac{-1}{3}}$ hr when $i = 90$ deg.

[4] Noticeable exceptions are provided by transit timing variations (TTV) measurements in systems with multiple transiting planets. In such cases (e.g. Kepler-9, Holman *et al.* 2010, and Kepler-11, Lissauer *et al.* 2011a), modeling the patterns of variation of the times of transit center induced by the mutual gravitational interaction between the systems' components allows us to obtain direct estimates of planets' radii and masses even in the absence of confirmation radial-velocity measurements (e.g. Agol *et al.* 2005, Holman and Murray 2005, Ragozzine and Holman 2010).

separately limited aspects of the physics of planet building using a "compartmentalized" approach. However, improvements are being made toward the definition of more robust theories capable of simultaneously explaining a large range of the observed properties of extrasolar planets, as well as of making new, testable predictions. To this end, help from future data obtained with a variety of techniques will prove invaluable. Planet search surveys, initially focused solely on planet discovery, are now being designed to put the emerging properties of planetary systems on firm statistical grounds and thus thoroughly test the theoretical explanations put forth to explain their existence. Furthermore, both NASA and ESA are now formulating strategies to establish a logical sequence of missions and telescope construction to optimize the pace of exoplanet discoveries and address key questions on the physical characterization and architecture of planetary systems.

As for the most successful of indirect detection techniques, Doppler surveys are extending their time baseline and/or are achieving higher-velocity precision (≤ 1 m/s, see for example Lovis *et al.* 2006, Pepe and Lovis 2008, Eggenberger and Udry 2010) to continue searching for planets at increasingly larger orbital distances (e.g. Fischer *et al.* 2007, Wittenmyer *et al.* 2011) and with increasingly smaller masses (e.g. Udry *et al.* 2007, Mayor *et al.* 2009). Ultimately, the limiting factor may not be the intrinsic stability of new-generation spectrographs, but rather the primary stars themselves, through astrophysical noise sources such as stellar surface activity, rotation, and acoustic p-modes. These problems are already severely limiting Doppler surveys from investigating the existence of giant planets orbiting stars departing significantly from our Sun in age, mass, and metal content. The difficulties inherent in precisely determining exoplanet masses with precision radial-velocity measurements in the presence of significant stellar activity is very clearly illustrated by the hot debate on the interpretation of the data collected for the CoRoT-7b super-Earth planet, whose mass, as of June 2011, varies anywhere between $\sim 1 M_{\oplus}$ and $\sim 10 M_{\oplus}$ (Hatzes *et al.* 2011 and references therein).

While ground-based wide-field photometric transit surveys are allowing us to unveil fundamental properties of strongly irradiated giant planets, the Kepler and CoRoT missions are designed to photometrically detect transiting Earth-sized planets in the habitable zone of solar-type host stars, providing the first measure of the occurrence of rocky planets and ice giants. The recent release of >1000 Kepler transiting planet candidates (awaiting confirmation spectroscopy), with radii even below the Earth's, has allowed for example the determination, as discussed at the beginning of this chapter, that $f_p \sim 13 \pm 1\%$ for super-Earths and Neptunes within 0.25 AU of solar-type stars (Howard *et al.* 2011). However, the host stars reside at typical distances beyond 250 pc, making imaging and spectroscopic follow-up of the planets difficult. The prospects for detailed characterization of giant planets and super-Earths transiting nearby solar-type as well as cooler (K- and M-type) stars are tied to the approval of proposed all-sky surveys in space (e.g. PLATO), and to the possible success of currently ongoing and planned ground-based photometric searches for transiting rocky planets around M dwarfs (e.g. MEarth, APACHE).

Gravitational microlensing surveys from the ground within the next decade have the potential to deliver a complete census of the cold-planet population down to $\sim 1 M_{\oplus}$ orbiting low-mass stars at separations $a > 1.5$ AU. Proposed microlensing observatories in space (e.g. WFIRST) could extend the census to planets of $\sim 1 M_{\oplus}$ with separations exceeding

1 AU. We note however that observations with this technique are non-reproducible and follow-up analyses are virtually impossible (the detected systems are typically located more than 1 kpc from the Sun), thus such findings will mostly have statistical value but will help little toward the physical characterization of planetary systems.

During the next 20 years, the prospects are becoming increasingly "bright" for the direct detection of exoplanets and the spectroscopic characterization of their atmospheres using techniques to spatially or temporally separate them from their parent stars. Data from upcoming and proposed observatories for visible-light, near- and mid-infrared imaging and spectroscopy and equipped with single- and multiple-aperture telescopes from the ground (e.g. VLT/SPHERE, ELT/EPICS) and in space (e.g. JWST, SPICA, EChO, Darwin and TPF-C/I avatars) will completely transform our view of the nature of planetary systems. On the path to imaging extrasolar Earths, we will encounter first extrasolar giant planets which, being brighter, are the natural technological and scientific stepping stones.

27.3 The potential of microarcsecond astrometry

27.3.1 Observable model

Similarly to the spectroscopic technique, astrometric measurements can detect the stellar wobble around the system barycenter due to the gravitational perturbation of nearby planets. The main observable (assuming circular orbits) is the "astrometric signature," i.e. the apparent semi-major axis of the stellar orbit scaled by the distance to the observer and the planet-to-star ratio:

$$\alpha = \left(\frac{M_p}{M_\odot}\right)\left(\frac{M_\odot}{M_\star}\right)\left(\frac{a_p}{1\,\mathrm{AU}}\right)\left(\frac{\mathrm{pc}}{d}\right)\,\mathrm{arcsec} \tag{27.1}$$

However, by reconstructing the orbital motion in the plane of the sky, astrometry alone can determine the entire set of seven orbital elements (see Chapters 23 and 24), thus breaking the $M_p \sin i$ degeneracy intrinsic to Doppler measurements and allowing an actual mass estimate for the companion to be derived.

Depending on the limiting astrometric precision desired to characterize planetary signatures, the astrometric observations may have to be corrected for a variety of effects that modify the apparent position of the target. Some of these effects include the mode of operation (wide-, narrow-angle, or global astrometry), the wavelength regime (visible or near-infrared), and the instrument (monolithic or diluted configuration) used to carry out the measurements. These corrections can be classical in nature or intrinsically relativistic, and can be due to (a) the motion of the observer (e.g. aberration), (b) secular variations in the target space motion with respect to the observer (e.g. perspective acceleration), or (c) the gravitational fields of massive bodies in the vicinity of the observer (see Sozzetti 2005, and Chapters 5, 7, and 19). Then, upon detection of one or more significant periodicities (through some version of a periodogram analysis) in the post-single-star fit residuals, the problem reduces to solving a non-linear least-squares problem with $5 + 7 \times n_p$ parameters (the five astrometric and seven orbital elements for every detected planetary signal). The process is carried out in the presence of observational errors (see Chapters 11, 14, and 19) and a variety of noise sources (atmospheric, such as turbulence and differential color refraction (see Chapter 9), and "astrophysical," such as astrometric jitter induced by stellar surface activity, or due to the environment, such as variable illumination of a protoplanetary disk). For multiple-component orbital fits, the inclusion of N-body integrators might have to be considered, in order to account for possibly significant dynamical interactions (see Chapter 6).

The state-of-the-art astrometric precision is nowadays set to ~ 1 milliarcsecond (mas) by Hipparcos and the Hubble Space Telescope Fine Guidance Sensor (HST/FGS) (see Chapters 2 and 24). By looking at Eq. (27.1), we realize how the magnitude of the perturbation induced by a 1-Jupiter-mass planet in orbit at 5 AU around a $1 M_\odot$ star at 10 pc from the Sun is $\alpha \simeq 500$ microarcsecond (μas). For the same distance and primary mass, a "hot Jupiter" with $a_p = 0.01$ AU induces $\alpha = 1$ μas, and an Earth-like planet ($a_p = 1$ AU) causes a perturbation $\alpha = 0.33$ μas. We then understand why, despite several decades of attempts (e.g. Strand 1943, Reuyl and Holmberg 1943, Lippincott 1960, van de Kamp 1963, Gatewood 1996, Han et $al.$ 2001, Pravdo and Shaklan 2009), and a few recent successes primarily thanks to HST/FGS astrometry (Benedict et $al.$ 2002, 2006, 2010; McArthur et $al.$ 2004, 2010; Bean et $al.$ 2007; Martioli et $al.$ 2010) and recent analyses of the re-reduced Hipparcos intermediate astrometric data (Sozzetti and Desidera 2010, Sahlmann et $al.$ 2011, Reffert and Quirrenbach 2011), astrometric measurements with mas precision have so far proved of limited utility when employed as either a follow-up tool or to independently search for planetary-mass companions orbiting nearby stars (for a review of the approach to planet detection with astrometry see for example Sozzetti 2005, 2010 and references therein). However, an improvement of two to three orders of magnitude in achievable measurement precision, down to the μas level, would allow this technique to achieve in perspective the same successes of the Doppler method, for which the improvement from the km/s to the m/s precision opened the doors for ground-breaking results in exoplanetary science. Indeed, μas astrometry is almost coming of age. Provided the demanding technological and calibration requirements to achieve the required level of measurement precision are met (e.g. Sozzetti 2005, and Chapters 5, 9, 14, and 19 in this book), future observatories at visible and near-infrared wavelengths, using both monolithic as well as diluted architectures from the ground (with VLTI/PRIMA, e.g. Launhardt et $al.$ 2008) and in space (with Gaia, Casertano et $al.$ 2008) or proposed ultra-high-precision observatories (Malbet et $al.$ 2010) such as NEAT (Malbet et $al.$ 2011) hold promise for crucial contributions to many aspects of planetary-system astrophysics (formation theories, dynamical evolution, internal structure, detection of Earth-like planets), in combination with data collected with other indirect and direct techniques. Fig. 27.1 shows the M_p–a diagram with the plotted present-day and achievable sensitivities of transit photometry and radial velocity, and with the expected detection thresholds for a space-borne instrument with 1 μas precision in narrow-angle mode and Gaia at 10 pc, 25 pc, and 200 pc, respectively. The presently known planets detected by the various methods are also shown, along with the predicted distribution of recent models (Ida and Lin 2008). At first glance, you could get the impression that the impact of astrometric measurements (except for those

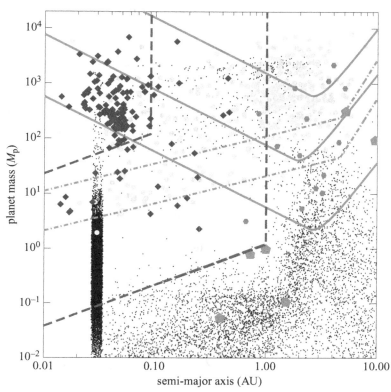

Fig. 27.1 Exoplanets discovery space in the M_p–a plane for the astrometric, Doppler, and transit techniques. Detectability curves are defined on the basis of a 3σ criterion for signal detection. The upper and middle solid curves are for Gaia astrometry with $\sigma_A = 10$ µas, assuming a $1M_\odot$. G dwarf primary at 200 pc and a $0.4M_\odot$ M dwarf at 25 pc, respectively, while the lower curve is for narrow-field astrometry of a $1M_\odot$ star at 10 pc with $\sigma_A = 0.6$ µas. For both Gaia and the narrow-angle astrometric instrument, survey duration is set to 5 yr. The radial-velocity curves (dashed-dotted lines) assume $\sigma_{RV} = 3$ m/s (upper curve) and $\sigma_{RV} = 1$ m/s (lower curve), $M_\star = 1M_\odot$., and 10-yr survey duration. For visible-light transit photometry (long-dashed curves), the assumptions are $\sigma_V = 5 \times 10^{-3}$ mag (upper curve) and $\sigma_V = 1 \times 10^{-5}$ mag (lower curve), signal-to-noise ratio (SNR) $= 9$, $M_\star = 1M_\odot$., $R_\star = 1R_\odot$., uniform and dense ($\gg 1000$ data points) sampling. The light-grey circles indicate the inventory of Doppler-detected exoplanets as of June 2011. Transiting systems are shown as dark-grey filled diamonds, while the grey hexagons are planets detected by microlensing. Solar System planets are also shown as large grey pentagons. The small black crosses represent a theoretical distribution of masses and final orbital semi-major axes from Ida and Lin (2008).

obtained by 1 µas precision differential astrometry around the nearest stars) may not bear great potential. However, the relative importance of different planet detection techniques should not be gauged by looking at their discovery potential per se, but rather in connection to outstanding questions to be addressed and answered in the science of planetary systems. Some of the most important issues for which µas astrometry will play a key role in the next decade are summarized below.

27.3.2   Planet formation and migration models

The evidence for a dependence of planet properties (orbital elements and mass distributions, and correlations among them) and frequencies on the characteristics of the parent stars (spectral type, age, metallicity, and binarity/multiplicity) constitutes one of the pivotal points in the comparison with predictions from competing models of planet formation and early evolution. However, the typical sample sizes of Doppler surveys are on the order of $\sim 10^3$ objects, sufficient to test only the most outstanding difference between the various populations. It is thus desirable to be able to provide as large a database as possible of stars screened for planets. High-precision astrometry, particularly with Gaia, has the potential to significantly refine our understanding of the statistical properties of extrasolar planets, thus helping to crucially test theoretical models of gas-giant planet formation and migration. For example, the Gaia unbiased and complete magnitude-limited census of $> 10^5$ F-G-K dwarfs screened for new planets out to 200 pc from the Sun will allow us to test the fine structure of giant-planet parameter distributions and frequencies, and to investigate their possible changes as a function of stellar mass with unprecedented resolution. In addition, astrometric measurements with Gaia of thousands of metal-poor stars and hundreds of young stars will instead allow us to probe specific predictions on giant-planet formation timescales and the role of varying metal content in the protoplanetary disk.

27.3.3   Multiple-planet systems

High-precision astrometry can provide meaningful estimates of the full three-dimensional geometry of any planetary system (without restrictions on the orbital alignment with respect to the line of sight) by measuring the mutual inclination angle between pairs of planetary orbits:

$$\cos i_{\text{rel}} = \cos i_{\text{in}} \cos i_{\text{out}} + \sin i_{\text{in}} \sin i_{\text{out}} \cos(\Omega_{\text{out}} - \Omega_{\text{in}}) \qquad (27.2)$$

where i_{in} and i_{out}, Ω_{in} and Ω_{out} are the inclinations and lines of nodes of the inner and outer planet, respectively. Coplanarity tests for hundreds of multiple-planet systems will be carried out with Gaia, and VLTI/PRIMA, and this, in combination with data available from Doppler measurements and transit timing, could allow us to discriminate between various proposed mechanisms for eccentricity excitation, thus significantly improving our comprehension of the role of dynamical interactions in the early as well as long-term evolution of planetary systems.

27.3.4   Direct detections of giant exoplanets

The first direct constraints on theoretical models of interiors and atmospheres of strongly irradiated gas-giant planets have been provided by the host of photometric and spectroscopic observations enabled by the special geometry of transiting hot Jupiters. Studies with high-contrast imaging instruments to spatially resolve cool, wide-separation giant planets around nearby stars have highlighted the strong dependence of the apparent brightness of a planet in reflected host-star light on orbit geometry, orbital phase, cloud cover, cloud composition, mass, and age.[5] In particular, accurate knowledge of all orbital parameters and actual mass are essential for understanding the thermophysical conditions on a planet and for determining its visibility. High-precision astrometric measurements (with Gaia, and VLTI/PRIMA) could then provide important supplementary data to aid in the understanding of direct detections of wide-separation extrasolar giant planets. For example, actual mass estimates and full orbital geometry determination for suitable systems (with typical separations $> 0.1''$) will inform direct imaging surveys (e.g. SPHERE/VLT, EPICS/E-ELT) and future spectroscopic characterization observatories (e.g. EChO) about the epoch and location of maximum brightness, in order to estimate optimal visibility, and will help in the modeling and interpretation of the phase functions and light curves of giant planets (the first prediction about where and when to look for the planet ε Eridani b was recently made by Benedict *et al.* 2006 using HST/FGS astrometry).

27.3.5 The hunt for other Earths

The most challenging achievement in exoplanet science will be the direct detection and characterization of terrestrial, habitable planets orbiting stars very close ($d \leq 25$ pc) to our Sun, searching for elements in their atmospheres that can be interpreted as "biomarkers" (Hitchcock and Lovelock 1967; Des Marais *et al.* 2002; Seager *et al.* 2005; Tinetti *et al.* 2007; Kaltenegger *et al.* 2007, 2010), implying the likely existence of a complex biology on the surface. Space-borne transit photometry carried out with CoRoT (Baglin *et al.* 2009) and Kepler (Borucki *et al.* 2009) will provide the strongest statistical constraints (including bona-fide detections) on the frequency $\eta_\oplus$ of Earth-sized planets orbiting within the habitable zone of Sun-like stars, but the typically large distances of the detected systems (several hundred parsecs) will make their further characterization highly unlikely. Astrometry of all nearby stars within 25 pc of the Sun with 1–10 µas precision (with Gaia in space and VLTI/PRIMA from the ground) will provide future direct-imaging and spectroscopic characterization observatories (e.g. Beichman *et al.* 2007, Tinetti *et al.* 2010) with essential supplementary data for the optimization of the target lists. For example, a comprehensive database of F-G-K-M stars astrometrically screened for Jupiter- and Saturn-mass companions out to several astronomical units (Sozzetti *et al.* 2003) would help to probe the long-term dynamical stability of their habitable zones, where terrestrial planets may have formed, and may be found, while systems containing astrometrically detected bona-fide terrestrial, (potentially) habitable planets (Malbet *et al.* 2010, Sozzetti 2011) will obviously constitute the highest-priority targets.

---

[5] The planet/host-star flux ratio at visible wavelengths is $\frac{F_p}{F_\star}(\Psi) = \left(\frac{R_p}{r}\right)^2 p\,\Phi(\Psi)$, where p is the geometric albedo, $r = a\frac{(1-e^2)}{1+e\cos\nu}$ is the orbital separation, and $\Phi(\Psi)$ is the phase function, which describes the relative flux at a phase angle Ψ to that at opposition. In general: $\Psi = \sin(\nu + \omega)\sin i \sin \Omega - \cos \Omega \cos(\nu + \omega)$. Depending upon orientation and eccentricity, the brightness of an extrasolar planet can be essentially constant (for $e \simeq 0.0$, $i \simeq 0°$, for example) or vary by over an order of magnitude (for $e \simeq 0.6$, $i \simeq 90°$, for example) along its orbit, and this can induce significant changes in the chemical composition of its atmosphere (e.g. from cloudy to cloud-free).

References

Agol, E., Steffen, J., Sari, R., and Clarkson, W. (2005). On detecting terrestrial planets with timing of giant planet transits. *MNRAS*, **359**, 567.

Alibert, Y., Mordasini, C., Benz, W., and Winisdoerffer, C. (2005). Models of giant planet formation with migration and disc evolution. *A&A*, **434**, 343.

Alibert, Y., Mordasini, C., and Benz, W. (2011). Extrasolar planet population synthesis. III. Formation of planets around stars of different masses. *A&A*, **526**, A63.

Baglin, A., Auvergne, M., Barge, P., *et al.* (2009). CoRoT: description of the mission and early results. *Proc. IAU Symp.*, **253**, 71.

Baraffe, I., Chabrier, G., and Barman, T. (2008). Structure and evolution of super-Earth to super-Jupiter exoplanets. I. Heavy element enrichment in the interior. *A&A*, **482**, 315.

Baraffe, I., Chabrier, G., and Barman, T. (2010). The physical properties of extra-solar planets. *Rep. Prog. Phys.*, **73**, 016901.

Barnes, R. and Quinn, T. (2004). The (in)stability of planetary systems. *ApJ*, **611**, 494.

Batygin, K., Stevenson, D. J., and Bodenheimer, P. H. (2011). Evolution of ohmically heated hot Jupiters. *ApJ*, **738**, a1.

Bean, J. L., McArthur, B. E., and Benedict, G. F. (2007). The mass of the candidate exoplanet companion to HD 33636 from Hubble Space Telescope astrometry and high-precision radial velocities. *ApJ*, **134**, 749.

Beaugé, C., Giuppone, C. A., Ferraz-Mello, S., and Michtchenko, T. A. (2008). Reliability of orbital fits for resonant extrasolar planetary systems: the case of HD82943. *MNRAS*, **385**, 2151.

Beaulieu, J.-P., Bennett, D. P., Fouqué, P., *et al.* (2006). Discovery of a cool planet of 5.5 Earth masses through gravitational microlensing. *Nature*, **439**, 437.

Beichman, C. A., Fridlund, M., Traub, W. A., *et al.* (2007). Comparative planetology and the search for life beyond the Solar System. In *Protostars and Planets V*, ed. B. Reipurth, D. Jewitt, and K. Keil. Tuscon, AZ: University of Arizona Press, p. 915.

Benedict, G. F., McArthur, B. E., Forveille, T., *et al.* (2002). A mass for the extrasolar planet Gliese 876b determined from Hubble Space Telescope Fine Guidance Sensor 3 astrometry and high-precision radial velocities. *ApJ Lett*, **581**, L115.

Benedict, G. F., McArthur, B. E., Gatewood, G., *et al.* (2006). The extrasolar planet ϵ Eridani b: orbit and mass. *ApJ*, **132**, 2206.

Benedict, G. F., McArthur, B. E., Bean, J. L., *et al.* (2010). The mass of HD 38529c from Hubble Space Telescope astrometry and high-precision radial velocities. *ApJ*, **139**, 1844.

Beuermann, K., Hessman, F. V., Dreizler, S., *et al.* (2010). Two planets orbiting the recently formed post-common envelope binary NN Serpentis. *A&A*, **521**, L60.

Bond, I. A., Udalski, A., Jaroszński, M., *et al.* (2004). OGLE 2003-BLG-235/MOA 2003-BLG-53: A planetary microlensing event. *ApJ Lett*, **606**, L155.

Borucki, W., Koch, D., Batalha, N., *et al.* (2009). Kepler: search for Earth-size planets in the habitable zone. *Proc. IAU Symp.*, **253**, 289-299.

Boss, A. P. (2001). Formation of planetary-mass objects by protostellar collapse and fragmentation. *ApJ Lett*, **551**, L167.

Boss, A. P. (2002). Stellar metallicity and the formation of extrasolar gas giant planets. *ApJ Lett*, **567**, L149.

Boss, A. P. (2006). Rapid formation of gas giant planets around M dwarf stars. *ApJ*, **643**, 501.

Boss, A. P. (2011). Formation of giant planets by disk instability on wide orbits around protostars with varied masses. *ApJ*, **731**, 74.

Burrows, A., Budaj, J., and Hubeny, I. (2008). Theoretical spectra and light curves of close-in extrasolar giant planets and comparison with data. *ApJ*, **678**, 1436.

Butler, R. P., Marcy, G. W., Vogt, S. S., *et al.* (2003). Seven new Keck planets orbiting G and K dwarfs. *ApJ*, **582**, 455.

Butler, R. P., Vogt, S. S., Marcy, G. W., *et al.* (2004). A Neptune-mass planet orbiting the nearby M dwarf GJ 436. *ApJ*, **617**, 580.

Butler, R. P., Wright, J. T., Marcy, G. W., *et al.* (2006). Catalog of nearby exoplanets. *ApJ*, **646**, 505.

Casertano, S., Lattanzi, M. G., Sozzetti, A., *et al.* (2008). Double-blind test program for astrometric planet detection with Gaia. *A&A*, **482**, 699.

Charbonneau, D., Brown, T. M., Burrows, A., and Laughlin, G. (2007). When extrasolar planets transit their parent stars. *Protostars and Planets V*, ed. B. Reipurth, D. Jewitt, and K. Keil. Tuscon, AZ: University of Arizona Press, p. 699.

Chauvin, G., Lagrange, A.-M., Zuckerman, B., *et al.* (2005). A companion to AB Pic at the planet/brown dwarf boundary. *A&A*, **438**, L29.

Collier Cameron, A. (2011). Statistical patterns in ground-based transit surveys. *Proc. IAU Symp.*, **276**, 129.

Cumming, A., Butler, R. P., Marcy, G. W., *et al.* (2008). The Keck planet search: detectability and the minimum mass and orbital period distribution of extrasolar planets. *PASP*, **120**, 531.

Des Marais, D. J., Harwitt, M. D., Jucks, K. W., *et al.* (2002). Remote sensing of planetary properties and biosignatures on extrasolar terrestrial planets. *Astrobiol.*, **2**(2), 153.

Durisen, R. H., Boss, A. P., Mayer, L., *et al.* (2007). Gravitational instabilities in gaseous protoplanetary disks and implications for giant planet formation. *Protostars and Planets V*, ed. B. Reipurth, D. Jewitt, and K. Keil. Tuscon, AZ: University of Arizona Press, p. 607.

Dvorak, R., Pilat-Lohinger, E., Bois, E., *et al.* (2010). Dynamical habitability of planetary systems. *Astrobiol.*, **10**, 33.

Eggenberger, A. and Udry, S. (2010). Detection and characterization of extrasolar planets through doppler spectroscopy. *EAS Pub. Ser.*, **41**, 27.

Endl, M., Cochran, W. D., Kürster, M., *et al.* (2006). Exploring the frequency of close-in jovian planets around m dwarfs. *AJ*, **649**, 436.

Fabrycky, D. C. (2010). Non-Keplerian dynamics of exoplanets. In *Exoplanets*, ed. S. Seager. Tuscon, AZ: University of Arizona Press, p. 217.

Fischer, D. A., Marcy, G. W., Butler, R. P., *et al.* (2003). A planetary companion to HD 40979 and additional planets orbiting HD 12661 and HD 38529. *AJ*, **586**, 1394.

Fischer, D. A. and Valenti, J. (2005). The planet–metallicity correlation. *ApJ*, **622**, 1102.

Fischer, D. A., Vogt, S. S., Marcy, D. W., *et al.* (2007). Five intermediate-period planets from the n2k sample. *ApJ*, **669**, 1336.

Ford, E. B. and Rasio, F. A. (2006). On the relation between hot Jupiters and the Roche limit. *ApJ Lett*, **638**, L45.

Ford, E. B. and Rasio, F. A. (2008). Origins of eccentric extrasolar planets: testing the planet–planet scattering model. *ApJ*, **686**, 621.

Fortney, J. J., Lodders, K., Marley, M. S., and Freedman, R. S. (2008). A unified theory for the atmospheres of the hot and very hot Jupiters: two classes of irradiated atmospheres. *ApJ*, **678**, 1419.

Fortney, J. J. and Nettelmann, N. (2010). The interior structure, composition, and evolution of giant planets. *Space Sci. Rev.*, **152**, 423.

Gatewood, G. (1996). Lalande 21185. *Bull. Am. Astron. Soc.*, **28**, 885.

Gaudi, B. S., Seager, S., and Mallen-Ornelas, G. (2005). On the period distribution of close-in extrasolar giant planets. *ApJ*, **623**, 472.

Gaudi, B. S., Bennett, D. P., Udalski, A., *et al.* (2008). Discovery of a Jupiter/Saturn analog with gravitational microlensing. *Science*, **319**, 927.

Goździewski, K., Migaszewski, C., and Konacki, M. (2008). A dynamical analysis of the 14 Herculis planetary system. *MNRAS*, **385**, 957.

Han, I., Black, D. C., and Gatewood, G. (2001). Preliminary astrometric masses for proposed extrasolar planetary companions. *ApJ Lett*, **548**, L57.

Hatzes, A. P., Fridlund, M., Nachmani, G., *et al.* (2011). The mass of CoRoT-7b. *ApJ*, **743**, 75.

Hinse, T. C., Michelsen, R., Jørgensen, U. G., Goździewski, K., and Mikkola, S. (2008). Dynamics and stability of telluric planets within the habitable zone of extrasolar planetary systems. Numerical simulations of test particles within the HD 4208 and HD 70642 systems. *A&A*, **488**, 1133.

Hitchcock, D. R. and Lovelock, J. E. (1967). Life detection by atmospheric analysis. *Icarus*, **7**, 149.

Holman, M. J. and Murray, N. W. (2005). The use of transit timing to detect terrestrial-mass extrasolar planets. *Science*, **307**, 1288.

Holman, M. J., Fabrycky, D. C., Ragozzine, D., *et al.* (2010). Kepler-9: a system of multiple planets transiting a Sun-like star, confirmed by timing variations. *Science*, **330**(6000), 51.

Howard, A. W., Marcy, G. W., Johnson, J. A., *et al.* (2010). The occurrence and mass distribution of close-in super-Earths, Neptunes, and Jupiters. *Science*, **330**, 653.

Howard, A. W., Marcy, G. W., Bryson, S. T., *et al.* (2011). Planet occurrence within 0.25 AU of solar-type stars from Kepler. *ApJ*, submitted (`eprint arXiv:1103.2541`).

Ida, S. and Lin, D. N. C. (2004a). Toward a deterministic model of planetary formation. I. A desert in the mass and semimajor axis distributions of extrasolar planets. *ApJ*, **604**, 388.

Ida, S. and Lin, D. N. C. (2004b). Toward a deterministic model of planetary formation. II. The formation and retention of gas giant planets around stars with a range of metallicities. *ApJ*, **616**, 567.

Ida, S. and Lin, D. N. C. (2005) Toward a deterministic model of planetary formation. III. Mass distribution of short-period planets around stars of various masses. *ApJ*, **626**, 1045.

Ida, S. and Lin, D. N. C. (2008). Toward a deterministic model of planetary formation. IV. Effects of type I migration. *ApJ*, **673**, 487.

Ida, S. and Lin, D. N. C. (2010). Toward a deterministic model of planetary formation. VI. Dynamical interaction and coagulation of multiple rocky embryos and super-Earth systems around solar-type stars. *ApJ*, **719**, 810.

Johnson, J. A., Aller, K. M., Howard, A. W., and Crepp, J. R. (2010). Giant planet occurrence in the stellar mass–metallicity plane. *PASP*, **122**, 905.

Johnson, J. A., Butler, R. P., Marcy, G. W., *et al.* (2007). A new planet around an M dwarf: revealing a correlation between exoplanets and stellar mass. *ApJ*, **670**, 833.

Jones, B. W. and Sleep, P. N. (2010). Habitability of exoplanetary systems with planets observed in transit. *MNRAS*, **407**, 1259.

Jones, H. R. A., Butler, R. P., Tinney, C. G., *et al.* (2003). An exoplanet in orbit around τ^1 Gruis. *MNRAS*, **341**, 948.

Jones, B. W., Underwood, D. R., and Sleep, P. N. (2005) Prospects for habitable "Earths" in known exoplanetary systems. *ApJ*, **622**, 1091.

Jones, H. R. A., Butler, R. P., Tinney, C. G., *et al.* (2006). High-eccentricity planets from the Anglo-Australian Planet Search. *MNRAS*, **369**, 249.

Kalas, P., Graham, J. R., Chiang, E., *et al.* (2008). Optical images of an exosolar planet 25 light-years from Earth. *Science*, **322**, 1345.

Kaltenegger, L., Traub, W. A., and Jucks, K. W. (2007). Spectral evolution of an Earth-like planet. *ApJ*, **658**, 598.

Kaltenegger, L., Selsis, F., Fridlund, M., *et al.* (2010). Deciphering spectral fingerprints of habitable exoplanets. *Astrobiol.*, **10**, 89.

Kasting, J. F., Whitmire, D. P., and Reynolds, R. T. (1993). Habitable zones around main sequence stars. *Icarus*, **101**, 108.

Latham, D. W., Rowe, J. F., Quinn, S. N., *et al.* (2011). A first comparison of Kepler planet candidates in single and multiple systems. *ApJ Lett*, **732**, L24.

Laughlin, G., Crismani, M., and Adams, F. C. (2011). On the anomalous radii of the transiting extrasolar planets. *ApJ Lett*, **729**, L7.

Launhardt, R., Henning, T., Queloz, D., *et al.* (2008). The ESPRI project: narrow-angle astrometry with VLTI-PRIMA. *Proc. IAU Symp.*, **248**, 417.

Lee, J. W., Kim, S.-L., Kim, C.-H., *et al.* (2009). The sdB+M eclipsing system HW Virginis and its circumbinary planets. *ApJ*, **137**, 3181.

Lineweaver, C. H. and Grether, D. (2003). What fraction of Sun-like stars have planets? *ApJ*, **598**, 1350.

Lippincott, S. L. (1960). The unseen companion of the fourth nearest star, Lalande 21185. *ApJ*, **65**, 349.

Lissauer, J. J., and Stevenson, D. J. (2007). Formation of giant planets. *Protostars and Planets V*, ed. B. Reipurth, D. Jewitt, and K. Keil. Tuscon, AZ: University of Arizona Press, p. 591.

Lissauer, J. J., Fabrycky, D. C., Ford, E. B., *et al.* (2011a). A closely packed system of low-mass, low-density planets transiting Kepler-11. *Nature*, **470**, 53.

Lissauer, J. J., Ragozzini, D., Fabrycky, D. E., *et al.* (2011b). Architecture and dynamics of Kepler's candidate multiple transiting planet systems. *ApJS*, **197**, a8.

Lovis, C., Mayor, M., Pepe, F., *et al.* (2006). An extrasolar planetary system with three Neptune-mass planets. *Nature*, **441**(7091), 305.

Malbet, F. (2011). High precision astrometry mission for the detection and characterization of nearby habitable planetary systems with the Nearby Earth Astrometric Telescope (NEAT). *Exp. Astron.*, in press (eprint arXiv:1107.3643).

Malbet, F., Sozzetti, A., Lazorenko, P., *et al.* (2010). Review from the Blue Dots Astrometry Working Group. *ASP Conf. Ser.*, **430**, 84.

Marley, M. S., Fortney, J., Seager, S., and Barman, T. (2007). Atmospheres of extrasolar giant planets. *Protostars and Planets V*, ed. B. Reipurth, D. Jewitt, and K. Keil. Tuscon, AZ: University of Arizona Press, p. 733.

Marois, C., Macintosh, B., Barman, T., *et al.* (2008). Direct imaging of multiple planets orbiting the star HR 8799. *Science*, **322**, 1348.

Marois, C., Zuckerman, B., Konopacky, Q. M., Macintosh, B., and Barman, T. (2010). Images of a fourth planet orbiting HR 8799. *Nature*, **468**, 1080.

Martioli, E., McArthur, B. E., Benedict, G. F., *et al.* (2010). The mass of the candidate exoplanet companion to HD 136118 from Hubble Space Telescope astrometry and high-precision radial velocities. *ApJ*, **708**, 625.

Mayer, L., Quinn, T., Wadsley, J., and Stadel, J. (2002). Formation of giant planets by fragmentation of protoplanetary disks. *Science*, **298**, 1756.

Mayor, M. and Queloz, D. (1995). A Jupiter-mass companion to a solar-type star. *Nature*, **378**, 355.

Mayor, M., Marmier, M., Lovis, C., *et al.* (2009). The HARPS search for southern extrasolar planets. XVIII. An Earth-mass planet in the GJ 581 planetary system. *A&A*, **507**, 487.

McArthur, B. E., Endl, M., Cochran, W. D., *et al.* (2004). Detection of a Neptune-mass planet in the ρ^1 Cancri system using the Hobby-Eberly Telescope. *ApJ Lett*, **614**, L81.

McArthur, B. E., Benedict, G. F., Barnes, R., *et al.* (2010). New observational constraints on the υ Andromedae system with data from the Hubble Space Telescope and Hobby-Eberly Telescope. *ApJ*, **715**, 1203.

Menou, K. and Tabachnik, S. (2003). Dynamical habitability of known extrasolar planetary systems. *ApJ*, **583**, 473.

Mordasini, C., Alibert, Y., Benz, W., and Naef, D. (2009). Extrasolar planet population synthesis. II. Statistical comparison with observations. *A&A*, **501**, 1161.

Muraki, Y., Han, C., Bennett, D. P., *et al.* (2011) Discovery and mass measurements of a cold, 10-Earth mass planet and its host star. *ApJ*, **741**, a22.

Oppenheimer, B. R., Kulkarni, S. R., and Stauffer, J. R. (2000). Brown dwarfs. *Protostars and Planets IV*, ed. B. Reipurth, D. Jewitt, and K. Keil. Tuscon, AZ: University of Arizona Press, p. 1313.

Papaloizou, J. C. B., Nelson, R. P., Kley, W., Masset, F. S., and Artymowicz, P. (2007). Disk–planet interactions during planet formation. *Protostars and Planets V*, ed. B. Reipurth, D. Jewitt, and K. Keil. Tuscon, AZ: University of Arizona Press, p. 655.

Pepe, F. A. and Lovis, C. (2008). From HARPS to CODEX: exploring the limits of Doppler measurements. *Physica Scripta*, **130**, 014007.

Pollack, J. B., Hubickyj, O., Bodenheimer, P., *et al.* (1996). Formation of the giant planets by concurrent accretion of solids and gas. *Icarus*, **124**, 62.

Pont, F., Husnoo, N., Mazeh, T., and Fabrycky, D. (2011). Determining eccentricities of transiting planets: a divide in the mass–period plane. *MNRAS*, **414**, 1278.

Pravdo, S. H. and Shaklan, S. B. (2009). An ultracool star's candidate planet. *ApJ*, **700**, 623.

Ragozzine, D. and Holman, M. J. (2010). The value of systems with multiple transiting planets. *ApJ*, submitted (`eprint arXiv:1006.3727`).

Reffert, S. and Quirrenbach, A. (2011). Mass constraints on substellar companion candidates from the re-reduced Hipparcos intermediate astrometric data: nine confirmed planets and two confirmed brown dwarfs. *A&A*, **527**, A140.

Reuyl, D. and Holmberg, E. (1943). On the existence of a third component in the system 70 Ophiuchi. *ApJ*, **97**, 41.

Ribas, I., and Miralda-Escudé, J. (2007). The eccentricity–mass distribution of exoplanets: signatures of different formation mechanisms? *A&A*, **464**, 779.

Sahlmann, J., Segransan, D., Queloz, D., *et al.* (2011). Search for brown-dwarf companions of stars. *A&A*, **525**, A95+.

Santos, N. C., Israelian, G., and Mayor, M. (2004). Spectroscopic [Fe/H] for 98 extra-solar planet-host stars. Exploring the probability of planet formation. *A&A*, **415**, 1153.

Seager, S. and Deming, D. (2010). Exoplanet atmospheres. *ARAA*, **48**, 631.

Seager, S., Turner, E. L., Schafer, J., and Ford, E. B. (2005). Vegetation's red edge: a possible spectroscopic biosignature of extraterrestrial plants. *Astrobiol.*, **5**, 372.

Sozzetti, A. (2005). Astrometric methods and instrumentation to identify and characterize extrasolar planets: a review. *PASP*, **117**, 1021.

Sozzetti, A. (2010). Detection and characterization of planetary systems with µas astrometry. *EAS Publ. Ser.*, **42**, 55.

Sozzetti, A. (2011). Astrometry and exoplanets: the Gaia era and beyond. *EAS Publ. Ser.*, **45**, 273.

Sozzetti, A. and Desidera, S. (2010). Hipparcos preliminary astrometric masses for the two close-in companions to HD 131664 and HD 43848. A brown dwarf and a low-mass star. *A&A*, **509**, A103.

Sozzetti, A., Casertano, S., Lattanzi, M. G., and Spagna, A. (2003). The Gaia astrometric survey of the solar neighborhood and its contribution to the target database for DARWIN/TPF. ESA Special Publication SP**539**, 605.

Silvotti, R., Schuh, S., Janulis, R., *et al.* (2007). A giant planet orbiting the "extreme horizontal branch" star V391 Pegasi. *Nature*, **449**, 189.

Sozzetti, A., Torres, G., Latham, D. W., *et al.* (2009). A Keck HIRES Doppler search for planets orbiting metal-poor dwarfs. II. On the frequency of giant planets in the metal-poor regime. *ApJ*, **697**, 544.

Strand, K. A. (1943). 61 Cygni as a triple system. *PASP*, **55**, 29.

Tabachnik, S. and Tremaine, S. (2002). Maximum-likelihood method for estimating the mass and period distributions of extrasolar planets. *MNRAS*, **335**, 151.

Tinetti, G., Vidal-Madjar, A., Liang, M.-C., *et al.* (2007). Water vapour in the atmosphere of a transiting extrasolar planet. *Nature*, **448**, 169.

Tinetti, G., Griffith, C. A., Swain, M. R., *et al.* (2010). Exploring extrasolar worlds: from gas giants to terrestrial habitable planets. *Faraday Discuss.*, **147**, 369.

Tremaine, S. and Dong, S. (2011). The statistics of multi-planet systems. *AJ*, **143**, a94.

Udry, S. and Santos, N. C. (2007). Statistical properties of exoplanets. *ARAA*, **45**, 397.

Udry, S., Mayor, M., and Santos, N. C. (2003). Statistical properties of exoplanets. I. The period distribution: constraints for the migration scenario. *A&A*, **407**, 369.

Udry, S., Bonfils, X., Delfosse, X., *et al.* (2007). The HARPS search for southern extra-solar planets. XI. Super-Earths (5 and 8 $M_\oplus$) in a 3-planet system. *A&A*, **469**, L43.

Valencia, D. (2011). Composition of transiting and transiting-only super-Earths. *Proc. IAU Symp.*, **276**, 181.

Valencia, D., Sasselov, D. D., and O'Connell, R. J. (2007). Detailed models of super-Earths: how well can we infer bulk properties? *ApJ*, **665**, 1413.

Valencia, D., Ikoma, M., Guillot, T., and Nettelmann, N. (2010). Composition and fate of short-period super-Earths. The case of CoRoT-7b. *A&A*, **516**, A20.

van de Kamp, P. (1963). Astrometric study of Barnard's star from plates taken with the 24-inch Sproul refractor. *ApJ*, **68**, 515.

Wittenmyer, R. A., Tinney, C. G., O'Toole, S. J., Jones, H. R. A., Butler, R. P., Carter, B. D., and Bailey, J. (2011). On the frequency of Jupiter analogs. *ApJ*, **727**, a102.

Wright, J. T., Upadhyay, S., Marcy, G. W., *et al.* (2009). Ten new and updated multiplanet systems and a survey of exoplanetary systems. *ApJ*, **693**, 1084.

28

Astrometric measurement and cosmology

RICHARD EASTHER

Introduction

Thanks to rapid progress in observational techniques and detectors, great progress has been made in understanding the large-scale properties of the Universe. Traditionally, parameters that describe the global properties of the Universe – its density, the relative fractions of its different constituents, and its expansion rate – were derived from sequences of largely independent observations. In the case of the Hubble constant, H_0, the struggle to develop the "distance ladder" is one of the great stories of twentieth-century astronomy. This effort is exemplified by the Hubble Space Telescope Key Project (Freedman *et al.* 2001), which returned the "headline number" $H_0 = 72 \pm 8$ km/s/Mpc. In the late 1990s, a different approach to cosmological-parameter estimation emerged: concordance cosmology. This is based on a *model* which describes the overall composition and evolution of the Universe – the simplest and best-known example being ΛCDM, or a universe dominated by cold dark matter and a "conventional" cosmological constant. These models contain a small number of free parameters, which are estimated simultaneously, usually via Monte Carlo Markov Chain (MCMC) techniques.[1]

These approaches represent opposite but complementary pathways to determining the overall properties of our Universe. The distance ladder makes few physical assumptions but relies on a carefully concatenated series of observations, from stellar distances to extragalactic astronomy. Conversely, the concordance model encodes explicit assumptions regarding the physical mechanisms governing the overall evolution of the Universe, but may be usefully constrained by a *single* high-quality cosmic microwave background (CMB) data set, or a combination of a small number of data sets (e.g. CMB, with large-scale structure and supernovae).

28.1 Cosmological parameters

At the time of writing, the "benchmark" concordance parameters are provided by the WMAP team's analysis of their five-year dataset, in conjunction with other survey information

---

[1] This discussion of concordance cosmology is adapted from Adshead and Easther 2008b.

Astrometry for Astrophysics: Methods, Models, and Applications, ed. William F. van Altena. Published by Cambridge University Press. © Cambridge University Press 2013.

Table 28.1 The parameters of the current concordance cosmology. The contribution of the i-th constituent of the overall energy density is measured by Ω_i, in units where the critical density corresponding to a spatially flat universe is unity. All current observations are consistent with flatness, so $\Omega_b + \Omega_{CDM} + \Omega_\Lambda \equiv 1$, and there are thus just six free parameters. The labels A_s and n_s define the primordial scalar or density perturbations, parameterized by Eq. (28.1).

Label	Definition	Physical origin	Value
Ω_b	Baryon fraction	Baryogenesis	0.0462 ± 0.0015
Ω_{CDM}	Dark matter	TeV scale physics (?)	0.233 ± 0.013
Ω_Λ	Dark energy	Quantum gravity (?)	0.721 ± 0.015
τ	Optical depth	First stars	0.084 ± 0.016
h	Hubble parameter	Cosmological epoch	0.701 ± 0.013
A_s	Amplitude	Inflation	$(2.45 \pm 0.09) \times 10^{-9}$
n_s	Spectral index	Inflation	0.960 ± 0.014

(Komatsu *et al.* 2009). The concordance cosmology contains just six free parameters, which are listed in Table 28.1. Most of these parameters encode information about new physics. Three parameters describe the present-day energy density of our Universe: baryonic matter, dark matter, and dark energy. Baryonic matter is of course familiar, but the *quantity* of baryonic matter in the Universe (relative to the photon number density) is not predictable using our present knowledge of fundamental theory. There is no non-astrophysical evidence for dark matter or dark energy, although the concordance cosmology implies that the *dark sector* comprises ∼95% of the current energy density, dominating both the dynamics of large-scale structure and the overall expansion of the Universe.[2] It is not unreasonable to expect that the properties of the (non-baryonic) dark matter involve Large Hadron Collider (LHC)-scale particle physics, whereas understanding dark energy could shed light on quantum gravity.

Looking at the remaining parameters, the optical depth τ parametrizes the time at which the first stars turn on, which re-ionizes the neutral hydrogen in the Universe. This process does not depend on unknown physics: τ is a free parameter due to the difficulty of building an *ab initio* understanding of the initial period of star formation. The primordial spectrum of density (or *scalar*) perturbations has the empirical characterization

$$P_s(k) = A_s(k_\star) \left(\frac{k}{k_\star} \right)^{n_s(k_\star)-1} \tag{28.1}$$

where A_s is the amplitude, n_s the spectral index, and $k_\star$ is a specified but otherwise irrelevant pivot scale. This spectrum is widely believed to have been fixed during a primordial inflationary phase (Baumann *et al.* 2009) – in which case A_s and n_s depend on the inflationary dynamics and thus the properties of (very) high-energy particle physics. The quantity most

---

[2] We might replace dark matter or dark energy by either modifying gravity on very large scales, or (in the case of dark energy) allowing the underlying spacetime geometry to differ from the homogeneous and isotropic background of the concordance model. These proposals – like the concordance model – will be subject to increasingly stringent tests as the data improve.

relevant to astrometry is h, the dimensionless Hubble parameter. The Hubble parameter is a strictly decreasing quantity in a Friedmann Robertson Walker universe, and is far from constant over cosmological timescales.[3] Consequently, h effectively measures the age of the Universe at the moment we observe it.

The concordance parameter set is not arbitrary, but is chosen to maximize the χ^2 per degree of freedom, or via Bayesian evidence (e.g. Trotta 2008). Upper limits can be set on many other parameters, and these would be added to the concordance set if they were detected in the data (Adshead and Easther 2008b, Liddle 2004). Crucially, adding parameters to the concordance model can *weaken* bounds on the existing parameters, since the overall predictions of a model often change very slowly as some combination of two or more parameters is held constant; a so-called *degeneracy*. Degeneracies are typically broken by fitting to data from more than one source, via *cosmic complementarity* (Eisenstein *et al.* 1999).

To explore the contribution that future astrometric experiments can make to the global cosmological data set, I focus on a single example – the degeneracy between the Hubble constant and the masses of the neutrino species. On the face of it, these quantities appear to be unrelated, but they are linked via their impact on the observed properties of the microwave background. The Standard Model of elementary particle physics contains three neutrino species. These species (and their antiparticles) are assumed to be in thermal equilibrium in the very early Universe. The temperature falls as the Universe expands, and the neutrinos – which only interact very weakly – eventually fall out of equilibrium.[4] This occurs when the temperature is somewhat higher than 1 MeV ($\sim 10^{10}$ K). Since neutrino masses are (at most) a fraction of an electron volt, they are ultra-relativistic in the early Universe (Yao *et al.* 2006). Thus, while thermal equilibrium holds, the number density of each species of neutrinos is the same as the number density of photons, up to constant factors that differentiate the photon's bosonic statistics from the neutrinos' fermionic statistics.

Electrons and positrons are also initially in equilibrium with radiation, and annihilate when the temperature of the Universe, and thus the energy of the typical photon, drops below the rest mass of the electron ($\sim$0.5 MeV). The electrons and positrons decay electromagnetically, creating pairs of photons – heating the photon gas, relative to the neutrinos. Taking into account this heating and the Fermi statistics of the neutrinos, it is found (e.g. Kolb and Turner 1990)

$$T_\nu = \left(\frac{4}{11} \right)^{1/3} T_\gamma \tag{28.2}$$

where $T_{\nu,\gamma}$ are the neutrino and photon temperatures. As usual, the number and energy densities of the neutrinos scale as T^3 and T^4, respectively. Consequently, the number density of *neutrinos* in the present-day Universe is comparable to the number density of *photons*, almost all of which are accounted for by the black-body radiation of the

---

[3] Looking at the Einstein equations, dH/dt is a negative number for any physically reasonable source of mass-energy, and is only zero in a de Sitter universe which is entirely dominated by a cosmological constant.

[4] I use "natural units," where Boltzmann's constant and the speed of light are both set to be equal to unity, and temperatures and masses can all be specified in terms of energies. These energies are conventionally written in terms of the electron volt, or eV, the energy gained by an electron moving through a potential difference of 1 volt (Kolb and Turner 1990).

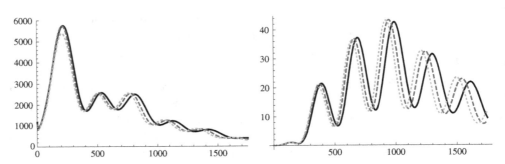

Fig. 28.1 We plot the CMB temperature ($\langle T T \rangle$, left) and the E-mode polarization ($\langle E E \rangle$, right) angular power spectra, computed with the central parameter values found by WMAP 5, and three neutrino species with Σm_ν of 0 (solid), 2 (dashed) and 4 (dotted) eV respectively. In the latter two cases, the neutrinos are massive at recombination, and the peaks move to lower multipoles, while the relative height of the first peak begins to decrease sharply with Σm_ν.

CMB. The neutrino background receives far less attention than the CMB, since these very low energy neutrinos are undetectable with present-day technology. The Universe has expanded dramatically since these backgrounds were laid down, and both have cooled dramatically. For the photons, T_γ is simply the measured black-body temperature of the CMB, 2.725 K $\approx 2.3 \times 10^{-4}$ eV. At least two of the neutrinos (and most likely all three) have small but non-zero masses. Crucially, it is not the masses themselves that have been measured, but the mass *differences* and these are substantially less than 1 eV – one million times smaller than the electron mass. Once the temperature of the neutrino background (or equivalently the CMB) becomes less than their rest mass (in eV), the neutrinos are non-relativistic.

As the Universe expands, the relative *number* densities of photons, neutrinos, and massive particles is essentially constant. Moreover, while the Universe is initially radiation-dominated, the contributions of radiation and matter become equal at the point of matter–radiation equality. From astrophysical data we deduce that matter–radiation equality occurs at a redshift of around 3200, and thus a temperature of 8700 K. At a redshift of slightly less than 1100 (or $T = 3000$ K) protons and electrons combine, and the Universe becomes transparent.[5]

Neutrinos make their presence felt in two ways. Firstly, the relic neutrino background substantially enhances the density of radiation – in the sense of effectively mass-less particles – beyond that provided by photons alone, delaying the moment of matter–radiation equality. Secondly, a neutrino species with a rest mass significantly greater than 0.25 eV moves non-relativistically during recombination, altering the detailed physics of the peaks in the microwave background power spectrum (e.g. Lesgourgues and Pastor 2006), as illustrated in Figure 28.1.

Given the quality of present data, neutrino mass constraints are usually written in terms of the total mass of all three neutrino species Σm_ν, with the individual masses assumed to

---

[5] 3000 K $\sim$ 0.25 eV, much less than the 13.6 eV required to ionize hydrogen. The Universe contains approximately a billion photons for every proton; however, it cools well below 13.6 eV and therefore the high-energy tail of the black-body distribution of photons cannot ionize hydrogen.

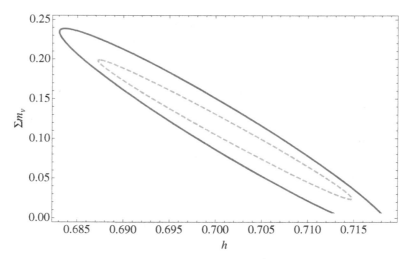

Fig. 28.2 Forecast $1 - \sigma$ errors in the $\Sigma m_\nu - h$ plane, where the present-day Hubble constant is $H_0 = 100\,h$ km/s/Mpc, assuming that the central value of $\Sigma m_\nu = 0.1$ eV. The outer ellipse is derived from the Ideal satellite of Verde *et al.* (2006) and Adshead and Easther (2008a), roughly equivalent to ambitious proposals for future missions (Bock *et al.* 2008). The forecasts are based on the usual concordance parameter set, plus r (the tensor-scalar ratio), α (the scale dependence in n_s) and Σm_ν. The Ideal satellite forecast assumes a low but non-zero signal-to-noise ratio in the detectors; the inner ellipse assumes a cosmic variance limited measurement up to $\ell = 1500$. Foregrounds are assumed to be perfectly subtracted.

be degenerate. The five-year WMAP data set (WMAPs) gives $\Sigma m_\nu < 1.3$ eV (95%), which tightens to $\Sigma m_\nu < 0.67$ eV (95%) with the addition of baryon acoustic oscillation (BAO) and supernovae data. For comparison, the forthcoming KATRIN experiment is projected to be sensitive to an electron neutrino mass of 0.2 eV, approximately one third of the *current* cosmological upper bound on Σm_ν. Moreover, the measured splittings of the neutrino masses tells us that $\Sigma m_\nu \gtrsim 0.05$ eV, so the astrophysical limit is now within an order of magnitude of the lower bound. Figure 28.1 shows that even a relatively small Σm_ν produces a detectable shift in the CMB power spectrum, and it is the precision of current CMB data that allows us to derive the bounds above. Crucially, this shift can be compensated for by adjustments to the Hubble constant, which is degenerate with Σm_ν (Lesgourgues and Pastor 2006). Forecasts (see Figure 28.2) suggest that a "next generation" CMB mission designed for excellent polarization sensitivity might make a $1 - \sigma$ "detection" if $\Sigma m_\nu = 0.1$ eV.

28.2 How astrometric measurements can help constrain cosmological parameters

Recall that astrometric measurements greatly improve the precision with which key "rungs" of the cosmological distance ladder can be measured via both direct trigonometric parallaxes of Cepheids, or "proper motion parallaxes" in nearby large galaxies (Unwin *et al.* 2007). Given that the distance ladder measures h independently of the concordance cosmology, values obtained in this way provide a vital consistency-check on the concordance model. Second, if astrometric measurements permit a high-precision measurement of h – which in this context means 1% or better – these results can then be included as *priors* when fitting to the concordance model and its extensions. For example, we see from Figure 28.2 that an independent measurement of h of this quality would break the degeneracy between h and σm_ν, allowing direct constraints on the neutrino sector to be derived from CMB measurements. Consequently, in addition to their immediate science goals, future astrometric programs promise to improve constraints on both cosmological models and, potentially, fundamental particle physics.

References

Adshead, P. and Easther, R. (2008a). Constraining inflation. *JCAP*, **0810**, 047.

Adshead, P. and Easther, R. (2008b). Neutrinos and future concordance cosmologies. *J. Phys. Conf. Ser.*, **136**, 022044.

Baumann, D., Jackson, M. G., and Adshead, P. (2009). CMBPol mission concept study: probing inflation with CMB polarization. *AIP Conf. Proc.*, **1141**, 10.

Bock, James, *et al.* (2008). The Experimental Probe of Inflationary Cosmology (EPIC): a mission concept study for NASA's Einstein Inflation Probe. Pasadena, CA: Jet Propulsion Laboratory. *arXiv: 0805.4207.*

Eisenstein, D. J., Hu, W., and Tegmark, M. (1999). Cosmic complementarity: joint parameter estimation from CMB experiments and redshift surveys. *APJ*, **518**, 2.

Freedman, W. L., Madore, B. F., Gibson, B. K., *et al.* (2001). Final results from the Hubble Space Telescope Key Project to measure the Hubble Constant. *ApJ*, **553**, 47.

Kolb, E. W. and Turner, M. S. (1990). *The Early Universe.* Addison-Wesley.

Komatsu, E., Dunkley, J., Nolta, M. R., *et al.* (2009). Five-year Wilkinson Microwave Anisotropy Probe (WMAP) observations: cosmological interpretation. *ApJS*, **180**, 330.

Lesgourgues, J. and Pastor, S. (2006). Massive neutrinos and cosmology. *Phys. Rep.*, **429**, 307.

Liddle, A. R. (2004). How many cosmological parameters? *MNRAS*, **351**, L49.

Trotta, R. (2008). Bayes in the sky: Bayesian inference and model selection in cosmology. *Contemp. Phys.*, **49**, 71.

Unwin, S. C., *et al.* (2008). Taking the measure of the Universe: precision astrometry with SIM PlanetQuest. *PASP*, **120**, 38.

Verde, L., Peiris, H., and Jimenez, R. (2006). Optimizing CMB polarization experiments to constrain inflationary physics. *JCAP*, **0601**, 019.

Yao, W. M., Amsler, C., Asner, D., and the Particle Data Group, *et al.* (2006). Review of particle physics. *J. Phys.*, **G33**.

Index

Printed in the United States
by Baker & Taylor Publisher Services